Equations of State

Theories and Applications

ACS SYMPOSIUM SERIES 300

Equations of State

Theories and Applications

K. C. Chao, EDITOR
School of Chemical Engineering, Purdue University

Robert L. Robinson, Jr., EDITOR
School of Chemical Engineering, Oklahoma State University

Developed from a symposium sponsored by
the Division of Industrial and Engineering Chemistry
at the 189th Meeting
of the American Chemical Society,
Miami Beach, Florida,
April 28–May 3, 1985

American Chemical Society, Washington, DC 1986

SEP/AE
CHEM

53139896

Library of Congress Cataloging-in-Publication Data

Equations of state.
(ACS symposium series; 300)

"Developed from a symposium sponsored by the Division of Industrial and Engineering Chemistry at the 189th meeting of the American Chemical Society, Miami Beach, Florida, April 28–May 3, 1985."

Includes bibliographies and indexes.

1. Equations of state—Congresses.

I. Chao, Kwang-Chu, 1925– . II. Robinson, Robert L., 1937– . III. American Chemical Society. Division of Industrial and Engineering Chemistry. IV. American Chemical Society. Meeting (189th: 1985: Miami Beach, Fla.)

QC173.4.E65E65 1986 541 86-1109
ISBN 0-8412-0958-8

PRINTED IN THE UNITED STATES OF AMERICA

SD
5/27/86

ACS Symposium Series

M. Joan Comstock, *Series Editor*

FOREWORD

The ACS SYMPOSIUM SERIES was founded in 1974 to provide a medium for publishing symposia quickly in book form. The format of the Series parallels that of the continuing ADVANCES IN CHEMISTRY SERIES except that, in order to save time, the papers are not typeset but are reproduced as they are submitted by the authors in camera-ready form. Papers are reviewed under the supervision of the Editors with the assistance of the Series Advisory Board and are selected to maintain the integrity of the symposia; however, verbatim reproductions of previously published papers are not accepted. Both reviews and reports of research are acceptable, because symposia may embrace both types of presentation.

CONTENTS

PREFACE

THE EQUATION-OF-STATE APPROACH to model and correlate fluid-phase equilibria has been emphasized more and more in the eight years since the first symposium on this topic. In 1979, we edited the volume entitled *Equations of State in Engineering and Research* (Advances in Chemistry No. 182), which was based on that symposium. Meanwhile, research activities have been continually robust, particularly in building new models and in extending established equations for new and improved applications.

The present volume is based on the April 1985 equation-of-state symposium held to present a comprehensive state-of-the-art view of progress in this area.

The term "equation of state" is used in a broad sense to include mathematical description of volumetric behavior, derived properties, mixture behavior, and phase equilibrium of fluids. The main thrust continues to be the description of fluid-phase equilibrium, a phenomenon of enduring interest because it is basic to mass transport and separation operations. At the present stage of development, nonpolar fluids are modeled almost exclusively with equations of state, and active research is extending to polar fluids.

The twenty-eight chapters in this volume, reporting work and progress on a broad front, are arranged in seven sections. Contributions are made by authors from diverse disciplines, including chemical engineers, physical chemists, and chemical physicists. The division of the volume into sections is not rigorous; some papers can fit easily into more than one section. Nevertheless, the arrangement should serve as a helpful guide to readers in their initial encounter with this substantial collection.

We greatly appreciate the encouragement and support of the Division of Industrial and Engineering Chemistry of the American Chemical Society in sponsoring the symposium. Special thanks go to the authors of the papers. In addition to the regular task of preparing their manuscripts, they have gone the extra mile by preparing them in a camera-ready form. Their care and devotion are clear from the quality of the finished product. Robin Giroux of the ACS Books Department worked with us throughout the development of the volume.

K. C. CHAO
School of Chemical Engineering
Purdue University
West Lafayette, IN 47907

ROBERT L. ROBINSON, JR.
School of Chemical Engineering
Oklahoma State University
Stillwater, OK 74078

December 2, 1985

EQUATIONS OF STATE AND SOLUTION THERMODYNAMICS

1

Equations of State and Classical Solution Thermodynamics

Survey of the Connections

Michael M. Abbott and Kathryn K. Nass

Department of Chemical Engineering and Environmental Engineering, Rensselaer Polytechnic Institute, Troy, NY 12180-3590

Many contemporary researches in equation-of-statery focus on the representation of properties of liquid mixtures. Here, the connection with experiment is made through excess functions, Henry's constants, and related quantities. We present in this communication a review and discussion of the apparatus linking the equation-of-state formulation to that of classical solution thermodynamics, and illustrate the key ideas with examples.

The correlation and prediction of mixture behavior are central topics in applied thermodynamics, important not only in their own right, but also as necessary adjuncts to the calculation of chemical and phase equilibria. Two major formalisms are available for representation of mixture properties: the PVTx equation-of-state formulation, and the apparatus of classical solution thermodynamics. It is well known that the two formalisms are related, that a PVTx equation of state in fact implies full sets of expressions for the quantities employed in the conventional thermodynamics of mixtures. Only with advances in computation, however, has it become possible to take advantage of these relationships, which are now used in building and testing equations of state.

The formulations differ in at least two major respects: in the choices of independent variables, and in the definitions of special functions used to represent deviations of real behavior from standards of "ideality". In the equation-of-state approach, temperature, molar volume, and composition are the natural independent variables, and the residual functions are the natural deviation functions. In classical solution thermodynamics, temperature, pressure, and composition are favored independent variables, and excess functions are used to measure deviations from "ideality". Thus translations from one formulation to the other involve both changes in independent variables and conversions between residual functions and excess functions.

0097-6156/86/0300-0002$10.75/0

Simple as all this sounds, we have found it frustrating not to have available a single source in which the connections between the two formulations are neatly laid out in purely classical terms. Particularly vexing is the lack of a flexible but clean notation. To meet our own needs, we have synthesized a system of definitions and notation, partially described in the following pages. The system seems adequate for both research and classroom use.

Deviation Functions

Rationale and Definitions. It is rarely practical to work directly with a mixture molar property M. For example, M may not be defined unambiguously, and may therefore not admit, even in principle, direct experimental determination. Thus, neither U nor S nor H nor G is defined at all, in the strict sense of the word. Both U and S are primitive quantities, and H and G are "defined" in terms of one or both of them. Moreover, property M by itself may not admit physical interpretations, except in a loose sense (e.g., entropy as a "measure of disorder"). For these and other reasons, one finds it convenient to introduce such quantities as residual functions and excess functions. These quantities, themselves thermodynamic properties, are examples of a general class of functions which we call deviation functions.

Deviation functions represent the difference between actual mixture property M and the corresponding value for M given by some model of behavior:

$$M(\text{deviation}) \equiv M(\text{actual}) - M(\text{model})$$

The choice of a model is to some extent arbitrary, but to be useful the model must have certain attributes. Its molecular implications should be thoroughly understood, so that deviation functions defined with respect to it can be given clean interpretations. Real behavior should approach model behavior in well-defined limits of state variables or substance types, so that the deviation functions have unambiguous zeroes. Finally, to facilitate numerical work, it is desirable that the properties of the model be capable of concise analytical expression.

Once a model is chosen, the conditions at which the comparison (real vs. model) is made must be specified. There are several possibilities here, but two are particularly felicitous. Thus, we may define deviation functions at uniform temperature, pressure, and composition:

$$M^D \equiv M - M^{mod}(T,P,x) \tag{1}$$

Here, the notation signals that mixture molar property M^{mod} for the model is evaluated at the same T,P, and composition as the actual mixture property M; superscript (capital) "D" identifies the deviation function as a constant - T,P,x deviation function. Alternatively, we may define deviation functions at uniform temperature, molar volume (or density), and composition:

$$\boxed{M^d \equiv M - M^{mod}(T,V,x)} \tag{2}$$

Here, M^{mod} is evaluated at the same T,V, and composition as the actual mixture property M; superscript (lower-case) "d" distinguishes this class of functions from that defined by Equation 1.

The two kinds of deviation function are related. Subtraction of Equation 1 from Equation 2 gives

$$M^d - M^D = M^{mod}(T,P,x) - M^{mod}(T,V,x)$$

whence we find that

$$\boxed{M^d = M^D + \int_{P^*}^{P} \left(\frac{\partial M^{mod}}{\partial P}\right)_{T,x} dP} \tag{3}$$

Here, pressure P^* is the pressure for which the mixture molar volume of the model has the same value V as that of the real solution at the given temperature and composition. According to Equation 3, deviation functions M^d and M^D are identical for those properties M for which M^{mod} is independent of pressure.

Residual Functions. The simplest model of mixture behavior is the ideal-gas mixture:

$$M^{mod} \equiv M^{ig}$$

Deviation functions defined with respect to the ideal-gas model are called residual functions, and are identified by superscript R or r. Thus, as special cases of Equations 1 and 2, we define

$$\boxed{M^R \equiv M - M^{ig}(T,P,x)} \tag{4}$$

and

$$\boxed{M^r \equiv M - M^{ig}(T,V,x)} \tag{5}$$

Residual functions M^R and M^r are related by Equation 3, with the assignments mod ≡ ig and $P^* = RT/V$. Thus

$$\boxed{M^r = M^R + \int_{\frac{RT}{V}}^{P} \left(\frac{\partial M^{ig}}{\partial P}\right)_{T,x} dP} \tag{6}$$

Ideal-gas properties U^{ig}, H^{ig}, C_V^{ig}, and C_P^{ig} are all independent of pressure. Hence

$$M^r = M^R \quad (M = U,H,C_V,C_P) \tag{7}$$

On the other hand, ideal-gas properties S^{ig}, A^{ig}, and G^{ig} are functions of pressure. In particular,

$$\left(\frac{\partial S^{ig}}{\partial P}\right)_{T,x} = -\frac{R}{P}$$

$$\left(\frac{\partial A^{ig}}{\partial P}\right)_{T,x} = \left(\frac{\partial G^{ig}}{\partial P}\right)_{T,x} = \frac{RT}{P}$$

and hence, by Equation 6,

$$S^r = S^R - R\ell nZ \quad (8)$$

$$A^r = A^R + RT\ell nZ \quad (9)$$

$$G^r = G^R + RT\ell nZ \quad (10)$$

where $Z \equiv PV/RT$ is the compressibility factor.

Expressions for either M^R or M^r are found from a PVTx equation of state by standard techniques: see e.g. Van Ness and Abbott (1). If the equation of state is explicit in pressure, then T,V (or molar density ρ), and composition are the natural independent variables and the M^r are the natural residual functions. If the equation of state is explicit in volume, then T,P, and composition are the natural independent variables and the M^R are the natural residual functions. Tables I and II summarize formulas for computing the M^r from a pressure-explicit equation of state, and the M^R from a volume-explicit equation of state. Conversion from M^R to M^r, or vice versa, is done by Equations 7 through 10. Note that V^r and P^R are identically zero.

Residual Function A^r: a Generating Function. Most realistic equations of state are explicit in pressure; T,V, and composition are the natural independent variables. These are also the canonical variables for the Helmholtz energy A, so the constant - T,V,x residual Helmholtz energy A^r plays a special role in equation-of-state thermodynamics. It can be considered a generating function, not only for the other constant - T,V,x residual functions, but also for the equation of state itself. The relevant working equations assume a pretty symmetry when the independent variables are chosen as reciprocal absolute temperature τ ("coldness" $\equiv T^{-1}$), reciprocal molar volume ρ (molar density $\equiv V^{-1}$), and composition. They are summarized in this form in Table III.

Application of these formulas may be illustrated by a simple example. We choose for this purpose the van der Waals equation of state, for which

$$A^r(\text{vdW}) = -RT\ell n(1-b\rho) - a\rho \quad (33)$$

where parameters a and b depend on composition only. The residual pressure P^r ($\equiv P-P^{ig} = P-\rho RT$) is found from Equation 27:

$$P^r(\text{vdW}) = \frac{b\rho^2 RT}{1-b\rho} - a\rho^2$$

so the equation of state is

$$P(vdW) = \frac{\rho RT}{1-b\rho} - a\rho^2 \quad (34)$$

The other residual functions follow from Equations 28 through 32. Two results are

$$S^r(vdW) = R\ln(1-b\rho) \quad (35)$$

$$U^r(vdW) = -a\rho \quad (36)$$

Equations 35 and 36 support the common interpretations of the hard-sphere (repulsive) term as representing an "entropic" contribution to the equation of state, and of the van der Waals "a" as an energy parameter. The material of this section can in fact be taken as a point of departure for the development of a classically-inspired generalized van der Waals theory, motivated by Equations 33 through 36, but unrestricted by the assumptions attendant to the original van der Waals equation of state (2).

Excess Functions. The conventional standard of mixture behavior for condensed phases is the ideal solution. Deviation functions reckoned against this model are called excess functions, and are identified by superscript E or e. We define

$$\boxed{M^E \equiv M - M^{id}(T,P,x)} \quad (37)$$

and

$$\boxed{M^e \equiv M - M^{id}(T,V,x)} \quad (38)$$

where superscript id identifies the ideal solution. Equations 37 and 38 are special cases of Equations 1 and 2, with the assignment mod ≡ id. Excess functions M^E and M^e are related by

$$\boxed{M^e = M^E + \int_{P^*}^{P} \left(\frac{\partial M^{id}}{\partial P}\right)_{T,x} dP} \quad (39)$$

which is a special case of Equation 3. Unlike the ideal-gas case, no simple general closed-form expression can here be written for P^*. By definition, P^* is in this case the pressure for which the ideal solution has the same molar volume V as the real solution at the given temperature and composition. Since

$$V^{id} = \sum_i x_i V_i$$

this pressure must be found as a solution to the equation

$$\sum_i x_i V_i(T,P^*) = V(T,P,x) \quad (40)$$

Clearly, numerical relation of M^e to M^E via Equations 39 and 40 requires equation-of-state information, both for the real pure com-

Table I. Residual Functions from a Pressure-Explicit Equation of State

$$P^r = \rho RT(Z-1) \tag{11}$$

$$U^r = -RT^2 \int_0^\rho \left(\frac{\partial Z}{\partial T}\right)_{\rho,x} \frac{d\rho}{\rho} \tag{12}$$

$$H^r = -RT^2 \int_0^\rho \left(\frac{\partial Z}{\partial T}\right)_{\rho,x} \frac{d\rho}{\rho} + RT(Z-1) \tag{13}$$

$$S^r = -R \int_0^\rho \left[T\left(\frac{\partial Z}{\partial T}\right)_{\rho,x} + Z-1\right] \frac{d\rho}{\rho} \tag{14}$$

$$A^r = RT \int_0^\rho (Z-1) \frac{d\rho}{\rho} \tag{15}$$

$$G^r = RT \int_0^\rho (Z-1) \frac{d\rho}{\rho} + RT(Z-1) \tag{16}$$

$$C_V^r = -RT \int_0^\rho \left[T\left(\frac{\partial^2 Z}{\partial T^2}\right)_{\rho,x} + 2\left(\frac{\partial Z}{\partial T}\right)_{\rho,x}\right] \frac{d\rho}{\rho} \tag{17}$$

$$C_P^r = C_V^r - R + R\left[Z + T\left(\frac{\partial Z}{\partial T}\right)_{\rho,x}\right]^2 \left[Z + \rho\left(\frac{\partial Z}{\partial \rho}\right)_{T,x}\right]^{-1} \tag{18}$$

Table II. Residual Functions from a Volume-Explicit Equation of State

$$V^R = \frac{RT}{P}(Z-1) \tag{19}$$

$$U^R = -RT^2 \int_0^P \left(\frac{\partial Z}{\partial T}\right)_{P,x} \frac{dP}{P} - RT(Z-1) \tag{20}$$

$$H^R = -RT^2 \int_0^P \left(\frac{\partial Z}{\partial T}\right)_{P,x} \frac{dP}{P} \tag{21}$$

$$S^R = -R \int_0^P \left[T\left(\frac{\partial Z}{\partial T}\right)_{P,x} + Z-1\right] \frac{dP}{P} \tag{22}$$

$$A^R = RT \int_0^P (Z-1)\frac{dP}{P} - RT(Z-1) \tag{23}$$

$$G^R = RT \int_0^P (Z-1)\frac{dP}{P} \tag{24}$$

$$C_P^R = -RT \int_0^P \left[T\left(\frac{\partial^2 Z}{\partial T^2}\right)_{P,x} + 2\left(\frac{\partial Z}{\partial T}\right)_{P,x}\right] \frac{dP}{P} \tag{25}$$

$$C_V^R = C_P^R + R - R\left[Z + T\left(\frac{\partial Z}{\partial T}\right)_{P,x}\right]^2 \left[Z - P\left(\frac{\partial Z}{\partial P}\right)_{T,x}\right]^{-1} \tag{26}$$

Table III. Residual Functions from A^r

$$P^r = \rho^2 \left(\frac{\partial A^r}{\partial \rho}\right)_{\tau,x} \tag{27}$$

$$S^r = \tau^2 \left(\frac{\partial A^r}{\partial \tau}\right)_{\rho,x} \tag{28}$$

$$U^r = \left[\frac{\partial(\tau A^r)}{\partial \tau}\right]_{\rho,x} \tag{29}$$

$$G^r = \left[\frac{\partial(\rho A^r)}{\partial \rho}\right]_{\tau,x} \tag{30}$$

$$H^r = A^r + \tau \left(\frac{\partial A^r}{\partial \tau}\right)_{\rho,x} + \rho \left(\frac{\partial A^r}{\partial \rho}\right)_{\tau,x} \tag{31}$$

$$C_V^r = -\tau^2 \left[\frac{\partial^2(\tau A^r)}{\partial \tau^2}\right]_{\rho,x} \tag{32}$$

ponents and for the real mixture. Hence the relation between M^e and M^E is not as "clean" as that between M^r and M^R.

Rough closed-form approximations to Equation 39, appropriate for applications to condensed phases, may be found however. We write Equation 39 as

$$M^e \approx M^E + \left(\frac{\partial M^{id}}{\partial P}\right)_{T,x}(P-P^*) \tag{41}$$

where the derivative is evaluated at the pressure P of the real mixture. An expression for P^* follows from Equation 41 by the assignment M = V, because $V^e = 0$ identically. Thus we find that

$$P^* \approx P - \frac{V^E}{\sum_i x_i \kappa_i V_i} \tag{42}$$

where κ_i is the isothermal compressibility of pure i:

$$\kappa_i \equiv - \frac{1}{V_i}\left(\frac{\partial V_i}{\partial P}\right)_T$$

All quantities on the right side of Equation 42 are evaluated at pressure P. According to Equation 42, the sign of V^E determines whether P^* is less than or greater than P.

Combination of Equations 41 and 42 gives

$$M^e \approx M^E + \left(\frac{\partial M^{id}}{\partial P}\right)_{T,x} \frac{V^E}{\sum_i x_i \kappa_i V_i} \tag{43}$$

which is the required approximation to Equation 39. Particular cases of Equation 43 are generated on specification of M and of the corresponding derivative $(\partial M^{id}/\partial P)_{T,x}$. Table IV summarizes expressions for this derivative in terms of the volumetric properties of the species composing the mixture. To demonstrate its application, let us take M = A. Then, by Equations 43 and 48 we find that

$$A^e \approx A^E + PV^E$$

But

$$G^E = A^E + PV^E$$

and hence we have rationalized the approximation

$$A^e \approx G^E$$

a result frequently used in molecular modeling of the constant - T,P,x excess Gibbs energy.

Table IV. Expressions for $\left(\frac{\partial M^{id}}{\partial P}\right)_{T,x}$

M	$\left(\frac{\partial M^{id}}{\partial P}\right)_{T,x} =$	
V	$-\sum_i x_i \kappa_i V_i$	(44)
U	$P\sum_i x_i \kappa_i V_i - T\sum x_i \beta_i V_i$	(45)
H	$\sum_i x_i V_i - T\sum x_i \beta_i V_i$	(46)
S	$-\sum_i x_i \beta_i V_i$	(47)
A	$P\sum_i x_i \kappa_i V_i$	(48)
G	$\sum_i x_i V_i$	(49)
C_P	$-T\sum_i x_i \beta_i^2 V_i - T\sum_i x_i \left(\frac{\partial \beta_i}{\partial T}\right)_P V_i$	(50)

Relations between Excess Functions and Residual Functions. Excess functions and residual functions are related. The relationships are most easily established through M^E and M^R. By Equations 4 and 37, we have

$$M^E \equiv M^R - (M^{id} - M^{ig}) \tag{51}$$

But

$$M^{id} = \sum_i x_i M_i - \mathcal{F} \sum_i x_i \ln x_i$$

and

$$M^{ig} = \sum_i x_i M_i^{ig} - \mathcal{F} \sum_i x_i \ln x_i$$

where

$$\mathcal{F} = \begin{cases} 0 & (M = U,H,C_P) \\ R & (M = S) \\ -RT & (M = A,G) \end{cases} \tag{52}$$

Hence Equation 51 becomes

$$M^E = M^R - (\sum_i x_i M_i - \sum_i x_i M_i^{ig})$$

or

$$\boxed{M^E = M^R - \sum_i x_i M_i^R} \tag{53}$$

Equation 53 is a basis for relations connecting M^E or M^e to M^R or M^r. It provides a link between excess functions and the equation of state and it suggests how physical interpretations applying to particular residual functions are carried over to the corresponding excess functions.

Experiment provides values of the constant - T,P,x excess functions M^E, whereas the constant - T,V,x residual functions M^r, particularly A^r, are most cleanly related to a pressure-explicit equation of state. By Equations 7 through 10, we have

$$M^R = M^r + \mathcal{F} \ln Z \tag{54}$$

with $\mathcal{F}$ defined by Equation 52. Combination of Equations 53 and 54 produces an expression relating M^E to M^r:

$$\boxed{M^E = M^r - \sum_i x_i M_i^r + \mathcal{F} \sum_i x_i \ln(V/V_i)} \tag{55}$$

Table V summarizes formulas for the constant - T,P,x excess functions H^E, S^E, G^E, and C_P^E in terms of A^r and its temperature derivatives. These recipes are useful for testing the abilities of a PVTx equation of state to represent liquid-mixture properties. Consider as an example the van der Waals equation of state, Equation 34, for which A^r is given by Equation 33. Application of Equations 56 through 59 gives

$$H^E(\text{vdW}) = -\left(a\rho - \sum_i x_i a_{ii}\rho_i\right) + PV^E(\text{vdW}) \tag{60}$$

$$S^E(\text{vdW}) = R\sum x_i \ln\left[\frac{\rho_i(1-b\rho)}{\rho(1-b_{ii}\rho_i)}\right] \tag{61}$$

$$G^E(\text{vdW}) = H^E(\text{vdW}) - TS^E(\text{vdW}) \tag{62}$$

$$C_P^E(\text{vdW}) = R\left\{\left[1 - \frac{2a}{RT}(1-b\rho)^2\rho\right]^{-1} - \sum x_i\left[1 - \frac{2a_{ii}}{RT}(1-b_{ii}\rho_i)^2\rho_i\right]^{-1}\right\} \tag{63}$$

Here, $V^E(\text{vdW})$ is the excess volume implied by the equation of state. Unsubscripted quantities are mixture properties, and subscripted quantities refer to the pure fluids.

Partial Properties

Rationale and Definitions. The partial-property concept is central to applied solution thermodynamics. First, it represents a formal (but arbitrary) basis for apportioning a mixture molar property M amongst the constituents of a phase. Second, it provides an elegant apparatus for describing infinitely-dilute solutions. Finally, it serves as a unifying concept in formulating mixture equilibrium problems, because the chemical potential and its relatives stand to the Gibbs energy and its relatives as partial properties: see Table VI.

The conventional definition of a partial property $\bar{M}_i$ is

$$\bar{M}_i \equiv \left[\frac{\partial(nM)}{\partial n_i}\right]_{T,P,n_j} \tag{64}$$

where by implication temperature, pressure, and composition are favored independent variables. This choice is entirely appropriate in the laboratory frame of reference, because temperature and pressure are the variables susceptible to precise measurement and control. There are instances however where it is useful to broaden the partial-property concept, to accommodate alternative choices of

Table V. Excess Functions and the Residual Helmholtz Energy

$$H^E = (A^r - \sum_i x_i A_i^{\,r}) - T\left[\left(\frac{\partial A^r}{\partial T}\right)_{V,x} - \sum_i x_i \left(\frac{\partial A_i^r}{\partial T}\right)_{V_i}\right] + PV^E \tag{56}$$

$$S^E = -\left[\left(\frac{\partial A^r}{\partial T}\right)_{V,x} - \sum_i x_i \left(\frac{\partial A_i^r}{\partial T}\right)_{V_i}\right] + R\sum x_i \ln(V/V_i) \tag{57}$$

$$G^E = (A^r - \sum_i x_i A_i^r) - RT\sum_i x_i \ln(V/V_i) + PV^E \tag{58}$$

$$C_P^E = -T\left[\left(\frac{\partial^2 A^r}{\partial T^2}\right)_{V,x} - \sum_i x_i \left(\frac{\partial^2 A_i^r}{\partial T^2}\right)_{V_i}\right] - T\left[\left(\frac{\partial P}{\partial T}\right)_{V,x}^2 \left(\frac{\partial P}{\partial V}\right)_{T,x}^{-1} - \sum_i x_i \left(\frac{\partial P}{\partial T}\right)_{V_i}^2 \left(\frac{\partial P}{\partial V_i}\right)_T^{-1}\right] \tag{59}$$

independent variables. Foremost among these is the application of pressure-explicit equations of state to the calculation of chemical and phase equilibria.

Table VI. Important Partial Properties

M	$\bar{M}_i$	
G	μ_i	(65)
G/RT	$\ln \lambda_i$	(66)
G^R/RT	$\ln \hat{\phi}_i$	(67)
$\Delta G/RT$	$\ln \hat{a}_i$	(68)
G^E/RT	$\ln \gamma_i$	(69)

Reis (3) discusses the generalization of the partial-property concept to other sets of independent variables, and Abbott (4) has extended the definition of Equation 64 to higher-order derivatives with respect to mole numbers. A result common to both kinds of generalizations is that the "summability feature" of the $\bar{M}_i$, viz.,

$$M = \sum_i x_i \bar{M}_i \qquad (70)$$

is in fact possessed by a very large number of "partial properties", in addition to $\bar{M}_i$.

The generalization of Equations 64 and 70 for choices of independent intensive variables other than T and P is quite simply rationalized. We outline a development here. Let the total property $M^t = nM$ of a phase be a function of the set of mole numbers $n_1, n_2, \ldots$, and of two arbitrary intensive variables X and Y. Then the total differential d(nM) corresponding to an arbitrary change of state is

$$d(nM) = \left[\frac{\partial(nM)}{\partial X}\right]_{Y,n} dX + \left[\frac{\partial(nM)}{\partial Y}\right]_{X,n} dY + \sum_i \left[\frac{\partial(nM)}{\partial n_i}\right]_{X,Y,n_j} dn_i \qquad (71)$$

where subscript n denotes constancy of all mole numbers and n_j denotes constancy of all mole numbers save n_i. The following identities apply to Equation 71:

$$d(nM) = ndM + Mdn$$

$$dn_i = ndx_i + x_i dn$$

$$\left[\frac{\partial(nM)}{\partial X}\right]_{Y,n} = n\left(\frac{\partial M}{\partial X}\right)_{Y,x}$$

$$\left[\frac{\partial(nM)}{\partial Y}\right]_{X,n} = n\left(\frac{\partial M}{\partial Y}\right)_{X,x}$$

where subscript x denotes constancy of all mole fractions. Additionally, let us define the generalized partial property $\hat{M}_i$ as

$$\boxed{\hat{M}_i \equiv \left[\frac{\partial(nM)}{\partial n_i}\right]_{X,Y,n_j}} \tag{72}$$

Combining the last six equations and collecting coefficients of n and of dn, we obtain

$$\left[dM - \left(\frac{\partial M}{\partial X}\right)_{Y,x} dX - \left(\frac{\partial M}{\partial Y}\right)_{X,x} dY - \sum_i \hat{M}_i dx_i\right] n + \left[M - \sum_i x_i \hat{M}_i\right] dn = 0$$

But quantities n and dn are independent and arbitrary; the two bracketed terms must therefore separately be zero, and we find that

$$dM = \left(\frac{\partial M}{\partial X}\right)_{Y,x} dX + \left(\frac{\partial M}{\partial Y}\right)_{X,x} dY + \sum_i \hat{M}_i dx_i \tag{73}$$

and

$$\boxed{M = \sum_i x_i \hat{M}_i} \tag{74}$$

Equation 73 is merely a special case of Equation 71, with n=1. Equation 74 is "new" however; it is the required extension of the "summability feature" of the $\bar{M}_i$ to other classes of partial properties.

Other analogs of the usual partial-property relations are found by straightforward mathematics. For example, according to Equation 74, the total differential dM is

$$dM = \sum_i x_i d\hat{M}_i + \sum_i \hat{M}_i dx_i$$

But this expression must be equivalent to Equation 73; comparing the two gives immediately a generalized Gibbs-Duhem equation:

$$\sum_i x_i d\hat{M}_i = \left(\frac{\partial M}{\partial X}\right)_{Y,x} dX + \left(\frac{\partial M}{\partial Y}\right)_{X,x} dY \tag{75}$$

which is merely an extension of the familiar

$$\sum_i x_i d\bar{M}_i = \left(\frac{\partial M}{\partial T}\right)_{P,x} dT + \left(\frac{\partial M}{\partial P}\right)_{T,x} dP \tag{76}$$

Relations among Partial Properties of Different Types. Sometimes it is necessary to convert one type of partial property to another. Let us define, analogous to $\hat{M}_i$, a second partial property $\hat{M}_i'$:

$$\hat{M}_i' \equiv \left[\frac{\partial(nM)}{\partial n_i}\right]_{X',Y',n_j} \tag{77}$$

where intensive variables X' and Y' may be different from intensive variables X and Y. We wish to relate partial property $\hat{M}_i'$ to partial property $\hat{M}_i$. By Equation 71 we find that

$$\left[\frac{\partial(nM)}{\partial n_i}\right]_{X',Y',n_j} = \left[\frac{\partial(nM)}{\partial X}\right]_{Y,n} \left(\frac{\partial X}{\partial n_i}\right)_{X',Y',n_j} + \left[\frac{\partial(nM)}{\partial Y}\right]_{X,n} \left(\frac{\partial Y}{\partial n_i}\right)_{X',Y',n_j} + \left[\frac{\partial(nM)}{\partial n_i}\right]_{X,Y,n_j}$$

which becomes on simplification

$$\boxed{\hat{M}_i' = \hat{M}_i + n\left(\frac{\partial M}{\partial X}\right)_{Y,x} \left(\frac{\partial X}{\partial n_i}\right)_{X',Y',n_j} + n\left(\frac{\partial M}{\partial Y}\right)_{X,x} \left(\frac{\partial Y}{\partial n_i}\right)_{X',Y',n_j}} \tag{78}$$

Equivalent statements of Equation 78 are possible. For example, the derivative $(\partial X/\partial n_i)_{X',Y',n_j}$ may be written as

$$\left(\frac{\partial X}{\partial n_i}\right)_{X',Y',n_j} = -\left(\frac{\partial X}{\partial Y'}\right)_{X',x} \left(\frac{\partial Y'}{\partial n_i}\right)_{X',X,n_j}$$

or as

$$\left(\frac{\partial X}{\partial n_i}\right)_{X',Y',n_j} = \frac{1}{n}(\hat{X}' - X)$$

The derivative $(\partial Y/\partial n_i)_{X',Y',n_j}$ may be similarly rewritten.

Only two classes of partial properties are of importance to us here: those defined at constant T and P ("laboratory" partial properties), and those defined at constant T and V ("equation-of-state" partial properties). Thus we define, as special cases of Equation 72,

$$\overline{M}_i \equiv \left[\frac{\partial(nM)}{\partial n_i}\right]_{T,P,n_j} \tag{64}$$

and

$$\widetilde{M}_i \equiv \left[\frac{\partial(nM)}{\partial n_i}\right]_{T,V,n_j} \tag{79}$$

Here and henceforth the tilde (~) is used to distinguish a constant - T,V partial property from the conventional constant - T,P variety. Note that $\widetilde{M}_i$ can equally well be considered a constant - T,ρ partial property, where ρ is molar density.

Both classes of partial properties obey the summability relation of Equation 74:

$$M = \sum_i x_i \overline{M}_i \tag{70}$$

and

$$M = \sum_i x_i \widetilde{M}_i \tag{80}$$

Each class of partial properties has its own Gibbs-Duhem equation:

$$\sum_i x_i d\overline{M}_i = \left(\frac{\partial M}{\partial T}\right)_{P,x} dT + \left(\frac{\partial M}{\partial P}\right)_{T,x} dP \tag{76}$$

and

$$\sum_i x_i d\widetilde{M}_i = \left(\frac{\partial M}{\partial T}\right)_{V,x} dT + \left(\frac{\partial M}{\partial V}\right)_{T,x} dV \tag{81}$$

Finally, by Equation 78, we find the following relations between the two classes of partial properties:

$$\widetilde{M}_i = \overline{M}_i - n\left(\frac{\partial M}{\partial V}\right)_{T,x}\left(\frac{\partial V}{\partial n_i}\right)_{T,P,n_j} \tag{82}$$

and

$$\overline{M}_i = \widetilde{M}_i - n\left(\frac{\partial M}{\partial P}\right)_{T,x}\left(\frac{\partial P}{\partial n_i}\right)_{T,V,n_j} \tag{83}$$

Equation 82 effects a conversion of laboratory partial properties to equation-of-state partial properties. Since

$$\overline{V}_i \equiv \left[\frac{\partial(nV)}{\partial n_i}\right]_{T,P,n_j} = V + n\left(\frac{\partial V}{\partial n_i}\right)_{T,P,n_j}$$

we can rewrite it as

$$\tilde{M}_i = \bar{M}_i + (V - \bar{V}_i)\left(\frac{\partial M}{\partial V}\right)_{T,x} \tag{84}$$

The conventional partial molar volume $\bar{V}_i$ is thus a key quantity for this type of conversion.

More common are applications requiring conversion of constant - T,V partial properties to constant - T,P partial properties. Here Equation 83 is appropriate; the analog of Equation 84 is

$$\bar{M}_i = \tilde{M}_i + (P - \tilde{P}_i)\left(\frac{\partial M}{\partial P}\right)_{T,x}$$

where

$$\tilde{P}_i \equiv \left[\frac{\partial(nP)}{\partial n_i}\right]_{T,V,n_j}$$

An equivalent result involving compressibility factors is

$$\bar{M}_i = \tilde{M}_i + \frac{RT}{V}(Z - \tilde{Z}_i)\left(\frac{\partial M}{\partial P}\right)_{T,x} \tag{85}$$

As an example of an application of this and earlier material, consider the following standard problem: to determine an expression for the component fugacity coefficient $\hat{\phi}_i$, convenient for use with a pressure-explicit equation of state. We know that $\ln\hat{\phi}_i$ is a constant - T,P partial property with respect to G^R/RT: see Equation 67, Table VI. Moreover, by Equations 10 and 16, we have the following recipe for G^R/RT:

$$\frac{G^R}{RT} = Z - 1 - \ln Z + \int_0^{\rho} (Z - 1)\frac{d\rho}{\rho}$$

Since density appears explicitly in this equation, an expression for $\tilde{G}_i^R/RT$ is readily found. It is

$$\frac{\tilde{G}_i^R}{RT} = \tilde{Z}_i\left(\frac{Z-1}{Z}\right) - \ln Z + \int_0^{\rho} (\tilde{Z}_i - 1)\frac{d\rho}{\rho}$$

All that remains is to relate $\bar{G}_i^R/RT$ (= $\ln\hat{\phi}_i$) to $\tilde{G}_i/RT$. Here, we apply Equation 85, with the assignment $M = G^R/RT$. The pressure derivative of G^R/RT is

$$\left[\frac{\partial(G^R/RT)}{\partial P}\right]_{T,x} = \frac{V^R}{RT} = \frac{V}{RT}\left(\frac{Z-1}{Z}\right)$$

Hence

$$\frac{\bar{G}_i^R}{RT} = \frac{\tilde{G}_i^R}{RT} + (Z-\tilde{Z}_i)\left(\frac{Z-1}{Z}\right)$$

and thus we find that

$$\ell n\hat{\phi}_i = Z - 1 - \ell nZ + \int_0^\rho (\tilde{Z}_i-1)\frac{d\rho}{\rho} \quad (86)$$

Equation 86 can of course be obtained by other procedures. In this example we have attempted to make full use of material presented in this and preceding sections.

Suppose that Z is represented by the van der Waals equation of state, Equation 34:

$$Z(\text{vdW}) = \frac{1}{1-b\rho} - \frac{a\rho}{RT} \quad (87)$$

Then

$$\tilde{Z}_i(\text{vdW}) = \frac{1}{1-b\rho} + \frac{(\tilde{b}_i-b)\rho}{(1-b\rho)^2} - \frac{\tilde{a}_i\rho}{RT}$$

and hence, by Equations 86 and 87,

$$\ell n\hat{\phi}_i(\text{vdW}) = \frac{\tilde{b}_i\rho}{1-b\rho} - \frac{(a+\tilde{a}_i)\rho}{RT} - \ell n(1-b\rho)Z \quad (88)$$

Here, quantities $\tilde{a}_i$ and $\tilde{b}_i$ are partial equation-of-state parameters:

$$\tilde{a}_i \equiv \left[\frac{\partial(na)}{\partial n_i}\right]_{T,\rho,n_j}$$

$$\tilde{b}_i \equiv \left[\frac{\partial(nb)}{\partial n_i}\right]_{T,\rho,n_j}$$

Explicit expressions for $\tilde{a}_i$ and $\tilde{b}_i$ require explicit expressions for the mixing rules for parameters a and b.

Mixture Fugacity Behavior

Fugacity Coefficients, Activity Coefficients, and Henry's Constants. Component fugacity coefficients are readily obtained from a PVTx equation of state. For developing and testing equations of state for phase-equilibrium applications, however, it is sometimes useful to deal directly with quantities conventionally used for description of the liquid phase, e.g., activity coefficients and Henry's constants. We review in the following paragraphs the connections among these measures of component fugacity behavior, and illustrate how they are determined from pressure-explicit equations of state.

The fugacity coefficient is defined as

$$\hat{\phi}_i \equiv \frac{\hat{f}_i}{x_i P} \tag{89}$$

where $\hat{f}_i$ is the fugacity of species i in solution. The activity coefficient γ_i is

$$\gamma_i \equiv \frac{\hat{f}_i}{x_i f_i^\circ} \tag{90}$$

where f_i° is the standard-state fugacity. Two standard states are popularly employed: Lewis-Randall ("Raoult's-Law") standard states, for which f_i° is the fugacity of pure i at the mixture T and P,

$$f_i^\circ\ (\mathrm{LR}) = f_i \tag{91}$$

and Henry's-Law standard states, for which

$$f_i^\circ\ (\mathrm{HL}) = \mathcal{H}_i \tag{92}$$

where Henry's "constant" $\mathcal{H}_i$ is defined as

$$\mathcal{H}_i \equiv \lim_{x_i \to 0} \frac{\hat{f}_i}{x_i} \tag{93}$$

the limit being taken at the mixture T and P.

If we write Equation 90 as

$$\gamma_i = \frac{\hat{f}_i}{x_i P} \frac{P}{f_i^\circ} = \hat{\phi}_i \frac{P}{f_i^\circ}$$

then we see that all that is required to "convert" a fugacity coefficient to an activity coefficient is an expression for the ratio

f_i°/P. For Lewis-Randall standard states, we have

$$\frac{f_i^\circ}{P} = \frac{f_i}{P} = \phi_i$$

where ϕ_i is the fugacity coefficient of pure i. Thus the conventional Lewis-Randall activity coefficient is

$$\boxed{\gamma_i = \frac{\hat{\phi}_i}{\phi_i}} \tag{94}$$

For Henry's-Law standard states, we have

$$\frac{f_i^\circ}{P} = \frac{\mathcal{H}_i}{P} = \frac{1}{P} \lim_{x_i \to 0} \frac{\hat{f}_i}{x_i}$$

But

$$\lim_{x_i \to 0} \frac{\hat{f}_i}{x_i} = P \lim_{x_i \to 0} \hat{\phi}_i \equiv P\hat{\phi}_i^{\infty}$$

where $\hat{\phi}_i^{\infty}$ is the fugacity coefficient at infinite dilution. Thus

$$\boxed{\mathcal{H}_i = P\hat{\phi}_i^{\infty}} \tag{95}$$

and the Henry's-Law activity coefficient γ_i^* is

$$\boxed{\gamma_i^* = \frac{\hat{\phi}_i}{\hat{\phi}_i^{\infty}}} \tag{96}$$

Conversion from fugacity coefficients to activity coefficients and Henry's constants is thus straightforward. One needs in addition to the component fugacity coefficient $\hat{\phi}_i$ one or another of its limiting values, viz.,

$$\phi_i \equiv \lim_{x_i \to 1} \hat{\phi}_i$$

or

$$\hat{\phi}_i^{\infty} \equiv \lim_{x_i \to 0} \hat{\phi}_i$$

For a pressure-explicit equation of state, both of these are found as limits of Equation 86.

Consider as an example the van der Waals equation of state, for which $\hat{\phi}_i$ is given by Equation 88. For pure i, this equation yields as a special case

$$\ell n\phi_i(\mathrm{vdW}) = \frac{b_{ii}\rho_i}{1-b_{ii}\rho_i} - \frac{2a_{ii}\rho_i}{RT} - \ell n(1-b_{ii}\rho_i)Z_i \tag{97}$$

where the doubly-subscripted parameters refer to pure i. Equations 94, 88, and 97 when combined produce an expression for the Lewis-Randall activity coefficient implied by the van der Waals equation.

Evaluation of $\hat{\phi}_i^\infty$ (and hence of $\mathcal{H}_i$ or γ_i^*) requires a little more care, because the state of infinite dilution for a species in a multicomponent mixture can in principle be defined in many ways. The natural definition of this state is as that state for which x_i approaches zero as the i-free mole fractions x'_j remain constant. (Here, $x'_j \equiv x_j/\sum_k x_k$, where j,k ≠ i.) By this definition, Equation 88 yields

$$\ell n\hat{\phi}_i^\infty(\mathrm{vdW}) = \frac{\tilde{b}_i^\infty\rho'}{1-b'\rho'} - \frac{(a'+\tilde{a}_i^\infty)\rho'}{RT} - \ell n(1-b'\rho')Z' \tag{98}$$

For the general multicomponent case, the primed quantities in Equation 98 are i-free mixture properties. Parameters $\tilde{a}_i^\infty$ and $\tilde{b}_i^\infty$ are partial equation-of-state parameters, evaluated at the same i-free composition as the mixture properties. For the binary case, say of infinitely dilute solute 1 in solvent 2, Equation 98 reduces to the simple result

$$\ell n\hat{\phi}_1^\infty(\mathrm{vdW}) = \frac{\tilde{b}_1^\infty\rho_2}{1-b_{22}\rho_2} - \frac{(a_{22}+\tilde{a}_1^\infty)\rho_2}{RT} - \ell n(1-b_{22}\rho_2)Z_2 \tag{99}$$

Here, the 1-free "mixture" is just pure solvent 2.

Partial Equation-of-State Parameters. Composition is introduced into many analytical engineering equations of state via "one-fluid theory", in which an equation of state for a mixture is assumed to have the same functional form as that for the pure species. Component mole fractions appear explicitly only in mixing rules for the equation-of-state parameters. As illustrated by the example just considered, evaluation of $\hat{\phi}_i$, γ_i, $\mathcal{H}_i$, or γ_i^* then requires expressions for the partial equation-of-state parameters. Letting π denote a generic equation-of-state parameter, we define, analogously as for any partial property,

$$\bar{\pi}_i \equiv \left[\frac{\partial(n\pi)}{\partial n_i}\right]_{T,P,n_j} \tag{100}$$

or

$$\tilde{\pi}_i \equiv \left[\frac{\partial(n\pi)}{\partial n_i}\right]_{T,\rho,n_j} \tag{101}$$

Usually, parameter π is taken to depend at most upon temperature and composition; in such cases the restrictions to constant P or ρ are superfluous in these definitions, and

$$\bar{\pi}_i = \tilde{\pi}_i = \left[\frac{\partial(n\pi)}{\partial n_i}\right]_{T,n_j}$$

Note however that Equation 101 accommodates the currently-popular concept of "density-dependent mixing rules". Since pressure-explicit equations of state are favored for engineering applications, we henceforth consider only $\tilde{\pi}_i$.

Development and testing of mixing rules is a major area of research in applied thermodynamics, and new formulations appear regularly: see other papers in this volume. For concreteness, and to illustrate procedures, we treat here only the familiar "van der Waals prescription", according to which parameter π is quadratic (or, as a special case, linear) in mole fraction:

$$\pi = \sum_{k}\sum_{\ell} x_k x_\ell \pi_{k\ell} \tag{102}$$

Application of Equation 101 to Equation 102 yields on rearrangement the simple result

$$\tilde{\pi}_i = 2\sum_k x_k \pi_{ki} - \pi \tag{103}$$

where π is the mixture parameter, given by Equation 102. In deriving Equation 103, we assume that the parameters remain unchanged on permutation of subscripts: $\pi_{ik} = \pi_{ki}$.

For pure i, Equation 103 yields the expected result:

$$\lim_{x_i \to 1} \tilde{\pi}_i = \pi_{ii}$$

To evaluate $\tilde{\pi}_i^\infty$ (for infinitely dilute i), we first introduce i-free mole fractions into Equations 103 and 102, obtaining

$$\tilde{\pi}_i = 2x_i\pi_{ii} + 2(1-x_i)\sum_k{}' x_k' \pi_{ki} - \pi$$

where

$$\pi = x_i^2 \pi_{ii} + 2x_i(1-x_i)\sum' x'_k \pi_{ki} + (1-x_i)^2 \sum_k' \sum_\ell' x'_k x'_\ell \pi_{k\ell} \qquad (104)$$

Combination of these equations gives

$$\tilde{\pi} = x_i(2-x_i)\pi_{ii} + (1-x_i)^2 \sum_k' \sum_\ell' x'_k x'_\ell (2\pi_{ki} - \pi_{k\ell}) \qquad (105)$$

Here, the primed mole fractions are i-free mole fractions, and the primed sums specifically exclude species i. Equation 105 is entirely equivalent to Equation 103; taking the limit $x_i \to 0$ we obtain directly an expression for $\tilde{\pi}_i$ at infinite dilution of i:

$$\tilde{\pi}_i^\infty = \sum_k' \sum_\ell' x'_k x'_\ell (2\pi_{ki} - \pi_{k\ell}) \qquad (106)$$

Similarly, we find from Equation 104 in the limit as $x_i \to 0$ an expression for the i-free mixture parameter π':

$$\pi' = \sum_k' \sum_\ell' x'_k x'_\ell \pi_{k\ell} \qquad (107)$$

Equations 102, 103, 106, and 107 complete the apparatus required to evaluate $\hat{\phi}_i$ and related quantities from a pressure-explicit equation of state, with mixing rules given by the van der Waals prescription. With respect to examples treated previously, they apply in particular to the van der Waals equation of state, where π is identified with parameters a and b.

Numerical Examples

We illustrate the use of preceding material with two numerical examples, both for the van der Waals equation of state. First, consider the van der Waals excess functions, for which expressions are given by Equations 60 through 63. Calculation of numerical values of these quantities requires values for the equation-of-state parameters. For binary mixtures containing species 1 and 2, six parameters are needed: a_{11}, b_{11}, a_{22}, b_{22}, a_{12}, and b_{12}. The first four are found from information on the pure components; the interaction parameters a_{12} and b_{12} are estimated from combining rules or, preferably, from mixture data.

Since the application is to the representation of excess functions for liquid mixtures, it is reasonable to determine the a_{ii} and b_{ii} from data on pure liquids. Various combinations of properties are employed for this purpose; we choose here the liquid/vapor saturation pressure P_i^{sat}, the molar density ρ_i^ℓ of the saturated liquid, and the molar heat of vaporization $\Delta H_i^{\ell v}$.

A molar property change of vaporization $\Delta M_i^{\ell v}$ of pure fluid i is defined as

$$\Delta M_i^{\ell v} \equiv M_i^v(T,P_i^{sat}) - M_i^{\ell}(T,P_i^{sat}) \tag{108}$$

where the terms on the right side denote molar properties of pure i as saturated vapor and as saturated liquid. Since temperature and pressure are uniform, $\Delta M_i^{\ell v}$ is just a difference between constant - T,P,x residual functions:

$$\boxed{\Delta M_i^{\ell v} = M_i^{R,v} - M_i^{R,\ell}} \tag{109}$$

Here, the terms on the right are constant - T,P,x residual functions for pure i as saturated vapor and as saturated liquid:

$$M_i^{R,v} \equiv M_i^v(T,P_i^{sat}) - M_i^{ig}(T,P_i^{sat})$$

$$M_i^{R,\ell} \equiv M_i^{\ell}(T,P_i^{sat}) - M_i^{ig}(T,P_i^{sat})$$

Residual functions are readily determined from a PVTx equation of state by procedures reviewed earlier. Hence, by Equation 109, so also are property changes of vaporization.

Consider the molar heat of vaporization $\Delta H_i^{\ell v}$. By Equation 109,

$$\Delta H_i^{\ell v} = H_i^{R,v} - H_i^{R,\ell}$$

and, by the definition of H,

$$H_i^R = U_i^R + PV_i^R$$

where, by Equation 7,

$$U_i^R = U_i^r$$

Thus, by the last three equations,

$$\boxed{\Delta H_i^{\ell v} = \left(U_i^{r,v} - U_i^{r,\ell}\right) + P_i^{sat}\left(\frac{1}{\rho_i^v} - \frac{1}{\rho_i^{\ell}}\right)} \tag{110}$$

Equation 110 expresses the molar heat of vaporization in a form convenient for use with a pressure-explicit equation of state. Constant - T,V,x residual internal energies are found from Equation 12, and P_i^{sat} is related to ρ_i^{ℓ} and ρ_i^v by the equation of state itself.

For the van der Waals equation of state, we have by Equations 110, 36, and 34 that

$$\Delta H_i^{\ell v}(\text{vdW}) = a_{ii}(\rho_i^{\ell} - \rho_i^{v}) + P_i^{sat}\left(\frac{1}{\rho_i^{v}} - \frac{1}{\rho_i^{\ell}}\right) \tag{111}$$

where

$$P_i^{sat} = \frac{\rho_i^{\ell}RT}{1-b_{ii}\rho_i^{\ell}} - a_{ii}(\rho_i^{\ell})^2 \tag{112}$$

and

$$P_i^{sat} = \frac{\rho_i^{v}RT}{1-b_{ii}\rho_i^{v}} - a_{ii}(\rho_i^{v})^2 \tag{113}$$

Given experimental values for P_i^{sat}, ρ_i^{ℓ}, and $\Delta H_i^{\ell v}$ at specified T, one solves Equations 111, 112, and 113 for ρ_i^{v}, a_{ii}, and b_{ii}. The value of ρ_i^{v} so obtained is merely an intermediary quantity; because the equal-fugacity requirement for liquid/vapor equilibrium is not invoked, ρ_i^{v} is not necessarily the "true" saturation vapor density implied by the equation of the state at the specified T. (The equal-fugacity requirement would provide a fourth equation; vapor pressure P_i^{sat} would then be treated as an unknown, to be determined along with ρ_i^{v}, a_{ii}, and b_{ii}.)

The above-described procedure when applied separately to pure 1 and to pure 2 provides values for a_{11}, b_{11}, a_{22}, and b_{22}. To find a_{12} and b_{12} we assume the availability of data for H^E and V^E, each at a single composition. The working equations follow from Equations 60, 34, and 102, applied to the liquid phase:

$$H^E(\text{vdW}) = -(a\rho - x_1a_{11}\rho_1 - x_2a_{22}\rho_2) + PV^E(\text{vdW}) \tag{114}$$

$$P(\text{vdW}) = \frac{\rho RT}{1-b\rho} - a\rho^2 \tag{34}$$

$$V^E(\text{vdW}) = \rho^{-1} - x_1\rho_1^{-1} - x_2\rho_2^{-1} \tag{115}$$

$$a = x_1^2a_{11} + x_2^2a_{22} + 2x_1x_2a_{12} \tag{116}$$

$$b = x_1^2b_{11} + x_2^2b_{22} + 2x_1x_2b_{12} \tag{117}$$

Here, liquid molar densities ρ_1, ρ_2, and ρ are found as solutions to Equation 34, under appropriate assignments for the equation-of-state parameters. Agreement with experiment is forced for H^E and V^E at the single states from which a_{12} and b_{12} are determined; calculations of H^E and V^E at other states, or of other excess functions at any states, constitute extrapolations. When compared with experiment, these extrapolations provide tests of the van der Waals mixing rules and (to a lesser degree) of the ability of this equation of state to represent properties of the liquid phase.

The literature abounds with such comparisons and it is not our purpose to survey them here. Instead we show in Figure 1 predicted values of the scaled excess functions V^E/x_1x_2, G^E/x_1x_2RT, H^E/x_1x_2RT, S^E/x_1x_2R, and $C_P{}^E/x_1x_2R$ for the system argon(1)/krypton(2). Pure-component parameters were estimated as described above from saturation data compiled by Vargaftik (5). Parameters a_{12} and b_{12} were estimated from equimolar values of H^E and V^E at zero pressure, as given by the correlations of Lewis et al. (6) for H^E at 116.9 K and Davies et al. (7) for V^E at 115.77 K.

Both pressure and temperature effects are illustrated in Figure 1. Particularly to be noted are the essential equivalence of the 0 bar and 1 bar isobars at 120 K. This justifies the frequent use of the zero-pressure liquid state in calculations of excess functions from equations of state. For example. one obtains for the van der Waals equation at zero pressure an explicit expression for the liquid density:

$$\rho^{\ell}(\text{vdW};P=0) = \frac{1}{2b}\left(1 + \sqrt{1 - \frac{4bRT}{a}}\right)$$

Moreover, the expression for the excess enthalpy simplifies to

$$H^E(\text{vdW};P=0) = -\left(a\rho - \sum_i x_i a_{ii}\rho_i\right)$$

Next we examine the composition dependence of Henry's constant $\mathcal{H}_{1;2,3}$ for solute species 1 in a mixed solvent containing species 2 and 3. Here, the solute-free mole fractions x_2' and x_3' are appropriate measures of composition. Subtleties of behavior are nicely displayed through the "excess" quantity $\ell n\mathcal{H}^E_{1;2,3}$, defined as

$$\ell n\mathcal{H}^E_{1;2,3} \equiv \ell n\mathcal{H}_{1;2,3} - x_2'\ell n\mathcal{H}_{1;2} - x_3'\ell n\mathcal{H}_{1;3} \qquad (118)$$

where $\mathcal{H}_{1;2}$ and $\mathcal{H}_{1;3}$ are Henry's constants for species 1 in pure solvents 2 and 3. The comparison in Equation 118 is made at uniform T,P, and composition, and $\ell n\mathcal{H}^E_{1;2,3}$ is identically zero if the three species form an ideal solution; however, $\ell n\mathcal{H}^E_{1;2,3}$ is not a true excess function as defined by Equation 37, because $\ell n\mathcal{H}_{1;2,3}$ is not a mixture molar property.

According to Equation 95, Henry's constant is proportional to the fugacity coefficient at infinite dilution. Combination of Equations 118 and 95 thus yields the general result

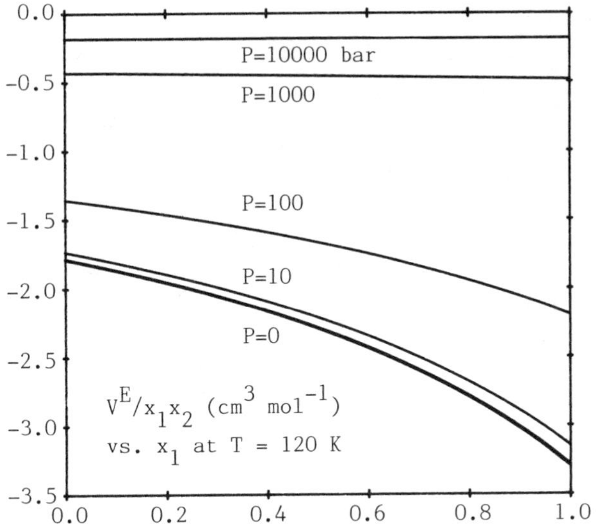

(A)

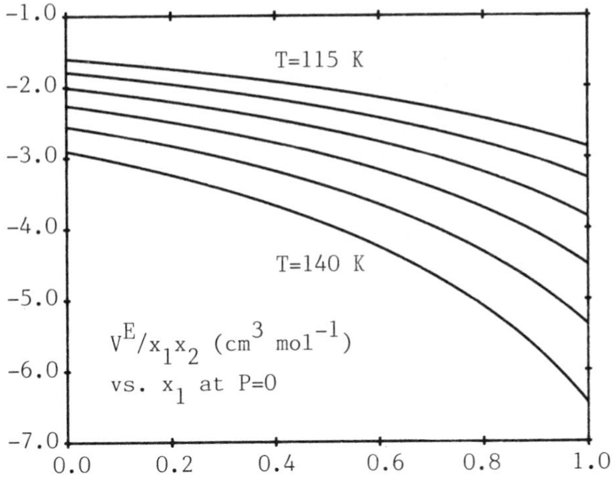

(B)

Figure 1 (A,B). Scaled van der Waals excess functions for liquid mixtures of argon(1) and krypton(2). (The 10,000-bar isobar is not shown in A).

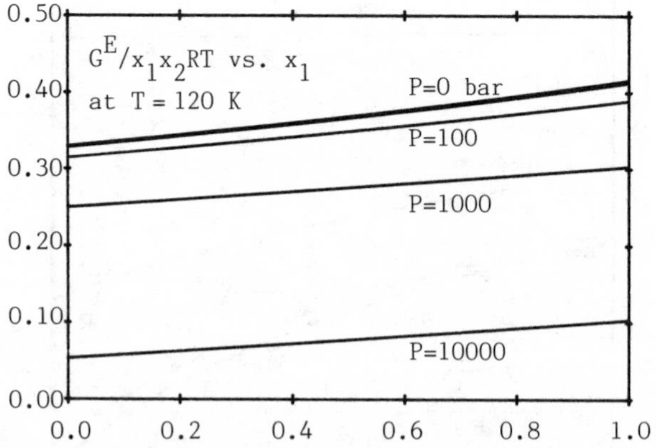

(C)

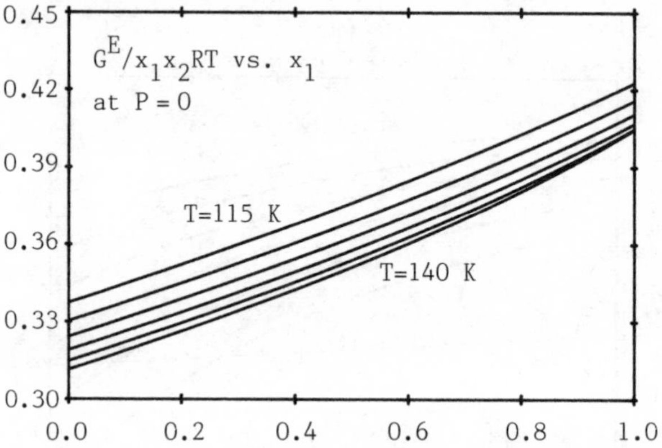

(D)

Figure 1 (C,D). Scaled van der Waals excess functions for liquid mixtures of argon(1) and krypton(2).

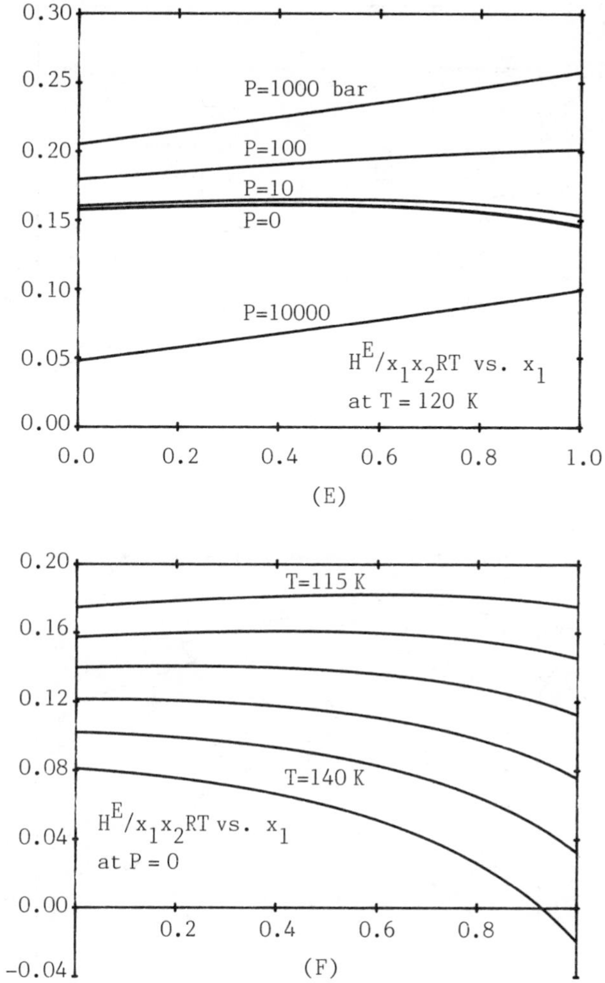

Figure 1 (E,F). Scaled van der Waals excess functions for liquid mixtures of argon(1) and krypton(2).

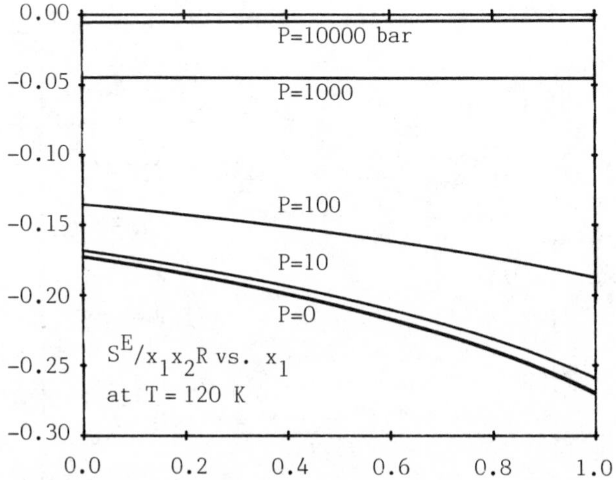

(G)

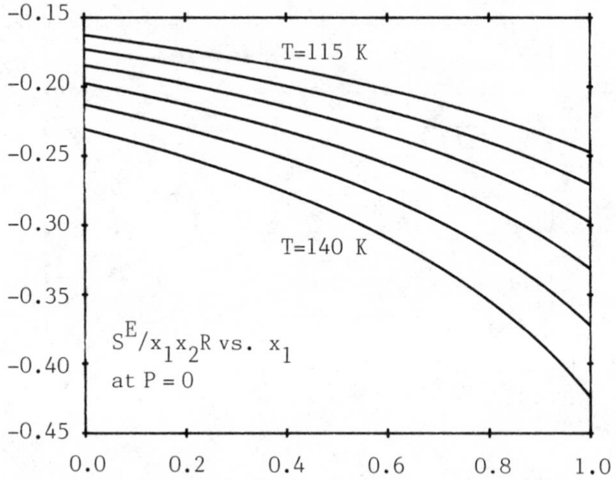

(H)

Figure 1 (G,H). Scaled van der Waals excess functions for liquid mixtures of argon(1) and krypton(2).

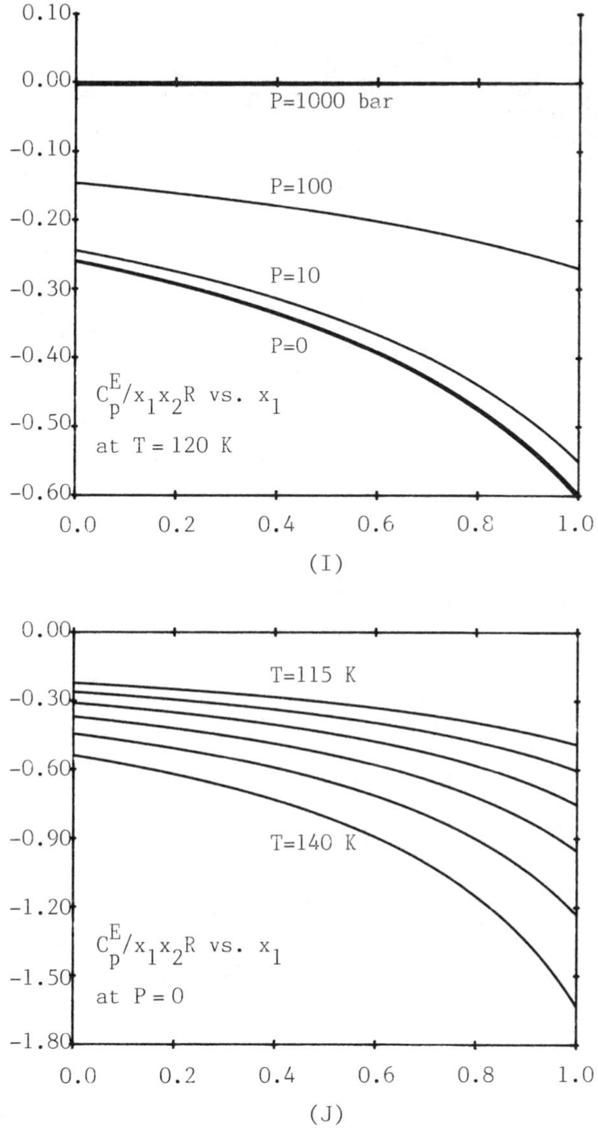

Figure 1 (I,J). Scaled van der Waals excess functions for liquid mixture of argon (1) and krypton (2).

$$\ln \mathcal{H}^E_{1;2,3} = \ln \hat{\phi}^\infty_{1;2,3} - x'_2 \ln \hat{\phi}^\infty_{1;2} - x'_3 \ln \hat{\phi}^\infty_{1;3} \tag{119}$$

For the van der Waals equation, $\ln \hat{\phi}^\infty_i$ is given by Equation 98. With quadratic mixing rules for parameters a and b, we find from Equations 98, 99, 106, and 107 the following expressions for the fugacity coefficients:

$$\ln \hat{\phi}^\infty_{1;2,3}(\mathrm{vdW}) = \frac{2(x'_2 b_{12} + x'_3 b_{13}) - b'}{1 - b'\rho'} \rho' - \frac{2(x'_2 a_{12} + x'_3 a_{13})\rho'}{RT} - \ln(1 - b'\rho')Z' \tag{120}$$

$$\ln \hat{\phi}^\infty_{1;2}(\mathrm{vdW}) = \frac{(2b_{12} - b_{22})\rho_2}{1 - b_{22}\rho_2} - \frac{2a_{12}\rho_2}{RT} - \ln(1 - b_{22}\rho_2)Z_2 \tag{121}$$

$$\ln \hat{\phi}^\infty_{1;3}(\mathrm{vdW}) = \frac{(2b_{13} - b_{33})\rho_3}{1 - b_{33}\rho_3} - \frac{2a_{13}\rho_3}{RT} - \ln(1 - b_{33}\rho_3)Z_3 \tag{122}$$

Solute-free mixture parameters a' and b' are given by

$$a' = (x'_2)^2 a_{22} + (x'_3)^2 a_{33} + 2x'_2 x'_3 a_{23} \tag{123}$$

$$b' = (x'_2)^2 b_{22} + (x'_3)^2 b_{33} + 2x'_2 x'_3 b_{23} \tag{124}$$

and liquid densities and compressibility factors are found as solutions to the equation of state.

For the present exercise, ten parameters are needed: a_{22}, b_{22}, a_{33}, b_{33}, a_{12}, b_{12}, a_{13}, b_{13}, a_{23}, and b_{23}. Ideally, the pure-component parameters would be estimated from liquid-phase data (e.g., as in the excess-function example), and the interaction parameters from liquid/vapor equilibria and gas-solubility data. We adopt for this example a more straightforward approach. Parameters a_{ii} and b_{ii} are estimated from critical constants T_{c_i} and V_{c_i} via the classical relations

$$a_{ii} = \frac{9}{8} RT_{c_i} V_{c_i} \tag{125}$$

$$b_{ii} = \frac{1}{8} V_{c_i} \tag{126}$$

Interaction parameters are determined from the conventional combining rules

$$a_{ij} = (1-k_{ij})(a_{ii}a_{jj})^{1/2} \tag{127}$$

$$b_{ij} = (1-\ell_{ij})(b_{ii}+ b_{jj})/2 \tag{128}$$

where parameters k_{ij} and ℓ_{ij} are pure numbers, of absolute value less than unity. We wish to illustrate the effects of varying k_{ij} and ℓ_{ij} on the "excess" quantity $\ell n \mathcal{H}^E_{1;2,3}$, i.e., the sensitivity of mixed-solvent Henry's constants to the numerical values of the interaction parameters.

The parametric study is done for a simulated ternary system at 300 K and 1 bar, in which hydrogen(1) is the solute, and n-heptane(2) and n-decane(3) compose the mixed solvent. Parameters for the pure fluids are obtained from Equations 125 and 126 and parameters a_{23} and b_{23} are fixed once and for all by setting $k_{23} = \ell_{23} = 0$ in Equations 127 and 128. Assignment of numerical values to k_{12}, ℓ_{12}, k_{13} and ℓ_{13} then permits calculation of $\ell n \mathcal{H}^E_{1;2,3}$ from Equations 119 through 124.

Numerical results are displayed on Figure 2. Figures 2A and 2B illustrate the effects of independently varying the energy interaction parameters; here, we have set $\ell_{12} = \ell_{13} = 0$. Figures 2C and 2D similarly show the effects of varying ℓ_{12} and ℓ_{13}, with $k_{12} = k_{13} = 0$. The results confirm that mixed-solvent Henry's constants, like excess functions for liquid mixtures, can serve as probes for assessing mixing rules and combining rules for PVTx equations of state.

Closure

Connections between the PVTx equation-of-state formalism and the conventional apparatus of classical solution thermodynamics are cleanly exposed through a few unifying concepts, e.g., generalized deviation functions, generalized partial properties, and component fugacity coefficients. We have found the notion of partial equation-of-state parameters to be particularly helpful, because it allows one to postpone questions relating to composition dependence until they really need to be addressed.

Much of the substance of this communication resides in definitions and generalizations, and in the summaries of working formulas collected in the tables. To keep the paper to a reasonable length, we have provided examples and illustrations for but a single

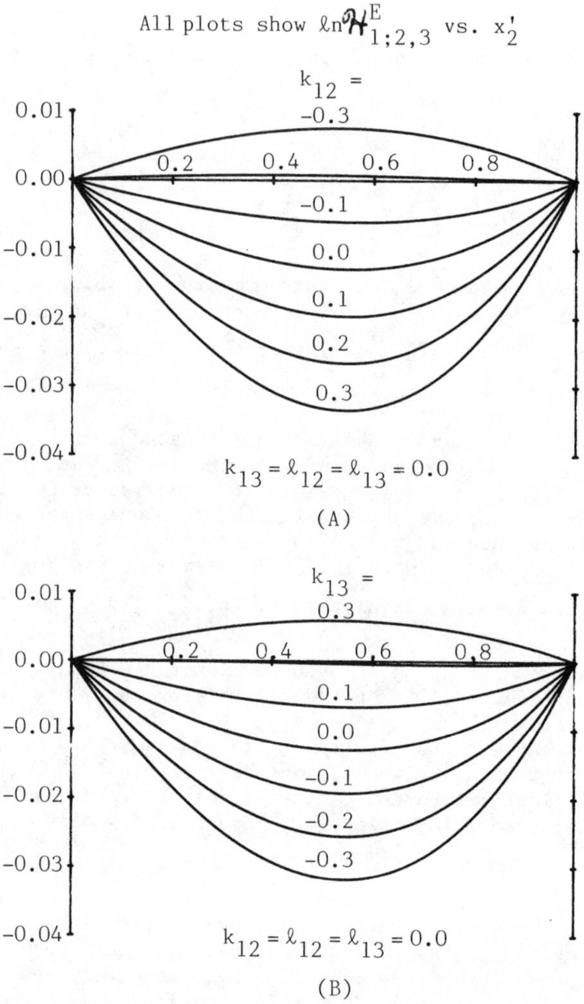

Figure 2 (A,B). van der Waals Henry's constants at 300K and 1 bar, for hydrogen(1) in mixed solvents containing n-heptane(2) and n-decane(3).

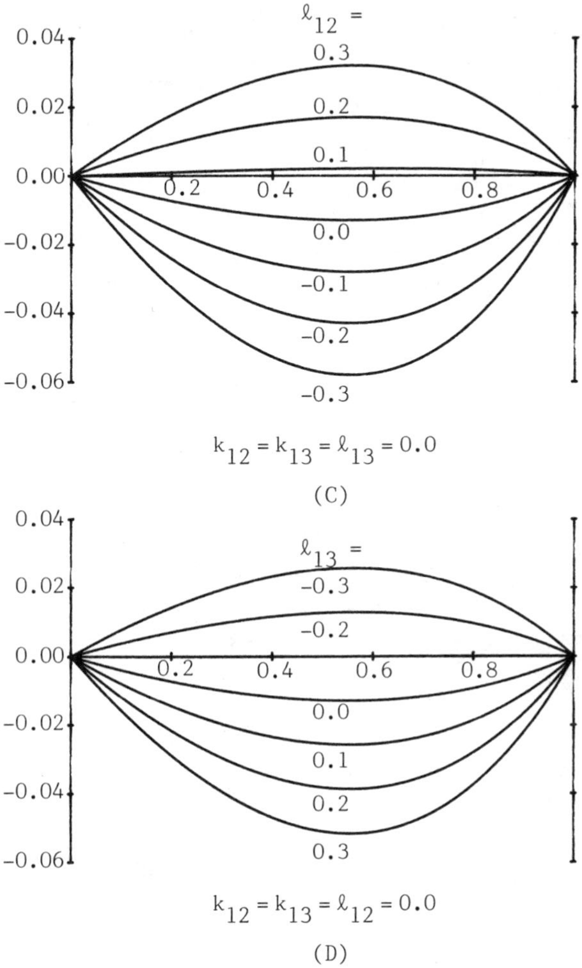

Figure 2 (C,D). van der Waals Henry's constants at 300K and 1 bar, for hydrogen(1) in mixed solvents containing n-heptane(2) and n-decane(3).

equation of state: the van der Waals equation. The principles and procedures are of course easily applied to other, more realistic equations of state.

Exercises in synthesis necessarily build on precedents, in this case too diffuse and numerous to cite in detail. We are however pleased to acknowledge as general sources of inspiration the published researches of P.T. Eubank, K.R. Hall, the late A. Kreglewski, M.L. McGlashan, K.N. Marsh, J. Mollerup, S.I. Sandler, R.L. Scott, K.E. Starling, and J. Vidal. To these and to other equation-of-state enthusiasts we acknowledge our indebtedness.

Acknowledgments

This work was partially supported by the National Science Foundation under Grant No. CPE-8311785. K.K.N. is pleased to acknowledge support provided by an AAUW Educational Foundation Fellowship, in the form of Conoco Engineering/Mobil Awards. Eugene Dorsi assisted with the calculations.

Legend of Symbols

a, b	=	parameters in van der Waals equation of state
$\hat{a}_i$	=	activity of species i
A	=	molar Helmholtz energy
C_P, C_V	=	molar heat capacities
f_i	=	fugacity of pure i
$\hat{f}_i$	=	fugacity of species i in solution
f_i°	=	standard-state fugacity of species i
G	=	molar Gibbs energy
H	=	molar enthalpy
$\mathcal{H}_i$	=	Henry's constant for species i
M	=	arbitrary molar, or intensive (e.g. M = P), property
M^d	=	constant - T,V,x deviation function
M^D	=	constant - T,P,x deviation function
M^e	=	constant - T,V,x excess function
M^E	=	constant - T,P,x excess function
M^r	=	constant - T,V,x residual function
M^R	=	constant - T,P,x residual function
M^t	=	total property $\equiv$ nM
$\hat{M}_i$	=	generalized partial property
$\bar{M}_i$	=	constant - T,P partial property
$\tilde{M}_i$	=	constant - T,V partial property
ΔM	=	constant - T,P,x property change of mixing

$\Delta M_i^{\ell v}$ = molar property change of vaporization of pure i

n = amount of substance ("mole number")

P = pressure

P_i^{sat} = liquid/vapor saturation pressure of pure i

R = universal gas constant

S = molar entropy

T = absolute temperature

U = molar internal energy

V = molar volume

x = mole fraction

X,Y = arbitrary intensive variables

Z = compressibility factor ≡ PV/RT

Greek Letters

β = volume expansivity ≡ $V^{-1}(\partial V/\partial T)_{P,x}$

γ_i = activity coefficient of species i

κ = isothermal compressibility ≡ $-V^{-1}(\partial V/\partial P)_{T,x}$

λ_i = absolute activity of species i

μ_i = chemical potential of species i

π = arbitrary equation-of-state parameter

ρ = molar density

τ = "coldness" ≡ T^{-1}

ϕ_i = fugacity coefficient of pure i

$\hat{\phi}_i$ = fugacity coefficient of species i in solution

Superscripts

id = denotes an ideal-solution property

ig = denotes an ideal-gas property

mod = denotes a model mixture property

∞ = denotes a property of a species at infinite dilution

Literature Cited

1. Van Ness, H.C.; Abbott, M.M. "Classical Thermodynamics of Nonelectrolyte Solutions: With Applications to Phase Equilibria," Appendix C, McGraw-Hill, New York, 1982.

2. Abbott, M.M.; Prausnitz, J.M. "Generalized van der Waals Theory: A Classical View," manuscript in preparation.

3. Reis, J.C.R. "Theory of Partial Molar Properties," J. Chem. Soc., Faraday Trans. II 1982, 78, 1595.

4. Abbott, M.M. "Higher-Order Partial Properties," seminar presented at the University of California, Berkeley, 21 Sept. 1983; unpublished notes.

5. Vargaftik, N.B. "Tables on the Thermophysical Properties of Liquids and Gases," 2nd Edition, Wiley, New York, 1975.

6. Lewis, K.L.; Lobo, L.Q.; Staveley, L.A.K. "The Thermodynamics of Liquid Mixtures of Argon + Krypton," J. Chem. Thermodynamics 1978, 10, 351.

7. Davies, R.H.; Duncan, A.G.; Saville, G.; Staveley, L.A.K. "Thermodynamics of Liquid Mixtures of Argon and Krypton," Trans. Far. Soc. 1967, 63, 855.

RECEIVED November 5, 1985

DATA, PHENOMENA, AND CRITIQUE

2

The Collinearity of Isochores at Single- and Two-Phase Boundaries for Fluid Mixtures

John S. Rowlinson[1], Gunter J. Esper[2,3] James C. Holste[2], Kenneth R. Hall[2], Maria A. Barrufet[2], and Philip T. Eubank[2]

[1]Physical Chemistry Laboratory, Oxford University, Oxford OX1 3QZ, United Kingdom
[2]Department of Chemical Engineering, Texas A&M University, College Station, TX 77843

Fluid isochores for mixtures of fixed overall composition generally change slope on passing across the isopleth (dew-bubble point curve, DBC) from the homogeneous to heterogeneous phase region on a pressure/temperature diagram. A thermodynamic proof is given which shows the isochores to be always collinear at the cricondentherm (or any temperature extremum), rather than at the mixture critical point. The proof agrees with a different, more mathematical proof given earlier by Griffiths. These thermodynamic proofs are supported by our new, high-precision density data for the CH_4/CO_2 and N_2/CO_2 equimolar binaries in addition to previously published measurements for $^3He/^4He$ and for H_2/CH_4 at Duke University and Rice University, respectively.

Collinearity of the isentropes at the cricondenbar is also demonstrated and supported by recent calculations using our CH_4/CO_2 data. Because fluid isochores for both pures and mixtures in the homogeneous region are approximately linear, most accurate equations of state (EOS) begin with this premise and then add correction terms for curvature. The present results contain thermodynamic constraints for such EOS.

Figure 1 illustrates a typical dew-bubble point curve (DBC) or isopleth for a binary mixture of fixed overall mol fraction, z_1. The mixture critical point (CP) is shown to lie between the point of maximum pressure, the cricondenbar (CB), and the point of maximum temperature, the cricondentherm (CT). However, if the slope of the critical locus, (dP_c/dT_c), is positive for a particular z_1 (1 = more volatile component), then the CP will lie outside the CB/CT gap. This occurrence is usual for $z_1 \geq 0.8$ causing a CP to the left of the

[3]Current address: Institut für Thermo- und Fluid-dynamik, Ruhr-Universität, D-4630 Bochum 1, Federal Republic of Germany.

0097-6156/86/0300-0042$06.00/0

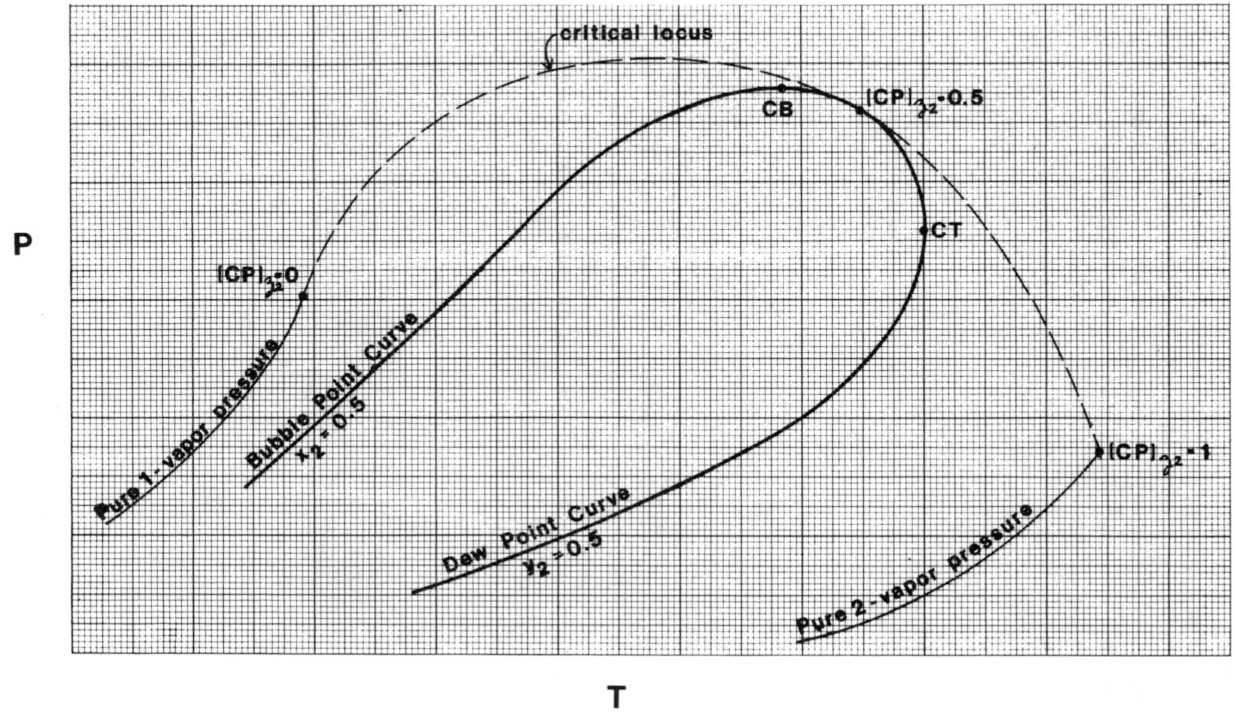

Figure 1. A pressure (P)/temperature (T) diagram showing the critical locus and DBC (isopleth) curve for a typical binary mixture of constant composition (here, equimolar). CB and CT are the cricondenbar and cricondentherm, respectively, for the mixture whereas CP denotes the critical point for the mixture as well as that of the two pure components.

CB but also happens for $z_1 \leq 0.1$ when $(dP_c/dT_c)|_{CP2} > 0$ causing a CP below the CT [1].

The saturation density increases monotonically as we traverse the DBC as in Figure 2 from CT to CP to CB. The low density isochores are shown to have a steeper slope on the two-phase side of the DBC. We prove that the limiting slopes at the DBC from the single and two-phase sides are identical at the CT. Above the CT, the isochoric slope from the single phase side is steeper. These results are independent of the location of the CP, which indeed fails to exist for some mixtures of fixed composition.

BASIC IDENTITIES

Imagine a fixed volume cell containing liquid of total volume V^ℓ and gas of total volume V^g. Then,

$$dV^\ell = V_T^\ell\ (dT) + V_P^\ell\ (dP) + \bar{V}_1^\ell\ (dn_1) + \bar{V}_2^\ell\ (dn_2) \tag{1}$$

where

$$V_T^\ell \equiv (\partial V^\ell/\partial T)_{P,n_1^\ell,n_2^\ell}\ ,\ V_P^\ell \equiv (\partial V^\ell/\partial P)_{T,n_1^\ell,n_2^\ell}$$

and $dn_i \equiv dn_i^\ell$. The addition of a similar equation for dV^g to Equation 1 provides

$$(V_T^\ell + V_T^g)\ dT + (V_P^\ell + V_P^g)\ dP = \Delta\bar{V}_1\ (dn_1) + \Delta\bar{V}_2\ (dn_2) \tag{2}$$

where $\Delta\bar{V}_i \equiv (\bar{V}_i^g - \bar{V}_i^\ell)$ is the difference of partial molar volumes between the two phases. Two further equations connecting the four variables (T, P, n_1^ℓ, n_2^ℓ) follow from the equilibrium conditions of $d\hat{\mu}_i^\ell = d\hat{\mu}_i^g$ [2]. First

$$d\hat{\mu}_1^\ell = -\bar{S}_1^\ell\ (dT) + \bar{V}_1^\ell\ (dP) + (\partial\hat{\mu}_1/\partial n_1)_{T,P,n_2^\ell}\ (dn_1) + (\partial\hat{\mu}_1/\partial n_2)_{T,P,n_1^\ell}\ (dn_2) \tag{3}$$

where the sum of the last two terms is also $(\partial\hat{\mu}_1/\partial x_1)_{T,P}\ (dx_1)$ with x_1 the mol fraction in the liquid phase. Equating a similar expression for $d\hat{\mu}_1^g$ to Equation 3 results in

$$\Delta\bar{S}_1\ (dT) - \Delta\bar{V}_1\ (dP) = (\partial\hat{\mu}_1/\partial y_1)_{T,P}\ (dy_1) - (\partial\hat{\mu}_1/\partial x_1)_{T,P}\ (dx_1) \tag{4}$$

where y_1 is the mol fraction in the vapor phase. The analogous equation for the less volatile component is

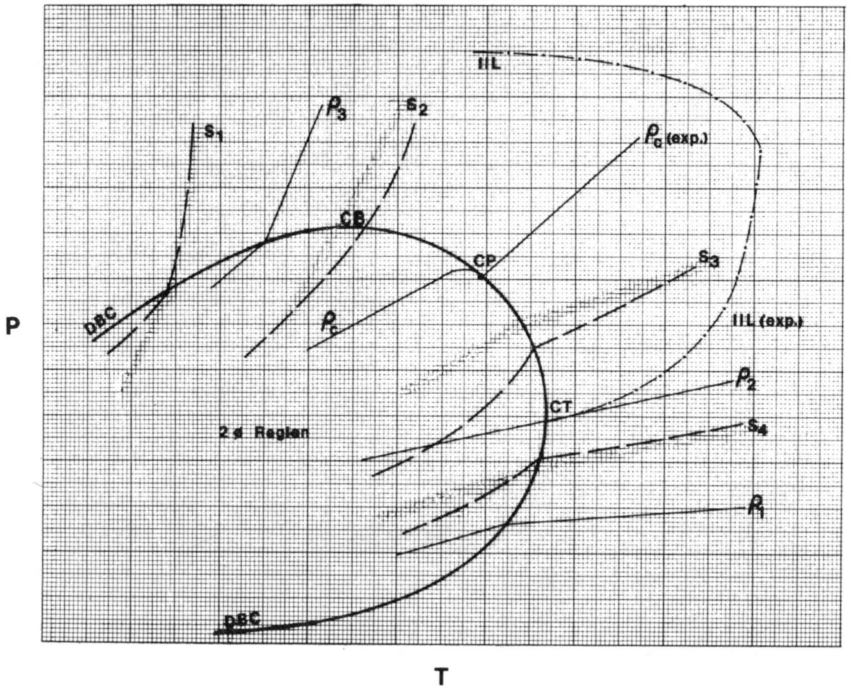

Figure 2. A qualitative diagram exhibiting the nature of isochores and isentropes in both the single and two-phase regions for a binary mixture of fixed composition. Note the collinearity of the isochores (lines of constant density, ρ) at the CT. The isentropes (lines of constant entropy, S) are collinear at the CB. Both the density and entropy are monotonic tracing the DBC and are subscripted in order of increasing magnitude. IIL is the isochoric inflection locus.

$$\Delta\bar{S}_2\ (dT) - \Delta\bar{V}_2\ (dP) = (\partial\hat{\mu}_2/\partial y_1)_{T,P}\ (dy_1) - (\partial\hat{\mu}_2/\partial x_1)_{T,P}\ (dx_1) \qquad (5)$$

Application of the Gibbs-Duhem equation to each phase yields $x_1\ (\partial\hat{\mu}_1/\partial x_1)_{P,T} = -\ x_2\ (\partial\hat{\mu}_2/\partial x_1)_{P,T}$ and $y_1\ (\partial\hat{\mu}_1/\partial y_1)_{P,T} = -y_2\ (\partial\hat{\mu}_2/\partial y_1)_{P,T}$. Multiplication of Equation (4) by x_1 and Equation (5) by x_2 followed by addition yields

$$\langle x_i\Delta\bar{V}_i\rangle\ (dP) - \langle x_i\Delta\bar{S}_i\rangle\ (dT) = (\Delta x/y_2)\ (\partial\hat{\mu}_1/\partial y_1)_{P,T}\ (dy_1) \qquad (6)$$

where $\langle x_i\Delta\bar{V}_i\rangle \equiv x_1\Delta\bar{V}_1 + x_2\Delta\bar{V}_2$ and $\Delta x \equiv (y_1 - x_1)$. Likewise, multiplication of Equation (4) by y_1 and Equation (5) by y_2 followed by addition provides

$$\langle y_i\Delta\bar{V}_i\rangle\ (dP) - \langle y_i\Delta\bar{S}_i\rangle\ (dT) = (\Delta x/x_2)\ (\partial\hat{\mu}_1/\partial x_1)_{P,T}\ (dx_1) \qquad (7)$$

Before combining Equations (2), (6) and (7), we modify Equation (2) by first noting that

$$x_2\ (dn_1/dT)_{V,n_1,n_2} = x_1\ (dn_2/dT)_{V,n_1,n_2} + n^{\ell}\ (dx_1^{\ell}/dT)_{V,n_1,n_2} \qquad (8)$$

and that a similar equation can be written for the gas phase. The reader is reminded that $dn_1 \equiv dn_1^{\ell}$ whereas $n_1 = n_1^{\ell} + n_1^{g}$. Multiplication of Equation (8) by y_1 and of the gas phase equation by x_1 followed by subtraction yields

$$(y_1 - x_1)\ (dn_1/dT) = n^{\ell}y_1\ (dx_1/dT) + n^{g}x_1\ (dy_1/dT) \qquad (9)$$

where all total derivatives indicate a constancy of total volume V, n_1 and n_2. Likewise,

$$(y_1 - x_1)\ (dn_2/dT) = n^{g}x_2\ (dy_1/dT) + n^{\ell}y_2\ (dx_1/dT) \qquad (10)$$

Equation (2) together with Equations (9) and (10) for (dn_1/dT) and (dn_2/dT), respectively, becomes

$$(dP/dT) = \frac{n^{\ell}(dx_1/dT)\langle y_i\Delta\bar{V}_i\rangle + n^{g}(dy_1/dT)\langle x_i\Delta\bar{V}_i\rangle - (V_T^{\ell}+V_T^{g})(\Delta x)}{(V_P^{\ell} + V_P^{g})\ (\Delta x)} \qquad (11)$$

Equation (6) shows that

$$(dy_1/dT) = [\langle x_i\Delta\bar{V}_i\rangle(dP/dT) - \langle x_i\Delta\bar{S}_i\rangle]\ y_2/(\Delta x)\ (\partial\hat{\mu}_1/\partial y_1)_{P,T} \qquad (12)$$

where

$$(\partial\hat{\mu}_1/\partial y_1)_{P,T} = \langle x_i\Delta\bar{V}_i\rangle(\partial P/\partial y_1)_T\ (y_2/\Delta x) = -\langle x_i\Delta\bar{S}_i\rangle(\partial T/\partial y_1)_P\ (y_2/\Delta x) \tag{13}$$

Substitution of Equations (12) and (13) into Equation (11) along with analogous equations for (dx_1/dT) and $(\partial\hat{\mu}_1/\partial x_1)_{P,T}$ results in

$(dP/dT) =$

$$\frac{n^{\ell}\langle y_i\Delta\bar{V}_i\rangle[(\frac{\partial x_1}{\partial T})_P+(\frac{\partial x_1}{\partial P})_T(\frac{dP}{dT})]+n^{g}\langle x_i\Delta\bar{V}_i\rangle[(\frac{\partial y_1}{\partial T})_P+(\frac{\partial y_1}{\partial P})_T(\frac{dP}{dT})]-(V_T^{\ell}+V_T^{g})(\Delta x)}{(V_P^{\ell}+V_P^{g})(\Delta x)}. \tag{14}$$

This equation is solved for (dP/dT) with the quality, $q \equiv [n^g/(n^g + n^{\ell})]$, introduced as well as $v_T^{\ell} \equiv (\partial v^{\ell}/\partial T)_{P,x_1} \equiv (V_T^{\ell}/n^{\ell})$, v_T^g, v_P^{ℓ} and v_P^g:

$-(dP/dT) =$

$$\frac{[(1-q)v_T^{\ell}+qv_T^{g}](\Delta x)+(1-q)\langle y_i\Delta\bar{V}_i\rangle(\partial x_2/\partial T)_P+q\langle x_i\Delta\bar{V}_i\rangle(\partial y_2/\partial T)_P}{[(1-q)v_P^{\ell}+qv_P^{g}](\Delta x)+(1-q)\langle y_i\Delta\bar{V}_i\rangle(\partial x_2/\partial P)_T+q\langle x_i\Delta\bar{V}_i\rangle(\partial y_2/\partial P)_T} \tag{15}$$

Because $(\partial P/\partial T)_{x_2} = -(\partial x_2/\partial T)_P\ (\partial P/\partial x_2)_T$, an alternate form of Equation (15) is

$$-(dP/dT) = \frac{[(1-q)v_T^{\ell}+qv_T^{g}](\Delta x)-(\partial P/\partial T)_{x_2}\cdot\eta^{\ell}-(\partial P/\partial T)_{y_2}\cdot\eta^{g}}{[(1-q)v_P^{\ell}+qv_P^{g}](\Delta x)+\eta^{\ell}+\eta^{g}} \tag{16}$$

where $\eta^{\ell} \equiv (1-q)\langle y_i\Delta\bar{V}_i\rangle(\partial x_2/\partial P)_T$ and $\eta^{g} \equiv q\langle x_i\Delta\bar{V}_i\rangle(\partial y_2/\partial P)_T$

PURE COMPONENT CHECK

Both η^{ℓ} and η^{g} are everywhere zero for a pure compound causing Equation (16) to reduce to

$$-(dP/dT) = [(1-q)\ v_T^{\ell} + qv_T^{g} - (\Delta v)\ (dq/dT)]/[1-q)v_P^{\ell} + qv_P^{g}] \tag{17}$$

For either pure components or mixtures of fixed overall composition, quality lines form a family of curves in the heterogeneous region issuing from the critical point. Any continuous path through the heterogeneous region ending at the CP results in $(dq/dT) = 0$ at the CP via straight line tangency. For any fixed value of q nearing the CP, Equation (17) contains $v_T^\ell \to v_T^g$ and $v_P^\ell \to v_P^g$ so that $(dP^\sigma/dT) = -(v_T^f/v_P^f)_c = (\partial P/\partial T)_{\rho_c}$. Here the homogeneous phase side derivaties v_T^f and v_P^f, (f = fluid), are both divergent but their ratio is identically the slope of the critical isochore at the CP. This collinearity of the vapor pressure curve with the critical isochore was known to van der Waals [3].

BINARY MIXTURE AT THE CRICONDENTHERM (CT)

At the CT, $(\partial P/\partial T)_{y_2}$ and $(\partial P/\partial y_2)_T$ are infinite while Δx, $(\partial T/\partial y_2)_P$ and $\langle x_i \Delta \bar{S}_i \rangle$ are finite so that $\langle x_i \Delta \bar{V}_i \rangle$ is zero from Equation (13). With q unity, $\eta^g = 0$ and $(\partial P/\partial T)_{y_2} \cdot \eta^g = -q \langle x_i \Delta \bar{V}_i \rangle \cdot (\partial y_2/\partial T)_P^{CT} = (1) \cdot (0) \cdot (f) = 0$, where f = finite (nonzero). Equation (16) then reduces to $-(dP/dT) = (v_T^g/v_P^g) = -(\partial P/\partial T)_\rho$ or collinearity is obtained for any extremum (maximum or minimum) in the temperature on the DBC. We have assumed here that the CT lies on the dew-point curve but in the unusual case where it is on the bubble-point curve the conclusions are identical. Figure 2 is a pressure/temperature diagram showing the qualitative behavior of fluid isochores for a binary mixture exhibiting a classical DBC including a vapor/liquid critical point. Figure 3 is the analogous qualitative diagram for equimolar CO_2/N_2 which has no CP nor CB but a CT and a minimum temperature (MT). As the density of the homogeneous phase increases it is termed first "gas", then "liquid" and, finally, once again "gas". Both the isochores ρ_3 and ρ_5 at CT and MT, respectively, are collinear.

ISOCHORIC COLLINEARITY PROOF OF GRIFFITHS

Levelt Sengers [4] has noted an earlier proof due to Griffiths which appeared as an appendix in the important article of Doiron, Behringer and Meyer [5], which contains their $^3He/^4He$ isochoric density measurements. Our previous discussions with engineers and chemists leads us to believe that this proof is known mostly to a limited number of physicists. We repeat here a backwards version of this terse, brilliant proof to call it to the attention of a wider audience. Compared to the proof given earlier, Griffiths' proof is more concise and mathematical. The reader can take his pick because these two different proofs lead to the same conclusion.

A standard mathematical argument is employed thrice to relate various quantities across the DBC. Let $\xi(\psi,\omega)$ be a continuous function whose derivatives are discontinuous along a curve in the (ψ,ω) plane separating regions I and II. The differential of ξ along this curve is

$$d\xi = (\partial\xi/\partial\psi)_\omega^I \, d\psi + (\partial\xi/\partial\omega)_\psi^I \, d\omega = (\partial\xi/\partial\psi)_\omega^{II} d\psi + (\partial\xi/\partial\omega)_\psi^{II} \, d\omega \qquad (18)$$

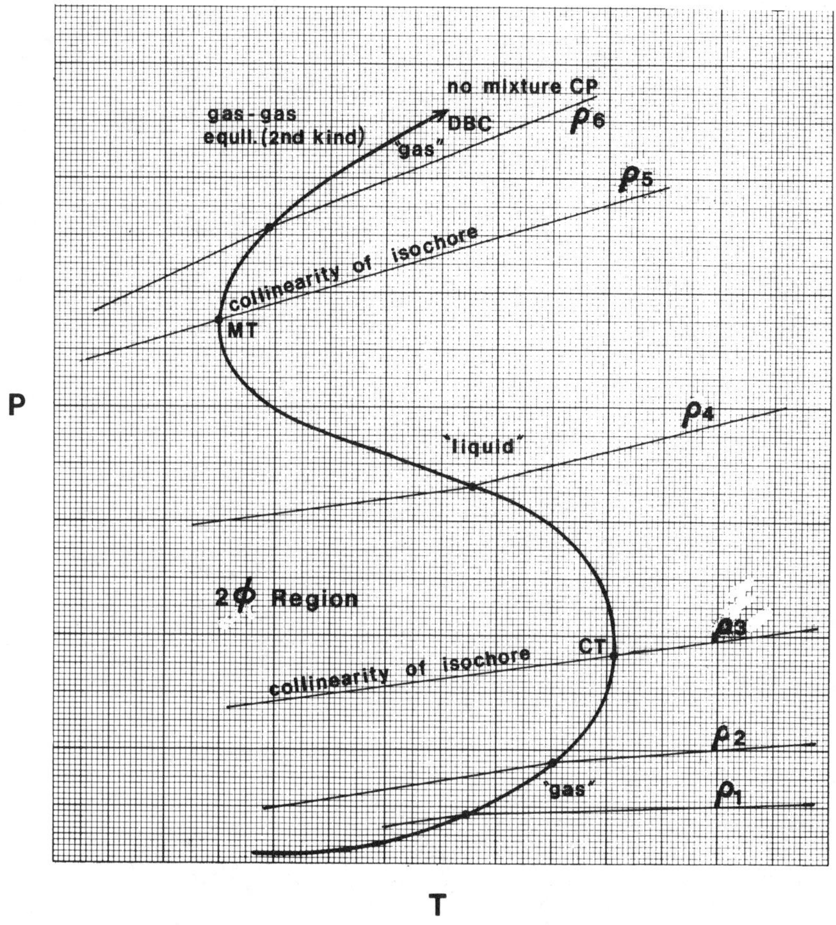

Figure 3. A qualitative diagram representing a binary mixture with no critical point (e.g., equimolar CO_2/N_2). Note the difference between isochoric slopes on either side of the DBC tracing the DBC from low to high densities.

where $(\partial\xi/\partial\psi)^{I}_{\omega}$ refers to the derivative at the DBC taken from the side of region I, etc. Then

$$[(\partial\xi/\partial\psi)^{I}_{\omega} - (\partial\xi/\partial\psi)^{II}_{\omega}]\, d\psi = -[(\partial\xi/\partial\omega)^{I}_{\psi} - (\partial\xi/\partial\omega)^{II}_{\psi}]\, d\omega \qquad (19)$$

or

$$\delta\, (\partial\xi/\partial\psi)_{\omega} = -\, (d\omega/d\psi) \cdot \delta\, (\partial\xi/\partial\omega)_{\psi} \qquad (20)$$

where δ indicates the difference, I-II. We are specially interested in $\delta(\partial P/\partial T)_{\rho,z_1}$, for which Equation (20) reads

$$\delta\, (\partial P/\partial \rho)_{T,z_1} = -\, (dT/d\rho)_{z_1} \cdot \delta\, (\partial P/\partial T)_{\rho,z_1} \qquad (21)$$

where the total derivative $(dT/d\rho)_{z_1}$ is taken along the DBC. This derivative is positive at low densities passing through zero at the CT to become negative at higher densities (see Figure 1). Griffiths proves that the left-hand side of Equation (21) is always nonnegative so that $\delta(\partial P/\partial T)_{\rho,z_1}$ must be negative below the CT, positive above the CT and thus zero itself at the CT.

To prove that $\delta(\partial P/\partial \rho)_{T,z_1} \geq 0$, Equation (20) is again applied but with $\xi \equiv \rho$, $\psi \equiv P$ and $\omega \equiv z_1$ at constant temperature:

$$\delta\, (\partial \rho/\partial P)_{z_1,T} = -\, (dz_1/dP)_T \cdot \delta(\partial\rho/\partial z_1)_{P,T} \qquad (22)$$

The derivative $(\partial\rho/\partial P)_{z_1,T} \geq 0$ for mechanical stability (see Ref. [2], p. 18). A Maxwell relation,

$$(\partial\rho/\partial z_1)_{P,T} = -\, \rho^2\, (\partial\Delta/\partial P)_{z_1,T} \qquad (23)$$

where $\Delta \equiv \hat{\mu}_1 - \hat{\mu}_2$, is then used to obtain

$$\delta(\partial\rho/\partial P)_{z_1,T} = \rho^2\, (dz_1/dP)_T \cdot \delta(\partial\Delta/\partial P)_{z_1,T} \qquad (24)$$

Equation (20) is applied a third time with $\xi \equiv \Delta$, $\psi \equiv P$ and $\omega \equiv z_1$ to replace $\delta(\partial\Delta/\partial P)_{z_1,T}$ with $\delta(\partial\Delta/\partial z_1)_{P,T}$:

$$\delta(\partial\rho/\partial P)_{z_1,T} = -\, \rho^2\, (dz_1/dP)^2_T \cdot \delta\, (\partial\Delta/\partial z_1)_{P,T} \qquad (25)$$

Now $(\partial\Delta/\partial z_1)_{P,T}$ is zero in the two-phase region (due to a phase rule constraint) and is nonnegative in the homogeneous region for material stability or $(\partial^2 G_m/\partial z_1^2)_{P,T} \geq 0$, where G_m is the molar Gibbs energy (see Ref. [2], p. 115). Hence, $\delta(\partial\rho/\partial P)_{z_1,T} \leqq 0$ or $\delta(\partial P/\partial\rho)_{z_1,T} \geqq 0$ because $(\partial\rho/\partial P)_{z_1,T} \geqq 0$ for mechanical stability. This completes the proof of Griffiths.

EXPERIMENTAL RESULTS

The 80.5% ^3He/19.5% ^4He mixture of Ref. 5 from Duke University showed collinear isochores at the CT for a system with the CP to the left of the CB on the P/T diagram. Figure 7 of Ref. 5 also showed the slope (dP/dT) to increase but weakly with temperature and with overall density in the heterogeneous region.

Later, isochoric measurements for a 20.05% H_2/79.95% CH_4 mixture at Rice University by Kobayashi and coworkers [6] illustrated collinearity at the CT for a system without a CB--at least not within 30 degrees of the CT temperature.

At the same time, measurements for CO_2 binaries in our laboratories at Texas A&M University showed isochoric collinearity for a nearly equimolar mixture of CO_2/CH_4 as seen in Figure 4. The CP for this mixture is not known exactly but has been estimated using a BACK equation of state (EOS). Later, we measured a nearly equimolar mixture of CO_2/N_2 resulting in the quantitative Figure 5, a dramatic illustration of isochoric collinearity at the CT. As discussed in a previous section, this mixture has no CP nor CB but a CT and a minimum temperature (MT) as shown by the qualitative Figure 3, based partly on necessarily less precise data at the higher densities.

COLLINEARITY OF ISENTROPES

In the fundamental equations of thermodynamics for pure components the variables are P, T, S and V. P and T are analogous potential functions of zero degree of mathematical homogeneity whereas S and V are analogous functions of first degree. The isentropic slope, $(\partial P/\partial T)_S$, is collinear with the vapor pressure curve at the CP for pure components as is the isochoric slope.

For binary mixtures the role of S and V can be reversed in any of the proofs given above with the result that isentropes are collinear at the CB. The qualitative nature of binary isentropes is illustrated in Figure 1. For mixtures, it may be said that "volume is prejudiced in favor of temperature whereas entropy favors pressure".

To confirm experimentally the collinearity of isentropes at the CB, we have taken the equimolar CO_2/CH_4 data of Figure 4 and calculated entropy increments on isotherms via the identity

$$(\Delta S_m)_T = -\int_{\rho_1}^{\rho_2} (\partial P/\partial T)_\rho \, (d\rho/\rho^2) \qquad (26)$$

To traverse temperature it is convenient to use the entropy residual function, $S_m^r(\rho,T)$:

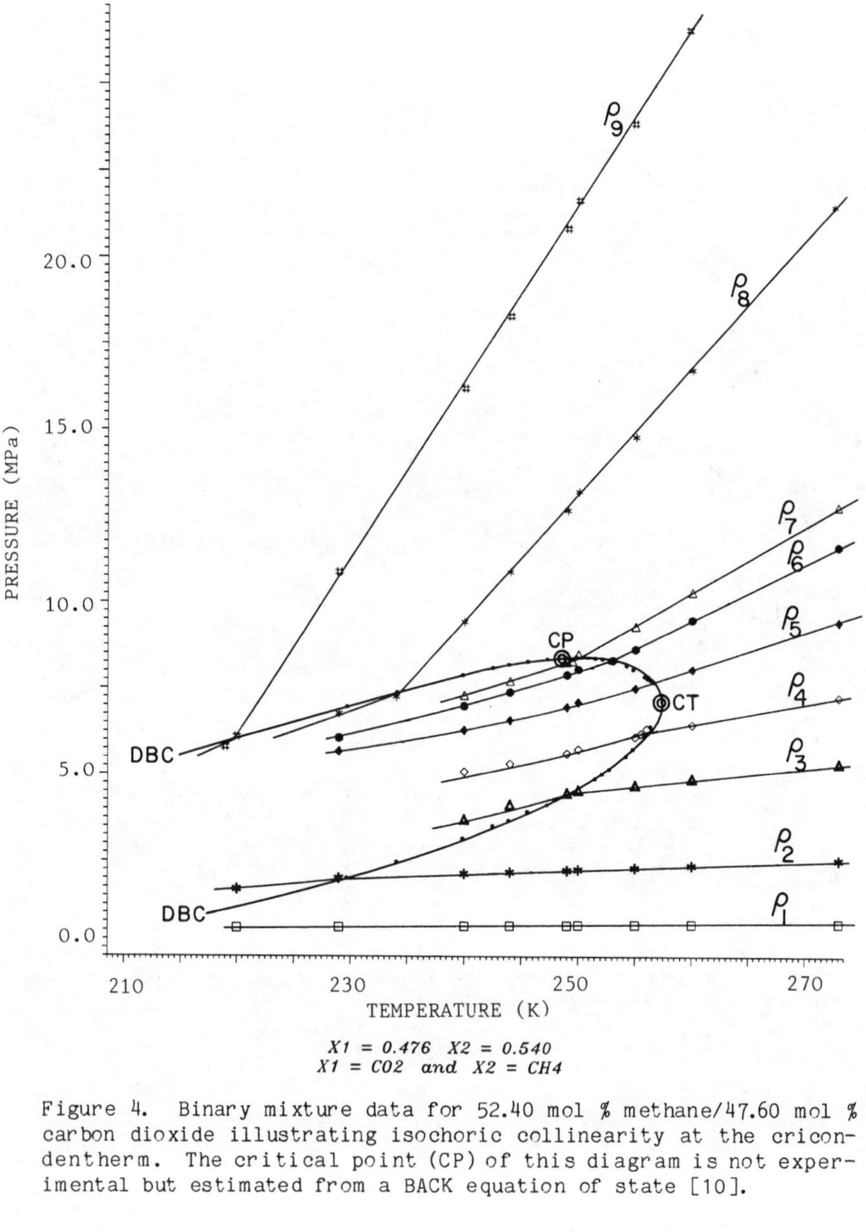

Figure 4. Binary mixture data for 52.40 mol % methane/47.60 mol % carbon dioxide illustrating isochoric collinearity at the cricondentherm. The critical point (CP) of this diagram is not experimental but estimated from a BACK equation of state [10].

$$(S_m^r/R) = T^{-1} \cdot \int_0^\rho (\partial Z_m/\partial T^{-1})_\rho \, (d\rho/\rho) + \int_0^\rho [(Z_m-1)/\rho]_T \, d\rho \qquad (27)$$

where Z_m ($\equiv P/\rho RT$) is the compressibility factor and $S_m^r \equiv S_m(T,\rho) - S_m^*(T,\rho)$ with S_m^* the perfect gas mixture value based upon a reference pressure and temperature, P_o and T_o, respectively. Like Equation (26), Equation (27) assumes isothermal integration. When the isotherm crosses the DBC, special precautions must be taken. Although we use Equation (27) for the calculations, Equation (26) is easier to examine. Imagine that our isotherm crosses the DBC first at a dew-point pressure or density, ρ_{DP}, and second at a bubble-point pressure or density, ρ_{BP}. Because of discontinuities of the derivatives, $(\partial P/\partial T)_\rho$ and $(\partial Z_m/\partial T^{-1})_\rho$, Equation (27) must be integrated in three separate steps: (1) from zero density to ρ_{DP}, (2) across the two-phase region from ρ_{DP} to ρ_{BP} and (3) from ρ_{BP} to a higher density in the compressed liquid. When the isotherm is supercritical, the bubble point is simply replaced by the upper dew point. Along the isotherm, S_m is a continuous function of density but $(\partial S_m/\partial \rho)_T$ suffers a discontinuity at the DBC.

Values of S_m calculated from Equation (27), see Table I, were then graphed as S_m versus pressure along isotherms for the CO_2/CH_4 binary. These isotherms must show a positive isobaric heat capacity, $C_{p,z_2} \equiv T\,(\partial S_m/\partial T)_{P,z_2}$, for thermal stability. The counter derivative, $(\partial S_m/\partial P)_{T,z_2} = -\rho^{-2} \cdot (\partial \rho/\partial P)_{T,z_2} \cdot (\partial P/\partial T)_{\rho,z_2}$, is usually negative but $(\partial P/\partial T)_{\rho,z_2}$ may be negative in the two-phase region near the CP. With these criteria in mind, some smoothing of our two-phase results were made on the sensitive S/P diagram, Figure 6. While our high-precision, P(ρ,T) measurements usually withstand well the differentiation and integration of Equation (27), such is not always the case in the two-phase region. Here we first have fewer data points and second are concerned about equilibrium when two phases are present in a blind cell without mixing capability.

Following this slight, judicious massaging of our isotherms on the S/P graph, Figure 6, a crossplot of isentropes was made on a P/T diagram--Figure 7. Although more data are desirable at lower entropies (i.e., higher pressures and lower temperatures), Figure 7 supports the notion of isentropic collinearity at the CB.

We also checked that $(\partial P/\partial T)_{S,z_2} > (\partial P/\partial T)_{\rho,z_2}$ at each (T,P) of Figures 4 and 7. The difference of these two slopes is $(\rho C_V/T) \cdot (\partial T/\partial V_m)_{P,z_2}$. Further, it can be shown that $(\partial^2 P/\partial T^2)_{S,z_2}$ is generally positive so that the isentropes of Figure 7 are concave upwards. Both pure methane and pure CO_2 exhibit this behavior with $(\partial P/\partial T)_{S_c,z_2}$ collinear with the vapor pressure curve at the CP.

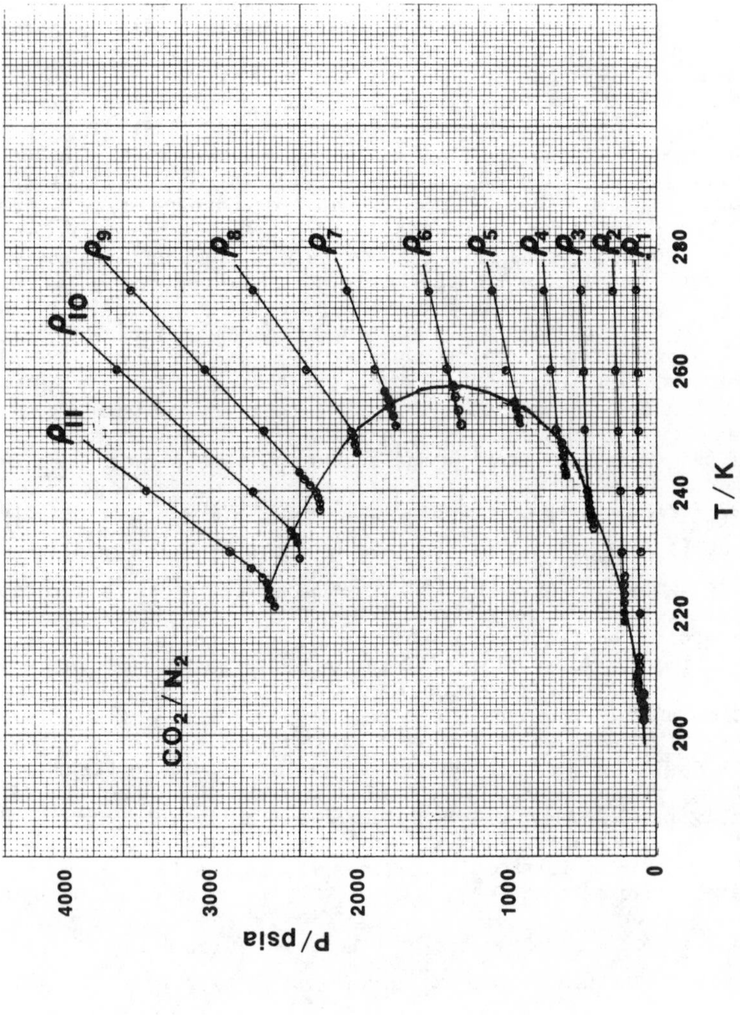

Figure 5. Binary mixture data for 44.70 mol% carbon dioxide/55.30 mol % nitrogen illustrating isochoric collinearity at the cricondentherm for a system known to exhibit the qualitative behavior of Figure 3 at the higher pressures and densities.

Table I: The calculation of Entropies for 52.40 mol% CH_4 and 47.60 mol% CO_2

Reference State: P_0 = 0.101325 MPa (1 Atm.), T_0 = 273.15 K

The entries are: pressure (Roman), (MPa), and dimensionless entropy (Italic), (S_m/R).

T(K)	Density (mol/m^3) 320.871	1246.85	3074.69	4799.85	7539.35	10509.1	11850.4	16560.7	18619.5
300.00	0.7817	2.8496	6.2404	8.8193	12.267	16.098	18.246	32.090	48.550
	2.080	*-3.550*	*-4.657*	*5.276*	*5.980*	*6.569*	*-6.807*	*-7.599*	*-7.956*
288.75	0.7511	2.7189	5.8710	8.1789	11.103	14.219	15.930	27.664	42.286
	2.207	*-3.679*	*-4.787*	*5.408*	*6.112*	*6.702*	*6.939*	*-7.729*	*-8.085*
273.20	0.7085	2.5362	5.3479	7.2619	9.4650	11.644	12.823	21.687	33.787
	-2.386	*-3.857*	*-4.967*	*5.589*	*6.297*	*6.890*	*-7.129*	*-7.920*	*-8.273*
260.08	0.6724	2.3802	4.8953	6.4597	8.0573	9.4941	10.306	16.774	26.764
	-2.539	*-4.012*	*-5.123*	*-5.745*	*-6.451*	*7.042*	*7.280*	*-8.069*	*-8.423*
256.61	0.6628	2.3386	4.7738	6.2429	7.6807	8.9282	9.6523	15.493	24.927
	-2.580	*-4.053*	*-5.164*	*5.786*	*6.492*	*7.178*	*-7.438*	*-8.205*	*-8.500*
256.30	0.6620	2.3350	4.7630	6.2236	7.6516	8.8780	9.5946	15.380	24.764
	-2.584	*-4.057*	*-5.167*	*-5.789*	*-6.887*	*7.162*	*7.874*	*-8.643*	*-8.937*
256.08	0.6612	2.3318	4.7537	6.2069	7.6266	8.8349	9.5450	15.282	24.625
	-2.588	*-4.060*	*-5.170*	*-5.792*	*7.336*	*8.141*	*8.405*	*-9.177*	*-9.471*
255.80	0.6606	2.3290	4.7455	6.1879	7.6044	8.7966	9.5010	15.196	24.501
	2.590	*-4.062*	*5.169*	*-5.823*	*7.353*	*-8.190*	*-8.454*	*-9.220*	*9.510*
254.78	0.6578	2.3167	4.7096	6.1055	7.5084	8.6310	9.3108	14.823	23.964
	-2.602	*-4.074*	*-5.182*	*-5.997*	*-7.461*	*-8.378*	*-8.652*	*-9.437*	*-9.728*
254.20	0.6562	2.3098	4.6892	6.0584	7.4535	8.5365	9.2023	14.610	23.659
	-2.609	*-4.081*	*5.189*	*6.066*	*-7.503*	*-8.447*	*-8.725*	*-9.563*	*9.864*
252.73	0.6521	2.2921	4.6372	5.9392	7.3147	8.2974	8.9287	14.072	22.886
	-2.627	*-4.099*	*5.207*	*6.462*	*-8.072*	*-8.938*	*-9.226*	*-10.04*	*-10.34*
252.10	0.6504	2.2845	4.6150	5.8884	7.2554	8.2297	8.8123	13.842	22.556
	-2.635	*-4.106*	*5.214*	*-6.530*	*-8.104*	*-8.946*	*-9.212*	*-10.04*	*-10.34*
251.50	0.6487	2.2773	4.5938	5.8398	7.1988	8.1650	8.7012	13.624	22.242
	2.642	*-4.114*	*-5.221*	*6.596*	*-8.135*	*-8.953*	*-9.233*	*-10.07*	*-10.37*
248.84	0.6414	2.2454	4.4790	5.6251	6.9487	7.8793	8.2954	12.681	20.856
	2.673	*-4.144*	*-5.316*	*6.903*	*-8.421*	*-9.189*	*-9.363*	*-10.08*	*-10.38*
245.50	0.6321	2.2049	4.2861	5.3540	6.6329	7.5183	7.9222	11.451	19.114
	-2.714	*-4.184*	*-5.602*	*-7.105*	*-8.518*	*-9.216*	*-9.370*	*-10.09*	*-10.46*
242.50	0.6237	2.1686	4.1134	5.1112	6.3499	7.1950	7.5878	10.374	17.561
	-2.750	*-4.221*	*-5.821*	*-7.246*	*-8.572*	*-9.223*	*9.373*	*-10.02*	*-10.51*
239.87	0.6164	2.1367	3.9621	4.8986	6.1022	6.9119	7.2951	9.4364	16.206
	-2.782	*-4.253*	*-5.999*	*-7.366*	*-8.625*	*-9.235*	*-9.378*	*10.07*	*-10.53*
234.08	0.6003	2.0662	3.6283	4.4292	5.5553	6.2871	6.6490	7.3829	13.236
	-2.853	*-4.323*	*-6.130*	*-7.323*	*-8.548*	*9.333*	*-9.588*	*-10.44*	*-10.80*

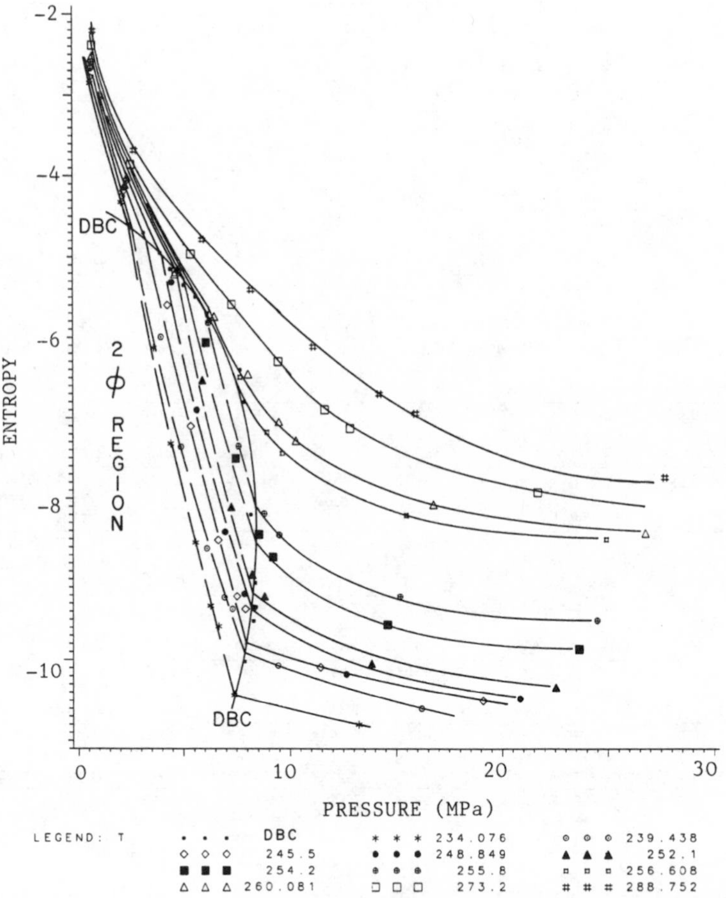

Figure 6. The calculated entropies for equimolar methane/carbon dioxide mixture (see caption of Figure 4). The slope of isotherms is $(-\rho/v_P)$ whereas the isobaric heat capacity, C_{px}, is roughly the average temperature multiplied by the isobaric entropy increment divided by the temperature increment.

LEGEND: ENTROPY

■ ■ ■ -10.1429	● ● ● -9.71429	◊ ◊ ◊ -9.42857
▲ ▲ ▲ -9.28571	♪ ♪ ♪ -8.71429	▫ ▫ ▫ -8.14286
* * * -7.85714	○ ○ ○ -7	△ △ △ -6.14286
# # # -5.85714	□ □ □ -3.85714	

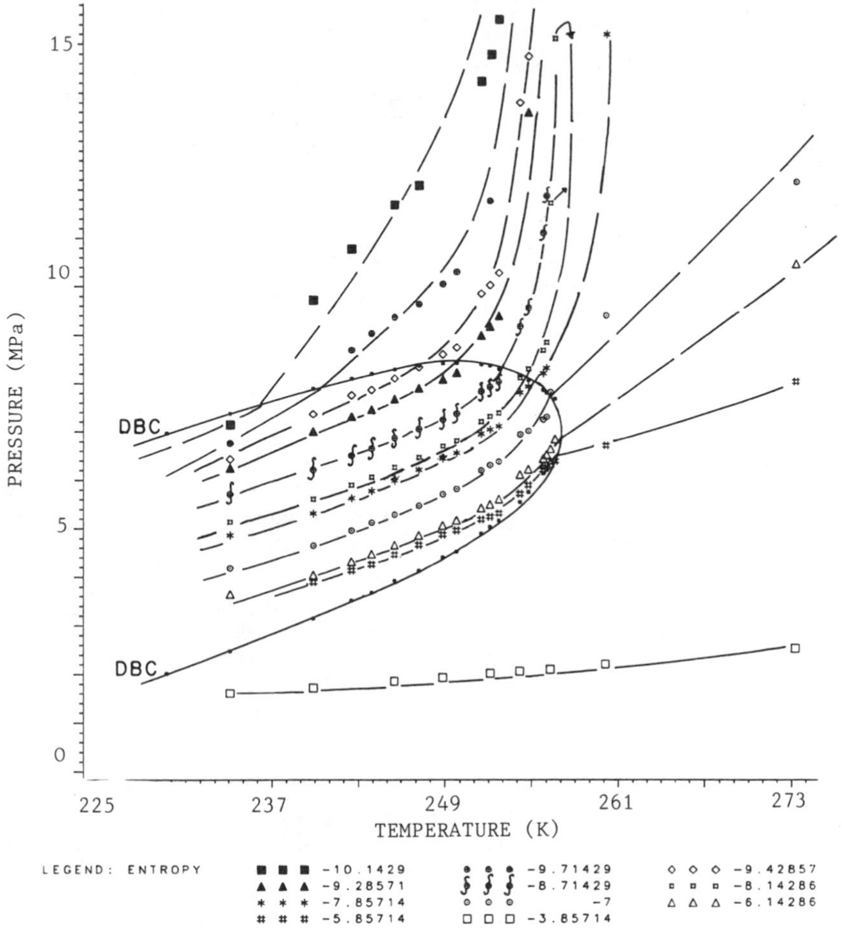

Figure 7. A pressure/temperature diagram illustrating isentropes (dashed lines) for the nearly equimolar methane/carbon dioxide mixture (see caption of Figure 4). Scatter in the calculated entropies prevents a definitive conclusion from these data alone of isentropic collinearity at the cricondenbar. However, these data are supportive of that proof and show that collinearity occurs on the upper part of the DBC between 237-255K.

THE ISOCHORIC INFLECTION LOCUS (IIL)

For pure components, it appears that the IIL always intersects the vapor pressure curve at the CP and again, for some compounds, at a liquid density roughly twice critical. Recent density data at Rice University (20.05 mol% H_2/79.95 mol% CH_4) and Texas A&M University (52.40 mol% CH_4/47.60 mol% CO_2 and 44.70 mol% CO_2/55.30 mol% N_2) suggest that the IIL may intersect the DBC at the CT for binary mixtures.

We know of no thermodynamic proof for this behavior whether for pure components or mixtures of fixed composition. A nonclassical proof might be required for pure compounds but if the mixture intersection is indeed the CT rather than the CP, then a classical proof should be possible. Considerable importance is attached to the IIL for not only is it the locus of the extremum of the heat capacity C_{v,z_2} but also its different qualitative behavior at densities above critical for pure compounds has been related to molecular effects, such as polarity [7].

DISCUSSION

Although we have presented proofs for (1) isochoric collinearity at the CT and (2) isentropic collinearity at the CB assuming a binary mixture, it is obvious that these results apply also to multicomponent mixtures because the derivatives are at constant overall composition. From another viewpoint, any multicomponent system may be considered a "pseudo-binary" and the above derivations repeated.

The derivative $(\partial P/\partial T)_{\rho,z_2}$ is important because a number of thermophysical property measurements are performed at constant volume and composition. For example, Nowak and Chan [8] have extended the "adiabatic non-flow saturation calorimeter method" to mixtures in the heterogeneous region. This method for measuring latent heats of vaporization and heat capacities was made famous by Osborne, Stimson and Ginnings (OSG) working primarily on water at NBS in the 1930's. The approximate working equations of Ref. 8 relate the apparent specific heat of the calorimeter to the desired isobaric heat capacity, C_{p,z_2}, for both the liquid and vapor at saturation. Because mixtures are difficult to duplicate, data are most conveniently taken along a series of isochores. In the homogeneous phase region, the density is lowered to the next isochore by simply exhausting part of the fluid or by expanding it into a second volume--as in the Burnett-isochoric method for the density measurements of Table I.

Finally, the present work has an important bearing upon mixture equations of state (EOS). Many physical EOS for both pure components and mixtures begin with straight isochores (van der Waals EOS) and then add correction terms to account for curvature [9]. It is an experimental fact that isochores are nearly linear in the homogeneous region on a pressure/temperature diagram for both pures and mixtures. The collinearity of isochores at the CT for mixtures provides a sensitive constraint for EOS parameters.

ACKNOWLEDGMENTS

We wish to thank Dr. Anneke Levelt Sengers for helpful discussions and for recalling the proof of Griffiths. The financial support of the National Science Foundation, Gas Processors Association, Petroleum Research Fund (ACS), Amoco Production Co. and Exxon Research and Engineering Co. are acknowledged. Special acknowledgment goes to the Gas Research Institute (Chicago) and to Dr. Frank Little, Project Manager, Thermodynamics. GRI provided the primary financial support for the CO_2/CH_4 and CO_2/N_2 data taken at Texas A&M University and for the data in the references cited from Rice University (Ref. 6 and 7).

LITERATURE CITED

[1] King, M.B., "Phase Equilibrium in Mixtures," Pergamon Press, Oxford, 1969.
[2] Rowlinson, J.S., and Swinton, F.L., "Liquids and Liquid Mixtures," Third Ed., Butterworths, London, 1982.
[3] Keesom, W.H., Communs. Phys. Lab. Univ. Leiden, (1901), No. 75.
[4] Levelt Sengers, J.M.H., on a visit to Texas A&M University to deliver the Lindsay Lecture "Dilute Near-Critical Mixtures", September 27, 1984.
[5] Doiron, T., Behringer, R.P., and Meyer, H., J. Low Temp. Phys., (1976), 24, 345.
[6] Magee, J.W. and Kobayashi, R., Adv. Cryog. Eng. (1984), 29, 943.
[7] Kobayashi, R., "Phase and Volumetric Studies for Methane/Synthesis Gas Separations: The Hydrogen-Carbon Monoxide-Methane System," Final Report to the Gas Research Institute (GRI-84/0084), Rice University, March 15, 1984; "Generalized Understanding Through Generalized Experiments," Lindsay Lecture (printed), Texas A&M University, Feb. 8, 1985.
[8] Nowak, E.S., and Chan, J.S., Proc. 7th Sym. Thermo. Prop., (1977), 538, ASME, New York.
[9] Nehzat, M.S., Hall, K.R., and Eubank, P.T., Adv. Chem. Series, (1979), 182, 109.
[10] Kreglewski, A., and Hall, K.R., Fluid Phase Equil., (1983), 15, 11.

RECEIVED November 8, 1985

3

The Equation of State of Tetrafluoromethane

J. C. G. Calado[1], R. G. Rubio[2], and W. B. Streett

School of Chemical Engineering, Cornell University, Ithaca, NY 14853

An experimental study of the P-V-T properties of tetrafluoromethane (CF_4) over wide ranges of temperature and pressure is used to test several semi-empirical equations of state and molecular theories. The experimental data have been correlated by the Strobridge equation, and comparisons are made with the Haar and Kohler, Deiters and BACK equations of state, as well as with the lattice gas model of Costas and Sanctuary, the variational inequality minimization (VIM) theory of Mansoori, and the perturbation theory of Gray and Gubbins.

We have recently completed a detailed experimental study of the equation of state of tetrafluoromethane, CF_4, covering a wide range of temperature and pressure (1).

Tetrafluoromethane is an attractive substance from both the technological and theoretical points of view. It is widely used as a low-temperature refrigerant (Freon 14) and a gaseous insulator, and its molecules offer an interesting theoretical study of the underlying intermolecular forces. Whilst its grosser features may be considered to be representable by a quasi-spherical (or globular) molecule, its thermodynamic behavior, especially in mixtures, displays a nonideality that is symptomatic of the anisotropy inherent in its microscopic interactions. For instance, a whole variety of phase diagrams arise when tetrafluoromethane is mixed with hydrocarbons (2). Even the existing low-density studies lead to contradictory conclusions about the intermolecular potential of CF_4. While in some cases a simple (12,6) Lennard-Jones potential is able to describe the second virial coefficient data (3), in other cases a spherical-shell model has been claimed to be necessary (4-6).

[1]Current address: Complexo I, Instituto Superior Técnico, 1096 Lisboa, Portugal.
[2]Current address: Departamento de Quimica Fisica, Facultad de Quimicas, Universidad Complutense, Madrid 28040, Spain.

0097-6156/86/0300-0060$06.00/0

Tetrafluoromethane seems to be an ideally placed molecule for this kind of study: on the one hand, it is fairly small, non-polar, highly symmetric (tetrahedral) and thus suitable for the testing of a wide variety of statistical theories and semi-empirical equations of state; on the other hand, since it has a relatively big octopole moment, it exhibits enough anisotropy to serve as a discriminator against cruder theoretical approaches. As such, tetrafluoromethane seems to fall between methane (CH_4) for which a (12, 6) Lennard-Jones potential seems to be adequate, and tetrachloromethane (CCl_4) for which a spherical-shell or site-site model is necessary.

Strong orientational correlations, which persist over several molecular diameters, have been found for tetrahedral molecules (7,8), and interlocking effects have been detected in tetrachloromethane molecules using Brillouin scattering techniques (9). These phenomena suggest that the spherical reference system, frequently utilized in perturbation and variational approaches (10) could fail for this kind of molecules. In addition, calculations carried out using the site-site distribution function formalism show that the disagreement with results from computer simulation is much larger than for diatomic molecules (11). Those interlocking and other effects should become more pronounced at higher densities, hence the need to extend the pressure range for which P-V-T data are available. Powles et al. (12) have pointed out that pressure and configurational energy data over wide ranges of density and temperature are necessary in order to improve intermolecular potential functions and to test theories.

There have been several experimental studies of the P-V-T properties of CF_4, but only up to about twice the critical density (ρ_c = 7.1 mol dm^{-3}). The most extensive and accurate measurements are those of Douslin et al. (4, 13) which cover the temperature range 273-623 K and pressures up to 400 bar. MacCormack and Schneider worked in the same temperature range but only with pressures up to 55 bar (14), while Lange and Stein (6) and Martin and Bhada (15) extended the measurements to lower temperature (203 K) and pressures up to 80 and 100 bar, respectively. Staveley and his co-workers measured both the orthobaric densities (16) and the saturation vapor pressure (17) of tetrafluoromethane. Thermodynamic properties of the saturated liquid, from the triple-point 89.56 K to the critical point 227.5 K, have been calculated by Lobo and Staveley (18), while Harrison and Douslin (19) calculated them for the compressed gas (temperatures 273-623 K and densities 0.75 - 11.0 mol dm^{-3}).

There was obviously a need for more data, especially in the low temperature, high-pressure region. We studied thirty-three isotherms in the temperature range 95-413 K and pressures up to 1100 bar, obtaining about one thousand and five hundred new P-V-T data points (1). Figure 1 shows the P - T regions for which data are now available. The P-V-T surface of tetrafluoromethane is now thus well defined over a wide range of temperature and pressure, from the dilute gas to the highly compressed liquid.

In this paper we examine the ability of several types of equations of state and molecular theories to predict the P-V-T properties of CF_4. Detailed tables of thermodynamic properties will be published elsewhere (20).

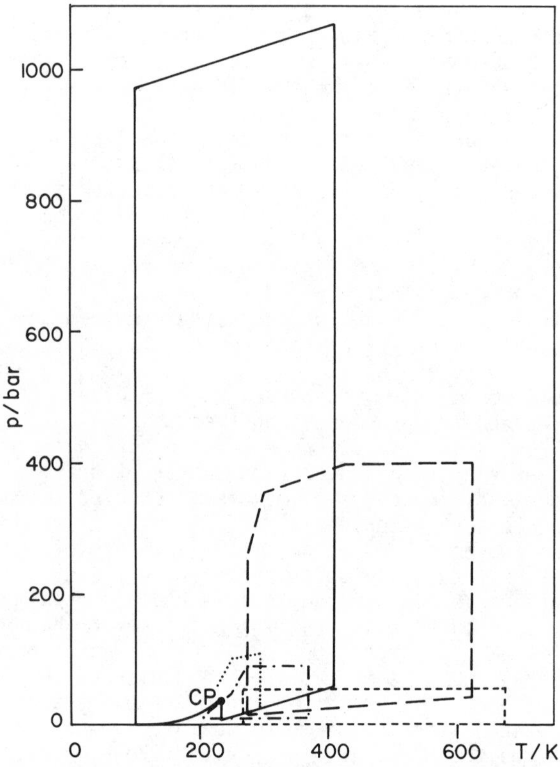

Figure 1. Pressure-temperature regions covered by different investigators : ——, this work; —— ——, Douslin *et al*. (4); -•-•- , Lange and Stein (6); ••••••••, Martin and Bhada (15); - - - -, MacCormack and Schneider (14)

Experimental

Since our study covered both low and moderately high temperatures, low and high densities, two different apparatuses were used. The measurements in the 95 to 333 K range were made in the gas-expansion type apparatus which has been utilized and described before (21). In this apparatus, the amount of substance contained in a calibrated 27.5 cm^3 cell, kept at a measured pressure and temperature, is determined by expansion into a large volume, followed by suitable P-V-T calculations (the pressure in the expansion volume is kept under 1.5 bar, to avoid any uncertainties in the equation of state of the gas). For the runs above 330 K a modification of the direct-weighing apparatus described by Machado and Streett (22) was used. Here the amount of substance is measured directly by weighing a full cell of approximately 100 cm^3 capacity.

Temperature control was achieved with a boiling liquid-type cryostat for the low-temperature apparatus (replaced by a simple water bath in the experiments above 270 K) and by a cascade-type oven in the higher temperature apparatus. Substances used in the boiling liquid type cryostat were nitrogen (95 K), argon (100 - 120 K), methane (130 - 152 K), tetrafluoromethane (160 - 200 K), ethane (210 - 245 K) and monochlorodifluoromethane (252 - 263 K). The temperature was controlled to within ± 0.02 K in the cryostat, and to within ± 0.002 K in both the liquid bath and the air oven. It was measured, in all cases, with a platinum resistance thermometer and referred to the IPTS - 68.

Pressures in the cells were measured with a Ruska dead-weight gage, (model 2450), the absolute accuracy being 0.1% or better, and the precision being about 0.01%.

The main source of error in both apparatuses lies in the imprecise knowledge of the volume of the systems. Details of the calibrating procedures have been given in previous papers (21, 22). We estimate that the average absolute error in density is about 0.1% for $\rho \geq 8$ mol dm^{-3}, 0.3% for $2 < \rho/\text{mol dm}^{-3} < 7$ and 0.4% for $\rho < 2$ mol dm^{-3}.

The CF_4 used in this work was from Linde (maximum purity 99.7%). It was purified by fractionation in a low-temperature column with a reflux ratio of 19/20. The final purity is estimated to be better than 99.99%.

Results

The over one thousand and five hundred data points were correlated by the Strobridge equation in the following form

$$p = RT\rho + (A_1RT + A_2 + A_3/T + A_4/T^2 + A_5/T^4)\rho^2 + (A_6RT + A_7)\rho^3 + A_8T\rho^4 + (A_9/T^2 + A_{10}/T^3 + A_{11}/T^4)\exp[A_{16}\rho^2]\rho^3 + (A_{12}/T^2 + A_{13}/T^3 + A_{14}/T^4)\exp[A_{16}\rho^2]\rho^5 + A_{15}\rho^6 \quad (1)$$

For the sake of completeness, the data of Douslin *et al.* (4) and Lange and Stein (6) were included in the correlation. Given their high quality, a statistical weight of two was ascribed to the data of Douslin *et al.*

The parameters of Equation 1 have been obtained using a method based on the Maximum Likelihood Principle as described by Anderson et al. (23). Values of the parameters are recorded in Table I, as well as their estimated uncertainties. Figure 2 shows the differences between experimental and calculated densities for several isotherms. Deviation plots for the other isotherms follow this same general trend: Equation 1 is able to represent the experimental results within their estimated errors, except for the low density region of a few isotherms. It is gratifying to note the good agreement obtained for the 227.53 K isotherm, which is very close to the critical one. Table II shows the density values of CF_4 calculated at round values of pressure and temperature from Equation 1.

In Figure 3 we compare the data of Douslin et al. (4) with the values generated by Equation 1. The agreement is, in general, within the combined errors of experiment and fitting. Martin and Bhada (15) have also reported large differences for the 273 K isotherm, from the equation they proposed.

The agreement between the different sets of data can be examined in Figure 4 where we plot the function B_V, defined by

$$B_V = (Z - 1)/\rho \qquad (2)$$

where Z is the compressibility factor, against pressure. As it has been pointed out by Douslin et al. (4) this plot provides an excellent test of the quality of the compressibility values, and the extrapolation to P = o gives the second virial coefficient, B. In the temperature range in which the three sets of results overlap, the agreement is very good. This is very encouraging since our experimental techniques were specially designed for high densities, whereas the other investigators were primarily concerned with the lower density region.

Equations of state and perturbation theory

With sixteen parameters Equation 1 is flexible enough to correlate PVT data over wide ranges of temperature and density, even in regions where $(\partial P/\partial \rho)_T$ is relatively large. It lacks sound theoretical foundation, although the format is that of an empirical modification of the virial equation of state. Looking at equations with a firmer theoretical basis is not only an intellectually rewarding exercise but also a serious attempt to develop more reliable and universal equations and improve on our present predictive abilities. The remainder of this paper will be devoted to an analysis and comparison of some of the most successful semi-empirical equations and theories proposed in the last few years.

The old Cartesian approach of dividing a complex problem into simpler, more manageable parts is of relevance here. Perturbation theory does just that. The reference or unperturbed system is a simple model whose properties have been fairly well understood. The real or complex system is recreated by adding successive layers of complexity (the perturbing terms) to the initial reference system. The genius or insight lies in finding a reference system

Table I. - Coefficients A_1 to A_{16}, their Estimated Error, and Gas Constant R for Equation 1

i			i		
1	0.041±0.001	$dm^3 mol^{-1}$	9	(5846±5)10	$bar\ dm^9\ K^2\ mol^{-3}$
2	1.76±0.04	$bar\ dm^6\ mol^{-2}$	10	$-(1.102±1)10^7$	$bar\ dm^9\ K^3\ mol^{-3}$
3	-1947±3	$bar\ dm^6\ K\ mol^{-2}$	11	$(6.566±0.007)10^8$	$bar\ dm^9\ K^4\ mol^{-3}$
4	$(1.347±0.002)10^5$	$bar\ dm^6\ K^2\ mol^{-2}$	12	-28.6 ± 0.4	$bar\ dm^{15}\ K^2\ mol^{-5}$
5	$(4.835±0.005)10^8$	$bar\ dm^6\ K^4\ mol^{-2}$	13	7970 ± 14	$bar\ dm^{15}\ K^3\ mol^{-5}$
6	$(0.53±0.02)10^{-2}$	$dm^6\ mol^{-2}$	14	$-(1.2011 ± 0.0004)10^6$	$bar\ dm^{15}\ K^4\ mol^{-5}$
7	-0.337±0.004	$bar\ dm^9\ mol^{-3}$	15	$(4.749 ± 0.002)10^{-5}$	$bar\ dm^{18}\ mol^{-6}$
8	$(4.00±0.05)10^{-5}$	$bar\ dm^{12}\ K^{-1}\ mol^{-4}$	16	-0.0040 ± 0.001	$dm^6\ mol^{-2}$

$R = 0.083144\ bar\ dm^3\ K^{-1}\ mol^{-1}$

Estimated Variance of Fit = 15

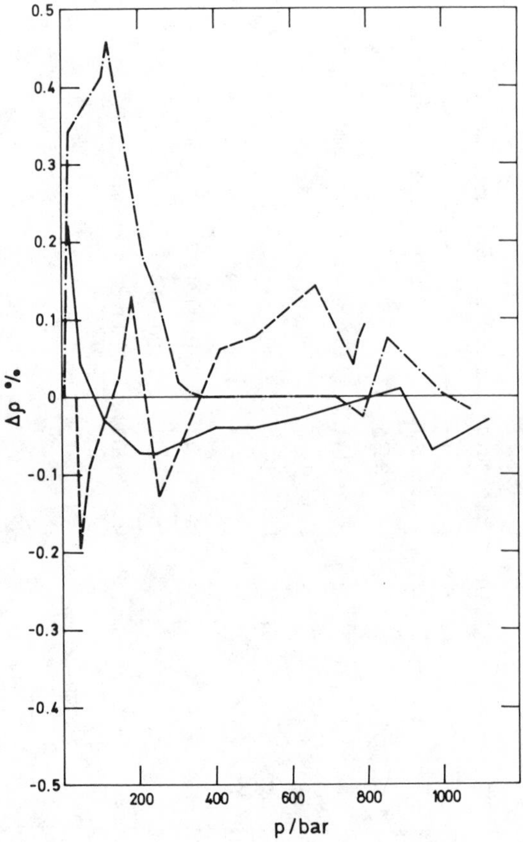

Figure 2. Density deviation plots $\Delta\rho = (\rho_e - \rho_c)/\rho_e \times 100$ for several isotherms. ρ_e is the experimental density and ρ_c its value calculated from Eq. (1). ———, 199.98 K; - - -, 227.53 K; -·-·-, 252.53 K;

Table II. Density values of CF_4 (in mol dm^{-3}) at round values of temperature and pressure, calculated from Equation 1.

T/k \ P/bar	5	10	15	25	50	100	200	300	400	500	600	700	800	900	1000	1100
95	21.054	21.063	21.072	21.090	21.134	21.218	21.376	------	------	------	------	------	------	------	------	------
100	20.807	20.817	20.826	20.845	20.892	20.981	21.148	21.302	21.446	------	------	------	------	------	------	------
120	19.739	19.752	19.765	19.790	19.852	19.970	20.188	20.385	20.567	20.735	20.892	21.039	21.179	21.311	21.437	21.557
140	18.587	18.605	18.623	18.659	18.745	18.908	19.198	19.454	19.684	19.893	20.085	20.264	20.431	20.588	20.736	20.877
160	17.330	17.357	17.385	17.438	17.565	17.796	18.190	18.521	18.810	19.067	19.299	19.512	19.709	19.892	20.063	20.225
180	0.368	15.924	15.970	16.058	16.260	16.606	17.152	17.583	17.943	18.256	18.533	18.783	19.011	19.221	19.417	19.599
200	0.323	0.703	1.180	14.332	14.716	15.287	16.069	16.632	17.081	17.458	17.785	18.074	18.336	18.574	18.793	18.997
220	0.288	0.613	0.987	2.003	12.576	13.746	14.927	15.667	16.221	16.671	17.052	17.384	17.680	17.947	18.192	18.417
240	0.261	0.545	0.858	1.607	5.778	11.812	13.716	14.687	15.366	15.897	16.336	16.713	17.045	17.341	17.610	17.857
260	0.238	0.492	0.765	1.377	3.541	9.346	12.445	13.700	14.520	15.138	15.638	16.061	16.429	16.755	17.049	17.316
280	0.220	0.450	0.693	1.218	2.839	7.092	11.158	12.724	13.691	14.399	14.962	15.430	15.834	16.189	16.507	16.795
300	0.204	0.415	0.635	1.100	2.436	5.679	9.938	11.781	12.891	13.687	14.309	14.823	15.261	15.644	15.985	16.293
320	0.190	0.386	0.587	1.007	2.162	4.801	8.800	10.897	12.131	13.006	13.685	14.240	14.712	15.121	15.485	15.811
340	0.179	0.361	0.547	0.931	1.958	4.207	7.953	10.090	11.419	12.363	13.092	13.685	14.187	14.621	15.005	15.349
360	0.168	0.339	0.513	0.868	1.798	3.774	7.206	9.367	10.762	11.760	12.532	13.159	13.687	14.144	14.547	14.907
380	0.159	0.320	0.483	0.813	1.668	3.442	6.592	8.728	10.161	11.200	12.006	12.661	13.213	13.690	14.110	14.485
400	0.151	0.303	0.456	0.767	1.559	3.177	6.084	8.166	9.165	10.682	11.515	12.194	12.766	13.259	13.694	14.082
420	0.144	0.288	0.433	0.725	1.466	2.960	5.660	7.674	9.121	10.205	11.057	11.755	12.343	12.852	13.299	13.698

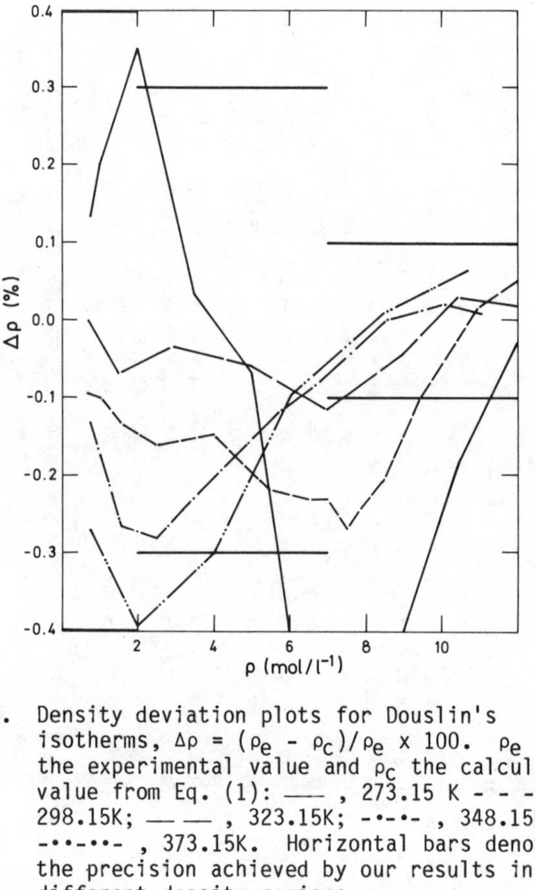

Figure 3. Density deviation plots for Douslin's isotherms, $\Delta\rho = (\rho_e - \rho_c)/\rho_e \times 100$. ρ_e is the experimental value and ρ_c the calculated value from Eq. (1): ——, 273.15 K - - - -, 298.15K; — —, 323.15K; -•-•-, 348.15K; -••-••-, 373.15K. Horizontal bars denote the precision achieved by our results in the different density regions.

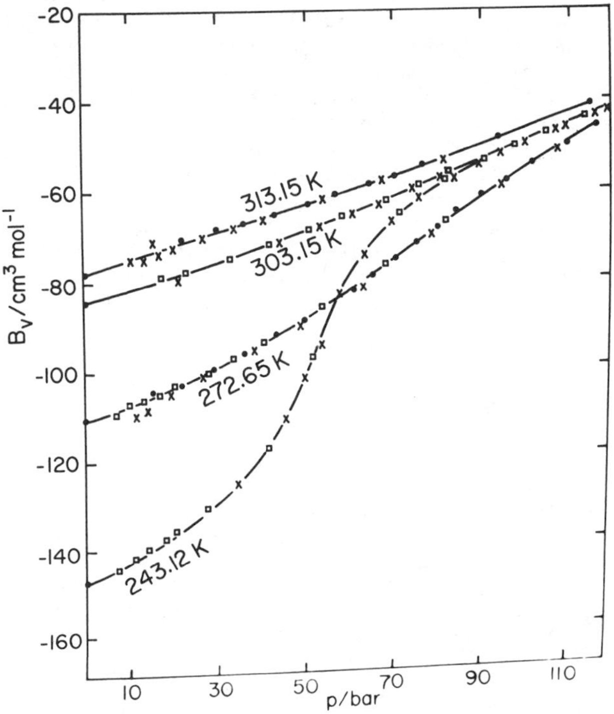

Figure 4. Values of B_V calculated from Equation 2: x, this work; ●, Douslin et al. (4); ◘, Lange and Stein (6). The points at P = o are the values of the second virial coefficient recommended by Dymond and Smith (24).

that is simple enough to allow its properties to be easily calculated, and yet close enough to the real system to avoid the need of many perturbing terms. Most of the calculations have been made using a spherical model (hard-sphere) as the reference system. Obviously in this case a perturbation theory is feasible if the molecules are not far from spherical. Recently Fischer (25) developed a perturbation theory which can effectively deal with anisotropic molecules, namely those of the two-center type (later also triangular, tetrahedral and octahedral molecules). The problem is that strongly anisotropic molecules are also usually polar, and the present theories are still unable to deal with both anisotropy and polarity. CF_4 is fortunately non-polar, but despite its high symmetry cannot be reasonably described as a spherical molecule. It is now well established that thermodynamic properties are markedly affected by the shape of the molecules.

Some of the more interesting equations of state can be obtained from perturbation theory. Indeed, the earliest of them, that due to van der Waals, can be derived from first-order perturbation theory with a reference system of hard-spheres (26). A common feature to many of these equations of state is its splitting, in true perturbational fashion, into two or more terms, accounting for different levels and types of complexity. One of those terms is usually built around a convex hard body. We will concentrate our attention, however, on what may be called the second generation of van der Waals - type equations. These are equations which retain a good approximation for the repulsive part (like that given by the Carnahan-Starling or Boublik-Nezbeda equations) while trying to improve on the attractive part of the potential.

The Haar and Kohler equation of state

A few years ago Haar et al. proposed a new approach to calculating P-V-T- data (27). The equation of state is split into two parts

$$P = P_B + P_R \tag{3}$$

where P_B is the so-called base equation (in the sense that it is a physically based expression) incorporating the effects of molecular repulsion and attraction, and P_R is a sum of residual terms, usually a series expansion in terms of ρ and T or some empirical function. For globular molecules a modified Carnahan-Starling equation is often used as P_B, but as many as twenty-six adjustable parameters are sometimes needed for P_R (28).

Kohler and Haar (29) showed that P_R would be a universal function for nonpolar fluids, provided that P_B takes into account the shape of the molecule. Recently, Moritz and Kohler came out with an empirical expression for P_R (30), leading to the following final form for the compressibility factor Z

$$Z = Z_h - \frac{a_0}{RT}\frac{y}{v^*} + y\{V' \exp\left[-\frac{y^2}{d_v^2}\right] + W \exp\left[-\frac{(y-y_{rd})^2}{d_w^2}\right]\} \tag{4}$$

Z_h is the compressibility factor for a hard-convex body (a tetrahedron for CF_4) like that given by the Boublik-Nezbeda equation of state (31), d_v and d_w are universal constants (0.098 and 0.128, respectively, according to reference 30) and $y = v^*\rho$ is the reduced density, with $v^* = N_A V^*$, V^* being the volume of the hard-body and N_A Avogadro's number. y_{rd} is the reduced density of the rectilinear diameter ρ_{rd} (the arithmetic mean of the densities of the orthobaric liquid and vapor)

$$\rho_{rd} = (\rho_L + \rho_G)/2 \qquad T < T_c \qquad (5)$$

For $T > T_c$, ρ_{rd} has a virtual value, obtained from the extrapolation of the rectilinear diameter curve.

Equation 4 has four parameters a_0, V', W and d ($d/2$ being the thickness of the smooth hard layer added to the hard body). For $T < 0.6\ T_c$, V' and W can be easily calculated from vapor-liquid equilibria using the second virial coefficient; thus, only d and a_0 (or a_0/v^*) need to be fitted at each temperature. The ratio a_0/v^* for CF_4 (~33 kJ mol^{-1}) is appreciably larger than for CH_4 (~20 kJ mol^{-1}), in qualitative agreement with the corresponding values of the solubility parameter.

The hard-core volume v^* is found to decrease with temperature, as expected, but following a different dependence law than that observed with the BACK equation (see later). Moritz and Kohler had reached a similar conclusion for methane (30). The observed density dependence for the van der Waals parameter a/v^*

$$\frac{a}{v^*} = RT\ (Z_h - Z)/y \qquad (6)$$

is also similar to that found for methane (30). Within the estimated uncertainties, we found that it is possible to describe the temperature dependence of all four parameters in Kohler's equation of state for CF_4, with simple functions incorporating only a few adjustable constants.

Figure 5 compares the results obtained with Equations 4 and 1. It is obvious that Kohler's equation fits the data better than the Strobridge equation below T_c, but the quality of the fitting worsens dramatically above T_c. This suggests that there is still room for improvement in the density dependence of a/v^*.

The Deiters equation of state

Another of the new, physically based equations of state is that derived by Deiters (32) from a square-well potential model of depth $(-\varepsilon)$

$$P = \frac{RT}{V}\left[1 + c\ c_0\ \frac{4y - 2y^2}{(1 - y)^3}\right] - \frac{RbcT}{V^2 w(y)}\left[\exp\ \frac{aw(y)}{cT} - 1\right] I_1(y) \qquad (7)$$

The equation has three adjustable parameters: the characteristic temperature a, the covolume b, and the correction factor for the number of density dependent degrees of freedom, c.

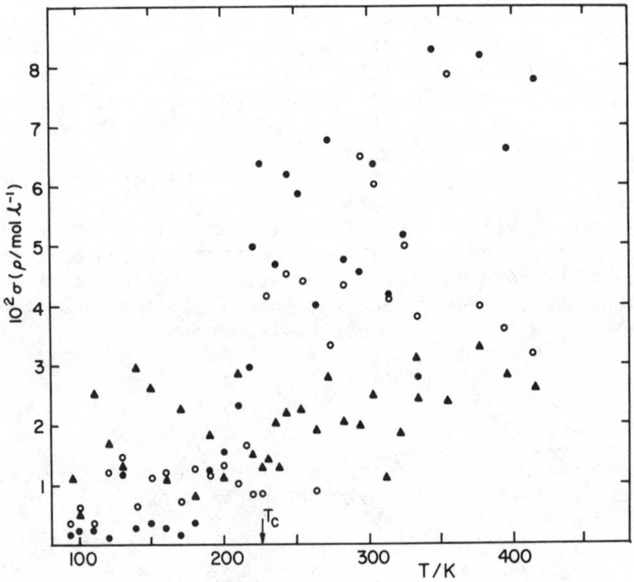

Figure 5. Mean standard deviations of density, σ, as a function of temperature. ▲, Strobridge equation; ●, Kohler equation; o, Deiters equation.

$$a = \varepsilon/k \quad (8)$$

$$b = N_A\ \sigma^3/\sqrt{2} \quad (9)$$

c can also be interpreted as a shape parameter (c = 1 for spherical molecules); c_0 = 0.6887 is a universal constant which accounts for the deviation of the real pair potential from the rigid-core model. $I_1(y)$ and w(y) are complicated functions of the density y and c, derived from statistical mechanics. w(y) may be described as an efficiency factor for the square well depth, and it takes into account the fact that the structure of the fluid is more and more determined by repulsive forces as density increases. The repulsive part of the Carnahan-Starling equation (33) has been retained as a good approximation for rigid spheres. Equation 7 can be modified, to account for three-body forces, by introducing an additive correction, $\lambda\rho$, to the reduced temperature, $\tilde{T}$ = cT/a (32).

We have used Equation 7 as P_B in Equation 3 and allowed the three parameters a, b and c to be temperature dependent. Figure 5 compares the results obtained with the Deiters equation with those given by Equations 1 and 4. For $T < T_c$ the highest deviations occur in the low density region, where the isotherms are the steepest. The results usually fall between those obtained with the Strobridge and Kohler equations.

Figures 6 and 7 show the fitted parameters of Deiters' equation as a function of temperature. It is interesting to note that both (c) and (a) are constant within their estimated errors, although they take distinct values below and above the critical temperature. Of course, the temperature dependence of any of the parameters is not a simple function in the critical region, so we cannot extrapolate over this region. The behavior found for parameter (b) is somewhat bizarre. In the range $T < T_c$, (b) follows the usual trend, i.e. it decreases with temperature (34); no explanation has been found for its apparent increase with temperature in the region $T > T_c$. It should be noted that similar behavior of the parameters a, b and c has been found for trifluoromethane CHF_3, whose study is now under way.

The calculations were repeated using a constant value for (c) throughout the entire temperature range (we used the more realistic 'high-temperature' value of Figure 6, c = 1.08), but letting both (a) and (b) float. Under these conditions parameter (a) is found to decrease monotonically from a value of about 235 K at 100 K to about 175 K at 400 K. Parameter (b) retains, however, the peculiar behavior displayed in Figure 7. The quality of the overall fitting also deteriorates when a constant value of (c) is used throughout, perhaps because packing effects become important at higher densities.

Using the Deiters equation as P_B (with the "constant" values of a and c, and b fitted to a Morse-like function) and an expression for P_R like that proposed by Haar et al. (27)

$$P_R = \sum_{n,j} C_{nj} \left(\frac{T_c}{T}\right)^j \rho^2 (1 - e^{-\alpha\rho})^n e^{-\alpha\rho} \quad (10)$$

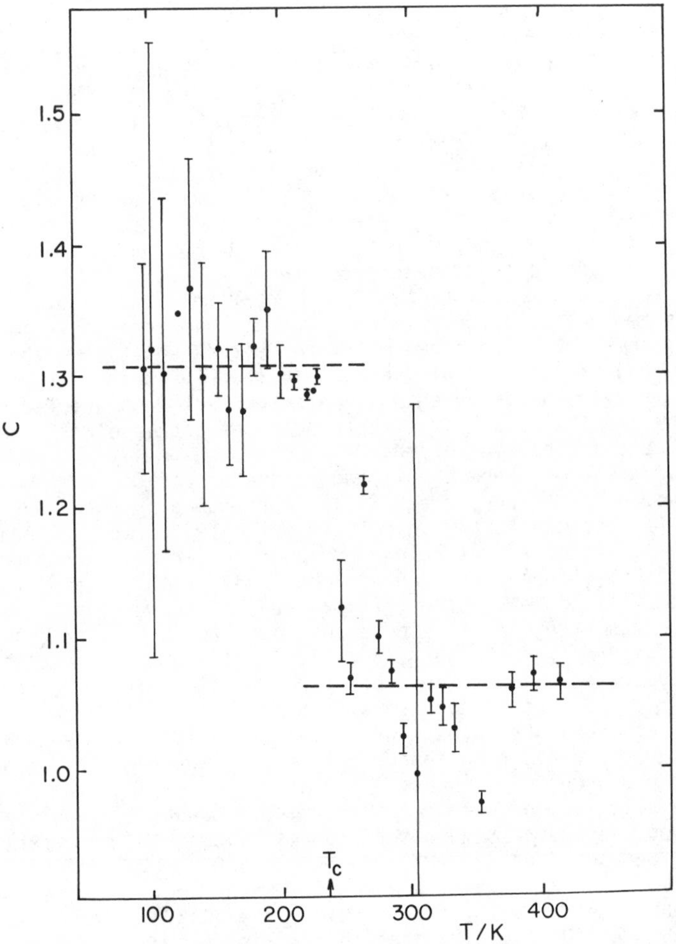

Figure 6. Parameter c in Deiters equation as a function of temperature (with error bars for estimated uncertainty).

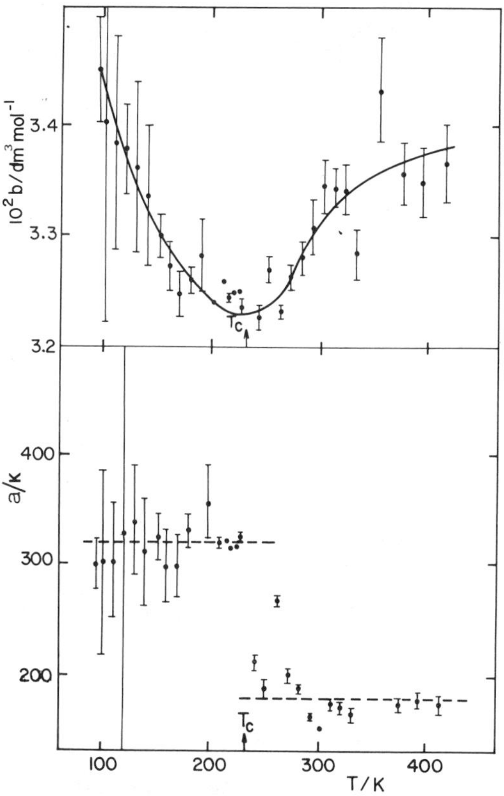

Figure 7. Parameters a and b in Deiters equation as a function of temperature (with error bars for estimated uncertainty).

We were unable, even with sixteen constants C_{nj}, to describe the whole set of P-V-T data with better accuracy than that provided by Equation 1. This could be due to the inadequacy of the residual terms used, but looking at Figure 5 suggests that an improvement in the low-temperature range can be obtained by substituting the Boublik-Nezbeda equation for the Carnahan-Starling equation in Deiters' model. This is confirmed by setting V' = W = 0 in the Kohler equation (4) (which uses the Boublik-Nezbeda equation as a reference) and finding that it leads to a better correlation of the experimental results than that obtained with the Deiters equation. Even more important, perhaps, that substitution would allow the separation of the contributions of external and internal degrees of freedom in the Deiters equation. One should then use data for molecules with vibrational degrees of freedom, instead of data on argon, as Deiters did.

The BACK equation of state

One of the best equations of state based on the generalized van der Waals model is the so-called BACK (from Boublik-Alder-Chen-Kreglewski) equation (35). It is an augmented hard-core equation which combines the Boublik expression for the repulsive part of the compressibility factor, with the polynomial developed by Alder *et al*. for the attractive part (36)

$$\frac{PV}{RT} = Z_h + Z_a \tag{11}$$

$$Z_h = \frac{1 + (3\gamma - 2)y + (3\gamma^2 - 3\gamma + 1)y^2 - \gamma^2 y^3}{(1 - y)^3} \tag{12}$$

$$Z_a = \sum_n^4 \sum_m^9 m D_{nm} \left(\frac{u}{kT}\right)^n \left(\frac{V^*}{V}\right)^m \tag{13}$$

Both the molecular hard core V* and the characteristic energy u are decreasing functions of temperature

$$V^* = V^{00} [\ 1 - C \exp(-3u^0/kT)]^3 \tag{14}$$

$$\frac{u}{k} = \frac{u^0}{k}\left(1 + \frac{\eta}{kT}\right) \tag{15}$$

Chen and Kreglewski (35) gave rules for calculating C and η, so that only three adjustable parameters remain in the BACK equation: V^{00}, γ and u^0/k. The fitting of our experimental data led to the following values: V^{00}=(0.0310 ± 0.0001)dm^3, γ = 1.099 ± 0.013 and u^0/k = (225 ± 2)K. The results are plotted in Figure 8. The agreement is satisfactory in the critical region, but it goes beyond the experimental error in the high density region. Besides, the value of γ does not clearly respect the tetrahedral geometry of the molecule, a fact which Moritz and Kohler also found for methane (30). As mentioned before, the temperature dependence of the hard-

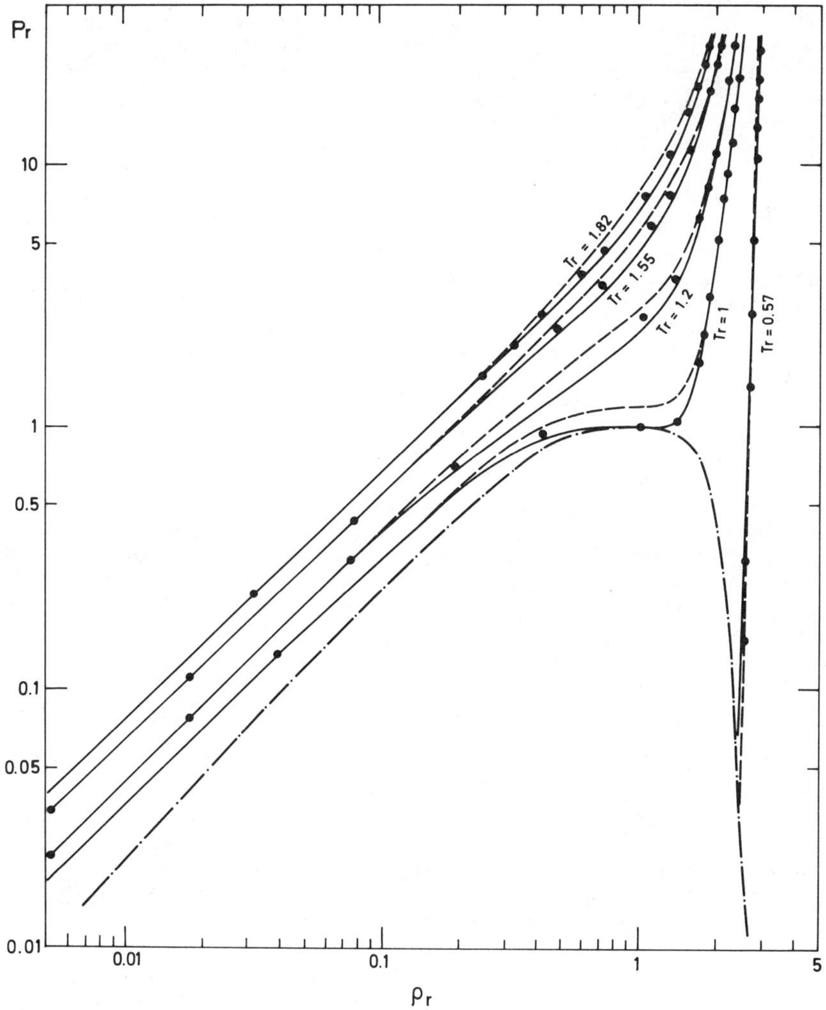

Figure 8. Several isotherms for CF_4 (in reduced variables). ●, experimental data; ——, calculated from BACK equation; - - -, calculated from lattice model; -•-•-, coexistence curve according to Lobo and Staveley (17).

core volume, as expressed by Equation 14, does not agree with the corresponding dependence in Kohler's equation. Use of the correlations proposed by Simnick et al. (37) for the calculation of parameters V^{00}, γ and u^0/k leads to even poorer agreement with the experimental results.

Taking the entire P-V-T range covered in this work, the quality of the fitting obtained with the BACK equation is considerably worse than that obtained with either the Kohler or Deiters equations, but this result should be set against the very small number of adjustable parameters utilized in the BACK equation.

Cell and Lattice Models

Although the cell and lattice models have been virtually abandoned for liquids, mainly because they over-estimate the degree of order, some of their modifications have been revived, with success, in the thermodynamic treatment of polymers. Recently Nies et al. (38) have shown that hole theories can be effective in reproducing the P-V-T behavior of a relatively simple fluid, ethylene. In the low-density region, they favor the lattice-gas model over the cell theory.

Sanchez and Lacombe (39) developed a theory of r-mers based upon a lattice model for liquids. They used a simplified Ising model, and were led to a four-parameter equation of state which successfully predicted a liquid-vapor phase transition. The four parameters were the non-bonded mer-mer interaction energy, ε^* (equivalent to a characteristic temperature), the close-packed -mer volume, v^*, the lattice coordination number z, and the number of mers per r-mer, r. Later Costas and Sanctuary (40, 41) removed r as an adjustable parameter and set it equal to the number of atoms, other than hydrogen, in the molecule. Their equation of state, in reduced variables, took the form

$$[\tilde{\rho}(1-\phi)/(1-\phi\tilde{\rho})]^2 + \tilde{\rho} + \tilde{T}\,[\ln(1-\tilde{\rho}) - (z/2)\ln(1-\phi\tilde{\rho})] = 0 \qquad (16)$$

where

$$\phi = (2/z)\,(1-1/r) \qquad (17)$$

Since they found that the calculated quantities were relatively insensitive to changes in z between the values 8 and 14, they chose to fix z equal to 12. Equation 16 became then a two-parametric equation.

In this work we used the Costas-Sanctuary equation of state in its three-parametric form. The adjusted values of the parameters were found to be $p^*/\text{bar} = 3338 \pm 123$, $T^*/K = 258 \pm 2$ and $v^*/l\ mol^{-1} = 0.0476 \pm 0.0002$. The results are plotted in Figure 8. It can be observed that the BACK equation gives a better overall description of the P-V-T properties of CF_4 than the Costas-Sanctuary equation, especially near the critical region. At very high or very low densities and pressures the two models give almost identical results. For high densities and $P_r < 0.8$ the agreement between the experimental values and those calculated from the lattice model is better than that obtained with the BACK equation, in particular near the coexistence curve. In the range $0.3 < \rho_r < 1.8$ the densities calculated with the BACK equation are too low, whereas those calculated from the Costas-Sanctuary equation are too high. In agreement with Costas and Sanctuary we have found that these conclusions still hold when the coordination number z changes from 8 to 12.

The VIM theory

In a previous paper we have already examined the ability of both variational and perturbation (see next section) theories to describe the thermodynamic properties of CF_4 (1). Here we reproduce the main conclusions in order to assess how these theories fare in comparison with more empirical equations of state.

The variational inequality minimization (VIM) theory as developed by Alem and Mansoori (42) led to a relatively simple equation of state for non polar fluids. They use a reference system of hard-sphere molecules whose diameter (σ) has been chosen in order to minimize the Helmholtz energy of the real system. The configurational entropy is then given by

$$S^C = -R\, y_{HS}\, (4 - 3y_{HS})/(1 - y_{HS})^2 \qquad (18)$$

Contrary to what Alem and Mansoori found for argon and methane, where σ was a linear function of density and inverse temperature, for tetrafluoromethane second order terms were needed. This reflects, perhaps, the higher anisotropy of CF_4.

Fitting of the P-V-T data led to the following values of the three adjustable parameters: $\varepsilon/k = (220 \pm 1)$ K; $\sigma/\text{nm} = 0.414 \pm 0.001$ and $10^3\nu/k = (8 \pm 1)\ \text{nm}^3\ \text{K}^{-1}$. ($\varepsilon$, σ) are the usual parameters in a Lennard-Jones type potential and ν is the coefficient of the Axilrod-Teller correction for the three-body potential (42). The final results are compared with the experimental values in Figure 9. The agreement is similar to that found by Alem and Mansoori for argon, but it is worse than that for methane. The theory underestimates the temperature coefficient of the density, leading to poor agreement in the low temperature region.

Perturbation theory

The perturbation theory of Gray, Gubbins and coworkers (10) has been extensively applied to molecular liquids and their mixtures. Since it uses a spherically-symmetric reference system, it stands the best chance of success when applied to molecules which exhibit high symmetry. A previous study (43) proposed a model for CF_4 which involved descriptions of the octopolar, anisotropic dispersion and charge overlap (shape) forces, in addition to a Lennard-Jones (n,6) potential

$$u_{CF_4/CF_4} = u_o^{(n,6)} + u_{\Omega\Omega}(336) + u_{ov}(303 + 033) + u_{disp.}(303 + 033) \qquad (19)$$

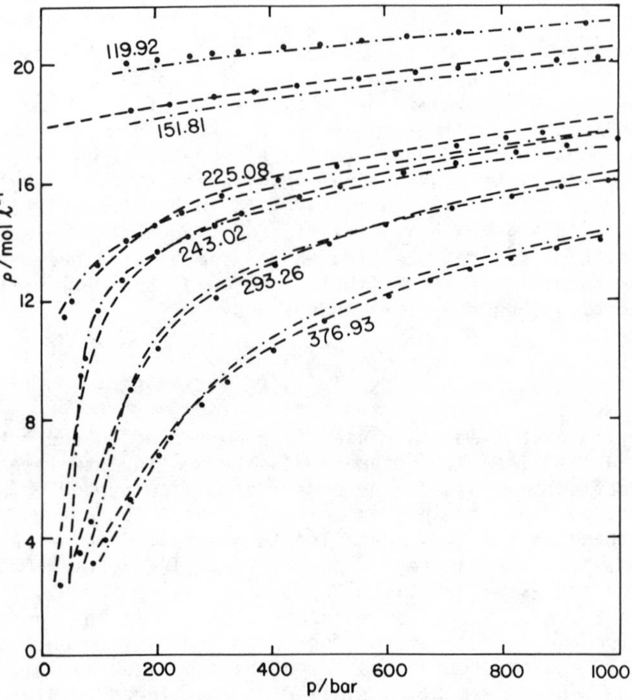

Figure 9. Comparison of experimental density values (•) with the predictions of the variational equation of state (-•-•-) and perturbation theory (- - - -), as a function of pressure, for different isotherms.

The potential was found to be extremely hard, with a repulsive exponent n = 20, perhaps reflecting the high electronegativity of fluorine atoms. Best values of the intermolecular parameters were ε/k = 232 K and σ = 0.4257 nm.

Use of the perturbation theory with the usual Padé approximant (1) led to the calculated densities shown in Figure 9. The agreement with experiment can be rated very good, the difference being, at most, 2% even at the highest pressures, where the reduced density is very close to the limit of validity of the reference equation (i.e. densities $\rho\sigma^3 < 1.0$).

Conclusions

The usual pattern is to have a new approach, be it an empirical equation of state or molecular theory, tested against a variety of systems, from pure fluids to mixtures, from simple molecules to nasty ones (to borrow an expression from John Prausnitz). Here we have followed a different route, viz. use the same body of experimental data to assess the goodness of several semi-empirical equations and of some theoretically based treatments. Obviously this latter approach can only be of value if the set of data is comprehensive enough to provide a strict test of the theory. In thermodynamic terms this means data over wide ranges of temperature and density, so that first order properties (temperature and pressure coefficents, for instance) can also be checked.

The first attempt at a molecular understanding of the properties of a fluid was that of van der Waals, with his famous equation of state

$$\frac{PV}{RT} = - \frac{a}{RTV} + \frac{1}{1-4y} \tag{20}$$

Although the first (attractive) term was soon recognized as a good approximation for the attractive field, the second was thought, even by van der Waals, to be a poor representation of the repulsive forces. In his Nobel address of 1910 he was still wondering if there was "a better way" of doing it, adding that "this question continually obsesses me, I can never free myself from it, it is with me even in my dreams". The problem was only solved more than fifty years later, when good approximations for the equation of state of a system of hard spheres were proposed. In the meantime, many people were busy trying to improve on the van der Waals equation by retaining its bad term while trying to correct the good one. The Redlich - Kwong equation of state is perhaps the most celebrated outcome of this approach.

$$\frac{PV}{RT} = - \frac{a}{(V+b)RT^{1.5}} + \frac{1}{1-4y} \tag{21}$$

No wonder it has generated over one hundred modifications, despite its own successes. John Prausnitz has said that Equation 21 is to applied thermodynamics what Helen of Troy has been to literature "... the face that launched a thousand ships."

In the 60's and 70's more successful members were added to the family of van der Waals equations: these combined the attractive part of the original Equation 20 or even Equation 21 with a good hard-sphere equation of state. The Carnahan-Starling equation, mentioned in this paper, is a good example.

In the last few years another type of equations of state has begun to emerge. They usually have a statistical mechanics basis and combine a good hard-sphere part (or some generalization to hard convex bodies) with a more flexible and sophisticated representation of the attractive field than that provided by the van der Waals equation. The Deiters and Kohler equations are good specimens of this second generation of generalized van der Waals equations of state.

Equations with a realistic molecular basis should be potentially more universal in their applicability and thus better suited to sound predictive methods.

There is obviously still room for progress. For instance, the Deiters equation could profit from a better description of shape effects, while the Kohler equation is perhaps deficient in its depiction of the attractive field. Perturbation theory seems, at present, to be more promising than the variational approach, especially if it succeeds in combining the ability to deal with polarity with the ability to take shape into account.

In any science, understanding must be synonymous with good quantitative agreement, and understanding really means the molecular level, even in chemical engineering. We owe it, after all, to van der Waals who, in his Nobel address, said: "It will be perfectly clear that in all my studies I was quite convinced of the real existence of molecules, that I never regarded them as a figment of my imagination". The road is long and arduous, but van der Waals pointed the way.

Glossary of Symbols

Latin Alphabet

A_i (i = 1 - 16) = coefficients of the Strobridge Equation (1)
a, a_0 = van der Waals parameters in Equation 4; charactic temperature in Deiters Equation (7)
B = second virial coefficient
B_v = function defined by Equation 2
b = covolume
C = parameter in Equation 14
C_{nj} = constants in Equation 10
c = shape parameter in Deiters Equation (7)
c_0 = universal constant in Deiters Equation (7)
d = twice the thickness of hard layer added to hard convex core
d_v,d_w = universal constants in Equation 4
I_1 (y) = a function of density in Deiters Equation (7)
k = Boltzmann's constant
N_A = Avogadro's constant
P = pressure
R = gas constant

r = number of mers per r-mer
S = entropy
T = absolute temperature
U = total energy
u, u^0 = characteristic energy in Equations 13-15
V = volume
V^{00} = volume parameter in Equation 14
V' = parameter in Equation 4
v^* = volume of one mole of hard bodies
V^* = volume of hard body
W' = parameter in Equation 4
$w(y)$ = efficiency factor for the square-well depth, Equation 7
y = reduced density
Z = compressibility factor ≡ PV/RT
z = lattice coordination number

Greek alphabet

α = coefficient in Equation 10
γ = shape parameter in Boublik Equation (12)
ε = intermolecular energy parameter (depth of potential)
η = parameter in Equation 15
ν = coefficient in Axilrod-Teller triple potential
ρ = molar density
σ = intermolecular energy parameter (molecular diameter); mean standard deviation
ϕ = parameter defined by Equation 17

Subscripts, superscripts

a = attractive
B = base (physically based)
c = configurational; calculated
e = experimental
G = gaseous
h = hard convex
HS = hard sphere
L = Liquid
R = Residual
r = reduced (by critical parameters)
rd = rectilinear diameter
Ω = octopole
~ = reduced (by characteristic parameters, other than critical)
* = characteristic

Literature Cited

1. Rubio, R.G; Calado, J.C.G.; Clancy, P.; Streett, W.B. J. Phys. Chem., in press.
2. Paas, R.; Schneider, G.M. J. Chem. Thermodynamics, 1979, 11 (3) 267.

3. MacCormack, K.E.; Schneider, W.G. J. Chem. Phys. 1951, 19 (7), 849.
4. Douslin, D.R.; Harrison, R.H.; Moore, R.T.; McCullough, J.P. J. Chem. Phys. 1961, 35 (4), 1357.
5. Kalfoglou, N.K.; Miller, J.G. J. Phys. Chem. 1967, 71 (5) 1256.
6. Lange Jr., H.B.; Stein, F.P. J. Chem. Eng. Data 1970, 15 (1), 56.
7. Montague, D.G.; Chowdhury, M.R.; Dore, J.C.; Reed, J. Mol. Phys. 1983, 50 (1), 1.
8. Narten, A.H. J. Chem. Phys. 1976, 65 (2) 573.
9. Asenbaum, A.; Wilhelm, E. Adv. Molec. Relax. Int. Proc. 1982, 22, 187.
10. Gray, C.G.; Gubbins, K.E. "Theory of Molecular Liquids"; Oxford University Press: London, 1984.
11. Monson, P.A. Mol. Phys. 1982, 47 (2), 435.
12. Powles, J.G; Evans, W.A.; McGrath, E.; Gubbins, K.G.; Murad, S. Mol. Phys. 1979, 38 (3), 893.
13. Douslin, D.R.; Harrison, R.H.; Moore, R.T. J. Phys. Chem. 1967, 71 (11), 3477.
14. MacCormack, K.E.; Schneider, W.G. J. Chem. Phys. 1951, 19 (7) 845.
15. Martin, J.J.; Bhada, R.K. AIChE J. 1971, 17 (3) 683.
16. Terry, M.J.; Lynch, J.T.; Bunclark, M.; Mansell, K.R.; Staveley, L.A.K. J. Chem. Thermodynamics 1969 1 (4) 413.
17. Lobo, L.Q.; Staveley, L.A.K. Cryogenics 1979, 19 (6) 335.
18. Lobo, L.Q.; Staveley, L.A.K. J. Chem. Eng. Data 1981, 26 (4) 404.
19. Harrison, R.H.; Douslin, D.R. J. Chem. Eng. Data 1966, 11 (3) 383.
20. Rubio, R.G.; Streett, W.B. J. Chem. Eng. Data submitted for publication.
21. Calado, J.C.G.; Clancy, P.; Heintz, A.; Streett, W.B. J. Chem. Eng. Data 1982, 27 (4) 376.
22. Machado, J.R.S.; Streett, W.B. J. Chem. Eng. Data 1983, 28 (2) 218.
23. Anderson, T.F.; Abrams, D.S.; Grens, E.A. AIChE J. 1978, 24 (1) 20.
24. Dymond, J.H.; Smith, E.B. "The Virial Coefficients of Pure Gases and Mixtures"; Clarendon Press: Oxford, 2nd. edition, 1980.
25. Fischer, J. J. Chem. Phys. 1980, 72 (10) 5371.
26. Henderson, D., in "Equations of State in Engineering and Research", Chao, K.C.; Robinson Jr., K.L. Eds.; ADVANCES IN CHEMISTRY SERIES No. 182, American Chemical Society, Washington, D.C., 1979.
27. Haar, L.; Gallagher, J.S.; Kell, G.S., Natl. Bur. Stand. (US) Internal Report No 81 - 2253, 1981.
28. Waxman, M. Gallagher, J.S J. Chem. Eng. Data 1983, 28 (2), 224.
29. Kohler, F.; Haar, L. J. Chem. Phys. 1981, 75 (1) 388.
30. Moritz, P.; Kohler, F. Ber. Bunsenges. Phys. Chem. 1984, 88 (8) 702.

31. Boublik, T.; Nezbeda, I. Chem. Phys. Lett. 1977, 46 (2) 315.
32. Deiters, U. Chem. Eng. Sci. 1981, 36 (7), 1139.
33. Carnahan, N.F; Starling, K.E. AIChE J. 1972, 18 (6) 1184.
34. Aim K.; Nezbeda, I. Fluid Phase Equil. 1983, 12 (3) 235.
35. Chen, S.S.; Kreglewski, A. Ber. Bunsenges. Phys. Chem. 1977, 81 (10) 1048.
36. Alder, B.J.; Young, D.A.; Mark, M.A. J. Chem. Phys. 1972, 56 (6) 3013.
37. Simnick, J.J.; Lin, H.M.; Chao, K.C. in "Equations of State in Engineering and Research", Chao, K.C.; Robinson Jr., R.L. Eds.; ADVANCES IN CHEMISTRY SERIES No. 182, American Chemical Society, Washington, D.C., 1979.
38. Nies, E.; Kleintjens, L.A.; Koningsveld, R.; Simha, R.; Jain, R.K. Fluid Phase Equil. 1983, 12 (1) 11.
39. Sanchez, I.C.; Lacombe, R.H. Nature (London) 1974, 252 (5482), 381.
40. Costas, M.; Sanctuary, B.C. J. Phys. Chem. 1981, 85 (21) 3153.
41. Costas, M.; Sanctuary, B.C. Fluid Phase Equil. 1984, 18 (1) 47.
42. Alem, A.H.; Mansoori, G.A. AIChE J. 1984, 30 (2) 468.
43. Lobo, L.Q.; McClure, D.; Staveley, L.A.K.; Clancy, P.; Gubbins, K.E.; Gray, C.G. J. Chem. Soc. Faraday Trans. 2, 1981, 77, 425.

RECEIVED November 5, 1985

4

Phase Equilibria for the Propane–Propadiene System from Total Pressure Measurements

Andy F. Burcham, Mark D. Trampe, Bruce E. Poling, and David B. Manley

Department of Chemical Engineering, University of Missouri–Rolla, Rolla, MO 65401

Total pressure measurements were made on the propane-propadiene system from 253.15 to 353.15K. The data were reduced using the Soave-Redlich-Kwong Equation of state with modified mixing rules containing several parameters. Corrections were made for the effects of known chemical impurities in the experimental system, and vapor pressures and relative volatilities were calculated.

The phase equilibrium behavior of the propane-propadiene system was studied using the total pressure method. This method has been applied to a number of systems of close boiling light hydrocarbons at the University of Missouri-Rolla (Steele et al., 1976; Martinez-Ortiz and Manley, 1978; Flebbe et al., 1982; Barclay et. al., 1982). The total pressure method consists of experimentally measuring the vapor pressures and total volume of two-phase mixtures under varying conditions of temperature and overall composition. These data were then reduced using a thermodynamically consistent set of equations to calculate phase compositions and densities.

In past studies, it has been necessary to use extremely pure materials in order to produce accurate results. For practical reasons this has limited the method to investigations concerning the relatively few chemicals which are easily purified. However, many chemicals present in industrial process mixtures cannot be studied in pure form because they react with themselves. In particular, light hydrocarbon diolefins and acetylenes tend to dimerize at process temperatures. One objective of this work was to extend the

0097–6156/86/0300–0086$06.50/0

previously developed technique to mixtures containing reactive components. Propadiene was chosen because, among other reasons, it is difficult to obtain in high purity; and it is moderately reactive. Therefore, it provides a good test case for the development of the necessary experimental and computational tools.

There are several ways to reduce the experimental total pressure data. In this work we chose to fit the data with the Soave-Redlich-Kwong (SRK) equation of state (Soave, 1972) with adjustable interaction parameters. The SRK equation was chosen because it has been used successfully for correlating hydrocarbon physical properties and because it was necessary to predict some properties in addition to correlating the observed pressures. Multicomponent mixtures were prepared with propadiene which contained quantitatively known impurities. The SRK interaction parameters for the minor constituents were estimated from the literature, but the parameters for the primary propane-propadiene binary were determined from the measured data. It was necessary to add additional interaction parameters in order to fit the data within the estimated experimental error. To avoid errors in calculated pure component vapor pressures and liquid densities, the SRK equation parameters were adjusted to yield accurate pure component vapor pressures. The Costald (Hankinson and Thompson, 1979) correlation was used to calculate the saturated liquid densities in the volume balance equation. Pure propadiene vapor pressures were estimated from the experimental data simultaneously with the interaction parameters in an iterative procedure.

Experiment

The equilibrium cell which holds about 6 cc of sample is shown in Figure 1. The chemicals of known composition were volumetrically metered into the evacuated cell, and their exact amounts were determined (to within 0.0001 grams) on an analytical balance. The cell was then placed in a thermostat controlled to within 0.01 °C, and the pressure was determined by balancing the hydrocarbon vapor pressure on the bottom of the stainless steel diaphragm with nitrogen on the top. The differential pressure null was determined to within 0.02 psia by the displacement transformer which had been previously calibrated. The nitrogen pressure was measured to within 0.01 psia with a Ruska dead weight gauge and Princo barometer. The temperature was measured to within 0.025 °C by a platinum resistance thermometer and Mueller bridge. Detailed descriptions of the equipment, operating procedures, and error analysis are given elsewhere (Barclay, 1980; Burcham, 1981, Barclay et al., 1982).

Research grade propane was obtained from Phillips Petroleum Company, which stated that infrared and mass spectrometer determinations showed the purity to be 99.99 mole percent. After careful degassing, the only impurity shown by gas chromatography was a trace amount of ethane. The propadiene was purchased from Columbia Inorganics Incorporated and had a stated purity of 99 weight percent. After careful degassing, analysis by gas chromatography and mass spectrometry identified the impurities as listed in Table I.

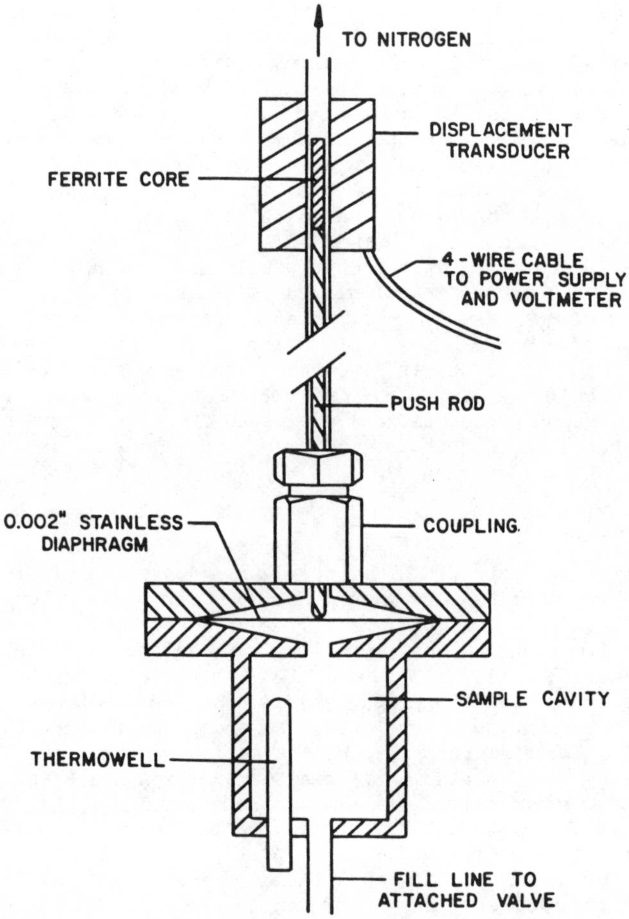

Figure 1. Sample cell and transducer

Table I. Propadiene Analysis

Compound	Mole %
Propadiene	96.84
Propene	2.11
Propyne	0.89
Propane	0.088
Ethene	0.047
Cyclopropane	0.031

Two cells were used, and a run consisted of filling the evacuated cells 25 to 60 percent full with degassed chemicals, inserting them into the bath, and measuring the pressure at different temperatures. The first set of measurements started at 80°C, proceeded down in 25°C steps to -20°C; a second set of measurements was then made from -20°C to 80°C. Comparison of the duplicated measurements provided a check on possible errors due to leaks, reactions, or non-equilibrium conditions.

The experimental results after adjustment to exact temperatures are given in Table II. The temperature adjustment procedure accounted for the difference between the actual bath temperature and the desired temperature. This was less than 0.15°C, and the correction (done with the Antoine equation) contributed negligible error.

Data Reduction

In order to compare with literature data the pure propane vapor pressures were correlated with the Goodwin equation (Goodwin, 1975).

$$P = P'EXP(AX + BX^2 + CX^3 + DX(1-X)^{1.5}) \qquad (1)$$

$$X = (1 - T'/T)/(1-T'/T_C)$$

The constants given in Table III, are regressed from the data of this study; and a comparison with recent literature data is shown in Figure 2. The scatter is typical for light paraffin vapor pressure data from different laboratories.

The thermodynamic state of a closed system at fixed temperature and specific volume is completely determined. Pressure can be measured, but not varied independently. Consequently, the pressure measurements can be used to determine appropriate parameters in the theoretical description of the system. This was accomplished by forming a system of equations relating the known and unknown

Table II. Temperature Corrected Pressure Data

Overall Mol Fraction Propane	Cell Volume cc	Propane grams	Propadiene (impure) grams	253.15 K Pressure Psia	278.15 K Pressure Psia	303.15 K Pressure Psia	328.15 K Pressure Psia	353.15 K Pressure Psia
0.00088	5.91	0.0000	1.4004	26.505	63.163	128.864	234.883	393.683
				26.502	63.169	128.897	234.847	393.801
0.1923	6.19	0.4009	1.5314	30.609	70.813	141.393	253.748	420.694
				30.608	70.800	141.437	253.750	420.639
0.3045	5.91	0.7216	1.4989	32.587	73.897	146.643	261.859	432.526
				32.590				
0.5071	5.91	0.46295	0.40935	34.339	77.865	153.263	272.067	447.469
				34.346	77.871	153.264	272.042	447.448
0.6293	6.19	0.5660	0.3033			155.711	275.819	453.110
0.7168	6.19	1.4471	0.5199	35.918	80.327	157.005	277.666	455.833
				35.956				
0.8782	6.19	1.8030	0.2274	36.040	80.824	157.632	278.380	456.782
				36.029	80.800	157.846	278.715	457.074
0.9459	5.91	1.7802	0.0926	36.092	80.805	157.158	277.697	455.680
				36.129	80.778	157.649	278.018	456.204
1.0000				35.505	79.971	156.627	276.631	454.489
				35.409	79.881	156.533	276.588	454.487
				35.481	79.935	156.472	276.621	454.494
				35.382	79.856	156.395	276.540	454.461
						156.522	276.591	454.239
						156.442	276.543	454.052

Table III. Goodwin Equation Constants for Propane and Propadiene

	Propane	Propadiene
A	3.26655	-1.57086
B	0.65753	9.00968
C	-0.18930	-3.35329
D	0.56757	5.35825
p', psia	14.696	14.696
T', K	231.1	238.7
T_c, K	369.82	393.0

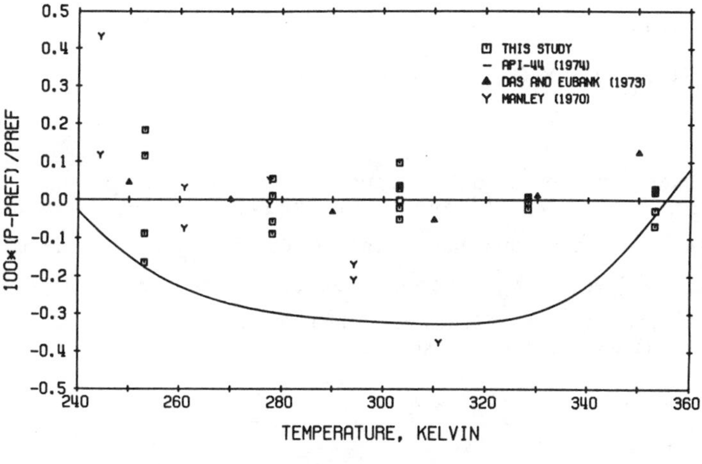

Figure 2. Comparison for propane vapor pressure

variables, and then solving for the unknown variables. For M components these equations are:

M Component Mass Balances

$$N_i - x_i N_x - y_i N_y = 0; \; i = 1,\ldots,M \qquad (2)$$

1 Total Volume Balance

$$V_T - N_x V_x - N_y V_y = 0 \qquad (3)$$

2 Summation Equations

$$1 - \sum_i x_i = 0; \; 1 - \sum_i y_i = 0 \qquad (4)$$

2 Equations of State

$$E(P,V_x,T) = 0; \; E(P,V_y,T) = 0 \qquad (5)$$

M Equilibrium Constraints

$$F_i(P,x_i,T) - F_i(P,y_i,T) = 0; \; i = 1,\ldots,M \qquad (6)$$

These equations contain only parameters and physically significant variables. When V_T, T, and N_i are known, and the parameters are specified; the equations can be solved for the 2M + 5 unknowns x_i, y_i, P, V_x, V_y, N_x, N_y. Comparison of the measured and calculated pressures provides verification of the specified parameters.

The SRK equation of state is:

$$P = RT/(V-b) - a/(V(V+b)) \qquad (7)$$

where

$$a = \Omega_a bRTF/\Omega_b \qquad (8)$$

$$b = \Omega_b RT_c/P_c \qquad (9)$$

$$\Omega_a = 0.42748 \qquad (10)$$

$$\Omega_b = 0.08664 \qquad (11)$$

where F is a parameter which forces equation 7 to reproduce pure component vapor pressures. Equations 7 through 11 correspond to the form described by Reid, et. al. (1977). Initially, the original mixing rules with one interaction parameter

$$a = \sum_{i}\sum_{j} x_i x_j (a_i a_j)^{1/2}(1-k_{ij}) \tag{12}$$

$$b = \sum_{i} x_i b_i \tag{13}$$

were applied; but, as described below, it was necessary to add additional parameters in order to fit the experimental data within its estimated uncertainty. Since the SRK equation predicts liquid specific volumes poorly, the more accurate Costald correlation (Hankinson and Thomson, 1979) was used in the total volume balance equation only. Because this is an extensive constraint, thermodynamic consistency is still satisfied.

At each temperature, pure component F parameters for equation (8) were determined so that pure component vapor pressures were accurately reproduced. For propane, the vapor pressure data of this study were used. For the impurities propene, propane, cyclopropane, and ethene, literature vapor pressure data were used (Bender, 1975; Vohra et. al., 1962; Heisig and Hurd, 1933; Lin et. al., 1970; Douslin and Harrison, 1968). The parameters are given in Table IV.

Interaction parameters, k_{ij} in Equation 12, for the minor impurities were considered relatively unimportant and were assigned a value of 0. The propene-propane parameter was determined from literature data (Manley and Swift, 1970) to be 0.0085, and the propene-propadiene parameter was estimated to be the same based on experience with C_4 hydrocarbon mixtures. The propane-propadiene parameter was estimated to be 0.028 from the experimental mixture data of this study.

Using the F parameters for all the components except propadiene, and the estimated interaction parameters; the theoretical-model equations were applied to the impure propadiene vapor pressure data. Pure propadiene F parameters and vapor pressures were calculated. The results are given in Table V. Goodwin equation parameters determined by a least squares fit to the pure vapor pressures are given in Table III, and a comparison with literature data is given in Figure 3. The relatively large scatter for these data is probably due to the difficulty in obtaining pure propadiene and in keeping it from dimerizing. The effect of accounting for the 3.26 mole percent impurity present in the propadiene used in this study was to reduce the estimated pure propadiene vapor pressure by about 1.5%.

Next, the propane-propadiene interaction parameter, k_{12}, for each data point was determinea. Each value of k_{12} is shown in Figure

Table IV. SRK F Parameters, Critical Properties, and Acentric Factors in this Study

	Propane	Propene	Propyne	Cyclopropane	Ethene	Propadiene
	Parameter F in Equation 8					
253.15 K	1.843929	1.798899	2.167369	2.032122	1.189297	
278.15 K	1.596573	1.558492	1.870922	1.769849	1.024429	
303.15 K	1.394985	1.362286	1.629764	1.554293	0.890463	
328.15 K	1.227697	1.199187	1.430623	1.375080	0.779921	
353.15 K	1.085501	1.059943	1.264037	1.223972	0.687514	
T_c,K	369.82	364.90	402.39	398.30	282.35	393.
P_c,psia	616.41	699.06	816.27	809.24	731.28	793.58
ω	0.152	0.148	0.218	0.132	0.086	0.1493

Table V. Propadiene Properties (Duplicate Data Points)

	253.15 K	278.15 K	303.15 K	328.15 K	353.15 K
Impure Propadiene	26.505	63.163	128.864	234.883	393.683
Vapor Pressures	26.502	63.169	128.897	234.847	393.801
Pure Propadiene	2.017308	1.749740	1.532509	1.352641	1.201258
F Parameters	2.017349	1.749706	1.532417	1.352645	1.201152
Pure Propadiene	25.882	62.053	127.083	232.230	389.897
Vapor Pressures	25.879	62.059	127.117	232.193	390.019

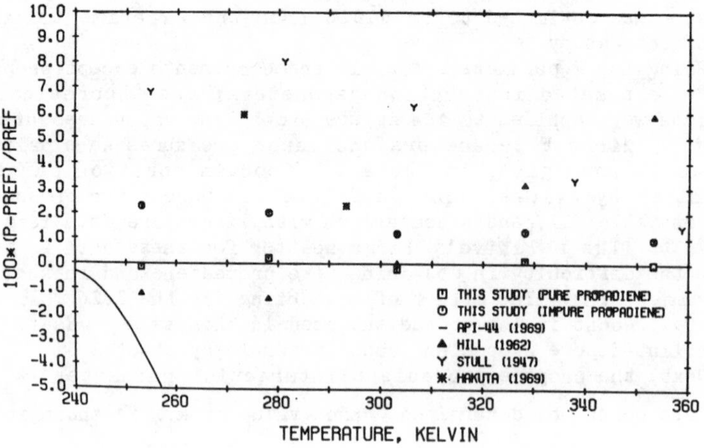

Figure 3. Comparison for propadiene vapor pressures

4. At all five temperatures, k_{12} increased steadily as the mole fraction propane increased. At each temperature, the value of k_{12} that minimized the sum of the squares of the pressure deviations was also determined. A typical pressure deviation plot (for 303.15K with k_{12} = 0.02792) is shown in Figure 5. This figure demonstrates a non-random error pattern consistent with the non-random composition dependence of k_{12}. This indicated a need for additional parameter(s) in the mixing rules for the SRK equation.

SRK Equation with Additional Interaction Parameters

Initially, a second parameter (β) was added to the standard SRK mixing rule in the form

$$k_{12} = k_{12}^{o} + \beta x_1/V \tag{14}$$

This equation introduces a linear composition dependence for the parameter k_{12}. The inverse volume factor was incorporated to force the mixing rule to reduce to a quadratic form in the low density limit. This is important if the equation is to be applied to the vapor as well as the liquid phase. Equation 14 led to satisfactory results, but two questions remained. How could equation 14 be extended to multicomponent mixtures, and what similarities were there between equation 14 and other recently proposed mixing rules? Mathias and Copeman (1983) recently published a modification of the Peng-Robinson equation (1976) with mixing rules based on local composition concepts. When the SRK equation is substituted for the Peng-Robinson equation, Mathias's logic leads to

$$P = \frac{RT}{V-b} - \frac{1}{V(V+b)} \left(a - \frac{\ln(1+b/V)}{bRT} \sum_i x_i a_{ci}^2 \left[\sum_j x_j d_{ji}^2 - \left(\sum_j x_j d_{ji}\right)^2\right]\right) \tag{15}$$

In equation 15, a and b are still given by equations 12 and 13. Two additional parameters per binary, d_{ij} and d_{ji} are introduced. For the system in this study values of d were to be introduced for only one of the binaries, namely, the propane-propadiene binary. When all d_{ij} values except d_{12} and d_{21} are zero, equation 15 may be written as

$$P = \frac{RT}{V-b} - \frac{1}{V(V+b)} \left(a - \frac{\ln(1+b/V)}{b} x_1 x_2 (a_1 a_2)^{1/2} (x_1 b_1 g_1 + x_2 b_2 g_2)\right) \tag{16}$$

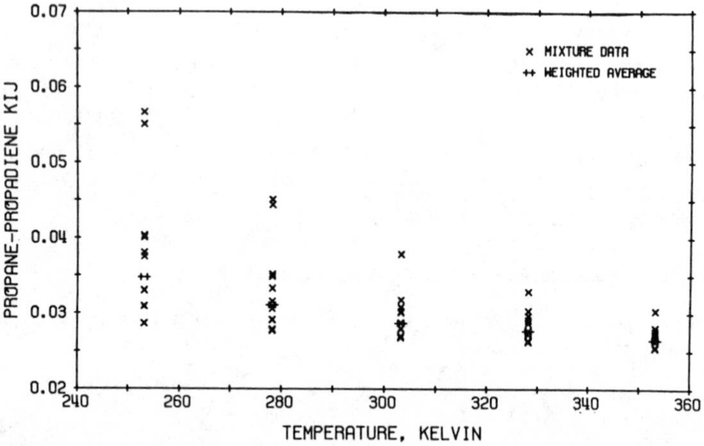

Figure 4. Propane-propadiene interaction parameters. k_{ij} values increase with increasing propane concentration.

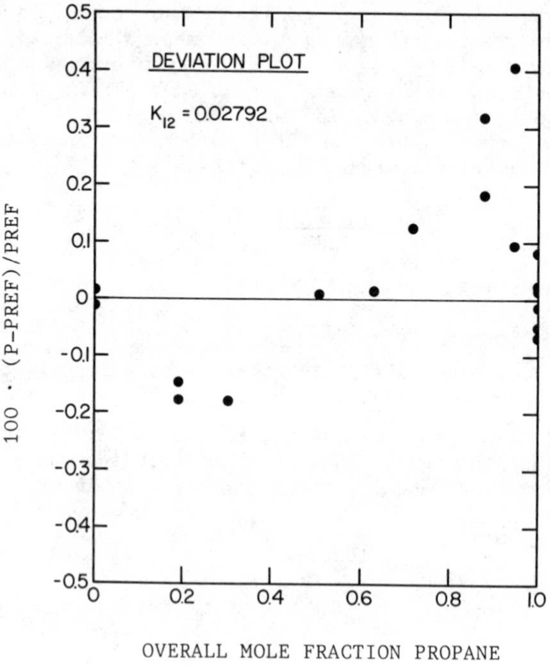

Figure 5. Mixture pressure deviations at 303.15K for one-parameter SRK fit.

where the two dimensionless parameters, g_1 and g_2 are given by

$$g_1 = \frac{a_{c1}^2 d_{21}^2}{(a_1 a_2)^{1/2} b_1 RT} \tag{17}$$

and

$$g_2 = \frac{a_{c2}^2 d_{12}^2}{(a_1 a_2)^{1/2} b_2 RT} \tag{18}$$

Equation (16) written for a binary mixture is

$$P = \frac{RT}{V-b} - \frac{1}{V(V+b)}[a_1 x_1^2 + a_2 x_2^2 + 2(a_1 a_2)^{1/2} x_1 x_2 (1-k_{12} - x_1 g_1 b_1 \frac{\ln(1+b/V)}{2b} - x_2 g_2 b_2 \frac{\ln(1+b/V)}{2b})] \tag{19}$$

In equations 15, 16, and 19, the g (or d) parameters primarily characterize the mixing behavior in the liquid phase, while k_{12} has an effect on both phases. When g_1 and g_2 are zero, equation 16 reduces to the original SRK equation. In equations 15, 16, and 19, the term ln(1+b/V) plays the same role as 1/V in equation 14. When g_2 is set equal to zero, equations 14 and 19 take on essentially the same composition dependence.

A general fugacity coefficient expression can be obtained from equation 15. However, since the extra parameters were used only for the propane-propadiene binary, it was more convenient to use equation 16. The fugacity coefficient expression that results from equation 16 is

$$\ln\phi_i = \frac{bi}{b}(z-1) - \ln(z - \frac{bP}{RT}) + \frac{a}{bRT}[\frac{bi}{b} - \frac{2a_i^{1/2}}{a}\sum_j x_j a_j^{1/2}(1-k_{ij})]\ln(1+b/V)$$
$$+ \frac{\ln(1+b/V)}{b^2 RT}(a_1 a_2)^{1/2}\{x_1 x_2 [(\frac{b_i}{V+b} - \frac{b_i}{b}\ln\frac{V+b}{V})(x_1 g_1 b_1 + x_2 g_2 b_2)$$
$$+ \frac{g_1 b_1 + g_2 b_2}{2}\ln\frac{V+b}{V}] + Q_i \ln\frac{V+b}{V}\} \tag{20}$$

For components 1 and 2,

$$Q_1 = x_1 x_2 g_1 b_1/2 + x_2(x_2 - x_1) g_2 b_2/2 \tag{21}$$

and

$$Q_2 = x_1x_2g_2b_2/2 + x_1(x_1-x_2)g_1b_1/2 \tag{22}$$

For all other components, $Q_i = 0$.

Results

The experimental data in Table II were recorrelated using equations 16 and 20 with g_1 set equal to zero. Values of the parameters that were obtained in this two parameter fit are listed in Table VII. Figure 6 shows the deviation plot for 303.15 K. The extra parameter has essentially removed the non- random deviation pattern demonstrated by Figure 5. The parameters in Table VII were used to calculate smoothed phase compositions, volumes, and relative volatilities, $\alpha = (y_1/x_1)/(y_2/x_2)$, of propane (1) in propadiene (2). These are shown in Tables VII-XI and Figure 7. The relative volatility curves are not linear with mole fraction. This non-linear behavior has been observed previously in close-boiling hydrocarbon mixtures (Flebbe et al., 1982; Barclay et al., 1982); characterization of this behavior is important in the design of separation processes. Hakuta et al. (1969) reported relative volatilities for the propane-propadiene system and comparisons to their results shown in Figures 8 and 9. A comparison to results reported by Hill et al. (1961) is shown in Table XII.

As discussed elsewhere (Barclay, 1980; Burcham, 1981; Barclay et al., 1982) estimated probable errors in the calculated liquid compositions and relative volatilities are ±0.0005 mole fraction and ±0.004 units respectively assuming that the pure material analysis is good and that calculated vapor and liquid densities are reasonably accurate. A comparison of the calculated saturated vapor and liquid compressibility factors for propane with published data (Das and Eubank, 1973) shows an error of +1% in $Z^V - Z^L$ at 253.15°K and +4% at 353.°K. Since these errors generate equivalent percent errors in $\ln\alpha$, the calculated relative volatilities may be off a maximum additional 0.0010 units at the pure propane conditions and 0.0100 units at the pure propadiene conditions. Experimental propadiene densities are not available and no effort was made to reduce this contribution to the total error.

As can be seen from equation 19, the binary interaction term when g_1 is zero is given by

$$1 - k_{12} - x_2g_2b_2 \frac{\ln(1+b/V)}{2b}$$

In this study, values of k_{12} were always positive and less than 0.04. Values of the term, $x_2g_2b_2 \ln(1+b/V)/2b$, were always positive and

Table VI. Binary Interaction Parameters for the Propane-Propadiene Binary

T,K	k_{12}	g_2
253.15	0.03329	0.1564
278.15	0.02956	0.1147
303.15	0.02775	0.0532
328.15	0.02694	0.0440
353.15	0.02599	0.0356

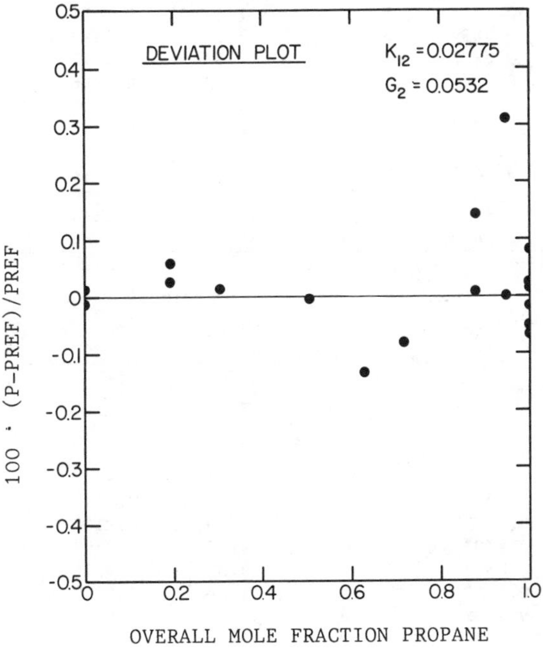

Figure 6. Mixture pressure deviations at 303.15K for two-parameter SRK fit.

Table VII. Calculated Propane(1)-Propadiene(2) VLE at 253.15 K. V^{LIQ} from Costald, Other Values from SRK Equation

X_1	Y_1	P,psia	V^{LIQ}, cm^3/g-mol	Z^{VAP}	Relative Volatility
0.0	0.0	25.88	63.81	0.9593	2.892
0.0500	0.1184	27.92	64.56	0.9558	2.552
0.1000	0.2019	29.43	65.31	0.9532	2.276
0.1500	0.2658	30.58	66.07	0.9511	2.051
0.2000	0.3180	31.50	66.83	0.9494	1.865
0.2500	0.3631	32.26	67.59	0.9479	1.710
0.3000	0.4038	32.91	68.36	0.9467	1.581
0.3500	0.4420	33.46	69.13	0.9455	1.471
0.4000	0.4788	33.95	69.90	0.9445	1.378
0.4500	0.5151	34.38	70.68	0.9436	1.298
0.5000	0.5516	34.75	71.47	0.9427	1.230
0.5500	0.5889	35.07	72.25	0.9419	1.172
0.6000	0.6272	35.33	73.04	0.9412	1.121
0.6500	0.6669	35.54	73.83	0.9406	1.078
0.7000	0.7082	35.70	74.63	0.9400	1.040
0.7500	0.7514	35.80	75.43	0.9395	1.008
0.8000	0.7967	35.84	76.24	0.9390	0.980
0.8500	0.8441	35.82	77.05	0.9386	0.956
0.9000	0.8938	35.75	77.86	0.9383	0.935
0.9500	0.9457	35.62	78.67	0.9380	0.917
1.0000	1.0000	35.44	79.49	0.9378	0.902

Table VIII. Calculated Propane(1)-Propadiene(2) VLE at 278.15 K. V^{LIQ} from Costald, Other Values for SRK Equation

X_1	Y_1	P,psia	V^{LIQ}, cm^3/g-mol	Z^{VAP}	Relative Volatility
0.0	0.0	62.06	67.21	0.9218	2.121
0.0500	0.0932	65.25	68.03	0.9172	1.952
0.1000	0.1673	67.80	68.86	0.9135	1.808
0.1500	0.2293	69.90	69.68	0.9103	1.686
0.2000	0.2833	71.66	70.52	0.9076	1.581
0.2500	0.3319	73.16	71.36	0.9051	1.490
0.3000	0.3769	74.46	72.20	0.9030	1.412
0.3500	0.4197	75.59	73.05	0.9010	1.343
0.4000	0.4611	76.58	73.90	0.8992	1.283
0.4500	0.5017	77.44	74.76	0.8976	1.231
0.5000	0.5422	78.19	75.62	0.8961	1.185
0.5500	0.5830	78.83	76.48	0.8947	1.144
0.6000	0.6243	79.36	77.36	0.8934	1.108
0.6500	0.6664	79.79	78.23	0.8922	1.076
0.7000	0.7097	80.11	79.11	0.8911	1.048
0.7500	0.7541	80.33	80.00	0.8902	1.022
0.8000	0.8000	80.45	80.89	0.8893	1.000
0.8500	0.8475	80.46	81.78	0.8885	0.980
0.9000	0.8965	80.38	82.69	0.8878	0.963
0.9500	0.9474	80.19	83.59	0.8872	0.947
1.0000	1.0000	79.91	84.50	0.8867	0.934

Table IX. Calculated Propane(1)-Propadiene(2) VLE at 303.15 K. V^{LIQ} from Costald, Other Values from SRK Equation

X_1	Y_1	P,psia	V^{LIQ}, cm^3/g-mol	Z^{VAP}	Relative Volatility
0.0	0.0	127.10	71.42	0.8676	1.667
0.0500	0.0771	131.59	72.33	0.8620	1.587
0.1000	0.1441	135.44	73.26	0.8571	1.515
0.1500	0.2040	138.77	74.19	0.8527	1.452
0.2000	0.2585	141.69	75.12	0.8487	1.395
0.2500	0.3093	144.25	76.07	0.8450	1.343
0.3000	0.3573	146.50	77.02	0.8416	1.297
0.3500	0.4033	148.49	77.97	0.8385	1.256
0.4000	0.4480	150.25	78.94	0.8357	1.218
0.4500	0.4919	151.78	79.91	0.8330	1.183
0.5000	0.5353	153.12	80.89	0.8306	1.152
0.5500	0.5785	154.26	81.87	0.8283	1.123
0.6000	0.6220	155.21	82.86	0.8262	1.097
0.6500	0.6658	155.99	83.86	0.8242	1.073
0.7000	0.7103	156.58	84.86	0.8224	1.051
0.7500	0.7556	157.00	85.87	0.8207	1.031
0.8000	0.8019	157.25	86.89	0.8192	1.012
0.8500	0.8493	157.32	87.92	0.8179	1.995
0.9000	0.8980	157.22	88.95	0.8166	0.979
0.9500	0.9482	156.95	89.99	0.8155	0.964
1.0000	1.0000	156.50	91.03	0.8146	0.950

Table X. Calculated Propane(1)-Propadiene(2) VLE at 328.15 K. V^{LIQ} from Costald, Other Values from SRK Equation

X_1	Y_1	P,psia	V^{LIQ}, cm^3/g-mol	Z^{VAP}	Relative Volatility
0.0	0.0	232.21	76.93	0.7939	1.462
0.0500	0.0690	238.75	78.00	0.7866	1.409
0.1000	0.1314	244.47	79.08	0.7799	1.361
0.1500	0.1887	249.53	80.17	0.7737	1.318
0.2000	0.2422	254.00	81.27	0.7679	1.279
0.2500	0.2930	257.97	82.38	0.7626	1.243
0.3000	0.3417	261.50	83.51	0.7577	1.211
0.3500	0.3889	264.62	84.64	0.7531	1.182
0.4000	0.4350	267.37	85.79	0.7488	1.155
0.4500	0.4805	269.77	86.94	0.7447	1.130
0.5000	0.5256	271.85	88.11	0.7410	1.108
0.5500	0.5706	273.62	89.29	0.7375	1.087
0.6000	0.6158	275.09	90.48	0.7342	1.068
0.6500	0.6612	276.26	91.69	0.7311	1.051
0.7000	0.7071	277.14	92.91	0.7283	1.035
0.7500	0.7537	277.74	94.14	0.7257	1.020
0.8000	0.8010	278.06	95.38	0.7234	1.006
0.8500	0.8491	278.10	96.63	0.7212	0.993
0.9000	0.8983	277.86	97.90	0.7192	0.981
0.9500	0.9485	277.36	99.19	0.7174	0.970
1.0000	1.0000	276.59	100.48	0.7158	0.960

TABLE XI. Calculated Propane(1)-Propadiene(2) VLE at 353.15 K. V^{LIQ} from Costald, Other Values from SRK Equation

X_1	Y_1	P,psia	V^{LIQ}, cm^3/g-mol	Z^{VAP}	Relative Volatility
0.0	0.0	389.96	84.99	0.6949	1.303
0.0500	0.0626	398.94	86.35	0.6849	1.268
0.1000	0.1208	406.97	87.73	0.6755	1.237
0.1500	0.1757	414.17	89.14	0.6667	1.208
0.2000	0.2281	420.64	90.57	0.6583	1.812
0.2500	0.2785	426.44	92.03	0.6503	1.158
0.3000	0.3276	431.63	93.51	0.6427	1.137
0.3500	0.3757	436.26	95.03	0.6355	1.117
0.4000	0.4230	440.36	96.57	0.6287	1.100
0.4500	0.4700	443.97	98.14	0.6221	1.084
0.5000	0.5167	447.09	99.74	0.6159	1.069
0.5500	0.5634	449.74	101.37	0.6100	1.056
0.6000	0.6102	451.94	103.04	0.6044	1.044
0.6500	0.6572	453.71	104.75	0.5992	1.032
0.7000	0.7046	455.04	106.50	0.5943	1.022
0.7500	0.7524	455.94	108.28	0.5896	1.013
0.8000	0.8007	456.43	110.12	0.5853	1.005
0.8500	0.8496	456.51	111.99	0.5814	1.997
0.9000	0.8990	456.18	113.92	0.5777	0.990
0.9500	0.9492	455.47	115.90	0.5744	0.983
1.0000	1.0000	454.37	117.94	0.5713	0.977

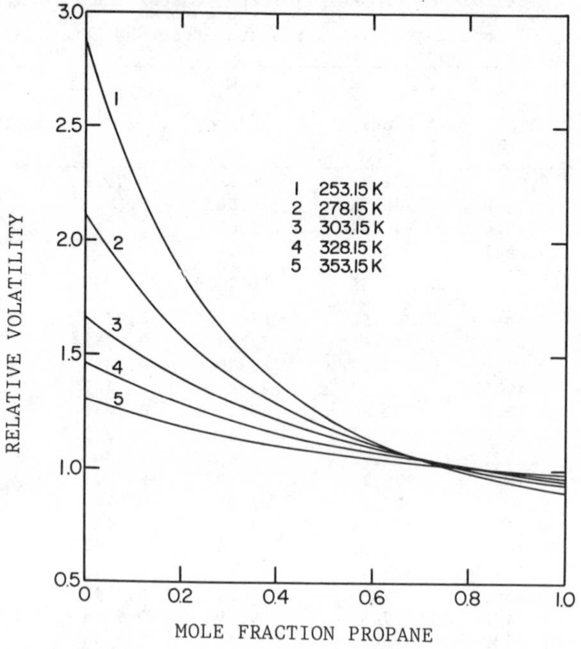

Figure 7. Relative volatility of propane propadiene.

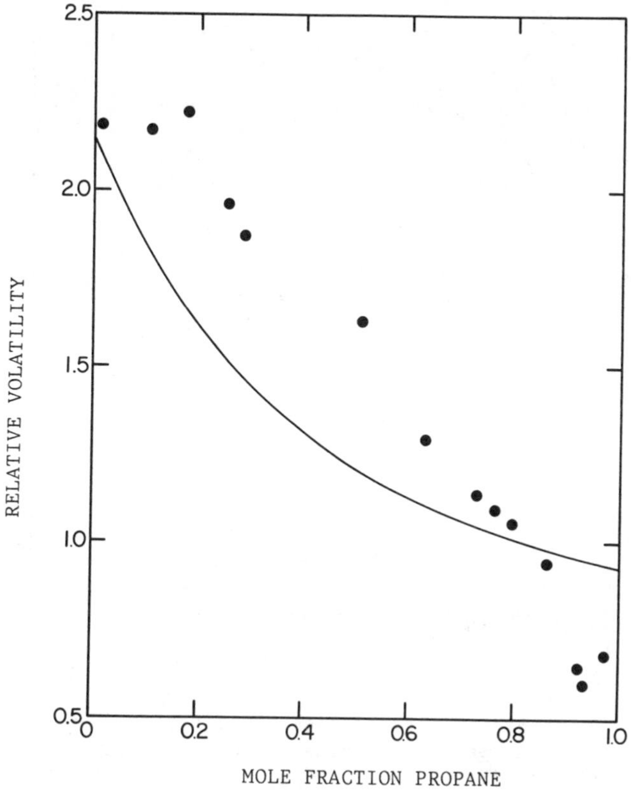

Figure 8. Comparison for relative volatilities at 273.15K. Line is calculated from results of this study, points are Hakuta's data (1969).

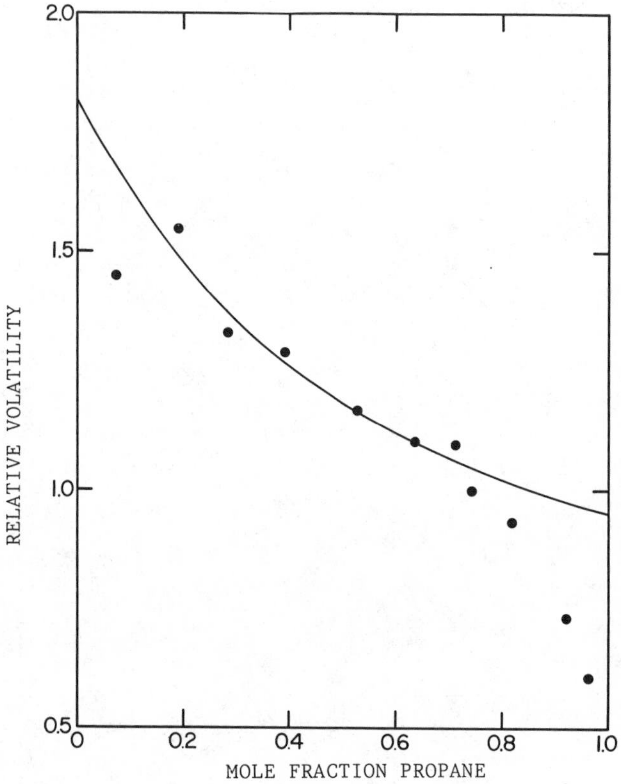

Figure 9. Comparison for relative volatilities at 293.25K. Line is calculated from results of this study, points are Hakuta's data (1969).

TABLE XII. Comparison of Experimental and Calculated Relative Volatilities of Propane in Propadiene

Liquid Phase Mole fractions					Relative Volatilities	
Propane	Propadiene	Propene	Propyne	Temperature	Experimental (Hill, 1961)	Calculated
0.912	0.013	0.058	0.017	305.37	0.976	0.967
0.924	0.014	0.047	0.015	305.37	0.973	0.965
0.026	0.935	0.027	0.012	305.37	1.441	1.584
0.012	0.933	0.037	0.018	305.37	1.464	1.595
0.010	0.941	0.036	0.013	305.37	1.464	1.601
0.919	0.012	0.055	0.014	333.15	0.971	.974
0.929	0.011	0.047	0.013	333.15	0.978	.973
0.014	0.959	0.016	0.011	333.15	1.287	1.401

less than 0.045 (the value for pure propadiene at 253.15K). At the four higher temperatures, this last term was never more than 0.026.

The data in this study were fit to various forms of equation 16, and none was judged to be as satisfactory as the form just reported, i.e. with $g_1 = 0$. For the case when all three parameters were fit, the three parameters proved to be highly correlated, the parameters did not vary smoothly with temperature, and the fit was improved by an insignificant amount. When other two-parameter combinations were tested, i.e., $k_{12} = 0$, or $g_2 = 0$, the fit to the data was better than with the one parameter model, but always worse than for the two parameter case when $g_1 = 0$.

When g_1 is zero, and g_2 is a positive number, the predicted relative volatility is increased at low propane mole fractions. When g_1 is a positive number, the effect is to lower the relative volatility predictions at high propane mole fractions.

Summary

New experimental total pressure data for the propane-propadiene system with several known minor impurities are presented. A computational method for reducing the data is developed and demonstrated. A modification of the SRK equation of state with an additional interaction parameter is applied. Calculated pure-propadiene vapor pressures and relative volatilities for the propane-propadiene system are given.

Acknowledgment

The authors acknowledge the University of Missouri-Rolla computer center for computer time. Andy Burcham acknowledges the Chemical Engineering Department at UMR for financial support.

Legend of Symbols

a,b	mixture constants in SRK equation; a_i, b_i are for pure i; a_{ci} is for pure i at its critical temperature
A,B,C,D,P',T'	constants in Goodwin vapor pressure equation
d_{ij}	binary iteraction parameter (see equation 15)
E	equation of state (see equation 5)
F_i	fugacity of i (see equation 6)
F	pure component parameter in SRK equation, see equation 8
g_1, g_2	binary interaction parameters (see equations 16-19). In this study $g_1 = 0$
k_{ij}	binary interaction parameter
k^o_{ij}	constants in equation 14

N	moles; N_x, moles in the liquid phase; N_y, moles in the vapor phase
P	pressure, psia; P_c, critical pressure, psia
Q_i	term in fugacity coefficient expression, see equations 19-21
T	temperature, K; T_c, critical temperature
V	molar volume ; V_x is molar volume of liquid, V_y is molar volume of vapor
V_T	total volume
X	temperature function in equation 1
x	mole fraction in the liquid phase in equations 2-6, in either phase in equations thereafter
y	mole fraction in the vapor phase
Z	compressibility factor; Z^V for vapor, Z^L for liquid
Greek	
α	relative volatility
β	constant in equation 14
ϕ_i	fugacity coefficient of i
Ω	a pure number, see equations 8-11

Literature Cited

1. **Am. Petr. Inst. Proj. 44**, Supplement 1974.
2. **Am. Petr. Inst. Proj. 44**, Supplement 1977.
3. Barclay, D. A., Ph.D. Thesis, University of Missouri-Rolla 1980.
4. Barclay, D. A.; Flebbe, J. L; Manley, D. B., **J. Chem. Eng. Data** 1982, **27**, 135.
5. Bender, E. **Cryogenics** 1975, **15**, 667.
6. Burcham, A. F., M. S. Thesis, University of Missouri, Rolla 1981.
7. Das, T. R.; Bubank, P. T. **Adv. Cryog. Eng.** 1973, **18**, 208.
8. Douslin, D. R.; Harrison, R. H. **J. Chem. Thermodyn.** 1976, **8**, 301.
9. Flebbe, J. L.; Barclay, D. A.; Manley, D. B. **J. Chem. Eng. Data** 1982, **27**, 405.
10. Goodwin, R. D. **J. Res. Nat. Bur. Stand.** 1975, **79A**, 71.
11. Hakuta, T.; Nagahama, K.; Hirata, M. **Bull. Jap. Petr. Inst.** 1969, **11**, 10.
12. Hankinson, R. W.; Thomson, G. H. **AIChE J.** 1979, **25**, 653.
13. Heisig, G. B; Hurd, C. D. **J. Amer. Chem. Soc.** 1933, **55**, 3485.
14. Hill, A. B.; McCormick, R. H.; Barton, P.; Fenske, M. R. **AIChE J.** 1961, **8**, 681.
15. Lin, D. C.; Silberberg, I. H.; McKetta, J. J. **J. Chem. Eng. Data** 1970, **15**, 483.
16. Manley, D. B.; Swift, G. W. **J. Chem. Eng. Data** 1971, **16**, 301.
17. Manley, D. B. **Hydro Proc.** 1972, **Jan.**, 113.
18. Martinez-Ortiz, J. A.; Manley, D. B. **Ind. Eng. Chem. Process Des. Dev.** 1978, **17**, 346.

19. Mathias, P. M.; Copeman, T. W. **Fluid Phase Equil.** 1983, **13**, 91.
20. Peng, D. Y.; Robinson, D. B. **Ind. Eng. Chem. Fundam.** 1976, **15**, 59.
21. Reid, R. C.; Prausnitz, J. M.; Sherwood, T. K. "The Properties of Gases and Liquids", 3rd ed., McGraw-Hill: New York 1977, 40.
22. Soave, G. **Chem. Eng. Sci.** 1972, **27**, 1197.
23. Steele, K.; Poling, B. E.; Manley, D. B. **J. Chem. Eng. Data** 1976, **21** 399.
24. Stull, D. R. **Ind. Eng. Chem.** 1947, **39**, 517.
25. Vohra, S. P.; Kang, T. L.; Kobe, K. A.; McKetta, J. J. **J. Chem. Eng. Data**, 1962, **7**, 150.

RECEIVED December 12, 1985

CRITICAL, NEAR-CRITICAL, AND SUPERCRITICAL STATES

5

Nonclassical Description of (Dilute) Near-Critical Mixtures

J. M. H. Levelt Sengers, R. F. Chang, and G. Morrison

Thermophysics Division, Center for Chemical Engineering, National Bureau of Standards, Gaithersburg, MD 20899

The organizers of this symposium have asked us to review the nonclassical equations of state and their applications in near-critical fluids and fluid mixtures. In this contribution, the emphasis will be on fluid mixtures. The behavior of fluid mixtures near the gas-liquid critical line has undergone a revival of interest for a variety of reasons. One is the appearance of a number of reports of large anomalies in properties such as apparent heats of dilution (1), apparent molar volumes (2) and apparent molar specific heats (3) of dilute salt solutions in near-critical steam, and of extraordinarily large enthalpies of mixing at supercritical pressures in mixtures with components of very different critical temperature (4). Another reason is the strong push for exploration of the supercritical regime in separation processes, a promising alternative to liquid extraction (5). Supercritical solubility is governed, in part, by the partial molar properties of the solute, which have been reported to behave anomalously. It seems therefore useful at this point to take stock and review: (1) what are the differences in prediction between classical and nonclassical equations for fluids and fluid mixtures; (2) how the nonclassical behavior is going to be meshed with classical behavior further away and (3) what types of nonclassical equations are available for fluid mixtures and what is their range of validity. The anomalous properties of dilute near-critical mixtures, which are of interest in custody transfer and in supercritical solubility, are discussed in the last part of this contribution.

Critical Behavior of Pure Fluids

This section defines the terminology and summarizes the differences between classical and nonclassical behavior of pure fluids in the briefest of terms. The material presented here is in standard notation and has been reviewed elsewhere (6, 7, 8).

Critical behavior is called classical (mean-field, or van der Waals-like) if the Helmholtz free energy a(V, T) can be expanded at the critical point in terms of $\delta V = V - V_c$ and $\delta T = T - T_c$, V being the volume, T the temperature and the subscript c denoting the critical value. If such an expansion exists, thermodynamic properties vary in accordance with simple power laws along given paths asymptotically near the critical point. The power laws and exponent values are summarized in Table 1 and the paths are indicated in Figure 1. Real fluids do not behave classically. Their coexistence curves and critical P-V isotherms are flatter than predicted by classical equations, and their specific heat C_V shows a weak divergence (Table 1). Renormalization-group theory explains how nonclassical exponent values result from the correlation of the critical fluctuations, and the theoretical exponent values for the universality class of 3D Ising-like systems have been confirmed in a number of delicate experiments, see, for instance (9), (10). The scaling laws (6 - 8, 11) that describe the thermodynamics of near-critical fluids can be viewed as a compact "packaging" of the power laws of Table 1. They imply the exponent equalities

$$2 - \alpha = \beta(\delta + 1) \; ; \; \gamma = \beta(\delta - 1) \qquad (1)$$

so that only two exponents are independent; these are, however, the same for all systems in the 3D Ising-like universality class. Likewise, only two amplitudes in the power-law expressions can be chosen independently; these are not universal. We will call the critical behavior shown by real fluids and predicted by renormalization-group theory nonclassical.

It is important to note the difference between strongly (γ-like) and weakly (α-like) diverging properties in the nonclassical case (12). Properties which are the same in coexisting phases are called "fields". Examples are pressure P, chemical potential μ and temperature T. Thus $\tilde{P}(\tilde{\mu}, \tilde{T})$, with $\tilde{P} = P/T$, $\tilde{\mu} = \mu/T$ and $\tilde{T} = 1/T$ is an example of a thermodynamic potential in which all variables are fields. First derivatives, the density $\rho = (\partial\tilde{P}/\partial\tilde{\mu})_{\tilde{T}}$ and the energy density $U\rho = (\partial\tilde{P}/\partial\tilde{T})_{\tilde{\mu}}$, are generally different in coexisting phases. Such first derivatives are called "densities". Other examples of densities include volume V, enthalpy H, entropy S, and composition x.

The strong divergences are obtained (classically and nonclassically) if a derivative is taken of a density with respect to a field while another field is kept constant, that is, in a direction intersecting the coexistence curve in Figure 1b. If, however, a density is kept constant at differentiation, one obtains the nonclassical weak divergences; a constant - density direction is parallel to the coexistence curve in Figure 1b. (12). Examples of strong and weak divergences in one-component fluids are listed in Table II.

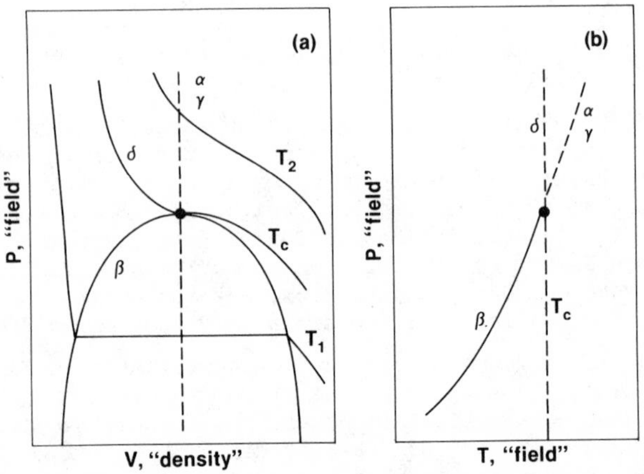

Figure 1 a, b. Paths along which critical exponents α-δ are defined in field-density (a) and in field-field space (b).

Table I
Critical exponent definitions and values
The paths are indicated in Figure 1

Property	Definition of critical exponent	Classical Value	Value for real fluids
Specific heat	$C_V \sim \lvert T-T_c\rvert^{-\alpha}$	$\alpha = 0$	$\alpha = 0.1$
Coexistence Curve	$\lvert \rho-\rho_c\rvert \sim \lvert T-T_c\rvert^{\beta}$	$\beta = 0.5$	$\beta = 0.33$
Compressibility	$K_T \sim \lvert T-T_c\rvert^{-\gamma}$	$\gamma = 1$	$\gamma = 1.24$
Critical isotherms	$\lvert P-P_c\rvert \sim \lvert \rho-\rho_c\rvert^{\delta}$	$\delta = 3$	$\delta = 4.8$

For details of the scaled form of the potential $\tilde{P}(\tilde{\mu}, \tilde{T})$ we refer to the literature (6 - 8). Conceptually, this form is construed as follows. The potential is split into a part that is analytic in $\tilde{\mu}$ and $\tilde{T}$ and a part that is scaled. The scaling variables are chosen as a reduced temperature difference $\tau = \Delta\tilde{T}$ in the weak direction along the coexistence curve and a reduced chemical potential difference $h = \Delta\tilde{\mu}$ in the strong direction intersecting this curve (Figure 2). The scaled part is written in algebraically closed form by means of Schofield's parametric variables, a distance-from-critical variable r and a contour variable θ. Two refinements are made. One goes by the name of "revised scaling": the gas-liquid asymmetry characteristic of real fluids is obtained by replacing the weak scaling variable by a linear combination of $\Delta\tilde{T}$ and $\Delta\tilde{\mu}$ (7, 13) which is equivalent with "tilting" the h axis in Figure 2. The other refinement is called "corrections to scaling" or "extended scaling". These corrections, introduced by Wegner, account for the difference between the critical Hamiltonian of the real fluid and the Hamiltonian characterizing the fixed point for 3D Ising-like systems [7].

PVT and thermal data for a number of pure fluids, including steam, ethylene and isobutane have been fitted accurately in the range of -0.01 to +0.1 in reduced temperature around the critical point by means of revised and extended scaling with one Wegner correction term (8).

Crossover from Classical to Scaled Behavior

Universal scaling behavior results when the behavior of a system is dominated by one length scale, that of the correlation length, which diverges at the critical point. When the system moves away from the critical point, other length scales begin to become important and the system gradually assumes the nonuniversal species-specific behavior that is described by classical equations. Only very recently, progress has been made with devising functions that scale near the critical point and are classical at some distance from it. There have been a number of attempts at empirically connecting "critical" and classical" regimes. Switch functions have been introduced to smoothly connect the two regimes (14, 15); great difficulty is, however, experienced with derived properties, because of anomalous contributions from higher derivatives of the switch function. Only if classical and scaled free energy are exceedingly close to each other in the range of the switch can these anomalous variations be suppressed. Hill reports good success with a switch function for steam (16).

A true crossover function is a function that behaves classically far from and is scaled close to the critical point. Fox transformed the variables in a classical thermodynamic potential in such a way that nonclassical behavior was obtained near the critical point (17). His method gives good results for PVT and coexistence curve data but has not been tried for higher derivatives such as C_V. White (18) has reported success with a

Table II. Strong and Weak Divergences

Strong (γ-type)	weak (α-type)	other (nondivergent)
second derivative in direction intersecting coexistence surface in field space	second derivative taken in coexistence, but not in critical surface in field space	second derivative along critical surface in field space
	one-component	
$K_T = -\frac{1}{V}\left(\frac{\partial V}{\partial P}\right)_T$	$K_S = -\frac{1}{V}\left(\frac{\partial V}{\partial P}\right)_S$	N.A.
$\alpha_P = -\frac{1}{V}\left(\frac{\partial V}{\partial T}\right)_P$		
$C_P = T\left(\frac{\partial S}{\partial T}\right)_P$	$C_V = T\left(\frac{\partial S}{\partial T}\right)_V$	
	two-component, $\Delta=\mu_1-\mu_2$	
$K_{T\Delta} = -\frac{1}{V}\left(\frac{\partial V}{\partial P}\right)_{T\Delta}$	$K_{Tx} = -\frac{1}{V}\left(\frac{\partial V}{\partial P}\right)_{Tx}$	$K_{Sx} = -\frac{1}{V}\left(\frac{\partial V}{\partial P}\right)_{Sx}$
$\alpha_{P\Delta} = \frac{1}{V}\left(\frac{\partial V}{\partial T}\right)_{P\Delta}$	$\alpha_{Px} = \frac{1}{V}\left(\frac{\partial V}{\partial T}\right)_{Px}$	
$C_{P\Delta} = T\left(\frac{\partial S}{\partial T}\right)_{P\Delta}$	$C_{Px} = T\left(\frac{\partial S}{\partial T}\right)_{Px}$	$C_{Vx} = T\left(\frac{\partial S}{\partial T}\right)_{Vx}$
$(\partial x/\partial\Delta)_{PT}$, osmotic susceptibility		
$(\partial x/\partial P)_{T\mu_2}$, supercritical solubility	$(\partial x/\partial P)_{TV}$	

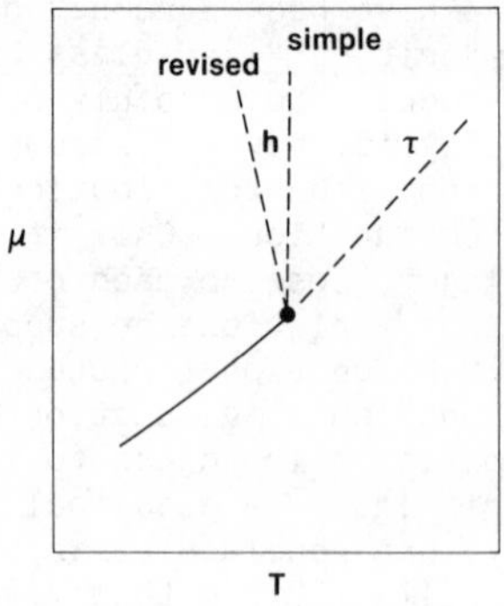

Figure 2. Directions of the strong (h) and weak (τ) scaling variables in the space of independent variables μ, T for a one-component fluid. In simple scaling, the h axis is vertical; in revised scaling, it is not.

systematic averaging procedure for fluctuations on various length scales for the van der Waals equation. Crossover from scaled to mean-field behavior was solved for the Landau-Ginzburg Hamiltonian for 3D Ising-like systems (19, 20) for both the symmetric and the asymmetric case, by Nicoll and Albright. The latter has constructed a crossover function that crosses from the scaled equation to the van der Waals equation in a range around the critical isochore (21).

Classical Critical Behavior of Fluid Mixtures

By classical behavior we mean again that the Helmholtz free energy $a(V, T, x)$, with x the mole fraction, can be expanded at the critical line in terms of its independent variables. The ideal-mixing term $RT[x \ln x + (1-x) \ln (1-x)]$ must be treated separately because it is not analytic at x=0 and x=1. The implications of classical behavior can be visualized by assuming, with van der Waals, that the mixture, at constant composition x, is in corresponding states with it components. In Figure 3a, b, we draw the equivalent of Figure 1a, b for the mixture, in reduced coordinates $P^* = P/P_c$, $V^* = V/V_c$ and $T^* = T/T_c$. For the mixture, P_c, ρ_c and T_c represent the pseudocritical parameters of the point at which the mixture would become mechanically unstable if it would remain homogeneous. In reality, the mixture becomes materially unstable before it reaches the pure-fluid coexistence curve. The curve of material instability, or dew-bubble curve, is indicated in Figure 3a. Inside this curve, the mixture splits into two phases of compositions other than x_1, and does not follow the pure-fluid isotherms. In a constant - x representation the critical point is not at the top and there are no tie lines in the plot. Figure 3a makes clear that the compressibility K_{Tx}, and therefore also α_{Px} and C_{Px}, are finite at the critical point. Thus, these mixture properties are not analogous to their strongly diverging counterparts in one-component fluids. Mixtures do show strong divergences; and example is the osmotic susceptibility $(\partial x/\partial \Delta)_{PT}$, with $\Delta = \mu_1 - \mu_2$ which diverges because of the criticality conditions:

$$(\partial \Delta/\partial x)_{PT} = (\partial^2 G/\partial x^2)_{PT} = 0 \; ; \; (\partial^2 \Delta/\partial x^2)_{PT} = (\partial^3 G/\partial x^3)_{PT} = 0 \qquad (2)$$

In Figure 3b we show the equivalent of Figure 1b for a mixture of constant composition; also indicated are the isobar that passes through the maximum on the phase boundary in Figure 3a and b, and the isotherm tangential to the phases boundary. In this example, the critical point is located between the two extrema in Figure 3b. This is the most common, but not the only case. In Figure 3b we also draw a few isochores. Note that the critical isochore changes slope at the phase boundary while the isochore passing through the temperature extremum does not change slope (22). By means of the law of corresponding states, van der Waals was able to predict almost all of the often-complex phase behavior of binary mixtures that has later been found experimentally. Van Konijnenburg and Scott (23) generated by computer all binary-mixture phase diagrams that follow from the van der Waals equation through the law of corresponding states.

Nonclassical Critical Behavior of Fluid Mixtures

The defects of a classical description are as obvious in fluid mixtures as in pure fluids. Classical models for the coexistence curves of mixtures are never flat enough at the top, since binary gas-liquid and liquid-liquid coexistence curves are cubic (24, 25). Griffiths and Wheeler (12) generalized the scaling laws for pure fluids to fluid mixtures by means of the concept of critical-point universality. Although, as we have seen, a one-component fluid and a mixture of constant composition do not have the same two-phase behavior, Griffiths and Wheeler postulated that they do if the mixture is considered at constant field. This principle is indicated schematically in Figure 4. In this figure the independent fields, for the scaled potential developed by Leung and Griffiths (26), are the weighted average activity $\bar{a} = c_1 a_1 + a_2$ and the normalized activity of the second component:

$$\zeta = c_2 a_2 / (c_1 a_1 + c_2 a_2) \quad (3)$$

Here c_1, c_2 are two constants related to the choice of zeropoints of chemical potentials, $a_i = \exp(\mu_i/RT)$, and ζ runs from 0 (pure component 1) to 1 (pure component 2). The variable ζ, rather than the customary composition variable x, connects the a-T diagrams of the two pure components. Whereas if x is the variable, a coexistence surface consists of two sheets, a dew and a bubble surface, it is single valued if ζ is the variable, since ζ is a field. A point on the critical line can now be approached in one of three ways. If two fields are kept constant, for instance ζ and the temperature, a path intersecting the coexistence curve is obtained. We expect this direction, indicated by h in Figure 4, to be "strong" just as in the pure fluid, Figure 2, and corresponding second derivatives to be strongly divergent. Several examples are given in Table II. Note that the osmotic susceptibility $(\partial x/\partial \Delta)_{PT}$ and the supercritical solubility $(\partial x/\partial P)_{T,\mu_2}$ in the presence of a solid or liquid phase of (almost) constant μ_2 are in this category of strong, γ-like divergences.

If one density is kept constant in the differentiation, for instance the volume or the entropy, then the critical point is approached along the coexistence surface, the weak direction indicated by τ in Figure 4, and second derivatives diverge weakly. Note, in Table 2, that K_{Tx}, C_{Px} and α_{Px} are in this category.

Finally, when two densities are held constant, the path to a point on the critical line becomes asymptotically parallel to this line. Since first derivatives such as volumes and entropies vary

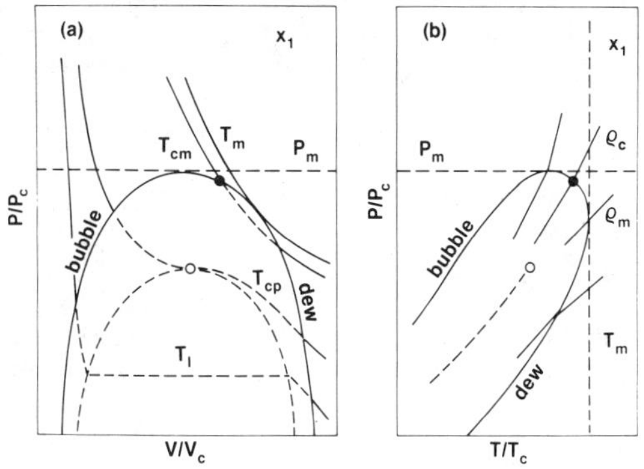

Figure 3 a, b. They are the equivalents of Figure 1a, b for a mixture of constant composition x, by virtue of the law of corresponding states. ● is the pure-fluid critical point, a pseudocritical point for the mixture. The mixture phase-separates on the dew-bubble curve and does not attain the states --- inside this curve. O is the real critical point of the mixture. T_m is the maximum temperature, P_m the maximum pressure, for the dew-bubble curve at x_1. In (b), several isochores are indicated. The critical isochore changes slope at the dew-bubble curve while the isochore through T_m does not.

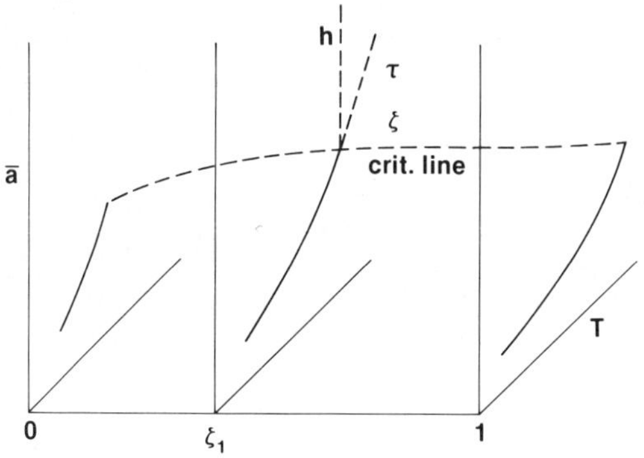

Figure 4. Coexistence surface and critical line in the space of independent variables $\bar{a}$ (average activity), $\zeta(=a_2/\bar{a})$ and T. The scaling variables h (strong) and τ (weak) are chosen in the plane of constant ζ. ζ is the third variable chosen along the critical line.

smoothly along the critical line, no divergences result if second derivatives are taken along this line, with the exception of x-derivatives in the limits $\zeta = 0$ or 1, to be discussed later.

In order to construct a scaled thermodynamic potential for the binary gas-liquid mixture, Leung and Griffiths (26) postulated that $\tilde{P}(\tilde{\bar{a}}, \tilde{T})$ at constant ζ is of the same form as $\tilde{P}(\tilde{a}_i, \tilde{T})$, or $\tilde{P}(\tilde{\mu}, \tilde{T})$, of the pure components (i = 1 or 2). Again, the potential is split into a regular and a scaled part. The scaled part, at constant ζ, is a function of the variables τ, along, and h, at an angle to the coexistence surface and is written in Schofield's parametric form, in complete analogy with the pure fluid. The nonuniversal scaling amplitudes and the analytic background are interpolated between those of the pure components, with ζ as the interpolation variable. The Leung-Griffiths equation, although phenomenogical in nature, algebraically tedious, restricted to continuous critical lines, and with limited predictive power because of the large number of adjustable parameters, is yet the only scaled equation available for modeling of gas-liquid mixtures near critical lines. It is analogous with simple scaling in one-component fluids and has not yet been generalized to include liquid-vapor asymmetry and corrections to scaling. It is therefore used with so-called apparent critical exponents which, by their departure from the true asymptotic values, accommodate empirically the corrections to scaling to a certain extent. The original Leung-Griffiths form has been used to describe one and two-phase properties, such as PVT and C_V, of mixtures of He^3 and He^4 (26, 27, 22). With minor modifications, it has been applied to one-and two-phase states of azeotropic mixtures such as CO_2-ethane (28) and CO_2-ethylene (29). Moldover, Rainwater and coworkers have modified the Leung-Griffiths equation in order to describe the liquid-vapor asymmetry in the two-phase region (30, 31). They have been very successful in describing dew-bubble curves of hydrocarbon mixtures, even with large differences in volatility, and at pressures down to half the critical pressure. Their model is presently limited to two-phase states.

The only other nonclassical model that has been applied in fluid mixtures is the decorated lattice gas (32, 33). The application has been mostly to the description of phase boundaries in binary liquids, some with closed-loop coexistence curves (34). Bartis and Hall applied such a model to gas-gas equilibria (35).

One of us (L.S., 36) constructed a scaled potential for binary liquid mixtures by a transformation of the scaling variables of the pure fluid to those of a binary mixture at constant pressure. In analogy with the Leung-Griffiths method, the two nonuniversal scaling amplitudes and the analytic background were made pressure-dependent. Recently, Johnson has used this potential to correlate excess enthalpies and volumes of a binary liquid mixture (37).

As of this date, no attempt has been made to solve the crossover problem for fluid mixtures. In view of the growing importance of supercritical fluid mixtures this appears to be an urgent task.

Experimental Evidence, or Lack of It, For Nonclassical Behavior in Mixtures

For all gross features of phase behavior of mixtures, classical equations do a reasonable job. There are not many mixtures for which we have the wealth of accurate experimental detail that we have for some of the pure fluids, so the shortcomings of classical equations are not as acutely felt. As we already mentioned, the most striking shortcoming of classical equations is in the representation of the coexisting phases near the critical point (24, 25) where nonclassical representations make a major improvement (17, 18, 25 - 36). Also, the strong anomaly of the osmotic susceptibility $(\partial x/\partial \Delta)_{PT}$ has been determined from the intensity of scattered light in liquid-liquid (38) and in gas-liquid (39) mixtures and the nonclassical exponent value was found for γ.

The situation with respect to the weak anomalies is far less clear-cut. Only the weak divergence predicted for the specific heat C_{Px} has been convincingly demonstrated in liquid-liquid (40) mixtures. Those in the expansion coefficient α_{Px} and the compressibility K_{Tx} have not been seen yet. In binary liquid mixtures, the reason is obvious. Since, according to the Pippard relations

$$\alpha_{Px}/K_{Tx} = dP/dT|_{CRL} \; ; \; C_{Px}/VT\alpha_{Px} = dP/dT|_{CRL} \tag{4}$$

where CRL denotes the critical line, and since $dP/dT|_{CRL}$ is very large in binary mixtures, the anomaly in C_{Px} has the best chance of being visible, followed by those in α_{Px} and in K_{Tx}, in this order.

In gas-liquid mixtures, the Leung-Griffiths model has been used to find the distance from the critical line where the weak divergence of K_{Tx} can be seen. For $He^3 - He^4$, this distance was estimated to be about 10^{-10} K (26), which is beyond experimental accessibility.

The prediction of a weakly diverging compressibility leads to an apparent paradox (41) in the case depicted in Figure 3, because the critical isotherm has to cross the phase boundary at the critical point with a horizontal slope (Figure 5). From the relation

$$dV/dP|_{\sigma,x} = (\partial V/\partial P)_{Tx} + (\partial V/\partial T)_{Px} \, dT/dP|_{\sigma,x} \tag{5}$$

where σ is the phase boundary, we note that if $(\partial V/\partial P)_{Tx}$ is to diverge to $-\infty$. while $(dV/dP)_{\sigma x}$ and $dT/dP|_{\sigma x}$ are finite and negative, then $(\partial V/\partial T)_{Px}$ has to diverge to $-\infty$. This means that in the one-phase region near the critical point, the expansion coefficient must pass through zero and diverge to $-\infty$. The negative expansion coefficient makes it appear that the isotherm enters the two-phase region while it is actually still in one phase, and vice versa on the other side of the critical point. Some indications of negative expansion coefficients have been seen in liquid-liquid mixtures near consolute points (42, 43) but never near gas-liquid critical lines. By using the Leung-Griffiths model, Wheeler and two of us (RFC and GM) have found that in 0.28 mole fraction CO_2 in ethane, the expansion coefficient changes sign at 10^{-22} in reduced temperature from the critical line! (41).

While several of the weak anomalies thus escape detection, some of the non-divergent properties actually appear to diverge. Thus the specific heat C_{Vx} of $He^3 - He^4$ behaves like the weakly-divergent specific heat C_V of pure fluids, while theory predicts it to remain finite (12). The Leung-Griffiths model explains that the same mechanism, the small size of a weakly-diverging property, is the cause of the apparent finite value of K_{Tx} and the apparent divergence of C_{Vx} (26).

A related theoretical prediction is that the isochore-isopleth, a curve on which two densities, V and x, are constant, should become confluent with the critical line and therefore with the constant - x phase boundary (Figure 5). The most detailed measurement of $(\partial P/\partial T)_{Vx}$ is that of Doiron et al. in $He^3 - He^4$ (22) in which the critical point was approached to 1 mK but $(\partial P/\partial T)_{Vx}$ remained finite. The Leung-Griffiths model predicts 10^{-11} K for the region in which $(\partial P/\partial T)_{Vx}$ begins to approach $dP/dT|_{CRL}$. In the $CO_2 - C_2H_6$ system, the isochoric slope has to become negative (Figure 5) - according to the Leung-Griffiths model, this happens at 10^{-22} in reduced temperature, the same point where the expansion coefficient turns negative (41).

It is obvious that for all practical purposes most of the predictions of nonclassical theory for the weak anomalies can be safely ignored. An exception is the weak divergence in C_{Px}. The departures from classical behavior for the strong divergences, however, are quite visible and need to be taken into account if accuracy is desired.

Dilute Near-Critical Mixtures: Classical Analysis

Dilute near-critical mixtures show a peculiarity that at a first glance is offensive to the thermodynamicist; partial molar properties of the solvent do not necessarily approach pure-solvent properties in the limit x=0 at the solvents critical point. This behavior was first experimentally studied by Krichevskii and coworkers (44, 45), and modeled by a classical equation by Rozen (46). Wheeler (47) gave a nonclassical treatment based on the decorated lattice gas.

The reason for this abnormal behavior is that the solvent critical point is the intersection of the locus of infinite dilution, $(\partial\Delta/\partial x)_{PT} \to -\infty$, and of the critical line, $(\partial\Delta/\partial x)_{PT} \to 0$; as a consequence the limiting values assumed by certain partial properties depend on the path of approach. This anomalous behavior is readily elucidated for the partial molar volume

$$\bar{V}_1 = V - x\,(\partial V/\partial x)_{PT} \tag{6}$$

since

$$(\partial V/\partial x)_{PT} = VK_{Tx}\,(\partial P/\partial x)_{VT} \tag{7}$$

In the limit $x \to 0$, K_{Tx} approaches the compressibility of the pure fluid which is strongly divergent. The limit of $(\partial P/\partial x)_{VT}$ is given by (44, 48)

$$[\lim_{x\to 0} (\partial P/\partial x)_{VT}]^c = dP/dx|^c_{CRL} - dP/dT|^c_{CXC}\; dT/dx|^c_{CRL} \tag{8}$$

where $dP/dx|^c_{CRL}$ are $dT/dx|^c_{CRL}$ are limiting slopes of the critical line in P-x and T-x space, respectively; the limit in (8), in general, is finite. Thus $(\partial V/\partial x)_{PT}$ diverges at the solvent's critical point. If the limit x=0 is taken first, it diverges strongly, as the compressibility. As a consequence, the partial molar volume of the solute

$$\bar{V}_2 = V + (1 - x)(\partial V/\partial x)_{PT} \tag{9}$$

diverges strongly, with a sign equal to that of $(\partial P/\partial x)^c_{VT}$ in (8). These diverging partial molar volumes of the solute have been reported by Khazanova and Sominskaya for $CO_2 - C_2H_6$ (45), by Benson et al. for NaCl in steam (2) and by Eckert et al. for naphthalene in ethylene (49). Close to the solvent critical point, $\bar{V}_1$ and $\bar{V}_2$ can be predicted from the solvent properties and the initial slope of the critical line (48, 50, 51).

In the classical treatment of dilute mixtures, the divergence of $(\partial V/\partial x)_{PT}$ along a path of choice is obtained by expanding $(VK_{Tx})^{-1}$ around the solvent's critical point, at which point it equals 0. We obtain (48):

$$VK^{-1}_{Tx} = a^c_{VVx}\, x + a^c_{VVT}\,(\delta T) + a^c_{VVVV}\,(\delta V)^2/2 \tag{10}$$

where subscripts V, x, T indicate repeated partial differentiation of the Helmholtz free energy a(V, T, x) with respect to the variable indicated, and where the superscript c indicated that the derivatives are evaluated at the solvent's critical point. Note that $(VK_{Tx})^{-1}$ is quadratic in volume because the criticality conditions imply that $a^c_{VV} = 0$, $a^c_{VVV} = 0$.

A typical path is the critical line (Figure 5). Here δT and δV are linear in x, so $(VK_{Tx})^{-1}$ is linear in x and $(\partial V/\partial x)_{PT}$ diverges as x^{-1}. The product $x(\partial V/\partial x)_{PT}$ in (6) is a finite constant and $\bar{V}_1$ does not approach V_c. A second path is the coexistence curve at $T=T_c$ (Figure 6). Here $(\delta V)^2 \sim x$ and $(\partial V/\partial x)_{PT}$ still diverges as 1/x, but with a different amplitude. On any isothermal path on which $(\delta V)^2$ varies more slowly than x, however, the term in $(\delta V)^2$ will dominate the critical behavior. One such path is the critical isotherm-isobar (Figure 6) on which $(\delta V)^3 \sim x$, so that $\bar{V}_2$ diverges as $x^{-2/3}$. On this path, $x(\partial V/\partial x)_{PT}$ approaches 0 and $\bar{V}_1 \rightarrow V_c$. Since the derivative $(\partial H/\partial x)_{PT}$ is asymptotically proportional to $(\partial V/\partial x)_{PT}$:

$$\lim_{x\to 0} (\bar{H}_2/\bar{V}_2) = T\,(dP/dT|^c_{CXC}) \tag{11}$$

we expect the same anomalies in partial enthalpies as in partial molar volumes (48, 50, 51) as was also noted by Wheeler (47). Strongly increasing slopes in the excess enthalpy curves of binary gas-liquid mixtures were reported by Christensen on isotherm-isobars near the critical points of each of the components (4, 50).

The above analysis is readily extended to higher-ordered derivatives such as the partial molar specific heat $\bar{C}_{P2} = (\partial C_P/\partial x)_{PT}$. This quantity is predicted to diverge much more strongly than the partial molar volume, namely as $(K_T)^2$ in the limit $x\to 0$ and as $(1/x)^2$ along the critical line. This extremely strong divergence of $\bar{C}_{P2}$ was experimentally found by Wood et al. for NaCl in steam (3).

The classical results are summarized in Table 3.

Dilute Near-Critical Mixtures; Nonclassical Analysis

We have investigated the nonclassical critical behavior of dilute mixtures (48) by writing

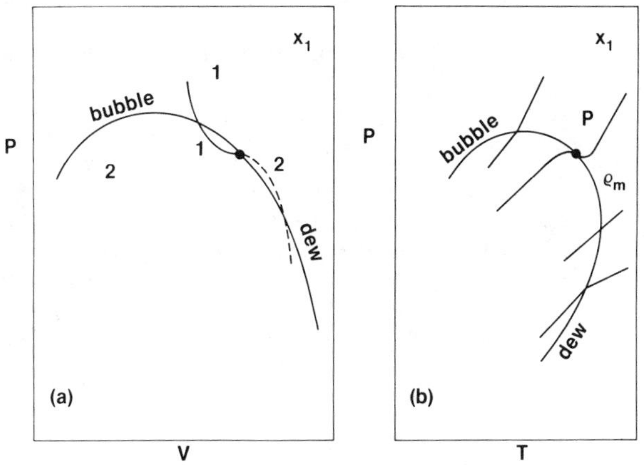

Figure 5. Nonclassical behavior near a mixture critical point. The compressibility is weakly divergent, so the critical isotherm is horizontal (a). The isochore-isopleth is confluent with the phase boundary and the critical line (b). If the slope of the phase boundary is is negative at the critical point, the expansion coefficient diverges to $-\infty$ and the slope of the isochore-isopleth is negative at the mixture's critical point.

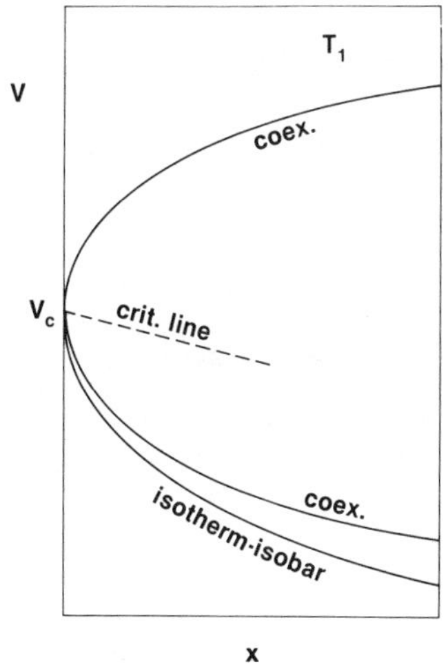

Figure 6. The paths in V-x space at the critical temperature of the solvent along which we have calculated the divergence of $\bar{V}_2$. The critical line is shown in projection.

$$(\partial V/\partial x)_{PT} = (\partial V/\partial \Delta)_{PT}/(\partial x/\partial \Delta)_{PT} \tag{12}$$

Both numerator and denominator, being second derivatives [of the potential $\mu_2(\Delta, P, T)$, (12)] in a direction intersecting the coexistence surface, diverge strongly at the critical line. As a consequence, $(\partial V/\partial x)_{PT}$ is finite at the critical line just as in the classical case, and not weakly divergent as stated in ref. (52). The osmotic susceptibility in the denominator, however, faces a dilemma when $x \to 0$ because it is zero at the pure-fluid axis and infinite at the critical line. By means of the Leung-Griffiths model we have been able to obtain the structure of this term. Schematically, it can be given, to the leading order in ζ and r, as

$$(\partial x/\partial \Delta)_{PT} \sim \zeta(\text{finite}) + \zeta^2(\text{strong}) \tag{13}$$

While the structure of the numerator is

$$(\partial V/\partial \Delta)_{PT} \sim \zeta(\text{strong}) \tag{14}$$

where (finite) denotes a thermodynamic derivative that remains finite, (strong) one that diverges strongly at the critical line. Thus the structure of $(\partial V/\partial x)_{PT}$ is

$$(\partial V/\partial x)_{PT} = \frac{(\text{strong})}{(\text{finite}) + \zeta(\text{strong})} \tag{15}$$

Schofield's r denotes the distance from the critical line at constant ζ, and the strong term can be written as proportional to $r^{-\gamma}$. A path to the solvent critical point can be characterized by stating how r goes to zero as ζ shrinks, or by the relation

$$r \sim \zeta^{\varepsilon} \tag{16}$$

If ε is small, ζ decreases much faster than r and the path is close to the pure-fluid axis. If ε is large, ζ decreases slower than r and the path is close to the critical line. Intermediate paths are obtained by varying the value of ε. Thus we may indicate the behavior of $(\partial V/\partial x)_{PT}$ schematically by

$$(\partial V/\partial x)_{PT} \sim \frac{r^{-\gamma}}{(\text{finite}) + \zeta r^{-\gamma}} \sim \frac{\zeta^{-\varepsilon\gamma}}{(\text{finite}) + \zeta^{-\varepsilon\gamma+1}} \tag{17}$$

Since for small ζ, $\zeta \sim x$, because the $RTx\ell nx$ term dominates the free energy, we obtain

$$(\partial V/\partial x)_{PT} \sim \frac{1}{(\text{finite})x^{\varepsilon\gamma} + x} \tag{18}$$

When ζ goes to 0 at finite r, we retrieve, from (15), the strong divergence of $(\partial V/\partial x)_{PT}$. If ζ and r both go to zero, there is a competition between the term in $x^{\varepsilon\gamma}$ and the term in x in (18). If ε is large,

$$\varepsilon > 1/\gamma,\ \varepsilon\gamma > 1,\ \varepsilon > 0.81 \tag{19}$$

then $(\partial V/\partial x)_{PT}$ behaves like 1/x and so does $\bar{V}_2$ (9). In this case $\bar{V}_1$ does not approach V_c. If ε is small

$$\varepsilon < 1/\gamma,\ \varepsilon\gamma < 1,\ \varepsilon < 0.81 \tag{20}$$

then $(\partial V/\partial x)_{PT}$ behaves like $1/x^{\varepsilon\gamma}$ and so does $\bar{V}_2$. In this case $\bar{V}_1$ does approach V_c. On the critical line (ε large) $\bar{V}_2$ diverges as 1/x. On the isothermal or isobaric coexistence curve h equals 0 and τ is linear in ζ to lowest order. Since τ is linear in r, we have ε=1 and $(\partial V/\partial x)_{PT}$ goes at 1/x.

On the critical isoterms-isobar, the pressure and temperature are to be kept constant and equal to those at the solvent's critical point. In this case both τ and h vary linearly with ζ for small ζ leading to $r^{\beta\delta} \sim \zeta$ or $\varepsilon = 1/\beta\delta$. It follows that

$$(\partial V/\partial x)_{PT} \sim 1/x^{\gamma/\beta\delta} = 1/x^{1-1/\delta} \tag{21}$$

and so does $\bar{V}_2$. The value of $1-1/\delta = 0.79$. Therefore x $(\partial V/\partial x)_{PT}$ $\rightarrow$ 0 and $\bar{V}_1 \rightarrow V_c$. The crossover from 1/x to slower-than 1/x behavior occurs at $\varepsilon = 1/\gamma = 0.81$ which is between the isothermal or isobaric coexistence curve (ε=1) and the critical isotherm-isobar (ε=0.64). We have also analyzed the behavior of the partial molar specific heat $\bar{C}_{P2}$. The results of our analysis are summarized in Table III.

The classical and nonclassical treatments of dilute mixtures show the same global features but differ in detail. The differences are due to the different values assumed by the strong exponents β, γ, δ and should therefore be experimentally detectable.

Impure Near-Critical Fluids

In the custody transfer of fluids, the density of the fluid is often obtained from measurements of pure P and temperature T and from the known equation of state. It is important to know what effect the presence of an impurity can have on the density at given P and T. If V indicates the molar volume, we may write

$$V(P, T, x) = V(P, T, 0) + (\partial V/\partial x)_{PT}x + \ldots \quad (22)$$

Since $(\partial V/\partial x)_{PT}$ is proportional to the solvent's compressibility, impurity effects become larger the larger the compressibility. The expansion (23) is not valid at the solvent's critical point. At T_c, P_c, we have $x(\partial V/\partial x)_{PT} \sim x^{1/\delta}$ (21). So even an impurity on the level of 10^{-5} in mole fraction can still modify the molar volume on the 10% level, since $1/\delta$ is about 0.2 for fluids.

Impurity effects can be drastic near T_c, P_c, because they may induce a phase transition. In Figure 7, we sketch how a nonvolatile impurity affects the molar volume of a near-critical fluid. If is obvious that impurity effects are highly nonlinear in concentration. Impurity effects in near-critical fluids can be calculated in a consistent but inaccurate way by means of classical corresponding-states treatments and their generalizations. They can be calculated consistently and accurately by means of the Leung-Griffiths model. In view of the complication of the latter approach it is tempting to model the pure fluid as accurately as possible with a multiparameter classical or a nonclassical thermodynamic surface, and then to calculate the impurity effects by means of corresponding states (53). Unfortunately, this idea is incorrect as can be appreciated from Figure 3. The nonclassical treatment of the major component modifies the behavior around the pure-fluid critical point. This part of the graph, however, is not reached by the mixture because it separates at the curve of material instability. The mixture therefore "sees" little or nothing at all of the nonclassical behavior characterizing the major component. The effect of combining a nonclassical description of the major component with classical corresponding states are more disturbing in dilute mixtures. To see this, we need the classical expressions for the initial slopes of the critical line, which are obtained from the classical expansion of the Helmholtz free energy and from the conditions of criticality. They are

$$\left.\frac{dT}{dx}\right|^{c}_{CRL} = \frac{(a^{c}_{Vx})^{2} - RT\, a^{c}_{VVX}}{RT\, a^{c}_{VVT}} \;;\; -\left.\frac{dP}{dx}\right|^{c}_{CRL} = a^{c}_{VT}\left.\frac{dT}{dx}\right|^{c}_{CRL} + a^{c}_{Vx} \quad (23)$$

the latter being the equivalent of (8). From (23) we derive for the initial slope of the critical line in P-T space

$$\left.\frac{dP}{dT}\right|_{CRL} = \left.\frac{dP}{dT}\right|^{c}_{CXC} - a^{c}_{VVT}\,\frac{RT}{a^{c}_{Vx} - RT\, a^{c}_{VVx}/a^{c}_{Vx}} \quad (24)$$

Table III
Critical Behavior in Dilute Mixtures

path	$\bar{V}_2$, $\bar{H}_2$	$\bar{K}_{T2}$, $\bar{C}_{P2}$	$\bar{V}_1$	$\lim_{x\to0} x\bar{V}_2$	$\lim_{x\to0} (\bar{V}_2/K_{Tx})$
			Classical		
isotherm-isobar	$1/x^{2/3}$	$1/x^{5/3}$	V_c	0	AV_c
isotherm-isochore	$1/x$	$1/x^2$	$\neq V_c$	$RT_c/(A-c_1/A)$	AV_c
isobar-isochore	$1/x$	$1/x^2$	$\neq V_c$	$RT_c/(A-c_2/A)$	AV_c
critical line	$1/x$	$1/x^2$	$\neq V_c$	RT_c/A	AV_c
			Nonclassical		
isotherm-isobar	$1/x^{1-1/\delta}$	$1/x^{2-1/\delta}$	V_c	0	AV_c
isotherm-isochore	$1/x$	$1/x^{3-\gamma}$	$\neq V_c$	RT_c/A	AV_c
isobar-isochore	$1/x$	$1/x^{3-\gamma}$	$\neq V_c$	RT_c/A	AV_c
critical line	$1/x$	---	$\neq V_c$	RT_c/A	0

$A = dP/dx|^c_{CRL} - dP/dT|^c_{CXC}\ dT/dx|^c_{CRL}$

$c_1 = RT_c a^c_{VVT}\ dT/dx|^c_{CRL}$

$c_2 = RT_c a^c_{VVT}\ dT/dP|^c_{CXC}\ dP/dx|^c_{CRL}$

$x \equiv x_2$

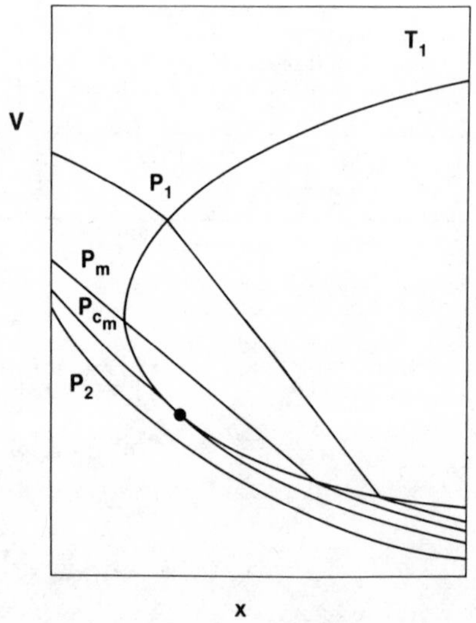

Figure 7. Impurity effects on extensive properties in dilute mixtures near the solvent's critical point are large and nonlinear in concentration. The plot shows the relation between molar volume and impurity concentration at fixed $T_1 > T_c$ and at four pressures: that of the mixture's critical point P_{cm}, $P_2 > P_{cm}$ and two pressures P_m, P_1 at which the impurity actually induces a phase transition.

where we have used $a^c_{VT} = -dP/dT|^c_{CXC}$, the limiting slope of the vapor pressure curve. If the major component has been treated nonclassically, or with a very accurate classical equation, the coefficient a^c_{VVT} will be zero or very small because $a^c_{VVT} = \partial(VK_T^{-1})/\partial T \sim \Delta T^{\gamma-1}$ and $\gamma > 1$. Thus, the initial slope of the critical line is forced to be the extension of the vapor pressure curve, which contradicts experiment. The combining of a nonclassical pure-fluid equation with corresponding states for the mixture induces a serious distortion in the phase diagram.

Conclusions

For one-component fluids, accurate nonclassical formulations of the critical region are available, incorporating the vapor-liquid asymmetry as well as corrections to scaling. Empirical techniques for switching from classical to critical behavior have been worked out. The fundamental solution of the crossover problem has advanced to the point that application to one-component fluids is near. For binary mixtures a nonclassical formulation is available for the one-and-two-phase regions mixtures with continuous critical lines. A variety of nonclassical models have been used to represent coexisting phases. No attempts have been made to devise crossover functions.

Dilute near-critical mixtures present some special anomalies not often encountered in thermodynamics. We describe these with both classical and nonclassical equations. We argue that a nonclassical thermodynamic description of the solvent can not be combined with a corresponding-states treatment of the dilute mixture. For incorporation of the nonclassical effects into practical thermodynamic surfaces of fluid mixtures, the solution of the crossover problem appears to be the most urgent task at hand.

Acknowledgment

We have profited from discussions with P.C. Albright and J.C. Wheeler.

References

1. Busey, R.H.; Holmes, H.F.; Mesmer, RE. J. Chem. Thermodynamics 1984, 16, 343.
2. Benson, S.W.; Copeland, C.S.; Pearson, D. J. Chem. Phys. 1953, 21, 2208. Copeland, C.S.; Silverman, J.; Benson, S.W. J. Chem. Phys. 1953, 21, 12.
3. Smith-Magowan, D.; Wood, R.H. J. Chem. Thermodynamics 1981, 13, 1047. Wood, R.H.; Quint, J.R. J. Chem. Thermodynamics 1982, 14, 1069. Gates, J.A.; Wood, R.H. J. Phys. Chem. 1982, 86, 4948.
4. Christensen, J.J.; Walker, T.A.C.; Schofield, R.S.; Faux, P.W.; Harding, P.R.; Izatt, R.M. J. Chem. Thermodynamics, 1984, 16, 445 and references therein.
   Wormald, C.J. Ber. Bunsenges. Phys. Chem. 1984, 88, 826 and references therein.

5. See for instance, "Chemical Engineering in Supercritical Fluid Conditions" (Paulaitis, M.E. et al., Eds). Ann Arbor Science Publishers, 1983.
6. Sengers, J.V.; Levelt Sengers, J.M.H. in "Progress in Liquid Physics" (Croxton, C.A. Ed.); Wiley: Chichester, U.K., 1978; Ch. 4, p. 103.
7. Levelt Sengers J.M.H.; Sengers, J.V. in "Perspectives in Statistical Physics" (Raveche, H.J. Ed.); North Holland Publ. Co: Amsterdam, Netherlands, 1981; Ch. 14, p. 239.
8. Sengers, J.V.; Levelt Sengers, J.M.H. Int. J. Thermophysics 1984, 5, 195.
9. Hocken, R.J.; Moldover, M.R. Phys. Rev. Letters 1976, 37, 29.
10. Greer, S.C. Phys. Rev. 1976, A14, 1770.
11. Widom, B. J. Chem. Phys. 1965, 43, 3898.
12. Griffiths, R.B.; Wheeler, J.C. Phys. Rev. 1970, A2, 1047.
13. Mermin, N.D.; Rehr, J.J. Phys. Rev. Letters 1971, 26, 1155. Phys. Rev. 1971, A4, 2408. Rehr, J.J.; Mermin, N.D. Phys. Rev. 1973, A8, 472.
14. Chapela, G.; Rowlinson, J.S. Faraday Trans. 1974, 70, 584.
15. Woolley, H.W. Int. J. Thermophysics 1983, 4, 51.
16. Hill, P.G. Proc. 10th International Conf. on Properties on Steam" Moscow, USSR, 1984; and Int. J. Thermophysics; to be published.
17. Fox, J.R. Fluid Phase Equilibria, 1983, 14, 45.
18. White, J.A.; Pustchi, M.E. Bull. Am. Phys. Soc., 1984, 30, 372; 1984, 29, 697; and references therein.
19. Nicoll, J.E. Phys. Rev. 1981, A24, 2203.
20. Nicoll, J.E.; Albright, P.C. Phys. Rev. 1985, B31, 4576.
21. Albright, P.C. Ph.D. Thesis, U. of Maryland, 1985.
22. Doiron, T.; Behringer, R.P.; Meyer, H. J. Low Temp. Phys. 1976, 24, 345.
23. Van Konynenburg, P.H.; Scott, R.L. Phil. Trans. Roy. Soc., 1980, 298, 495.
24. Kolasinska, G.; Moorwood, R.A.S.; Wenzel, H. Fluid Phase Equilibria 1983, 13, 121.
25. Bijl, H.; de Loos, Th.W.; Lichtenthaler, R.N. Fluid Phase Equilibria 1983, 14, 157.
26. Leung, S.S.; Griffiths, R.B. Phys. Rev. 1973, A8, 2670.
27. Wallace Jr, B; Meyer, H. Phys. Rev., 1972, A5, 953.
28. Chang, R.F.; Doiron, T. Int. J. Thermophysics 1983, 4, 337. Chang, R.F.; Levelt Sengers, J.M.H.; Doiron, T.; Jones, J. J. Chem. Phys. 1983, 79, 3058.
29. D'Arrigio, G.; Mistura, L.; Taraglia, P. Phys. Rev. 1975, A12, 2578.
30. Moldover, M.R.; Gallagher, J.S. in "Phase Equilibria and Fluid Properties in the Chemical Industry" (Storvick, T.S. and Sandler, S.I., Eds.). ACS Symposium Series 60, 1977; Ch. 30, p. 498.
31. Rainwater, J.C.; Moldover, M.R. in "Chemical Engineering in Supercritical Fluid Conditions" (Paulaitis, M.E., et al., Eds.), Ann Arbor Science Publishers, 1983; p. 199.
32. Wheeler, J.C. Ann. Rev. Phys. Chem., 1977, 28, 411 and references therein.
33. Wheeler, J.C. J. Chem. Phys. 1975, 62, 4332.

34. Andersen, G.R.; Wheeler, J.C. J. Chem. Phys., 1978, 69, 2082; J. Chem Phys., 1979, 70, 1326.
35. Bartis, J.T.; Hall, C.K. Physica, 1975, 78, 1.
36. Levelt Sengers, J.M.H. Pure and Applied Chemistry 1983, 55, 437.
37. Ewing, M.B.; Johnson, K.A.; McGlashan, J.L. J. Chem. Thermodynamics 1985, 17, 513; and to be published in the same Journal.
38. Chang, R.F.; Burstyn, H.; Sengers, J.V. Phys. Rev. 1979, A19, 866.
39. Giglio, M; Vendramini, A. Optics Comm. 1973, 9, 80.
40. Bloemen, E.; Thoen, J.; Van Dael, W. J. Chem. Phys. 1981, 75, 1488.
41. Wheeler, J.C.; Morrison, G.; Chang, R.F., accepted, J. Chem. Phys.
42. Morrison, G.; Knobler, C.M. J. Chem. Phys. 1976, 65, 5507.
43. Clerke, E.A.; Sengers, J.V.; Ferrell, R.A.; Battacharjee, J.K. Phys. Rev., 1983, 27, 2140.
44. Krichevskii, I.R. Russ. J. Phys. Chem. 1967, 41, 1332.
45. Khazanova, N.E.; Sominskaya, E.E. Russ. J. Phys. Chem. 1971, 45, 1485.
46. Rozen, A.M. Russ. J. Phys. Chem. 1976, 50, 837.
47. Wheeler, J.C. Ber. Bunsenges, Phys. Chem. 1972, 76, 308.
48. Chang, R.F.; Morrison, G.; Levelt Sengers, J.M.H. J. Phys. Chem. Letter 1984, 88, 3389.
49. Eckert, C.A.; Ziger, D.H.; Johnston, K.P.; Ellison, T.K. Fluid Phase Equilibria 1983, 14, 167.
50. Morrison, G.; Levelt Sengers, J.M.H.; Chang, R.F.; Christensen, J.J. in "Proceedings of the Symposium on Supercritical Fluids", (J. Penninger, Ed.). In press, Elsevier, Netherlands.
51. Levelt Sengers, J.M.H.; Morrison, G.; Chang, R.F. Fluid Phase Equilibria 1983, 14, 19.
52. Gitterman, M; Procaccia, I. J. Chem. Phys. 1983, 78, 2648.
53. Hastings, J.R.; Levelt Sengers, J.M.H.; Balfour, F.W. J. Chem. Thermodynamics 1980, 12, 1009.

RECEIVED November 8, 1985

6

Prediction of Binary Critical Loci by Cubic Equations of State

Robert M. Palenchar[1], Dale D. Erickson[2], and Thomas W. Leland[2]

[1]**Olefins Technology Division, Exxon Chemical Company, Baton Rouge, LA 70821**
[2]**Department of Chemical Engineering, Rice University, Houston, TX 77251**

This paper compares reduced forms of five different cubic equations of state of the van der Waals family as to their ability to predict critical loci of binary mixtures. For the calculation of binary critical loci, the constants for each equation were expressed as functions of composition by the same mixing rules which involve two unlike pair coefficients evaluated by fitting to the experimental critical temperature and critical pressure at the same single composition.

Only the critical temperature and critical pressure loci of van Konynenburg and Scott's Type I systems are predicted quantitatively over their entire composition range by the equations examined in this way. Critical volumes in all systems and critical temperature and pressure loci in Type II systems are only qualitatively predicted. Unlike pair coefficients evaluated at one point on a critical line in a two-phase region do not produce satisfactory predictions in portions of a critical locus where a third phase appears. There are no major differences in accuracy among any of the equations in predicting Type I critical loci. However, the Teja equation gave best predictions of Type I critical temperatures and pressures. The Peng-Robinson equation gave best predictions of Type I critical volumes. In Class 1, Type II and in all Class 2 systems, the two-constant Soave and Redlich-Kwong equations generally predict critical temperatures and critical pressures better than the more elaborate three-constant equations and two constant equations whose form has a larger deviation from that of the original van der Waals equation.

The accurate prediction of the properties of equilibrium phases and vapor-liquid equilibria in mixtures near their critical is an

0097–6156/86/0300–0132$07.00/0

important unsolved problem. The basic difficulty is that there is at present no accurate non-analytic equation of state for mixtures. The use of typical engineering type analytical equations of state with essentially empirical mixing rules fails in the near critical region. The nature of this failure is well understood. The divergence of the partial molar volume of a solute as it becomes infinitely dilute when the system approaches the solvent critical conditions has recently been explained clearly by Chang, Morrison, and Levelt Sengers (1). At finite concentrations the vapor and liquid phase properties predicted by reduced analytical equations using pseudo-criticals approach each other too rapidly as the system approaches its critical conditions. When determined from reduced equations in terms of pseudo-criticals these properties tend to predict critical conditions which lie below the true critical locus. The traditional method of dealing with this problem has been to stop phase equilibrium calculations well below the critical locus and to make an empirical extrapolation to an estimate of the true critical properties of the system.

It may be possible to overcome these problems with corresponding states methods which use a properly scaled reference fluid equation. Pseudo-critical relations at present cannot accurately predict the second and third derivatives of the Gibbs free energy with respect to composition which are needed to define the critical of a mixture. Furthermore, additional research is needed to determine the best way to map the critical region of a mixture on to the critical region of a pure scaled reference without destroying the effective representation of multicomponent equilibrium phases which is currently possible below the near critical region. Although research with these and other methods for improving predictions in the critical region is continuing, a common need with any method under investigation is a procedure for estimating the true critical locus of the mixture.

For this reason, this paper examines a feature of the classical van der Waals family of equations which enables them to give a very good representation of the critical loci of simple mixtures by using properly assigned unlike pair interaction coefficients. This procedure may cause properties near the mixture critical to be predicted poorly, but the critical locus itself is described remarkably well.

During the last twenty years a great deal of work has been done to study the prediction of binary critical loci in this manner (2-7). Using the original van der Waals equation, van Konynenburg and Scott (8) qualitatively predicted nine different types of binary critical loci. These different types are identified by the behavior of critical properties in a projection of the P, T, x surface of the system on to a pressure - temperature plane. These types are summarized in Table I.

Thermodynamics of the Critical Point

The thermodynamic conditions for the existence of a critical point were first derived by Gibbs (9) more than a hundred years ago. The defining equations for binary system critical points are as follows:

$$\left(\frac{\partial^2 G}{\partial x_1^2}\right)_{P,T} = 0 \quad (1)$$

$$\left(\frac{\partial^3 G}{\partial x_1^3}\right)_{P,T} = 0 \quad (2)$$

When working with pressure explicit equations of state it is more convenient to express the above equations in terms of their relationship to the Helmholtz free energy, which depend on the measured variables P, V, T, and x. This relationship has been developed by Redlich and Kister (10). The Redlich and Kister form for a binary system is:

$$\left(\frac{\partial^2 G}{\partial x_1^2}\right)_{P,T} = \frac{RT}{x_1 x_2} + \int_V^\infty \left(\frac{\partial^2 P}{\partial x_1^2}\right)_{V,T} dV - \left(\frac{\partial P}{\partial x_1}\right)_{V,T} \left(\frac{\partial V}{\partial x_1}\right)_{P,T} = 0 \quad (3)$$

$$\left(\frac{\partial^3 G}{\partial x_1^3}\right)_{P,T} = \frac{RT(x_1-x_2)}{x_1^2 x_2^2} + \int_V^\infty \left(\frac{\partial^3 P}{\partial x_1^3}\right)_{V,T} dV - \left(\frac{\partial P}{\partial x_1}\right)_{V,T} \left(\frac{\partial^2 V}{\partial x_1^2}\right)_{P,T} - \left(\frac{\partial V}{\partial x_1}\right)_{P,T}^2 \left(\frac{\partial^2 P}{\partial V \partial x_1}\right)_T - 2\left(\frac{\partial V}{\partial x_1}\right)_{P,T} \left(\frac{\partial^2 P}{\partial x_1^2}\right)_{V,T} = 0 \quad (4)$$

Equations of State Examined

The critical loci criteria in Equations (3) and (4) can be evaluated with an equation of state relating the variables P,V,T and x. For this work five reduced equations of state of the van der Waals family were used. These equations contain two, and sometimes three, constants which are universal functions of the reduced temperature, critical constants, and the acentric factor, ω.

Redlich - Kwong (11):

$$P = \frac{RT}{V-b} - \frac{a}{T^{0.5}V(V+b)} \quad (5)$$

$$a = 0.4278\ R^2\ T_c^{2.5}/P_c$$

$$b = 0.0867\ R\ T_c/P_c$$

Soave (12):

$$P = \frac{RT}{V-b} - \frac{a(T)}{V(V+b)} \quad (6)$$

$$a(T) = 0.42747\ \alpha(T)\ R^2\ T_c^2/P_c$$

$$b = 0.08664\ R\ T_c/P_c$$

$$\alpha(t) = \{1+m(1-Tr^{0.5})\}^2$$

$$m = 0.480 + 1.574\omega - 0.176\omega^2$$

Adachi (13):

$$P = \frac{RT}{V-b} - \frac{a(T)}{V(V+c)} \tag{7}$$

$$a(T) = A_o \; Z_c\{1+\alpha(1-Tr^{0.5})\}^2 R^2 \; T_c^{\,2}/P_c$$

$$b = B \; Z_c \; R \; T_c/P_c$$

$$c = C \; Z_c \; R \; T_c/P_c$$

$$\alpha = 0.479817 + 1.55553\omega - 0.287787\omega^2$$

$$A_o = B \; (1+C)^3/(1-B)^3$$

$$B = 0.260796 - 0.0682692\omega - 0.0367338\omega^2$$

$$C = [\{(4/B)-3\}^{0.5}-3]/2$$

$$Z_c = [1/(1-B)]-A_o/(1+C)$$

Peng and Robinson (14):

$$P = \frac{RT}{V-b} - \frac{a(T)}{V(V+b)+b(V-b)} \tag{8}$$

$$a(T) = 0.45724 \; \alpha(T) \; R^2 \; T_c^{\,2}/P_c$$

$$b = 0.07780 \; R \; T_c/P_c$$

$$\alpha(T) = \{1+\kappa(1-Tr^{0.5})\}^2$$

$$\kappa = 0.37464 + 1.54226\omega - 0.26992\omega^2$$

Teja (15):

$$P = \frac{RT}{V-b} - \frac{a(T)}{V(V+b)+c(V-b)} \tag{9}$$

$$a(T) = \Omega_a \; \alpha(T) \; R^2 \; T_c^{\,2}/Pc$$

$$b = \Omega_b \; R \; T_c/P_c$$

$$c = \Omega_c \; R \; T_c/P_c$$

$$\Omega_a = 3Z_c^{\,2} + 3(1-2Z_c)\Omega_b + \Omega_b^{\,2} + 1-3Z_c$$

Ω_b equals the smallest positive root of:

$$\Omega_b^{\,3} + (2-3Z_c)\Omega_b^{\,2} + 3Z_c^{\,2}\Omega_b - Z_c^{\,3} = 0$$

$$\Omega_c = 1 - 3Z_c$$

$$Z_c = 0.329032 - 0.076799\omega + 0.0211947\omega^2$$

$$\alpha(T) = \{1+F(1-Tr^{0.5})\}^2$$

$$F = 0.452413 + 1.30982\omega - 0.295937\omega^2$$

loci predicted by various equations using two unlike pair coefficients obtained from a single data point are illustrated in Figures 1-13. The errors reported in Table II for all the critical properties are obtained by the procedure shown below for the critical temperature:

$$\text{Average \% Relative Absolute Error} = \frac{1}{N}\sum_j\left(\frac{|(T_c)_{calc} - (T_c)_{expt}|}{(T_c)_{expt}}\right)_j(100) \qquad (15)$$

The subscript j in Equation (15) indicates a data point and N represents the total number of data points.

Unlike Pair Interaction Coefficients

Table III shows the values of the unlike pair coefficients ζ_{12} and λ_{12} in Equations (10) and (11) obtained by fitting to the experimentally measured critical temperature and critical pressure in an equi-molar, or nearly equi-molar, mixture. Because of the empirical nature of the equations of state in which these coefficients are used, their actual relationship to any effective pair potential is rather obscure. The name "unlike pair interaction coefficients" is used in an ideal sense only. Furthermore, it is important to note that these values apply only to the projected critical loci and bear little resemblance to their optimal values obtained by fitting other portions of the P-T-x surface.

Even when fitting critical loci alone, a significantly better correlation could obviously be obtained with any of the equations tested by optimizing either ζ_{12} when λ_{12} is set at 1.0, or by optimizing both ζ_{12} and λ_{12} together, over the entire composition range. This, however, obscures the relative degree of composition dependence of these unlike pair coefficients in the various equations of state and is less revealing of the shape of the critical locus predicted by the analytical form of any particular equation of state. However, among the equations with the same mixing rules with unlike pair coefficients evaluated in the same way from the same single data point within the same group of data points, the equation which produces the lowest average absolute error for the entire group of these data points should produce the best correlation when its unlike pair coefficients are optimized for all data points in the group.

In a few cases it was not possible to obtain an accurate prediction of both the critical pressure and the critical temperature with any combination of ζ_{12} and λ_{12} values. These cases are indicated with a * in Table III and the values presented are those giving the best prediction of T_c and P_c.

The policy of fitting the unlike pair coefficients at a single data point gives information regarding the relative degree of composition independence in the values of these coefficients among the various equations of state tested. Furthermore, this policy also gives information about the shape of the critical loci predicted by the various equations. This is illustrated by Figures 10 and 11 where the critical temperatures and critical pressures of

TABLE I

van Konynenburg and Scott Classification of Binary Critical Loci

Class 1

Type	Description
I.	One continuous gas-liquid critical line connecting the two pure component critical points C1 and C2. C1 represents the critical of the component with the lowest critical temperature and C2 is the critical of the other component.
I-A.	The same as I, with the addition of a negative azeotrope.
II.	Two critical lines: One is a vapor-liquid critical line between the pure component criticals C1 and C2. The second is a liquid-liquid critical line representing the merger of two liquids with limited miscibility in the presence of a solid phase. Because of the limited miscibility, this critical line persists to immeasurably high pressures so that its origin may be regarded as occurring at an infinitely large pressure. From this infinite pressure this liquid-liquid critical line connects at lower pressure with an upper critical end point (UCEP), representing the high pressure termination point of a three-phase liquid-liquid-solid line. At this termination point the two liquid phases become identical. The origin of this liquid-liquid line at an infinite critical pressure is designated with the symbol C_m by van Konynenburg and Scott (8).
II-A.	The same as II, but with the addition of a positive azeotrope.

Class 2

Type	Description
III-HA.	Two critical lines, but no continuous critical line runs between C1 and C2. The first of the critical lines is a vapor-liquid critical line from C1 to the high temperature termination point (UCEP) of a liquid-liquid-gas three phase line where the liquid richest in the component with the lowest critical temperature and the gas phase become identical. The other critical line connects one of two possible low temperature origins to the critical C2 of the component with the higher critical temperature. The portion of this second critical line near its low temperature origin is a liquid-liquid critical line which changes to a vapor-liquid critical line before approaching C2. The two points of origin for this second critical line identify two sub-types of Type III-HA defined as IIIa-HA and IIIb-HA as follows:

Continued on next page

Table I Continued

IIIa-HA. In Type IIIa-HA the two liquids which merge along the low temperature portion of the second critical line have limited miscibility so that their merger requires increasingly higher critical pressures as the critical temperature is lowered. The low temperature origin of this liquid-liquid critical line in this case may be regarded as occurring at an infinite critical pressure and is given the symbol C_m by van Konynenburg and Scott (8).

IIIb-HA. The second low temperature point of origin for the critical line to C2 occurs at the high pressure termination point, or UCEP, of a three phase liquid-liquid-solid line at a point where the two liquid phases become identical in the presence of the solid. Type IIIb-HA was not considered by van Konynenburg and Scott because its origin at the UCEP of a three phase liquid-liquid-solid line cannot be represented by the original van der Waals equation.

In Type IIIa-HA or Type IIIb-HA the entire three phase liquid-liquid-gas line lies at pressures above the vapor pressure curves of both of the two pure components. This condition represents what is called "hetero-azeotropic" behavior and is designated by the symbol "HA" in the classification of this type of critical behavior.

III. The same as III-HA except that the three phase liquid-liquid-gas line ending with the UCEP lies between the vapor pressure curves of the two pure components. Like Type III-HA, Type III may be subdivided into two sub-types IIIa and IIIb, depending upon the origin of the critical line to C2. Type IIIa originates at C_m and Type IIIb originates at the UCEP terminating a three phase liquid-liquid-solid line. Because the three phase liquid-liquid-gas line does not lie outside the two pure component vapor pressure curves, there is no hetero-azeotropic behavior in Type IIIa or Type IIIb systems.

IV. Three critical lines: A vapor-liquid critical line runs from C1 to the high temperature UCEP termination of a liquid-liquid-gas three phase line where a liquid and the gas phase become identical. In this case, however, this liquid-liquid-gas line is discontinuous. As the temperature is lowered, it ends at a LCEP where the two liquid phases merge. At temperatures below this LCEP there is a short range where the liquids are completely miscible and only a liquid and gas phase are present. At still lower temperatures, there is a second UCEP where two liquid phases are again critical. Below this second, or low temperature, UCEP the liquid-liquid-gas three phase line continues with decreasing temperature until it ends with the formation of a solid phase at a quadruple point.

A second critical line runs from the LCEP to C2. Near the LCEP it is a liquid-liquid critical line and in aproaching C2 it changes to a vapor-liquid critical line.

Table I Continued

The third critical line is a liquid-liquid critical locus between the second or low temperature UCEP and a C_m point at an immeasurably high pressure.

V. Two critical lines: In this system there is a single three phase liquid-liquid-gas line ending at a UCEP where one of the liquids and the gas phase become identical. As the temperature is lowered, this three phase line stops at a LCEP where the two liquid phases merge.

One of the critical lines is a gas-liquid critical locus from C1 to the UCEP. The second critical line runs from the LCEP to C2 and consists of a liquid-liquid critical line near the LCEP which changes to a vapor-liquid critical line when approaching C2.

V-A. The same as Type V but with the addition of a negative azeotrope.

Class 3

The van Konynenburg and Scott classification includes a Class 3 behavior which is exhibited by very complex mixtures with strong specific interactions, usually involving hydrogen bonds, between the components. These systems have LCSTs where there is a minimum in a T-x coexistence curve. Systems in this class cannot be represented by equations of state of the van der Waals family.

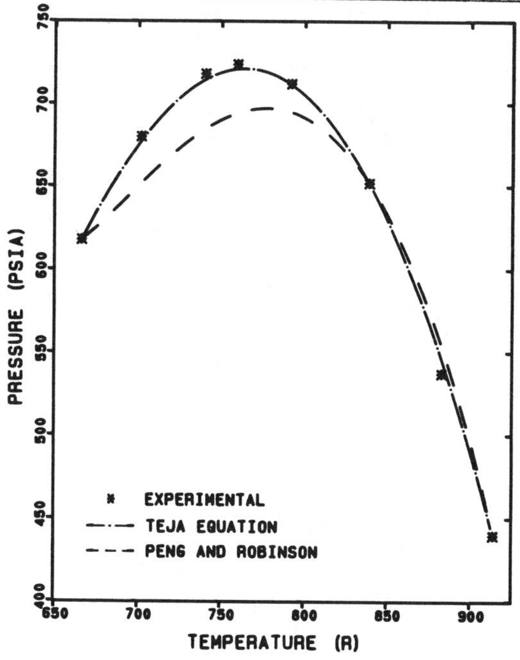

Figure 1. Critical loci of propane-hexane system.

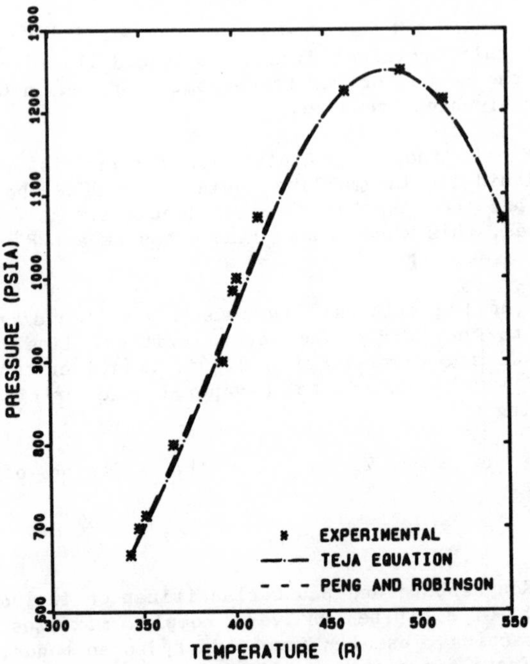

Figure 2. Critical loci of methane-carbon dioxide system.

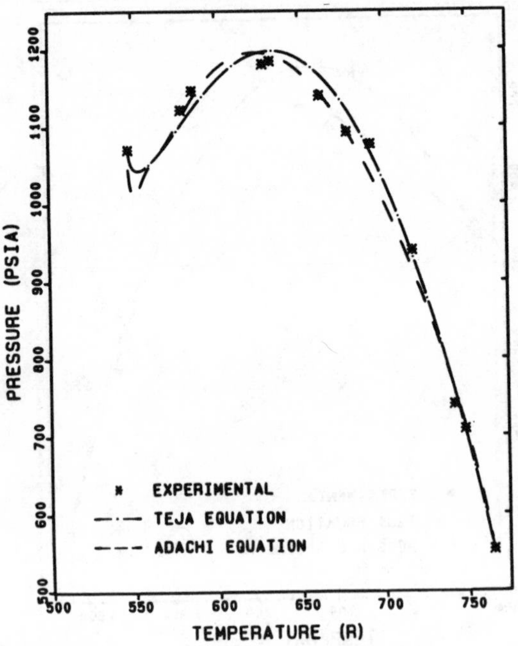

Figure 3. Critical loci of butane-carbon dioxide system.

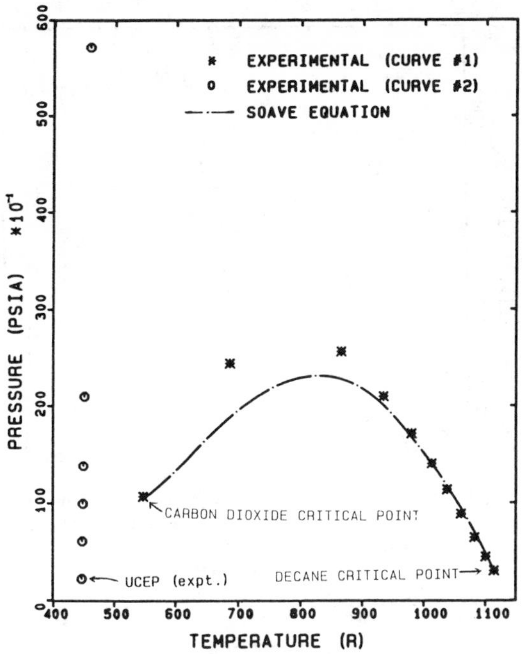

Figure 4. Critical loci of decane-carbon dioxide system.

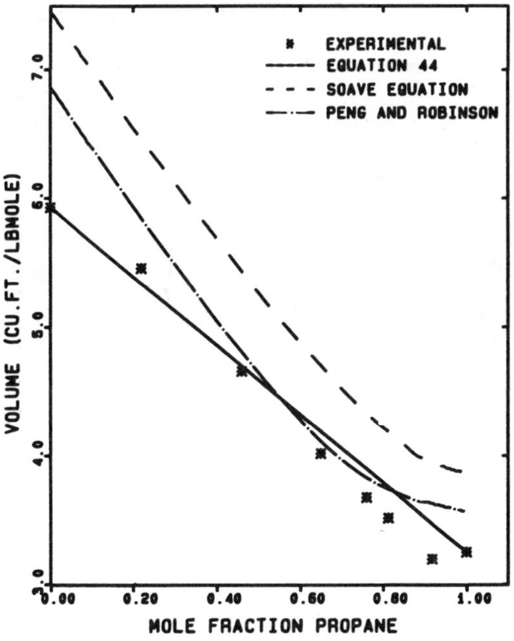

Figure 5. Critical loci of propane-hexane system.

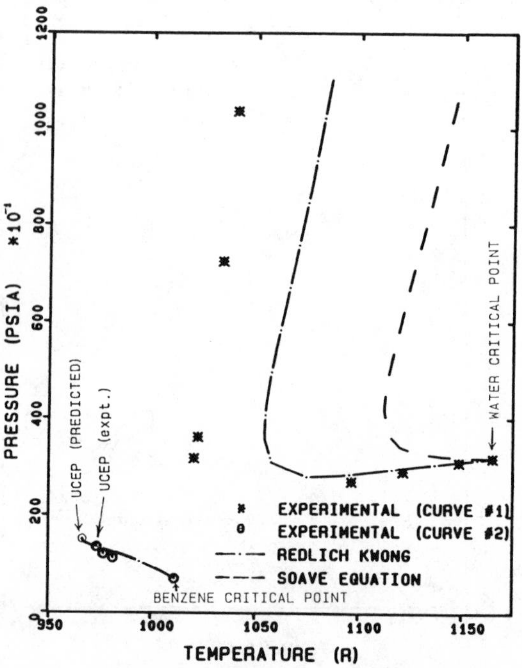

Figure 6. Critical loci of benzene-water system.

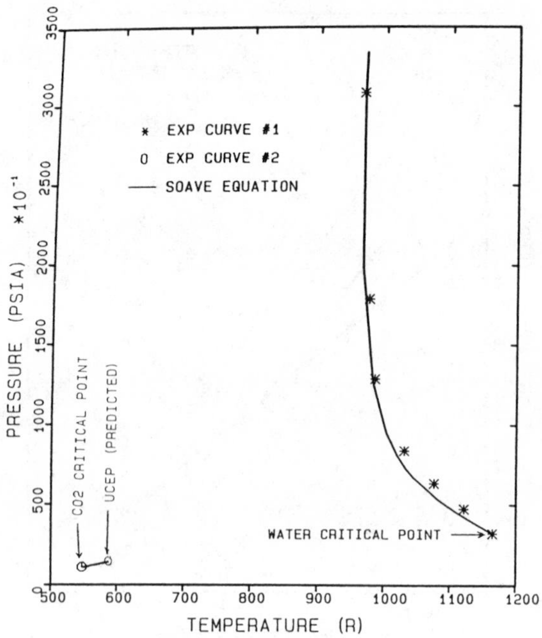

Figure 7. Critical loci of water-carbon dioxide system.

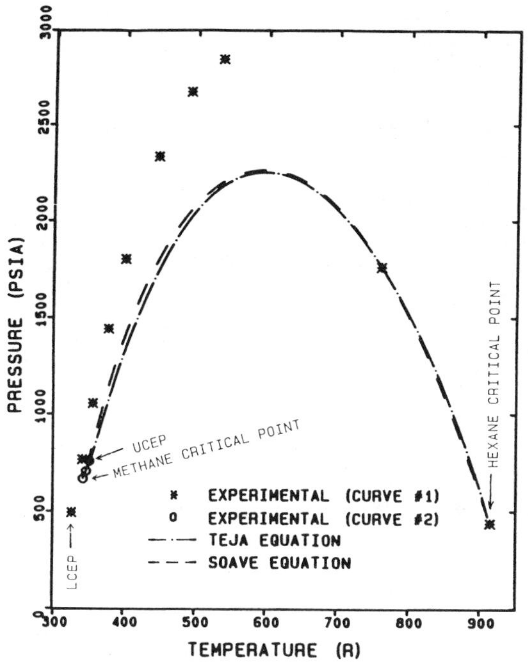

Figure 8. Critical loci of methane-hexane system.

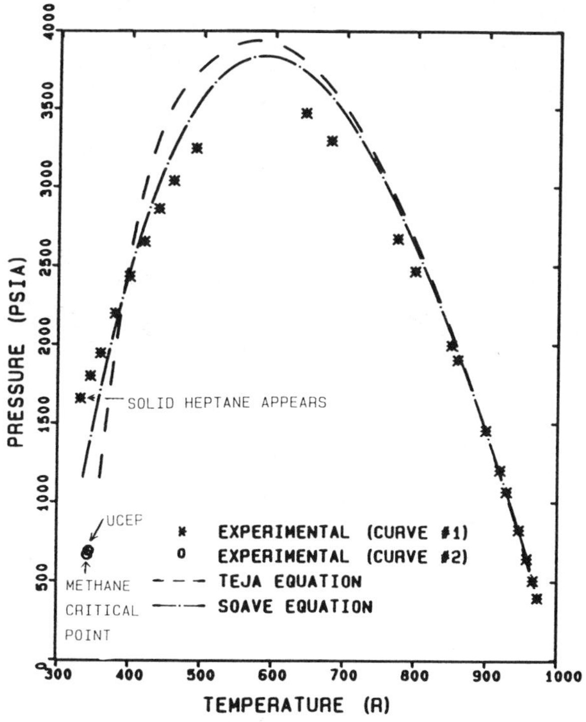

Figure 9. Critical loci of methane-heptane system.

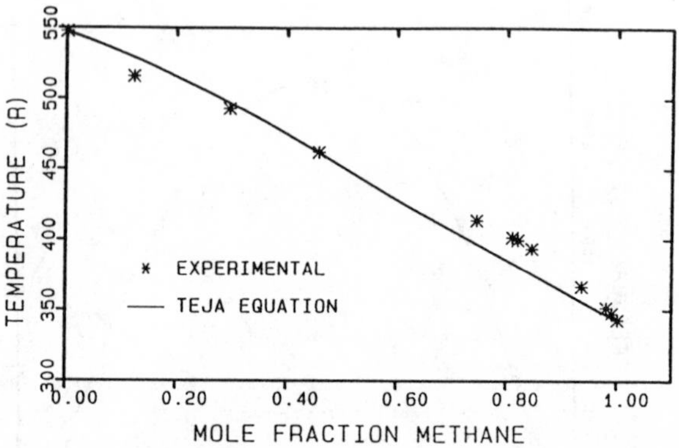

Figure 10. Critical temperature of methane-carbon dioxide system.

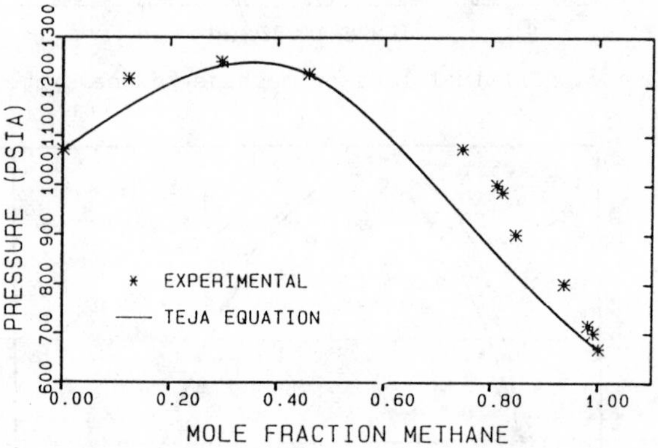

Figure 11. Critical pressure of methane-carbon dioxide system.

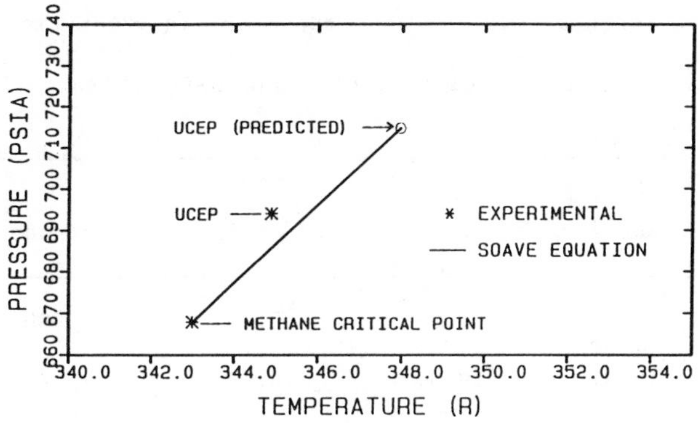

Figure 12. Critical loci of methane-heptane system near methane critical point.

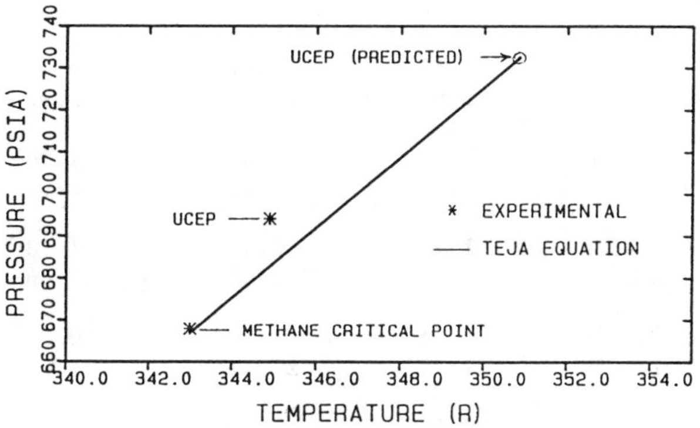

Figure 13. Critical loci of methane-heptane system near methane critical point.

TABLE II

Average Percent Relative Absolute Error in Predicting Binary Critical Loci by Cubic Equations of State

System		Redlich Kwong	Soave	Adachi	Peng and Robinson	Teja
propane	(1)	0.37	0.08	0.14	0.80	0.05
hexane	(2)	1.09	0.49	0.73	2.52	0.33
	(3)	18.31	19.51	16.78	7.70	16.78
methane/	(1)	2.53	2.18	2.88	3.64	2.16
CO_2	(2)	6.32	5.86	7.99	8.58	5.82
butane/	(1)	0.77	0.54	1.18	1.28	0.51
CO_2	(2)	2.62	1.77	2.96	2.47	1.56
	(3)	17.00	20.96	28.06	7.99	13.38
decane/	(1)	.81	1.67	2.81	3.63	2.33
CO_2	(2)	10.23	9.63	16.72	27.92	29.72
benzene/	(1)	.96	.77	.69	1.18	1.35
water	(2)	6.25	7.81	7.93	8.32	7.78
water/	(1)	1.38	1.14	6.61	1.28	2.02
CO_2	(2)	21.66	16.62	224.55	35.84	27.70
methane/	(1)	10.17	4.91	17.16	10.02	6.14
hexane	(2)	12.44	19.42	29.91	11.13	22.67
methane/	(1)	2.92	2.43	9.36	5.01	4.90
heptane	(2)	11.49	3.39	24.31	5.72	8.30
	(3)	12.31	18.89	23.50	8.65	22.23

(1) percent relative absolute error in critical temperature prediction
(2) percent relative absolute error in critical pressure prediction
(3) percent relative absolute error in critical volume prediction (if available)

TABLE III

Values of fitted parameters γ_{12} and

ζ_{12} for each Cubic Equation of State

system	Redlich Kwong	Soave	Adachi	Peng and Robinson	Teja
propane/(1)	0.9244	1.0097	0.9094	0.8320	0.9897
hexane (2)	0.9111	0.9734	0.8846	0.8245	0.9771
methane/(1)	0.9558	1.0524	1.0990	0.7476	1.0465
CO_2 (2)	0.8765	0.9222	0.8871	0.7893	0.9250
butane/ (1)	0.9182	0.9990	1.2607	0.6600	0.9578
CO_2 (2)	0.7567	0.8390	0.9713	0.7475	0.8294
decane/ (1)	1.2319	1.3268	-.2422	.4455	1.400*
CO_2 (2)	.5940	0.9352	-1.0090	1.5599	.700
benzene/(1)	0.8015	0.8019	.7257	.7210	.7562
water (2)	0.6556	0.5980	.5536	.7694	.6110
water/ (1)	1.0546	1.0446	1.000*	.6260	1.2307
CO_2 (2)	0.8828	1.0407	1.000	.95154	1.2927
methane/(1)	.9950	1.1310	.5000*	.6091	1.0682
hexane (2)	.9545	1.1441	.6000	.7347	0.9661
methane/(1)	.9093	1.0237	-.0096	.3793	0.8861
heptane (2)	.7319	0.9661	-.2814	.6732	0.7446

(1) Calculated value for λ_{12}
(2) Calculated value for ζ_{12}
* Values for λ_{12} and ζ_{12} do not fit the midpoint data exactly

The constants a, b, and c in the above equations are calculated entirely from pure component properties. To extend the equations to mixtures, the following mixing rules were used:

$$a = [\sqrt{a_1}x_1 + \sqrt{a_2}x_2]^2 + 2x_1x_2(\xi_{12}-1)\sqrt{a_1a_2} \quad (10)$$

$$b = [b_1x_1 + b_2x_2] + 2x_1x_2(\lambda_{12}-1)\left(\frac{b_1+b_2}{2}\right) \quad (11)$$

$$c = c_1x_1 + c_2x_2 \quad (12)$$

The constants in these mixture equations can be evaluated entirely from pure component properties alone by equating the parameters ζ_{12} and λ_{12} to unity. Equations (10) and (11) then become the mixing rules of the original van der Waals equation. The ζ_{12} and λ_{12} parameters other than unity, must be determined by fitting the equation to actual experimental data for the mixture.

Equations (3) and (4), together with an equation of state make up a system of non-linear equations in three unknowns: T_c, P_c and V_c. For a given composition, the free energy equations were solved for the critical temperature and critical volume of the system. The critical pressure was then determined from the equation of state. To calculate the values for λ_{12} and ζ_{12}, used in this work, the following functions were calculated at $x_1 \simeq 0.50$:

$$F_1 = |(T_c)_{exp} - (T_c)_{calc}| \quad (13)$$

$$F_2 = |(P_c)_{exp} - (P_c)_{calc}| \quad (14)$$

If experimental data at $X_1 = 0.50$ were unavailable, then the closest experimental value to $x_1 = 0.50$ was used. Values of λ_{12} and ζ_{12} were calculated by using a damped Newton Raphson method, until the functions F_1 and F_2 were less than 0.001. In most cases initial starting values of $\lambda_{12} = 1.0$ and $\zeta_{12} = 1.0$ converged satisfactorily. In other cases much lower initial trials were required. The values of λ_{12} and ζ_{12} obtained from data at $x_1 \simeq 0.5$ were then used to calculate the entire critical locus from $x_1 = 0.0$ to $x_1 = 1.0$.

Systems Selected for Study

The P-T projections of the P-T-x critical surface of eight different binary systems were selected for study. The systems selected represent six of the nine different critical behavior types discussed by van Konynenburg and Scott (8) and summarized in Table I. For each of these projected critical loci, the critical temperature, critical pressure, and critical volume were calculated from the five cubic equations of state in Equations (5) - (9).

As indicated in Table I, the Class 1 Type I projections are characterized by a single continuous critical line connecting the pure component critical temperatures. Binary mixtures which exhibit Class 1 Type I behavior have pure component critical temperatures are in the range $T_{c1}/T_{c2} < 2$ (8). The Class 1 Type I systems studied are:

(1) Propane and n-Hexane, Kay (16)
(2) Methane and Carbon Dioxide, Donnelly and Katz (17)
(3) n-Butane and Carbon Dioxide, Poettmann and Katz (18) and Redlich, Kister, and Lacy (19)

More complicated Class 1 systems are those of Class 1 Type II, which exhibit not only a continuous gas-liquid critical line between the pure component criticals, but also a liquid-liquid critical line from C_m to the UCEP termination of a three phase liquid-liquid-solid line, as described in Table I. One system of Class 1 Type II was studied in this work. This system is:

(4) n-Decane and Carbon Dioxide, Reamer and Sage (20)

Class II behavior is exhibited in binary systems in which the critical temperature of one of the pure components is usually more than twice the critical temperature of the other. In Class II systems there is no continuous critical line connecting the two pure component critical points. Four different examples of Class 2 systems were studied in this paper. These include a Class 2 Type IIIa-HA system in which the liquid-liquid critical line between C2 and C_m, as described in Table I, actually passes through a minimum temperature before diverging to the immeasurably high pressure at C_m. It is therefore characterized as a Type IIIa-HAm system. The example chosen was:

5) Benzene and Water, Rebert & Kay (21), and Schneider (22)

Class 2 Type III systems, as indicated in Table I, are the same as Type III-HA except that the three phase liquid-liquid-gas line lies between the two pure component vapor pressure curves so that there is no hetero-azetropic behavior, as discussed by Rowlinson (23). Two sample systems representing Type IIIa and Type IIIb were selected. These are:

6) Carbon Dioxide - Water, Takenouchi (24), Type IIIa
7) Methane-n-Heptane, Kohn (27); Chang, Hurt & Kobayashi (28); and Reamer, Sage and Lacey (29), Type IIIb

The final critical behavior studied in this work is that of Class 2 Type V. In this case there are two critical lines, one from the pure component critical Cl to a UCEP and a second from the other pure component critical C2 to a LCEP, as described in Table I. One example of this behavior was examined. This was:

8) Methane-n-Hexane, Shim & Kohn (25) and Poston & McKetta (26)

Results and Critical Loci Predictions

The average percent relative absolute error in predicting the selected binary critical loci is shown in Table II and the critical methane-carbon dioxide mixtures are plotted as a function of composition as predicted by the Teja equation. The critical values were fitted exactly at a methane mole fraction of 0.457. As shown in Figures 10 and 11, the equation generally follows the trend of the data but predicts critical temperatures slightly too high at mole fractions less than 0.457 and slightly too low at larger mole fractions. The same trends are exhibited by the other equations and the shapes of the predicted loci are similar.

Accuracy of the critical loci predictions is considerably enhanced by the use of two unlike pair parameters, instead of a single ζ_{12} parameter only, as is a common practice. This is particularly true when the members of the binary mixture have large dissimilarities in size and character. Usually, the assignment of $\zeta_{12} = 1.0$ and $\lambda_{12} = 1.0$ predicts the critical temperature much more accurately than the critical pressure. The critical pressure in the cases studied here, is always extremely low, frequently by more than 30%. These large Pc errors can be remedied by changing ζ_{12}, usually decreasing its value. Unfortunately, when the calculated and experimental Pc values are brought into good agreement by this process, the error in the calculated Tc is made larger than the initial result obtained with both λ_{12} and ζ_{12} equal to 1.0 .

Discussion

As indicated in Table II, the critical temperatures and critical pressures in systems of Class 1 Type I are predicted accurately by all equations of the van der Waals family. Best results for the predicted critical temperatures and pressures of these systems were obtained with the Teja equation. It is important to note that this comparison applies only to critical loci tested by the procedures described here. It does not imply that this ranking is valid over the entire P-T-x diagram for these systems.

The excellent results for the Class 1 Type I systems gives some support to the mixing rules adopted in Equations (10)-(11). The results indicate that the two fitted unlike pair coefficients in these mixing rules resemble true molecular properties in that the same values are valid over the entire composition range.

A remarkable feature of the van der Waals family of equations is the ability of all the equations tested to predict the local minimum in the critical locus of the n-butane-carbon dioxide system at high concentrations of carbon dioxide, as shown in Figure 3. This phenomenon is well established experimentally for Class 1 Type I systems involving carbon dioxide. Morrison and Kincaid (30) have shown experimentally that this minimum exists both for the n-butane-carbon dioxide and also for the propane-carbon dioxide systems.

It is clear from Table II that the excellent accuracy of the critical temperature and critical pressure predictions for Class 1 Type I systems does not extend at all to the critical volumes. A common feature of the van der Waals family of equations is that they predict pure component critical compressibility factors which are too large for components whose molecules contain more than one atom. This results in critical volume predictions which are always too high. The best prediction of critical volumes is given by the Peng-Robinson equation which predicts a critical compressibility factor of 0.307 (5). Although this value is higher than the critical compressibility factor of the components in the test systems, it is the lowest among the equations tested.

For a system belonging to Class 1, Type I, a more accurate method of predicting its mixture critical volume V_{cm} is to use a simple linear average in the form:

$$V_{cm} = x_1 V_{c1} + x_2 V_{c2} \tag{16}$$

Equation (16) gives a relative absolute error of 3.56% for the propane-hexane system and 6.5% for n-butane-carbon dioxide. These errors are much less than any of those in Table II obtained with a van der Waals type equation of state for these systems. Results of Equation (15) for the propane-hexane system are shown graphically in Figure 5.

Unfortunately, Equation (16) is not an effective method for predicting critical volumes in systems of more complicated types. It obviously fails for Class 2 systems where there is no continuous critical line between the pure component criticals. For the Class 2 Type V system methane-n-heptane, for example, Equation (16) predicts a critical volume which is generally worse than the equation of state predictions shown in Table II for the critical volume of this system.

All equations tested predict critical loci in Class 2 systems and in Class 1 Type II systems with much less accuracy than for the Class 1 Type I systems. In contrast to the results with Type I systems, the two-constant equations of state with the smallest deviation from the form of the original van der Waals equation, such as the Redlich-Kwong and Soave equations, give better predictions of critical temperature and critical pressure in Class 1 Type II and in Class 2 systems than the more elaborate three-constant equations and equations which deviate farther from the original van der Waals form.

An example of a Class 1 Type II system is shown in Figure 4 for the n-decane-carbon dioxide system, where predictions of the Soave equation are illustrated. The critical line between the pure component criticals was fit exactly at a mole fraction of carbon dioxide = 0.5 by the unlike pair coefficients shown in Table III. At lower carbon dioxide concentrations the predictions are excellent, as shown in Figure 4. At carbon dioxide mole fractions greater than 0.5, approaching the pure carbon dioxide critical, the predicted critical pressures are much too low. No predictions at all are possible for the liquid-liquid critical line from the UCEP to higher pressures.

Among the Class 2 systems, the benzene-water system is an example of Type III-HAm behavior, where the "m" indicates a minimum in the critical line between the critical of water (C2) and the high pressure point at C_m. The results obtained with the Soave and Redlich-Kwong equations, which gave the best results for this system, are shown in Figure 6. Although the fitting point at 0.4895 m.f. benzene lies on the vapor-liquid critical line between the UCEP and the benzene critical (C1), the predictions are qualitatively correct along the other critical line between C2 and C_m. It is remarkable also that the Redlich-Kwong equation shows the required minimum in this critical line, although it does not occur at the precisely correct pressure and temperature values.

Critical temperature and pressure predictions for the Class 2 Type IIIa system carbon dioxide-water are shown in Figure 7 for the Soave equation. Because there is no critical locus at x_{CO2} = 0.5 the unlike pair coefficients for the Soave equation were obtained at x_{CO2} = 0.321, corresponding to the highest pressure data point on the critical line between C_m and the critical point of water, C2. With coefficients evaluated at this concentration, the Soave equation can predict both parts of the critical line, although the UCEP is not located precisely so that results along the critical line from C1 to UCEP are only qualitatively predicted.

Among the other equations, it was possible to find unlike pair coefficients which could fit both the critical pressure and critical temperature at this same CO_2 composition only for the Teja equation. The nearest composition where most of the others could be fit to both the critical pressure and temperature was the data point in Figure 7 at 0.270 m.f. CO_2, corresponding to a critical pressure of 12,836 psia, although the predictions at other compositions are poor, particularly for the steeply rising pressure. For all equations except the Soave and Teja equations the errors in Table II are those obtained when using unlike pair coefficients fitted at 0.270 m.f. CO_2. Only for the Soave equation could the unlike pair coefficients, which were fitted exactly to the single data point at 0.321 m.f. CO_2, produce accurate pressure predictions along the critical line from C_m to C2 at other compositions. For the other equations, reasonable predictions over the entire critical line cannot be obtained from unlike pair coefficients fitted at a single composition.

Critical behavior of the Type IIIb methane-n-heptane system has been explained in detail by Chang, Hurt and Kobayashi (28). Calculated results are shown in Figure 9. Accurate results are obtained only at low methane concentrations along the critical line from the heptane critical (C2) to the UCEP at the termination point of a liquid-liquid-solid line when the two liquid phases become identical. Reasonable results are obtained only along this critical line between the heptane critical and a methane concentration of about 0.7 mole fraction at a temperature of 775°R. At higher methane concentrations the calculated critical pressures are much too large. Reasonable results are also obtained along the very short second critical line from the methane critical to the UCEP at the limit of the liquid-liquid-gas line.

Results for the Class 2 Type V methane-n-hexane system are shown in Figure 8. This system is predicted only qualitatively by the procedure used here for evaluating the unlike pair coefficients. On the critical line between the LCEP and the hexane critical (C2), methane concentrations above the fitting point of x_{CH4} = 0.625 at a critical temperature of 762°R give predicted pressures which are much too low with all the equations. It is possible to calculate some points on the gas-liquid critical line between the UCEP and the methane critical (C1) but the accuracy is extremely poor with the calculated critical pressure much too small. Reasonable results are obtained only at low methane concentrations near the hexane critical. Accuracy deteriorates rapidly as methane concentrations approach those where three phase conditions occur between the LCEP and the UCEP when the unlike pair coefficients are evaluated at conditions far from the three phase region.

Predictions of Class 2 critical loci by equations of the van der Waals family are limited by their inability to predict critical end points of three phase lines. As an example, Figures 12 and 13 show the experimental and predicted UCEP conditions for the termination of the liquid-liquid-solid line in the Type IIIb methane-n-heptane system. Results obtained with the Soave and Teja equations are shown. None of the equations tested here produced accurate UCEP values.

Conclusions

The results of this study may be summarized as follows:

1. Critical temperatures and critical pressures of Class 1, Type I, binary systems can be predicted successfully by all cubic equations of state of the van der Waals family tested in this paper. The mixing rules used in each equation studied here are those of the original van der Waals equation with an added correction. This correction is for the deviation of the contributions of unlike pair interactions from the form of these contributions assumed in the original van der Waals equation. It is described by two adjustable unlike pair interaction coefficients.

2. In most, but not all, of the cubic equations tested the unlike pair interaction coefficients required for Class 1, Type I systems do not depend appreciably on composition, temperature, or pressure. This is shown by the fact that when these coefficients are evaluated by fitting them to a single experimental data point for an approximately equi-molar mixture, the critical temperature and pressure at all other compositions on the critical locus are predicted with an average absolute relative error of less than 1%. Best accuracy in predicting Class 1, Type I critical temperatures and pressures in this way was obtained by using the Teja equation.

3. Critical volumes predicted at all compositions, when the unlike pair coefficients are evaluated to predict the critical temperature and pressure at a single data point, are always too large for all the equations tested. For all Class 1, Type I systems

the Peng-Robinson equation gave the best prediction of the critical volume with the unlike pair coefficients fitted to the critical temperature and pressure at a single data point. In Class 1, Type I systems a simple linear average of the pure component critical volumes gave a significantly better prediction of the mixture critical volume than any of the equations of state tested.

4. In Class 1, Type II systems and in all Class 2 systems, all the equations predict critical temperatures and pressures with much less accuracy than for Type I systems when using the test procedures of this study. The results, showed that critical temperature predictions are much more accurate than those for critical pressure.

5. In Class 1, Type II, and in all Class 2 systems, the region of a critical locus near the formation conditions for a second liquid phase or a solid phase cannot be predicted reliably when using unlike pair coefficients evaluated at a data point on a vapor-liquid or liquid-liquid critical locus far removed from the point of appearance of a third phase.

6. In Class 1, Type II, and in all Class 2 systems, either the Redlich - Kwong or the Soave equation shows the least variation of the unlike pair coefficients with composition, temperature, and pressure and gives best predictions of critical properties at all compositions from pair coefficients evaluated at a single data point. For these more complicated systems, it appears that the simpler two-constant equations, like the Redlich - Kwong and Soave, give better critical loci predictions than the more elaborate equations whose form deviates farther from that of the original van der Waals equation.

Acknowledgments

The authors appreciate the support of this research at Rice University by the Gas Research Institute.

The authors also appreciate the helpful suggestions made by Dr. J.M.H. Levelt Sengers.

Literature Cited

1. Chang, R. F.; Morrison, G.; Levelt Sengers, J.M.H. J. Phys. Chem. 1984, 88, 3389.

2. Joffe, J.; and Zudkevitch, D. Chem. Eng. Prog. Symp. S. 1967, 63, 43.

3. Spear, R. R.; Robinson, R.L.; Chao, K.C. Ind. Eng. Chem. Fundam. 1969, 8, 2.

4. Hissong, D. W.; Kay, W. B. AIChE J. 1970, 16, 580.

5. Peng, D.-Y.; Robinson, D. B. AIChE J. 1977, 23, 137.

6. Teja, A. S.; Rowlinson, J. S. Chem. Eng. Sci. 1973, 28, 529.

7. Teja, A. S. Smith, R. L.; Sandler, S. I. Fluid Phase Equilibria 1983, 14, 265.

8. van Konynenburg, P. H.; Scott, R. L. Phil. Trans. Roy. Soc. Lond. 1980, 298, 495.

9. Gibbs, J. W. "Collected Works"; Longmans, Green, and Company; New York, 1928, Vol. 1, 129.

10. Redlich, O.; Kister H. Ind. Eng. Chem. 1948, 40, 345.

11. Redlich, O.; Kwong, J. N. S. Chem. Rev. 1949, 44, 233.

12. Soave, G. Chem. Eng. Sci. 1972, 27, 1197.

13. Adachi, Y.; Lu B. C.-Y.; Sugie, H. Fluid Phase Equilibria 1983, 13, 133.

14. Peng, D.-Y; Robinson, D. B. Ind. Eng. Chem. Fundam. 1976, 15, 59.

15. Teja, A. S.: Patel, N. C. Chem. Eng. Sci. 1982, 37, 463.

16. Kay, W. B., J. Chem. Eng. Data 1971, 16, 137.

17. Donnelly, H. G.; Katz, D. L. Ind. Eng. Chem. 1954, 46, 511.

18. Poettmann, F. H.; Katz, D. L. Ind. Eng. Chem. 1945, 37, 59.

19. Redlich, O.; Kister H.; Lacey, W. N. Ind. Eng. Chem. 1949, 841, 475.

20. Reamer, H. H.; Sage, B. H. J. Chem. Eng. Data 1963, 8, 508.

21. Rebert, C. J.; Kay, W. B. AIChE J. 1959, 5, 285.

22. Schneider, G. M. Ber. Bunsenges. Physik. Chem. 1972, 76, 325.

23. Rowlinson, J. S. "Liquids and Liquid Mixtures" 2nd Edition, Butterworth and Co., 1969, 203-216.

24. Takenouchi, S.; Kennedy, G. C. Am. J. Sci. 1964, 262, 1055.

25. Shim, J.; Kohn, J. P. J. Chem. Eng. Data 1962, 7, 3.

26. Poston, R. S.; McKetta, J. J. J. Chem. Eng. Data 1966, 11, 362.

27. Kohn, J. P. AIChE J. 1961, 7, 514.

28. Chang, H. L.; Hurt, L. J.; Kobayashi, R. AIChE J. 1966, 12, 1212.

29. Reamer, H. H.; Sage, B. H.; and Lacey, W. N. Chem. Eng. Data S. 1956, 1, 29.

30. Morrison, G.; Kincaid, J. M. AIChE J. 1984, 30, 257.

RECEIVED November 8, 1985

7

Equation of State for Supercritical Extraction

J. S. Haselow, S. J. Han, R. A. Greenkorn, and K. C. Chao

School of Chemical Engineering, Purdue University, West Lafayette, IN 47907

Nine equations of state are evaluated regarding their ability to describe supercritical extraction. Experimental data on 31 binary mixture systems are compared with calculations from the equations of Redlich-Kwong, Soave, Peng-Robinson, Schmidt-Wenzel, Harmens, Kubic, Heyen, Cubic Chain-of-Rotators, and Han-Cox-Bono-Kwok-Starling. Interaction constants of the equations determined from the experimental data in the course of the evaluation are reported.

Supercritical extraction has received much attention recently for its many applications and potential applications. Supercritical carbon dioxide is used to decaffeinate coffee, denicotinize cigarettes, and extract spices (1,2). Carbon dioxide flooding has assumed major importance in petroleum production (3). Extraction of coal with a supercritical solvent has been the subject of investigation by the British Coal Board (4,5), Maddocks et al. (6), Blessing and Ross (7), Ross and Blessing (8), and Vasilakos et al. (9). Kerr McGee Oil Company developed supercritical extraction for the deashing of coal liquefaction reactor products in the ROSE process (10).

The design and operation of a supercritical extraction process requires the estimation of solute solubility in a supercritical solvent. Equation of state can be useful to satisfy this need. In this work, equations are screened regarding their applicability to the calculation of supercritical solubility, and the applicable equations are evaluated regarding their ability to quantitatively describe experimental solubility data.

0097–6156/86/0300–0156$07.00/0

The Equations of State

From the large number of equations of state that have been proposed in the literature, we have found only a small number of them to show promise of being useful for the estimation of supercritical solubility. A severe limitation is the availability of the equation constants. Equations that require their constants to be determined by fitting extensive experimental data on the pure substances to the equation are not useful, for rarely are experimental data available for the complex solutes of interest. We are thus forced to discard complex equations such as the Benedict-Webb-Rubin (11), Jacobsen (12), Goodwin (13), Perturbed-Hard-Chain (14), and Chain-of-Rotators (15).

For detailed evaluation, we have selected nine equations from among those that have been in use in phase equilibrium calculations and newer equations that show promise. Eight of these are cubic equations, and one (Han-Cox-Bono-Kwok-Starling) is complex. But all have been generalized to express the equation constants in terms of critical properties, which are known for a large number of substances. Group contribution (16,17) and other methods (18,19) are widely used for the estimation of the critical properties where experimental values have not been reported.

Table I presents the nine equations of state that are studied in this work. Table II shows the critical properties of the substances that are used in the calculations. For many higher molecular weight compounds T_c , p_c , and ω were estimated, T_c and p_c by the method of Lydersen (18) and ω by the method of Edmister (19).

Mixing rules described in the original papers are used in the calculations reported here for the Redlich-Kwong, HCBKS, Peng-Robinson, Kubic, Heyen, and CCOR equations. No interaction constant was employed by Soave in the mixing rules for his equation. We introduced an interaction constant k_{12} to the constant a, as we found it quite necessary. The Schmidt-Wenzel and Harmens-Knapp equations contained no mixing rules at all. We employ the classical one-fluid mixing rules to extend these equations to mixtures as follows:

$$\Theta_m = \sum_i \sum_j y_i y_j \Theta_{ij} \tag{1}$$

Where Θ may be either a or b, and the cross parameter Θ_{ij} is given by

$$a_{ij} = (1-k_{ij})(a_{ii}a_{jj})^{1/2} \tag{2}$$

and

$$b_{ij} = (b_{ii} + b_{jj})/2 \tag{3}$$

These are, in fact, the same rules employed in the other equations studied in this work.

Table I. Equations of State

1. Redlich-Kwong equation
Source: Redlich, O., and Kwong, J.N.S., Chem. Rev. (1949) 44 , 223.

$$p=\frac{RT}{(v-b)}-\frac{a}{T^{0.5}v(v+b)}$$

2. HCBKS equation
Sources: Cox, K.W., Bono, J.L., Kwok, Y.C., and Starling, K.E., Ind. Eng. Chem. Fundamen. (1971) 10 , 245. Starling, K.E., and Han, M.S., Hydrocarbon Processing (1972), 4 , 129.

$$p=\rho RT+(B_oRT-A_o-\frac{C_o}{T^2}+\frac{D_o}{T^3}-\frac{E_o}{T^4})\rho^2$$

$$+(bRT-a-\frac{d}{T})\rho^3+\alpha(a+\frac{d}{T})\rho^6$$

$$+\frac{c\rho^3}{T^2}(1+\gamma\rho^2)\exp(-\gamma\rho^2)$$

3. Soave equation
Source: Soave, G., Chemical Engineering Science (1972) 27 , 1197.

$$p=\frac{RT}{(v-b)}-\frac{a(T)}{v(v+b)}$$

4. Peng-Robinson equation
Source: Peng, D., and Robinson, D.B., Ind. Eng. Chem. Fund. (1976) 15 , 59.

$$p=\frac{RT}{(v-b)}-\frac{a(T)}{[v(v-b)+b(v-b)]}$$

5. Schmidt-Wenzel equation
Source: Schmidt, G., and Wenzel, H., Chemical Engineering Science (1980) 35 , 1503.

$$p=\frac{RT}{(v-b)}-\frac{a}{v^2+ubV+wb^2}$$

Table I Continued

6. Harmens-Knapp equation

   Source: Harmens, A., and Knapp, H., Ind. Eng. Chem. Fundam. (1980) 19 , 291.

$$p=\frac{RT}{(v-b)}-\frac{a}{v^2+cbV-(c-1)b^2}$$

7. Kubic equation

   Source: Kubic, W.L., Fluid Phase Equilibria (1982) 9 , 79.

$$p=\frac{RT}{(v-b)}-\frac{a}{(v+c)^2}$$

8. Heyen equation

   Source: Heyen, G., "A Cubic Equation of State with Extended Range of Application" in "Chemical Engineering Thermodynamics", Ann Arbor Science. (Ann Arbor), (1983).

$$p=\frac{RT}{(v-b)}-\frac{a}{v^2+(b+c)v-bc}$$

9. CCOR equation

   Source: Kim, H., Lin, H.M., and Chao, K.C., Ind. Eng. Chem. Fundam., (1985) in press.

$$p=\frac{RT(1+0.77b/v)}{(v-0.42b)}+\frac{0.055c^R RTb/v}{(v-0.42b)}-\frac{a}{v(v+c)}-\frac{bd}{v(v+c)(v-0.42b)}$$

Table II. Critical Properties Used and Sources

Compound	Tc(K)	Pc(atm)	ω	Vc(cm^3/mol)	Ref.
Carbon Dioxide	304.2	72.8	0.225	94.0	a
Ethylene	282.4	49.7	0.085	129.0	a
Benzoic Acid	752.0	45.0	0.620	341.0	a
Phenanthrene	890.0	32.5	0.429	644[d]	b
2,3-Dimethylnaphthalene	785.0	31.748	0.42403	522[d]	c
2,6-Dimethylnaphthalene	777.0	31.796	0.42013	516[d]	c
Naphthalene	748.4	40.0	0.302	410.0	a
Ethane	305.4	48.2	0.091	148.0	a
Fluorene	821[e]	29.5[e]	0.407[f]	590.0[d]	
1,4-Naphthoquinone	792[g]	40.7[g]	0.575[g]	257.0[d]	
Acridine	883[g]	31.5[g]	0.498[g]	399[d]	
Phenol	694.2	60.5	0.440	229	a
Hexamethylbenzene	752[e]	23.5[e]	0.498[f]	659[d]	
Triphenylmethane	863[e]	22.1[e]	0.576[f]	785[d]	
Pyrene	936[e]	25.7[e]	0.494[f]	751[d]	
Trifluoromethane	299	48.2	0.275	133	a
Chlorotrifluoromethane	302	38.7	0.180	180	a
Hexachloroethane	698.4[e]	32.97[e]	0.255[h]	470[d]	

[a] Reid, R.C., Prausnitz, J.M., and Sherwood, T.K., "The Properties of Liquids and Gases", 3rd ed.; McGraw-Hill: New York, 1977.

[b] GPA Research Report RR-30, "High Temperature V-L-E Measurements for Substitute Gas Components"; Tulsa, Oklahoma, 1978.

[c] Driesbach, R.R., "Physical Properties of Chemical Componds", American Chemical Society, Washington, D.C., 1955.

[d] Estimated by $Zc = 0.291-0.008\omega, Vc = ZcRTc/Pc$.

[e] Estimated by Lydersen's method.

[f] Estimated by Edmister's method.

[g] Estimated by method of Joback (R.C. Reid, Personal Communication).

[h] Perry, R.H., and Chilton, C.H., "Chemical Engineers Handbook", 5th ed.; McGraw-Hill: New York, 1973, for vapor pressures, definition of acentric factor.

Calculation of Supercritical Solubility

Solubility of a solid solute in a supercritical solvent is given by

$$y_1 = p_1^o \Phi_1^o \left\{ \exp[V_s(p-p_1^o)/RT] \right\} / (p\Phi_1) \tag{4}$$

where the symbols are defined in Glossary of Symbols. In Equation 4 it is assumed that the solvent is not dissolved in the condensed phase, which is valid when the condensed phase is a solid. We exclude solubility data of liquids. Equation 4 additionally assumes that the solid phase is incompressible. The fugacity coefficient of the saturated pure vapor of the solute Φ_1^o is set to be equal to 1 in view of the small vapor pressures.

The calculation of solubility y_1 according to Equation 4 requires the vapor pressure p_1^o and molal volume V_s of the solid solute to be known. Table III shows the sources from which vapor pressures of the solutes were obtained. Table IV shows the solid volume data that were used in the calculations. Since data were not available at the various temperatures of interest and since the effect of inaccuracy of solid volume is small, variation of the solid volume with temperature was ignored in the calculations.

The variation of supercritical solubility with pressure is given by p and Φ_1 in Equation 4. The characteristic supercritical effect of solute-solvent interaction is expressed by Φ_1. The calculation of Φ_1 by an equation of state, and the use of it in Equation 4 to calculate y_1 for comparison with experimental data constitutes the test of the equation of state.

Calculation of Φ_1 by an equation of state follows the standard thermodynamic procedure described in textbooks. The calculation is sensitively dependent on the value of the interaction constant(s). A value of k_{ij} of an equation of state is determined for each solute + solvent system from experimental solubility data for the best fitting of the data. The objective function

$$\Omega = \sum_{i=1}^{n} [\ln(y_{i,exp}/y_{i,cal})]^2 \tag{5}$$

is minimized in searching for the k_{ij} value. Here n stands for number of solubility data points. The k_{ij} thus determined is employed in a final round of calculations to give the calculated solubilities. The quality of the calculated supercritical solubility is expressed in terms of an average % deviation of the equation of state from the data given by

Table III. Vapor Pressure Data Sources

Compound	Reference
2,3-Dimethylnaphthalene	a
2,6-Dimethylnaphthalene	a
Phenanthrene	b
Benzoic Acid	b
Hexachloroethane	c
Naphthalene	d,e
Fluorene	f
1,4-Naphthoquinone	g
Acridine	g
Phenol	h
Hexamethylbenzene	i
Triphenylmethane	j
Pyrene	h

[a] Osborn, A.G., and Douslin, D.R., J. Chem. Eng. Data (1975) 20 , 229.

[b] deKrulf, C.G., and van Grunkel, C.H.D., presented at the Quatrieme Conference Interationale de Thermodynamique Chimique, 26 au 30, Montpellier, France, 1975.

[c] Sax, N.I., "Dangerous Properties of Industrial Materials", Van Nostrand Reinhold; Princeton, 1979.

[d] Diepen, G.A., and Scheffer, F.E.C., J. Am. Chem. Soc. (1948) 70 , 4085.

[e] Fowler, L., Trump, W., and Vogler, C., J. Chem. Eng. Data (1968) 13 , 209.

[f] Johnston, K.P., Ziger, D.H., and Eckert, C.A., Ind. Eng. Chem. Fundam. (1982) 21 , 191.

[g] Schmitt, W.J., and Reid, R.C., presented at Annual AIChE Meeting, San Francisco, CA (1984).

[h] Weast, R.C., "Handbook of Chemistry and Physics", 55th ed.; Chemical Rubber Company: U.S.A., 1975.

[i] Ambrose, D., Lawrenson, I.J., and Sprake, C.H.S., J. Chem. Thermodyn. (1975) 8 , 503.

[j] Aihara, B., Chem. Soc., Japan, in Weast, op. cit.

Table IV. Solid Volume Data and Sources

Compound	T°	$V_s(cm^3/gmole)$	Reference
Benzoic Acid	15	154.60	a
Hexachloroethane	20	113.22	a
Naphthalene	20	125.03	a
Phenanthrene	4	181.9	a
2,3-Dimethylnaphthalene	20	155.76	a
2,6-Dimethylnaphthalene	20	155.76	a
Phenol	20	89.0	a
Fluorene	0	138.18	a
1,4-Naphthoquinone	-	111.2	b
Acridine	-	178.3	b
Hexamethylbenzene	25	152.66	a
Triphenylmethane[c]	20	245	a
Pyrene	23	159.13	a

[a] Weast, R.C., "Handbook of Chemitry and Physics", 55th ed.; Chemical Rubber Co.: U.S.A. (1975).

[b] Schmitt, W.J., and Reid, R.C., presented at AIChE Annual Meeting, San Francisco, CA (1984).

[c] The density of diphenylmethane was used to calculate V_s

$$\% \text{ deviation} = 100\left\{\exp\left[(\Omega/n)^{1/2}\right] - 1\right\} \tag{6}$$

The use of logarithm in eq. 5 in preference to the sum of squares of relative deviations avoids the bias that would be produced when the calculated values are either very large or very small when compared to data.

Data and Results

Table V presents the sources of experimental supercritical solubility data for the 31 systems studied. Many more systems have been measured. Reid (20) made an extensive compilation. But limitations on available vapor pressure or solid volume data prevented some of them from being analyzed here.

Table VI presents the summary of the comparison of the equation of state in 31 parts. In each part, experimental y's of the solute in a supercritical solvent are compared with calculations with the nine equations of state. The average % deviation is shown for each equation. Interaction constants are presented, one for each equation of state, except for the CCOR equation for which two interaction constants k_{a12} and k_{c12} are given in this order.

The calculated results appear to depend on the source of the critical constants. The deviations are significantly larger for those solutes whose critical constants have been estimated by Lydersen's and Edmister's methods. The 31 systems studied are, therefore, arranged into two groups: the 14 systems with known T_c, p_c, and ω are shown in parts 1 through 14 in Table VI and the results are summarized in Table VII; the remaining 17 systems with estimated T_c, p_c, and ω are shown in parts 15 through 31 in Table VI. The overall summary comparison for all 31 systems is presented in Table VIII.

For systems of known critical constants, the Redlich-Kwong appears to give the best overall description of supercritical solubility, showing an average deviation of 17% in Tble VII. The HCBKS, Peng-Robinson, and Heyen equations belong together showing average percent deviations in the middle twenties. Soave, Schmidt-Wenzel and CCOR equations form the next group with average percent deviations in the lower thirties. Harmen-Knapp equation follows with a somewhat higher percent deviation of 41. The Kubic equation, showing a very large deviation, does not even qualitatively describe supercritical solubility. The cross interaction constant of the Kubic equation is not given by Equation 2, but by a different formula, which causes the failure.

Table V. Supercritical Solubility Data Sources

Mixture System (Solvent; Solute)	Reference
Carbon Dioxide, Benzoic Acid	a
Ethylene, Benzoic Acid	a
CO_2, Hexachloroethane	a
Ethylene, Phenanthrene	a
CO_2, Phenanthrene	a
Ethylene, 2,3-Dimethylnaphthalene	a
Ethylene, 2,6-Dimethylnaphthalene	a
CO_2, 2,3-Dimethylnaphthalene	a
CO_2, 2,6-Dimethylnaphthalene	a
Ethylene, Phenanthrene	a
Ethane, Naphthalene	b
Ethane, Phenanthrene	b
Ethane, Triphenylmethane	b
CO_2, Triphenylmethane	b
Ethylene, Hexamethylbenzene	b
CO_2, Hexamethylbenzene	b
Ethylene, Fluorene	b
CO_2, Fluorene	b
CO_2, Pyrene	b
Ethylene, Pyrene	b
CO_2, Phenol	c
CHF_3, Naphthalene	d
$CClF_3$, Naphthalene	d
CO_2, 1,4-Naphthoquinone	d
CHF_3, 1,4-Naphthoquinone	d
$CClF_3$, 1,4-Naphthoquinone	d
Ethane, 1,4-Naphthoquinone	d
CO_2, Acridine	d
Ethane, Acridine	d
CHF_3, Acridine	d

[a] Kurnick, R,T., Holla, S.J., and Reid, R.C., J. Chem. Eng. Data (1981) 26, 47.

[b] Johnston, K.P., Ziger, D.H., and Eckert, C.A., Ind. Eng. Chem. Fundamen. (1982) 21, 191.

[c] VanLeer, R.A., and Paulaitis, M.E., J.Chem. Eng. Data (1980) 25, 257.

[d] Schmitt, W.J., and Reid, R.C., presented at AIChE Annual Meeting, San Francisco, CA, (1984).

Table VI. Deviations of Equation of State from Solubility Data

	1. Benzoic acid in Carbon Dioxide		2. Benzoic acid in Ethylene		3. Naphthalene in Trifluorochloromethane	
Equation	k_{12}	av. % dev.	k_{12}	av. % dev.	k_{12}	av. % dev.
Redlich-Kwong	-.1797	16.8	-.2686	11.3	.0145	10.5
HCBKS	.0118	21.5	-.0381	36.5	.0571	11.2
Soave	.0433	43.2	-.0080	19.6	.0959	17.7
Peng-Robinson	.0381	36.8	-.0118	14.8	.0983	16.4
Schmidt-Wenzel	.0256	27.9	-.0557	31.2	.0907	27.6
Harmens-Knapp	.0156	40.2	-.0633	37.9	.0883	35.2
Kubic	-.1257	409.8	-.1696	153.9	-.0582	139.7
Heyen	-.0509	10.3	-.0845	23.1	.0644	21.2
CCOR*	.3796	19.0	.3969	13.4	.3644	16.5
	.5227		.5178		.4351	

	4. Phenanthrene in Ethylene		5. Phenanthrane in Carbon Dioxide		6. 2,3-dimethylnaphthalene in Ethylene	
Equation	k_{12}	av. % dev.	k_{12}	av. % dev.	k_{12}	av. % dev.
Redlich-Kwong	-.0820	19.4	.0094	9.9	-.1108	30.2
HCBKS	.0256	26.5	.0571	18.5	.0207	25.7
Soave	.0782	28.4	.1371	37.5	-.0546	39.1
Peng-Robinson	.0734	20.5	.1284	30.0	.0509	29.6
Schmidt-Wenzel	.0658	36.1	.1384	31.9	.0457	30.0
Harmens-Knapp	.0519	44.3	.1246	45.7	.0357	37.2
Kubic	-.2374	1611.1	-.1935	13523.2	-.1585	439.8
Heyen	-.0207	23.4	.0283	14.8	-.0270	24.7
CCOR	.4422	34.8	.4724	21.5	.3572	110.0
	.6242		.6694		.5103	

	7. 2,6-dimethylnaphthalene in Ethylene		8. 2,3-dimethylnaphthalene in Carbon Dioxide		9. 2,6-dimethylnaphthalene in Carbon Dioxide	
Equation	k_{12}	av. % dev.	k_{12}	av. % dev.	k_{12}	av. % dev.
Redlich-Kwong	-.1045	18.7	-.0045	12.8	-.0031	9.1
HCBKS	.0181	16.0	.0506	26.1	.0495	19.4
Soave	.0544	40.0	-.1197	41.9	.1170	33.7
Peng-Robinson	.0509	31.1	.1132	36.5	.1108	28.5
Schmidt-Wenzel	.0443	25.6	.1284	22.9	.1257	20.6
Harmens-Knapp	.0332	33.0	.1184	32.4	.1159	29.9
Kubic	-.1561	467.4	-.1222	1638.6	-.1197	1733.9
Heyen	-.0218	17.6	.0308	23.3	.0319	12.9
CCOR	.3894	56.5	.4261	32.2	.4248	31.2
	.5493		.6043		.6054	

* First constant is k_{a12}, second k_{c12}.

Table VI Continued

Equation	10. Naphthalene in Ethane k_{12}	av. % dev.	11. Phenanthrene in Ethane k_{12}	av. % dev.	12. Naphthalene in Carbon Dioxide k_{12}	av. % dev.
Redlich-Kwong	-.0519	29.9	-.0533	23.4	.0308	16.7
HCBKS	.0343	31.0	.0357	29.2	.0557	21.9
Soave	.0405	32.8	.0831	23.4	.1208	38.3
Peng-Robinson	.0405	30.6	.0807	17.1	.1146	32.9
Schmidt-Wenzel	.0471	44.3	.1032	51.0	.1032	21.8
Harmens-Knapp	.0433	47.1	.0945	60.2	.0934	33.7
Kubic	-.0796	84.3	-.2510	10600.3	-.0218	202.2
Heyen	.0083	20.7	-.0007	35.5	.0481	16.2
CCOR	.2932	31.9	.4285	30.2	.3518	60.6
	.3888		.5979		.4815	

Equation	13. Naphthalene in trifluoromethane k_{12}	av. % dev.	14. Phenol in Carbon Dioxide k_{12}	av. % dev.	15. Fluorene in Ethylene k_{12}	av. % dev.
Redlich-Kwong	.0232	14.3	-.0308	17.5	-.1121	33.0
HCBKS	.0606	14.7	.0495	23.4	.0183	32.6
Soave	.1083	18.8	.1007	16.2	.0433	57.8
Peng-Robinson	.1070	17.8	.0983	17.4	.0419	43.0
Schmidt-Wenzel	.0896	39.5	.1059	32.9	.0405	47.2
Harmens-Knapp	.0820	53.3	.1032	39.2	.0294	58.5
Kubic	-.0332	156.9	-.0408	22.7	-.2245	10224.6
Heyen	.0644	36.2	.0533	37.2	-.0457	29.6
CCOR	.3543	19.7	.3010	3.0	.3922	118.4
	.4552		.3462		.5843	

Equation	16. Hexachloroethane in Carbon Dioxide k_{12}	av. % dev.	17. 1,4-Naphthoquinone in Carbon Dioxide k_{12}	av. % dev.	18. 1,4-Naphthoquinone in Ethane k_{12}	av. % dev.
Redlich-Kwong	.0734	18.7	-.1135	19.0	-.0519	27.6
HCBKS	.0595	17.2	.0332	26.4	.0357	33.4
Soave	.1371	26.6	.0907	36.8	.1260	16.3
Peng-Robinson	.1322	21.4	.0834	30.6	.1233	15.8
Schmidt-Wenzel	.1322	9.5	.0758	68.7	.1308	55.5
Harmens-Knapp	.1295	12.6	.0633	88.9	.1270	64.8
Kubic	.0218	71.6	-.1371	1270.0	-.1797	2494.8
Heyen	.0820	11.2	-.0194	42.2	.0571	26.2
CCOR	.3795	37.4	.3849	35.1	.4330	23.1
	.5192		.5292		.5067	

Continued on next page

Table VI Continued

	19. 1,4-Naphthoquinone in Trifluoromethane		20. 1,4-Naphthoquinone in Trifluorochloromethane		21. Acridine in Carbon Dioxide	
Equation	k_{12}	av. % dev.	k_{12}	av. % dev.	k_{12}	av. % dev.
Redlich-Kwong	-.1623	86.3	-.0433	31.1	-.0045	20.1
HCBKS	.0294	85.4	.0519	31.7	.0582	37.1
Soave	.0509	51.1	.1371	20.3	.1433	64.2
Peng-Robinson	.0481	56.3	.1357	22.0	.1357	51.7
Schmidt-Wenzel	.0180	147.9	.1260	63.8	.1485	36.7
Harmens-Knapp	.0045	173.3	.1208	77.3	.1346	54.1
Kubic	-.1332	742.2	-.1384	749.2	-.2336	11211.1
Heyen	-.0370	132.1	.0671	47.1	.0256	19.1
CCOR	.3323	69.5	.4248	15.1	.4836	26.7
	.4414		.4579		.6632	

	22. Acridine in Ethane		23. Acridine in Trifluoromethane		24. Acridine in Trifluorochloromethane	
Equation	k_{12}	av. % dev.	k_{12}	av. % dev.	k_{12}	av. % dev.
Redlich-Kwong	-.0533	72.4	-.0332	34.4	-.0145	33.3
HCBKS	.0381	58.4	.0620	24.6	.0609	33.8
Soave	.1045	43.4	.1284	29.7	.1360	12.2
Peng-Robinson	.0997	43.9	.1233	21.7	.1333	19.2
Schmidt-Wenzel	.1146	106.8	.1083	96.0	.1360	74.0
Harmens-Knapp	.1059	117.3	.0921	124.9	.1260	91.6
Kubic	-.2561	87120.0	-.1948	41167.0	-.1284	12516.7
Heyen	.0183	54.9	.0256	70.0	.0544	48.4
CCOR	.4337	40.0	.4423	37.5	.4500	7.7
	.5805		.5581		.5553	

	25. Fluorene in Carbon Dioxide		26. Hexamethylbenzene in Ethylene		27. Hexamethylbenzene in Carbon Dioxide	
Equation	k_{12}	av. % dev.	k_{12}	av. % dev.	k_{12}	av. % dev.
Redlich-Kwong	-.0107	28.3	-.1246	28.5	-.0183	43.5
HCBKS	.0495	39.6	.0094	38.4	.0457	37.8
Soave	.1108	59.3	.0571	59.1	.1333	77.0
Peng-Robinson	.1059	49.1	.0533	41.6	.1257	63.4
Schmidt-Wenzel	.1184	18.0	.0658	55.6	.1409	28.0
Harmens-Knapp	.1083	28.6	.0557	66.7	.1295	33.4
Kubic	-.1609	27634.7	-.1911	6485.9	-.0772	1920.7
Heyen	.0142	35.6	-.0471	31.5	.0232	41.4
CCOR	.4387	44.8	.4009	73.3	.4422	33.8
	.6431		.5930		.6393	

Table VI Continued

	28. Triphenylmethane in Ethane		29. Triphenylmethane in Carbon Dioxide		30. Pyrene in Ethylene	
Equation	k_{12}	av. % dev.	k_{12}	av. % dev.	k_{12}	av. % dev.
Redlich-Kwong	-.1159	15.5	-.0481	43.1	-.0921	32.6
HCBKS	.0169	15.6	.0470	46.8	.0495	19.3
Soave	.0807	33.6	.1284	85.1	.0959	90.4
Peng-Robinson	.0720	25.5	.1195	64.7	.0945	68.5
Schmidt-Wenzel	.0969	39.5	.1561	28.5	.0571	35.6
Harmens-Knapp	.0858	52.4	.1433	43.5	.0343	51.9
Kubic	-.2423	6066.4	-.2412	525956.7	-.2059	5907.1
Heyen	-.0370	28.1	-.0194	45.1	-.0495	39.0
CCOR	.4486	28.3	.4922	62.4	.4798	25.2
	.6280		.6494		.6719	

	31. Pyrene in Carbon Dioxide	
Equation	k_{12}	ave. % dev.
Redlich-Kwong	.0180	114.2
HCBKS	.0671	163.8
Soave	.1748	244.8
Peng-Robinson	.1658	202.1
Schmidt-Wenzel	.1696	45.5
Harmens-Knapp	.1534	43.5
Kubic	-.2250	1697621.0
Heyen	.0343	104.0
CCOR	.5498	34.0
	.7319	

Table VII. Summary of Equation of State Calculations for 14 Systems with Known Tc, Pc, and ω

Equation of State	Av. % Deviation
Redlich-Kwong	17
HCBKS	24
Soave	31
Peng-Robinson	26
Schmidt-Wenzel	32
Harmens-Knapp	41
Kubic	> 10,000
Heyen	25
CCOR	34

Table VIII. Summary of Equation of State Calculations for all 31 Systems

Equation of State	Av. % Deviation
Redlich-Kwong	34
HCBKS	38
Soave	51
Peng-Robinson	43
Schmidt-Wenzel	49
Harmens-Knapp	60
Kubic	> 10,000
Heyen	40
CCOR	38

The results shown in Table VIII include all mixtures. For seventeen of these mixtures the critical properties of the solutes have to be estimated. There is a general increase in the deviation of the calculated solubility from the experimental value for all equations when the relatively uncertain critical properties are employed in the calculations. The distinction among the equations becomes blurred at the same time. Thus, the Redlich-Kwong, HCBKS, and CCOR equation appear together with percent deviations in the thirties. The Peng-Robinson, Schmidt-Wenzel, and Heyen equations form the next group with percent deviations in the forties. The total summary for all of the 31 mixtures in Table VIII appears less meaningful than the partial summary of Table VII.

Representation of the solubility data by an equation of state is examined in detail in Figures 1 through 10 for five equations - the Redlich-Kwong, HCBKS, Peng-Robinson, Heyen, and CCOR.

Calculations by the R-K equation are compared with experimental data on benzoic acid solubility in CO_2 in Figure 1. This mixture is chosen for illustration owing to the pure component properties being known, and the pressure range of the data being quite large. However, the temperature range is narrow, which is common for supercritical extraction experiments. Figure 1 shows that the equation does not describe the variation with temperature very well; the calculated temperature dependence is too weak. The calculated pressure dependence is qualitatively correct but misses the fine points.

Figure 2 shows the Redlich-Kwong equation in comparison with data on fluorene solubility in ethylene. This mixture is chosen for illustration for the large temperature range of the data. The critical properties of fluorene are estimated. The calculation describes the lower pressure data well, but departs from the data significantly at 200 bars and above.

The Heyen equation is compared with the same two mixtures in Figures 3 and 4, and the Peng-Robinson equation in Figures 5 and 6. The HCBKS equation is likewise shown in figures 7 and 8.

The CCOR equation is shown in Figure 9 for phenol in carbon dioxide, and in Figure 10 for acridine in $CClF_3$. The critical properties of phenol are known, but those of acridine are estimated. The CCOR equation calculations agree with data about equally well for both mixtures. The comparison shown in the two figures illustrates that the CCOR calculations are quite independent of the source of critical properties, estimated or known. This is also shown in Tables VII and VIII.

It seems clear from this work that the quantitative representation of supercritical solubility by the currently available equations of state with their classical one fluid mixing rules is not adequate. In view of the many applications and potential applications of supercritical extraction, there is a need for new development of an equation and mixing rules for the improved representation of supercritical extraction.

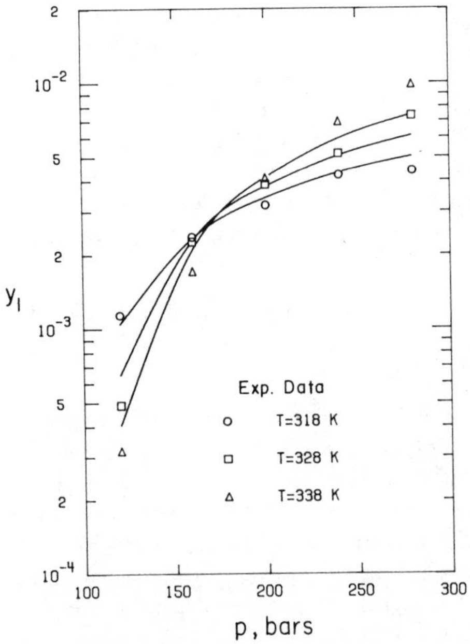

Figure 1. Solubility of Benzoic Acid in CO_2 by the Redlich-Kwong Equation

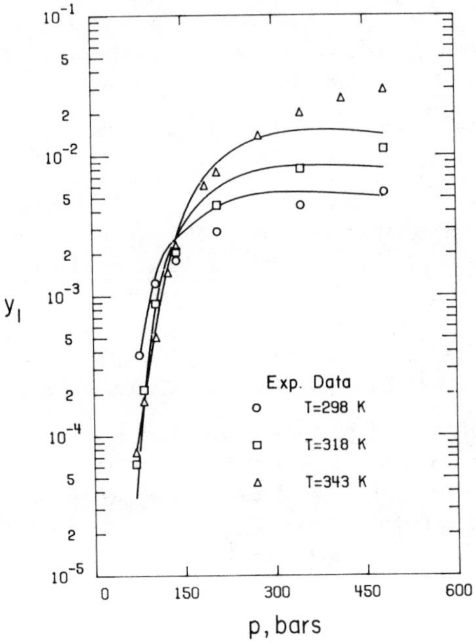

Figure 2 Solubility of Fluorene in Ethylene by the Redlich-Kwong Equation

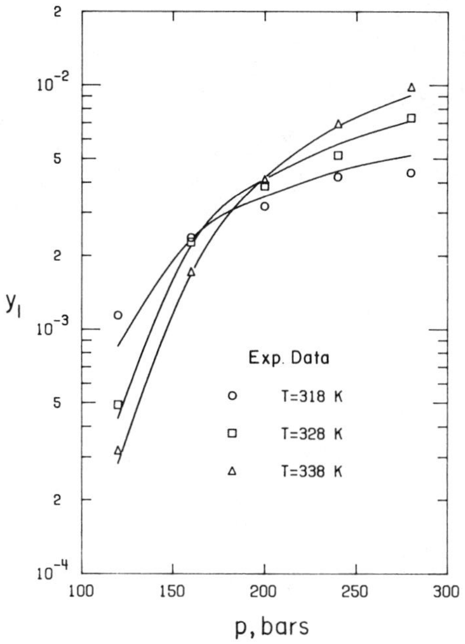

Figure 3. Solubility of Benzoic Acid in CO_2 by the Heyen Equation

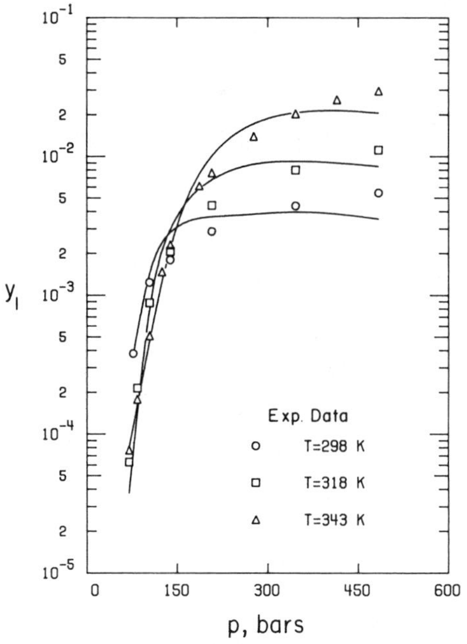

Figure 4. Solubility of Fluorene in Ethylene by the Heyen Equation

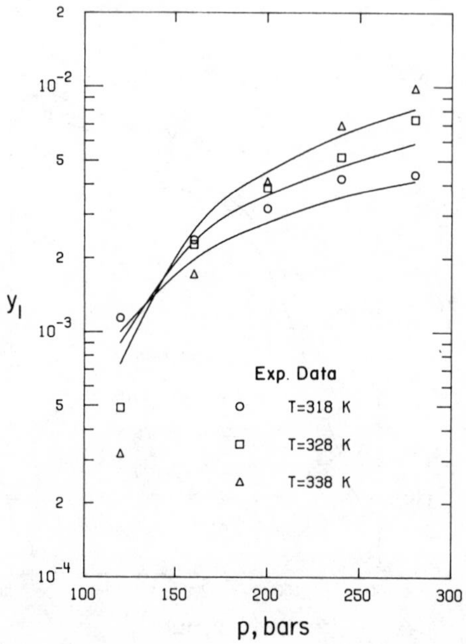

Figure 5. Solubility of Benzoic Acid in CO_2 by the Peng-Robinson Equation

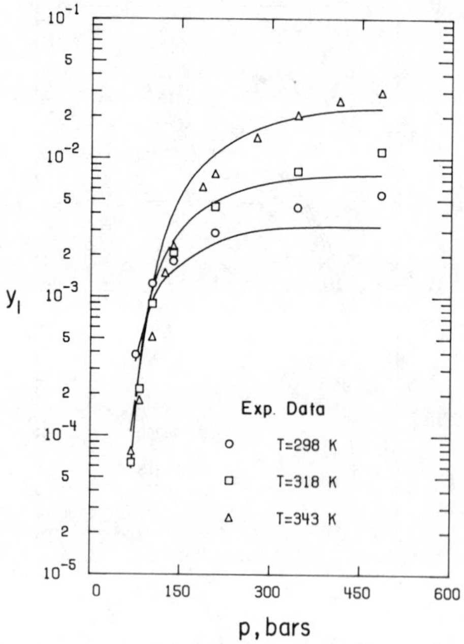

Figure 6. Solubility of Fluorene in Ethylene by the Peng-Robinson Equation

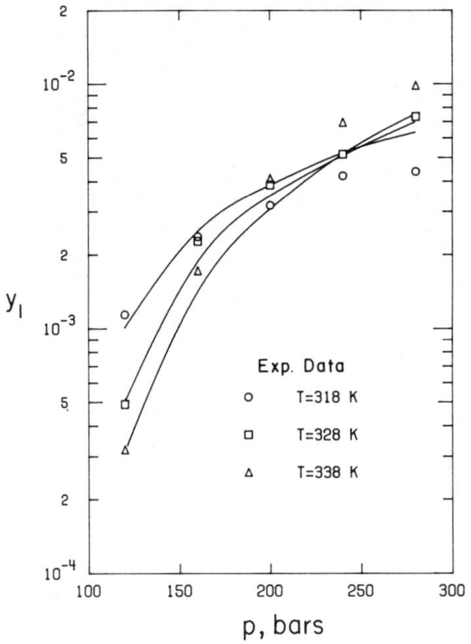

Figure 7. Solubility of Benzoic Acid in CO_2 by the HCBKS Equation

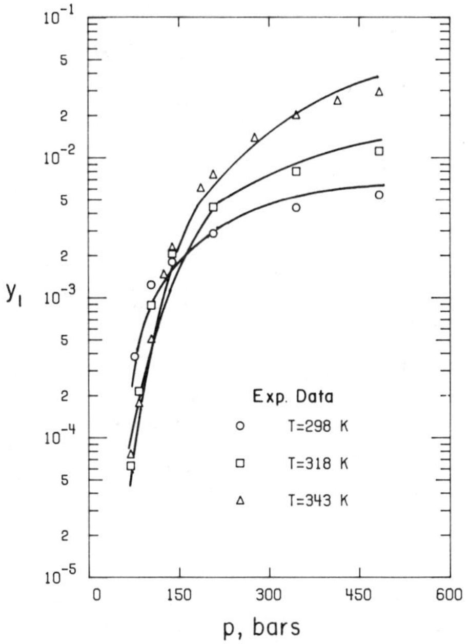

Figure 8. Solubility of Fluorene in Ethylene by the HCBKS Equation

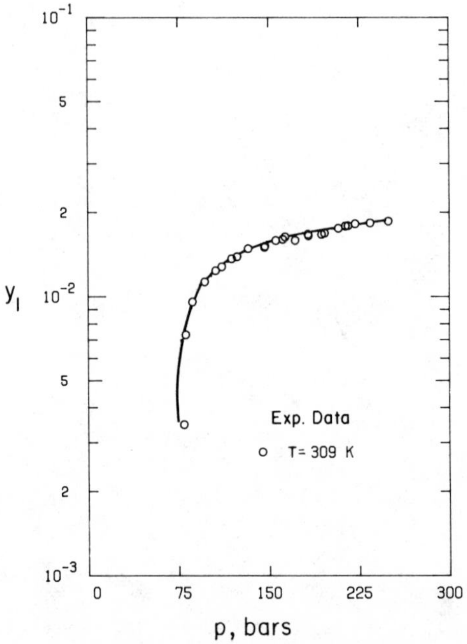

Figure 9. Solubility of Phenol in CO_2 by the CCOR Equation

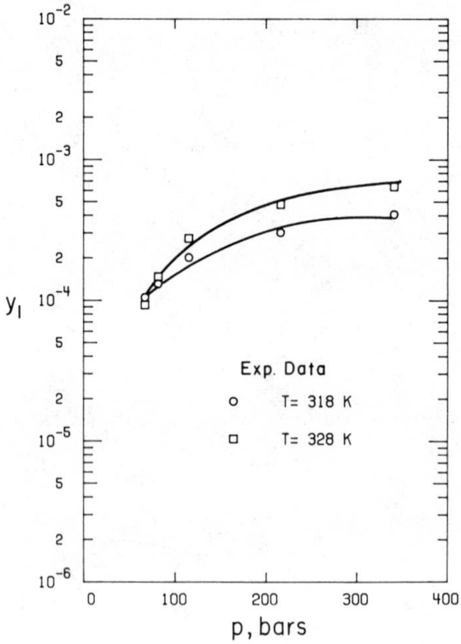

Figure 10. Solubility of Acridine in $CClF_3$ by the CCOR Equation

Glossary of Symbols

a, b, c, c^r, d	equation of state parameters
k_{ij}	binary interaction parameter
n	number of data points in a binary system
p	pressure
p^o	vapor pressure
R	gas constant
T	temperature
u	equation of state parameter
V	molar volume
V_s	solid molar volume
w	equation of state parameter
y	fluid phase mole fraction

Greek Letters

ρ	molar density
ϕ	fugacity coefficient of fluid phase
ϕ^o	fugacity coefficient of saturated vapor
ω	acentric factor
Ω	objective function

Subscripts

1	solid component
2	fluid component
c	critical property

Acknowledgment

Funds for this research were provided by the Department of Energy through contract DE-FG22-83PC60036.

Literature Cited

1. Zosel, K. Chem. Int. Ed. (1978) 17, 702.
2. Hubert, P.; Vitzthum, O.G. Agnew. Chem. Int. Ed. (1978) 17, 710.
3. Irani, C.A.; Funk, E.W. Recent Developments in Separation Science Volume 3, part A, p. 171; CRC Press, West Palm Beach, 1977.
4. Bartle, K.C., Martin, T.G.; Williams, D.F. Fuel 54, 226.
5. Martin, T.G.; Williams, D.F. Phil. Trans. Roy. Soc. London (1981) A300 , 183.
6. Maddocks, R.R., Gibson, J.; Williams, F. Chem. Eng. Progress (1979) 75, 16, 49.
7. Blessing, J.E.; Ross, D.S. Organic Chemistry of Coal ,p. 171 et. seq., Am. Chem. Soc., 1978.
8. Ross, D.S.; Blessing, J.E. Fuel (1979) 58, 423.
9. Vasilakos, N.P., Dobbs, J.M.; Parisi, A.S. Ind. Eng. Chem. Process Des. Dev. (1985) 24, 121.
10. Adams, R.M., Knebel, A.M.; Rhodes, D.E. Hydrocarbon Processing (1980), May 150.
11. Benedict, M., Webb, G.B.; Rubin, L.C. J. Chem. Phys. (1940) 8, 334.
12. Jacobsen, R.B. Ph.D. Thesis, Washington State University, Pullman, WA (1972).
13. Goodwin, R.D. "The Thermodynamic Properties of Methane from 90 to 500K at Pressures to 700 Bar", NBS Tech. Note 653 (1974).
14. Donohue, M.C.; Prausnitz, J.M. AIChE J. (1978) 24, 849.
15. Chien, C.H., Greenkorn, R.A.; Chao, K.C. AIChE J. (1983) 29, 560.
16. Klincewicz, K.M.; Reid, R.C. AIChE J. (1984) 30, 1, 137.
17. Lydersen, A.L. "Estimation of Critical Properties of Organic Compounds", Univ. Wis. Coll. Eng. Exp. Stn. Rep. 3, Madison, WI, April, 1955.
18. Edmister, W.C. Pet. Refiner. (1958) 37, 4, 173.
19. Lin, H.M.; Chao, K.C. AIChE J. (1984) 30, 981.
20. Reid, R.C. "Creativity and Challenges in Chemical Engineering", University of Wisconsin, Madison, WI (1982).

RECEIVED November 5, 1985

MOLECULES AND MODELS

8

The Generalized van der Waals Partition Function as a Basis for Equations of State

Mixing Rules and Activity Coefficient Models

S. I. Sandler, K.-H. Lee, and H. Kim

Department of Chemical Engineering, University of Delaware, Newark, DE 19716

The generalized van der Waals (GVDW) partition function provides a basis for understanding the molecular level assumptions underlying presently used equations of state, their mixing rules including the new local composition (density dependent) models, and activity coefficient models. Further, our Monte Carlo computer simulation results for mixtures of square-well fluids provide a method of testing these assumptions, many of which are found to be incorrect. Using the combination of the GVDW partition function and computer simulation results, we have formulated new equations of state and local composition models which are simple and accurate.

Thermodynamic modelling of real fluids and mixtures, using both equations of state and activity coefficients, is an area in which there is a strong temptation to introduce empiricism. The various modifications of the van der Waals equation, which now number in the hundreds, is one example of this. Statistical mechanics, on the other hand, leads to the detailed calculation of the properties of model fluids with very simple intermolecular potential functions, but few generalizations to real fluids. For the last several years we have been pursuing a different approach wherein we start from a firm statistical mechanical basis, but seek results of general validity, rather than to calculate numbers for an idealized interaction potential model. In particular, we are seeking answers to such questions as:

(1) Why are present equations of state and their mixing rules not applicable to mixtures of molecules of greatly differing functionality or size?, and

(2) What is the molecular basis for local composition and density dependent mixing rules, and how can such rules be improved? This communication is a progress report on our efforts.

The basic idea in our work is the use of the rigorous generalized van der Waals partition function, which we consider in the following sections, to understand the molecular level assumptions imbedded in presently used equations of state, their

0097–6156/86/0300–0180$06.50/0

mixing rules, and activity coefficient models, and then to test these assumptions using computer simulation. In addition, we have been using the results of our simulations to formulate new and better molecular thermodynamic models. Our progress to date will be reviewed here.

Pure Fluids

The starting point for our study of pure fluids is a new form of the generalized van der Waals partition function we have derived [1]:

$$Q(N,V,T) = \frac{1}{N!}\,\Lambda^{-3N}(q_r q_v q_e)^N\, V_f^N \exp\left(\frac{-N\Phi}{2kT}\right) \qquad (1)$$

where Q is the canonical partition function for N molecules in a volume V at temperature T, q_r, q_v and q_e are the single particle rotational, vibrational and electronic partition functions, respectively, Λ is the de Broglie wave-length and k is the Boltzmann constant. Also, V_f is the free volume defined such that

$$V_f^N = Z^{HC}(N,V) \qquad (2)$$

where Z^{HC} is the configurational integral for hard core molecules. The configuration integral for spherical particles is

$$Z(N,V,T) = \int_V \cdots \int e^{-u(\underline{r}_1,\ldots\underline{r}_N)/kT}\, d\underline{r}_1 \ldots d\underline{r}_N \qquad (3)$$

where $u(\underline{r}_1, \underline{r}_2, \ldots)$ is the intermolecular potential function when a molecule is located at position vector $\underline{r}_1$, a second molecule is located at position vector $\underline{r}_2$, etc. [The extension to nonspherical particles is considered briefly in reference 1.] Finally, E^{CONF}, the energy of interaction, for a pair-wise additive system is given by

$$E^{CONF} = \frac{N^2}{2V} \int u(r)g(r;N,V,T)d\underline{r} \qquad (4)$$

and the mean potential

$$\Phi = -\frac{2kT}{N} \int_{T=\infty}^{T} \frac{E^{CONF}}{kT^2}\, dT \qquad (5)$$

is the free energy of bringing the system from $T=\infty$ (where only hard-core repulsive forces are important) to the temperature of interest. In these last equations, u(r) is the two-body intermolecular potential and g(r;N,V,T) is the two-body radial distribution function, which is a function of density, temperature and intermolecular separation distance.

Once the partition function is known, all thermodynamic properties can be found. For example,

$$P(N,V,T) = kT\left(\frac{\partial \ln Q}{\partial V}\right)_{N,T} \qquad (6)$$

and

$$A(N,V,T) = -kT\ln Q \qquad (7)$$

The first of these equations provides a method of obtaining the volumetric equation of state from the partition function, while from the second the fundamental equation of state in the sense of Gibbs is obtained.

The importance of equations (1-5) is that they focus attention on the two quantities which are needed for thermodynamic model-building, the temperature and density dependence of the free volume V_f and the configurational energy, E^{CONF}. Indeed, these are central to obtaining good equations of state, as will be seen shortly. As an aside, we note that for the square-well potential

$$u(r) = \begin{cases} \infty & r<\sigma \\ -\varepsilon & \sigma\leq r\leq R\sigma \\ 0 & r>R\sigma \end{cases} \qquad (8)$$

we have that

$$E^{CONF} = \frac{-N^2\varepsilon}{2V}\int_{\sigma}^{R\sigma} g(r;N,V,T)d\underline{r} = \frac{-N\varepsilon}{2} N_c(N,V,T) \qquad (9)$$

and

$$\Phi = \frac{-2kT}{N}\int_{T=\infty}^{T} \frac{E^{CONF}}{kT^2}\,dT = \varepsilon kT\int_{T=\infty}^{T} \frac{N_c(N,V,T)}{kT^2}dT \qquad (10)$$

where

$$N_c(N,V,T) = \frac{N}{V}\int_{\sigma}^{R\sigma} g(r;N,V,T)d\underline{r}$$

is the number of molecules in the potential well of central molecule; we will refer to this quantity as the coordination number. Thus, in general, what is needed for applied thermodynamic modelling is the temperature and density dependence of the configuration energy; for the special case of the square-well molecule, temperature and density dependence of the coordination number suffices. This comment has special relevance to mixture local composition models, which will be considered later.

To illustrate how the generalized van der Waals partition function may be used, we restrict our attention here to the square-well fluid; the extension to real fluids is simply accomplished as described in reference 1. For the square-well fluid, the hardcore free volume is best given by the Carnahan and Starling [8] expression

$$V_f = V \cdot \exp\left(\frac{-\eta(4-3\eta)}{(1-\eta)^2}\right) \tag{11}$$

with $\eta = \pi\rho\sigma^3/6$, though the less accurate van der Waals expression

$$V_f = V-b \tag{12}$$

with $b = 4\eta$ has been incorporated into many cubic equations of state used in engineering.

There has been little consideration given to the coordination number in the thermodynamic literature. Implicit in the van der Waals model is the assumption that the coordination number, that is the number of molecules interacting with a central molecule, is linearly proportional to the density,

$$N_c = \frac{N}{V} C = \rho C \tag{13}$$

where $C = \int_{\sigma}^{R\sigma} g(r;N,V,T)d\underline{r}$ is assumed to be independent of temperature and density. In this case,

$$E^{CONF} = -\frac{N^2 \varepsilon C}{2V}, \quad \Phi = -\frac{N\varepsilon C}{V}, \quad \text{and } P = P^{HS} - \frac{aN^2}{V^2} \tag{14}$$

where $a = C\varepsilon/2$. Using the critical point conditions, we find that

$$C = \frac{27}{32}\left(\frac{kT_c}{\varepsilon}\right)^2 \left(\frac{\varepsilon}{P_c}\right).$$

At low densities, we have that

$$\lim_{\rho \to 0} g(r;N,V,T) = e^{-u(r)/kT} \tag{15}$$

Assuming that this equation is valid at all densities for the square-well fluid yields

$$N_c = \frac{N}{V}\frac{4\pi\sigma^3}{3}(R^3-1)e^{\varepsilon/kT} \tag{16}$$

and an equation of state whose volume dependence is like the van der Waals equation, but with the a parameter given by the temperature dependent function

$$a = \frac{2\pi}{3}\sigma^3 (R^3-1)kT(e^{\varepsilon/kT}-1) \tag{17}$$

The coordination number models built into other equations of state can also be obtained, and several are shown in Table I. The first three, the van der Waals, Redlich-Kwong [3] and Peng-

Robinson [4] equations, are representative of commonly used cubic equations of state, and the fourth is a twenty-seven constant equation of the extended virial form developed by Alder et al. [2] to fit their computer simulation data for the square-well fluid. It is interesting to note that it is this equation, with slightly modified parameters, that is used in the perturbed hard chain theory of Beret and Prausnitz [5] and Donohue and Prausnitz [6], and largely gives rise to its complexity. The remaining two equations will be discussed shortly.

It is now of interest to compare several of these coordination number models with data for the square-well fluid. We have obtained such data from our own Monte Carlo simulations described elsewhere [7]. Some of these data, together with the predictions of the various cubic equation of state models, are shown as a function of reduced temperature kT/ε and reduced density $\rho\sigma^3$ in Figure 1. This figure is for an R = 1.5 square-well fluid, and we have used the estimates of Alder et al [2] for the critical properties of the square-well fluid

$$\frac{kT_c}{\varepsilon} = 1.260, \quad \frac{V_c}{N\sigma^3} = 3.006 \quad \text{and} \quad \frac{P_c\sigma 3}{\varepsilon} = 0.120$$

to interrelate the square-well and critical parameters. As we can see from the figure, none of the coordination number models built into existing cubic equations of state are very good.

Based on a lattice-gas model, we have developed the simple expression [7]

$$N_c = \frac{Z_m V_o \exp(\varepsilon/2kT)}{V + V_o \{\exp(\varepsilon/2kT)-1\}} \tag{18}$$

where Z_m is the close-packed coordination number (Z_m = 18 for R = 1.5) and $V_o = N\sigma^3/\sqrt{2}$. Figure 1 also shows the coordination number predictions for this model, which are better than any of the ones considered previously. Using this coordination number model in the generalized van der Waals partition function we obtain

$$\frac{P}{NkT} = \frac{P^{HS}}{NkT} - \frac{Z_m V_o(e^{\varepsilon/2kT}-1)}{V(V+V_o[e^{\varepsilon/2kT}-1])} \tag{19}$$

Note that the density dependence of the attractive term in this new equation is like the Redlich-Kwong equation, but the denominator is temperature dependent. Thus, at low density, N_c should be proportional to exp (ε/kT) while in equation (18) it is proportional to exp $(\varepsilon/2kT)$, which from our data appears appropriate at high density. Therefore, an empirical modification of Eq. (18) is

$$N_c = \frac{Z_m V_o \exp(\varepsilon/\alpha kT)}{V + V_o\{\exp(\varepsilon/\alpha kT)-1\}} \qquad \text{where } \alpha = \frac{V}{V-V_o} \tag{20}$$

TABLE I. Coordination Number Models for the Square-Well Fluid in Various Equations of State

Equation of State	N_c
van der Waals	$\frac{27}{32}\left(\frac{kT_c}{\varepsilon}\right)^2 \cdot \left(\frac{\varepsilon}{Pc\sigma^3}\right) \cdot \rho\sigma^3$
Redlich-Kwong	$\frac{3(0.4278)}{0.0867}\left(\frac{kT_c}{\varepsilon}\right)^{1.5}\left(\frac{\varepsilon}{kT}\right)^{0.5} \ln[1+0.0867\left(\frac{\varepsilon}{Pc\sigma^3}\right)\left(\frac{kT_c}{\varepsilon}\right)\rho\sigma^3]$
Peng-Robinson	$\frac{0.45724}{0.07789\sqrt{2}}\left(\frac{kT_c}{\varepsilon}\right)(1+\kappa)[1+\kappa\{1-\sqrt{\frac{T}{T_c}}\}] \ln[\frac{1+(\sqrt{2}+1)\Delta}{1-(\sqrt{2}-1)\Delta}]$ where $\kappa = 0.37464 + 1.54226\omega - 0.26992\omega^2$ and $\Delta = 0.07780\left(\frac{kT_c}{\varepsilon}\right)\left(\frac{\varepsilon}{Pc\sigma^3}\right)(\rho\sigma^3)$
Alder	$-2\sum_{n=1}\sum_{m=1}\frac{nA_{nm}}{(\sqrt{2})^m}\left(\frac{\varepsilon}{kT}\right)^{n-1}(\rho\sigma^3)^m$
Eqn. (18)	$\frac{Z_m V_o \exp(\varepsilon/2kT)}{V + V_o\{\exp(\varepsilon/2kT)-1\}}$
Eqn. (20)	$\frac{Z_m V_o \exp(\varepsilon/\alpha kT)}{V + V_o\{\exp(\frac{\varepsilon}{\alpha kT})-1\}}$ where $\alpha = \frac{V}{V-V_o}$

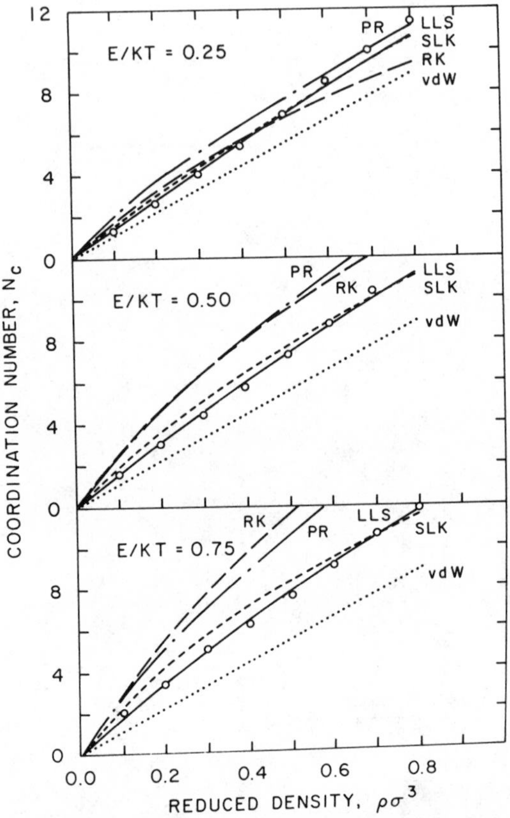

Figure 1. Predictions of coordination number for square-well fluid:
(o) Monte Carlo simulation; (····) van der Waals; (—— ——) Redlich-Kwong; (—— - ——) Peng-Robinson; (——) Eqn. (18); (----) Eqn. (20)

The predictions of this model are also shown in Figure 1, where they are found to be in very good agreement with computer simulation. Using this expression we obtain

$$\frac{P}{NkT} = \frac{P^{HS}}{NkT} - \frac{Z_m}{2} \left\{ \frac{V_o \ln[1 + \frac{V_o}{V} (e^{\varepsilon/\alpha kT} - 1)]}{(V-V_o)^2} \right.$$

$$\left. + \frac{V_o[(1 - \frac{\varepsilon}{kT}\frac{V_o}{V})e^{\varepsilon/\alpha kT} - 1]}{(V-V_o)[V + V_o(e^{\varepsilon/\alpha kT} - 1)]} \right\} \tag{21}$$

To test the accuracy of these and other equations of state, we compare, in Table II, the predicted compressibility factor for the square-well fluid with the data of Alder et al. [2]. For this comparison we have used the Carnahan and Starling [8] expression for the hard-sphere pressure, and restricted our attention to the vapor and liquid one-phase regions of the Alder simulation data. We have also included in Table II the predictions of the equations of Aim and Nezbeda [9] and Ponce and Renon [10], both of which are meant to describe the square-well fluid. It is interesting to note that equation (19) which is one of the simplest equations considered, is also the most accurate. Its accuracy is further demonstrated in

Table II. Absolute Average Deviation in the Compressibility Factor of the Square-Well Fluid as Predicted by Various Equations of State

Equation of State	Average Absolute Deviation
van der Waals	8.380
Redlich-Kwong	1.332
Peng-Robinson	2.159
Alder, et al.	0.380
Aim-Nezbeda w/o 3 body	0.418
Aim-Nezbeda with 3 body	0.323
Ponce-Renon	0.378
This work, Eqn. (19)	0.240
This work, Eqn. (21)	0.269

Figure 2 where the compressibility factor along several isotherms for various equations of state are plotted as a function of density together with the results of our Monte Carlo simulations.

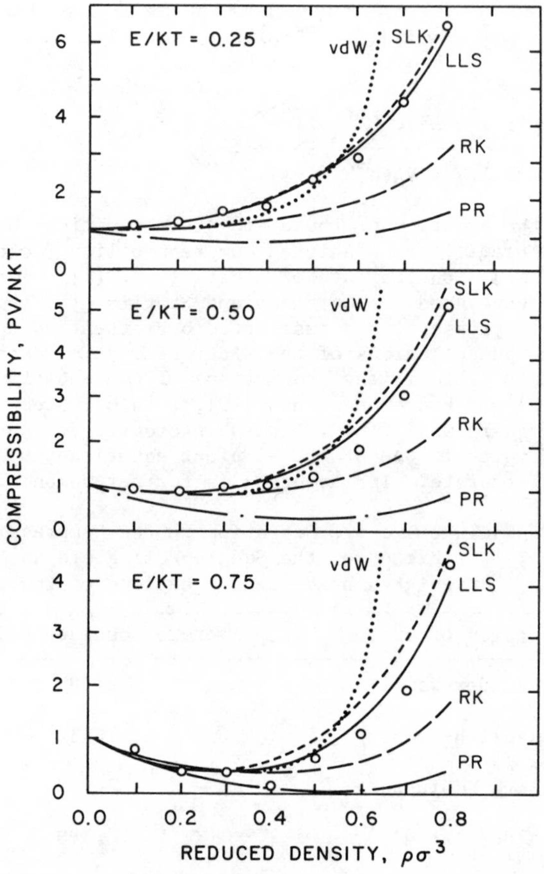

Figure 2. Predictions of compressibility for square-well fluid: (o) Monte Carlo simulation; (••••) van der Waals; (—— ——) Redlich-Kwong; (—— - ——) Peng-Robinson; (——) Eqn. (19); (----) Eqn. (21)

Based on our results, two conclusions can be drawn. First, is that there is an advantage to using the generalized van der Waals partition function as a basis for developing equations of state. This is demonstrated by the fact that by using this approach, we could develop simple equations with no adjustable parameters, that are much more accurate than the empirical or semitheoretical equations for the square-well fluid that had been used heretofore. Second, from this analysis, we have developed an equation of state which, as a result of its accuracy, is an obvious candidate for generalization to real fluids.

Mixtures and Local Compositions

Mixture behavior, and especially local composition effects and density dependent mixing rules, are of considerable practical interest. Here the generalized van der Waals partition function is of great utility in understanding the assumptions contained in presently used models for activity coefficients and equation of state mixing rules, and as a basis for improving upon them.

The extension of the generalized van der Waals partition function of the previous section to mixtures yields [1]

$$Q(N_1,N_2,\ldots V,T) =$$

$$\prod_i \left[\frac{1}{N_i!}\left(\frac{q_r q_v q_e}{\Lambda^3}\right)_i^{N_i}\right] V_f^N(T,V,N_1,N_2,\ldots)\exp\left(\frac{-N\Phi(T,V,N_1,N_2\ldots)}{2kT}\right) \tag{22}$$

where

$$\Phi(N_1,N_2,\ldots V,T) = \frac{-2kT}{N}\int_{T=\infty}^{T} \frac{E^{CONF}(N_1,N_2,\ldots,V,T)}{kT^2}dT \tag{23}$$

with

$$E^{CONF}(N_1,N_2,\ldots,V,T) = \sum_i \sum_j E_{ij}^{CONF}(N_1,N_2,\ldots,V,T) \tag{24}$$

and

$$E_{ij}^{CONF}(N_1,N_2,\ldots,V,T) = \frac{N_i N_j}{2V}\int u_{ij}(r)g_{ij}(N_1,N_2,\ldots,V,T;r)d\underline{r}$$
$$= \frac{N^2}{2V} x_i x_j \int u_{ij}(r)g_{ij}(N_1,N_2,\ldots,V,T;r)d\underline{r} \tag{25}$$

The form of Eqn. (22) is useful for the study of mixture equations of state and parameter mixing rules. For the study of excess free energy or activity coefficient models, it is more convenient to use the difference between Equation (22) and the analogous equations for the pure components to obtain the expression below for mixing at constant temperature and total volume (so that $V = \sum V_i$)

$$A^{EX}_{T,V} = A_{MIX}(T,V, N_1,N_2,\ldots)-\sum_i A_i(T,V_i,N_i)-kT\sum_i N_i \ln x_i$$

$$= -kT\ln\left[\frac{Q(T,V,N_1,N_2,\ldots)}{\pi_i Q_i(T,V_i,N_i)}\right] -kT\sum_i \ln\left(\frac{N_i}{N}\right)^{N_i}$$

$$= -kT\ln\left[\frac{Z(T=\infty,V,N_1,N_2,\ldots)}{\pi_i\left[Z_i(T=\infty,V_i,N_i)\left(\frac{N}{N_i}\right)^{N_i}\right]}\right] \tag{26}$$

$$- kT \int_{T=\infty}^{T} \left[\frac{E^{CONF}(T,V,N_1,N_2\ldots)-\sum_i E_i^{CONF}(T,V_i,N_i)}{kT^2}\right] dT$$

or

$$\frac{A^{EX}_{T,V}}{T} = \left.\frac{A^{EX}_{T,V}}{T}\right|_{T=\infty} - \int_{T=\infty}^{T} \left[\frac{E^{CONF}(T,V,N_1,N_2\ldots)-\sum_i E_i^{CONF}(T,V_i,N_i)}{T^2}\right] dT \tag{27}$$

The first term on the righthand side of this equation is the excess free energy of mixing when only repulsive forces are important. This term accounts for size and shape effects, and may be modelled by the Flory-Huggins [11], Guggenheim-Staverman [12] or other suitable expressions. The second term is the excess free energy of mixing for bringing the system from $T=\infty$ to the temperature of interest at constant composition and total volume. This term is the contribution of the soft portions of the intermolecular potential to the free energy of mixing.

As shown by Hildebrand [13], at low and moderate pressures, $A^{EX}_{T,V} \simeq G^{EX}_{T,P}$, where $G^{EX}_{T,P}$ is the excess Gibbs free energy change on mixing at constant temperature and pressure. It is from $G^{EX}_{T,P}$ that activity coefficient expressions are most easily obtained.

We want to stress that, in Equations (22-27), it is the configurational energies

$$E_{ij}^{CONF} = \frac{2\pi N_i N_j}{V} \int_0^\infty u_{ij}(r)g_{ij}(r)r^2 dr \tag{28}$$

that are important, rather than the (ambiguously defined) species coordination numbers

$$N_{ij} = \frac{4\pi N_i}{V} \int_0^{L_{ij}} g_{ij}(r)r^2dr \tag{29}$$

or local mole fractions

$$x_{ij} = \frac{N_{ij}}{\sum_i N_{ij}} \tag{30}$$

There are two problems with the species coordination numbers or mole fractions. First, since the cut-off distance L_{ij} is somewhat arbitrary, neither N_{ij} nor x_{ij} are uniquely defined. Second, it is the clearly defined configurational energies which appear in the generalized van der Waals analysis, and, in general, these cannot be gotten from the local compositions. It is only for square-well molecules, with its well-defined range, for which

$$E_{ij}^{CONF} = \frac{-N_j}{2} N_{ij}\varepsilon_{ij} \tag{31}$$

with

$$N_{ij} = \frac{N_i}{V} \int_{\sigma_{ij}}^{R\sigma_{ij}} g_{ij}(N_1,N_2,\ldots,V,T;r)d\underline{r} \tag{32}$$

that the local species-species coordination numbers (or <u>local compositions)</u> are unambiguous and contain equivalent information to the configurational energies. This is an important point since most local composition studies in the past for the Lennard-Jones 6-12 fluid [14, 15, 16] have concentrated on the species coordination numbers or local mole fractions, rather than the configurational energies, which are really needed.

A very useful characteristic of Equations (22-25) is that they are in a form which makes it possible to determine the applied thermodynamic model which results from any molecular level assumption. We demonstrate this here by considering a mixture of square-well molecules. In this case, Equation (31) is applicable, and the local coordination numbers, N_{ij}, are useful in thermodynamic modelling.

Clearly, different models for the local coordination number will give different equations of state, mixing rules, and activity coefficient models. For example, if we were, in analogy with the pure fluid van der Waals model, to assume that

$$N_{ij} = \frac{N_i}{V} C \tag{33a}$$

where C is a constant (which is reasonable only for similar size molecules), or even

$$N_{ij} = \frac{N_i}{V} C_{ij} \tag{33b}$$

where C_{ij} can have a different numerical value for 1-1, 2-2, 1-2, etc., interactions, we find that the van der Waals equation of state is again obtained with the attraction parameter given by

$$a = \sum \sum x_i x_j a_{ij} \tag{34}$$

Equation (34) is referred to as the van der Waals one-fluid mixing rule. Since it is known from statistical mechanics [17] that the second virial coefficient depends quadratically on mole fraction (and not on surface fraction or volume fraction), an important boundary condition on any cubic equation of state mixing rule is that the a and b parameters for the mixture at low density reduce to the one-fluid form.

Proceeding, we could instead assume that

$$N_{ij} = \frac{N_i}{V} C_{ij} \, e^{\varepsilon_{ij}/kT} \tag{35}$$

As indicated in the previous section, Equation (35) is the exact low density expression for the local coordination number in a square-well fluid if

$C_{ij} = \frac{4\pi}{3} \sigma_{ij}{}^3 (R_{ij}{}^3 - 1)$. Assuming this equation is valid for both the pure fluid and the mixture, we find that once again the van der Waals equation of state and one-fluid mixing rules apply, except that in this case

$$a_{ij} = \frac{kT}{2} C_{ij} (e^{\varepsilon_{ij}/kT} - 1) \tag{36}$$

instead of being a constant.

Continuing further, we find that while different assumptions of the type

$$N_{ij} = \frac{N_i}{V} f(N, V, T, \varepsilon_{ij}, \sigma_{ij}) \tag{37}$$

where f is any function of N, V, T and the potential parameters of all species, but independent of composition, lead to different forms of the attractive term in the equation of state, in each case the van der Waals one-fluid mixing rule of Equation (34) applies at all densities.

Defining N_{ci} to be the total coordination number for a species i molecule in the mixture, from Equation (33a) we have (for similar size molecules) that

$$N_{ci} = N_{ii} + N_{ji} = \frac{N}{V} C = \overset{\circ}{N}_{ci} \tag{38}$$

so that

$$N_{ii} = x_i \overset{\circ}{N}_{ci} \qquad N_{ji} = x_j \overset{\circ}{N}_{ci} \tag{39}$$

where $\overset{\circ}{N}_{ci}$ is the coordination number for pure species i molecules at the same density. Using these results in Equation (27) yields

$$A^{EX}_{T,V} = T \cdot \left.\frac{A^{EX}_{T,V}}{T}\right|_{T=\infty} + Nx_1x_2\theta \tag{40}$$

where $\Theta = \frac{1}{2}\overset{\circ}{N}_{c1}(\varepsilon_{11}-\varepsilon_{12}) + \frac{1}{2}\overset{\circ}{N}_{c2}(\varepsilon_{22}-\varepsilon_{12})$. For molecules of similar size and shape, the first term on the right-hand side of Equation (40) vanishes, and we have the one constant Margules expansion.

If, on the other hand, Equation (33b) is assumed, we then find that each species coordination number is a linear function of mole fraction,

$$N_{cj} = \overset{\circ}{N}_{cj} + x_i\delta_j \tag{41}$$

that

$$N_{ij} = x_iN_{cj} \qquad N_{jj} = x_jN_{cj} \tag{42}$$

and that

$$A^{EX}_{T,V} = T \cdot \left.\frac{A^{EX}_{T,V}}{T}\right|_{T=\infty} + Nx_1x_2\Theta' + Nx_1x_2(x_1-x_2)\Phi \tag{43}$$

with $\Theta' = \Theta - \frac{1}{4}\,[\delta_1(\varepsilon_{11}+\varepsilon_{21}) + \delta_2(\varepsilon_{22}+\varepsilon_{12})]$ and $\Phi = \frac{1}{4}\,[\delta_2(\varepsilon_{22}-\varepsilon_{12})-\delta_1(\varepsilon_{11}-\varepsilon_{21})]$. This is the two-constant Margules expansion if the T=∞ term is neglected, or an augmented two-constant Margules expansion if the Flory-Huggins

$$\left.\frac{A^{EX}_{T,V}}{T}\right|_{T=\infty} = k\sum N_i \ln \frac{\phi_i}{x_i} \tag{44}$$

or similar expressions are used for $\left.\frac{A^{EX}_{T,V}}{T}\right|_{T=\infty}$. In Equation (44), $\phi_i = x_i\nu_i/\sum x_j\nu_j$, where ν_i is some measure of the molecular volume of species i molecules.

If we assume that Equation (35) is applicable to each species in the mixture, then Equation (41) is again obtained, and the two-constant Margules expansion results, though with slightly different expressions for the Margules coeffcients. Indeed, whenever we start from local coordination number models of the type we have been considering, the excess Helmholtz (or Gibbs) free energy is always of the Margules form, and the equation of state mixing rules are of the van der Waals one-fluid form at all densities.

We now consider another class of local composition assumptions. Suppose, instead of Equations (33) we assume

$$\frac{N_{ij}}{N_{jj}} = \frac{N_i}{N_j}\frac{C_{ij}}{C_{jj}} = \frac{x_i}{x_j}\frac{C_{ij}}{C_{jj}} \tag{45}$$

where C_{ij} is some constant. [A common choice for the constants C is some measure of the molecular volume ν, so that $C_{ii} = C_{ij} = \nu_i$, and $C_{jj} = C_{ji} = \nu_j$. In this case the ratio of the local coordination numbers, that is the local composition ratio, is equal to the ratio of volume fractions.] By itself, Equation (45) is insufficient, since each species coordination number (i.e., the separate N_{ij} and N_{jj}) is needed in the generalized van der Waals partition function analysis. Implicit in many activity coefficient models is the <u>additional and quite separate assumption</u> that the total coordination number for each species in the mixture, $N_{cj} = N_{ij} + N_{jj}$, is independent of mole fraction, and therefore equal to the pure component coordination number. That is, the assumption is

$$N_{cj} = \overset{\circ}{N}_{cj} \tag{46}$$

Equation (46) has it origin in lattice theory where a molecule is presumed to have a fixed coordination number; it may be valid for similar size molecules at high density, but is incorrect for molecules of different size, or at low and moderate densities.

Solving Equations (45 and 46) yields

$$N_{ij} = \frac{x_i \Lambda_{ij} \overset{\circ}{N}_{cj}}{x_j + x_i \Lambda_{ij}} \quad \text{and} \quad N_{jj} = \frac{x_j \overset{\circ}{N}_{cj}}{x_j + x_i \Lambda_{ij}} \tag{47}$$

where $\Lambda_{ij} = C_{ij}/C_{jj}$. [Equations for N_{ji} and N_{ii} are gotten from the expressions above by an interchange of indices.] Assuming that $\Lambda_{ij} = \nu_i/\nu_j$ yields

$$G^{EX}_{T,P} = A^{EX}_{T,V} = T \cdot \left.\frac{A^{EX}_{T,V}}{T}\right|_{T=\infty} + \frac{N}{2}\frac{x_i \nu_i x_j \nu_j}{x_i \nu_i + x_j \nu_j}\,\Delta \tag{48}$$

where

$$\Delta = \frac{(\varepsilon_{ii} - \varepsilon_{ji})}{\nu_i}\overset{\circ}{N}_{ci} + \frac{(\varepsilon_{jj} - \varepsilon_{ij})}{\nu_j}\overset{\circ}{N}_{cj}$$

If the T=∞ term is set equal to zero, Equation (48) is the excess free energy expression which gives rise to the van Laar and Hildebrand-Scatchard Regular Solution activity coefficient models.

Alternatively, using Equation (44) the Flory-Huggins activity coefficient model is obtained. One could instead choose the molecular surface areas for the C_{ij}'s instead of molecular volumes, to obtain a variant of these models in which surface area fractions rather than volume fractions appear.

If, instead, we use Equation (47) and the assumption that the total coordination number is a linear function of density in Equation (22) we obtain the following equation of state mixing rule for the a parameter

$$a = (\sum x_i \nu_i) \cdot \sum \sum \phi_i \phi_j \frac{a_{ij}}{\nu_j} \tag{49}$$

This mixing rule does not satisfy the low density van der Waals one-fluid boundary condition (unless $\nu_i = \nu_j$), which is not surprising since Equation (47) on which it is based is incorrect at low density.

Next consider yet another local composition model

$$\frac{N_{ij}}{N_{jj}} = \frac{N_i}{N_j} \frac{C_{ij}}{C_{jj}} e^{(\varepsilon_{ij}-\varepsilon_{jj})/kT} \tag{50}$$

which is the ratio form of Equation (35). If we choose $C_{ii} = C_{ij} = \nu_i$, we have the local composition assumption of Wilson. If one further assumes that the total coordination number is constant (Equation (46)) then Equations (47) are again obtained, but with

$$\Lambda_{ij} = \frac{C_{ij}}{C_{jj}} e^{(\varepsilon_{ij}-\varepsilon_{jj})/kT} \tag{51}$$

Using these results in Equation (27) leads to the Wilson activity coefficient model; when used in Equation (22) with the assumption that the total coordination number is a linear function of density we obtain an a parameter mixing rule in the form of Equation (49) which, as already mentioned, does not satisfy the low density one-fluid mixing rule boundary condition.

While the various local composition models considered so far lead to different equation of state mixing rules and activity coefficient models, none result in density dependent mixing rules which have been of much interest lately. From the analysis using the generalized van der Waals partition function, it is evident that density dependent mixing rules can only result from a density dependent local composition model. Two such density dependent mixing rules have been suggested recently. The first, due to Whiting and Prausnitz [18] is of the form

$$\frac{N_{ij}}{N_{jj}} = \frac{N_i}{N_j} \frac{C_{ij}}{C_{jj}} e^{(\varepsilon_{ij}-\varepsilon_{jj})N_{cj}/2kT} \tag{52}$$

where the total coordination number for each species $N_{cj} = N_{ij} + N_{jj}$ is assumed to be linearly dependent on density, but independent of mole fraction. This local composition model leads to the following mixing rule

$$a = \sum_{i=1} x_i \{ \sum_{j=1} x_j \frac{a_{ji} b_{ii}}{b_{ji}} \exp \left(\frac{2a_{ji}\xi_i}{RTb_{ji}}\right) / \sum_{k=1} x_k \exp \left(\frac{2a_{ki}\xi_i}{RTb_{ki}}\right) \} \quad (53)$$

where b is a size parameter and $\xi_i = b_{ii}N/4V$. Also an activity coefficient model is obtained which is similar to the three parameter Wilson equation. However, this model should not be expected to be very good since it predicts that the effect of attractive forces (the exponential term in Equation (53)) vanishes at low density (where $N_{cj} \to 0$), and is largest at high density. This is opposite to what should be expected since Equation (50) is correct at low density, and as attractive forces are of little importance at high densities, the exponential term should approach unity (rather than increase in value) in this limit.

The recent "practical" local composition model of Hu et al. [19]

$$N_{ij} = \frac{4\pi}{3}(R_{ij}^3-1)\sigma_{ij}^3 \frac{N_i}{V} \exp[\alpha\varepsilon_{ij}/kT] \quad (54)$$

where

$$\alpha = 0.60-0.58 \left(\rho\sum_i x_i\sigma_{ii}^3\right)^{0.1865}$$

has a density dependence which is qualitatively in agreement with the observations above. This local composition model leads to the density dependent mixing rule given in reference 19. Many of the local composition models discussed above are summarized in Table III.

Which Local Composition Model Is Best?

In the previous section we considered numerous local composition models, and it is reasonable to ask which is best? To answer this question we have been calculating local compositions in mixtures of square-well fluids using Monte Carlo simulation and integral equation methods. We report briefly some of our simulation results here.

In Figure 3 we have plotted the quantity

$$\frac{N_{21}}{N_{11}} \frac{N_1}{N_2} \text{ and } \frac{N_{12}}{N_{22}} \frac{N_2}{N_1}$$

as a function of density and composition for a mixture of square-well molecules of equal diameter σ, but different well depths. The points represent our simulation results, and the arrows are the exact low density limits. For completely random mixing, both these ratios would be unity, which is clearly not the case. Since the molecules are of equal diameter, $\nu_1 = \nu_2$, and the local composition ratios of Equation (45) also reduce to a unity at all densities. The Wilson local composition model of Equation (50) is independent of density, and is seen to be correct at zero density, but to

Table III

Local Coordination Number Model	Equation of State Mixing Rule	Free Energy
$N_{ij} = \frac{N_i}{V} C$ $N_{jj} = \frac{N_j}{V} C$	vdW 1-fluid	1-constant Margules
$N_{ij} = \frac{N_i}{V} C_{ij}$ $N_{jj} = \frac{N_j}{V} C_{jj}$	vdW 1-fluid	2-constant Margules
$N_{ij} = \frac{N_i C_{ij}}{V} e^{\varepsilon_{ij}/kT}$ $N_{jj} = \frac{N_j C_{jj}}{V} e^{\varepsilon_{jj}/kT}$	vdW 1-fluid	2-constant Margules
$\frac{N_{ij}}{N_{jj}} = \frac{N_i}{N_j} \frac{\nu_i}{\nu_j}$ $N_{ij} + N_{jj} = N^{\circ}_{cj}$	$a = (\sum x_i \nu_i) \sum\sum \phi_i \phi_j \frac{a_{ij}}{\nu_j}$ if $N^{\circ}_{cj} = \frac{N}{V} C_j$	van Laar, Scatchard-Hildebrand Regular Solution and Flory-Huggins
$\frac{N_{ij}}{N_{jj}} = \frac{N_i}{N_j} \frac{\nu_i}{\nu_j} e^{(\varepsilon_{ij}-\varepsilon_{jj})/kT}$ $N_{ij} + N_{jj} = N^{\circ}_{cj}$	$a = (\sum x_i \nu_i) \sum\sum \phi_i \phi_j \frac{a_{ij}}{\nu_j}$ if $N^{\circ}_{cj} = \frac{N}{V} C_j$	Wilson

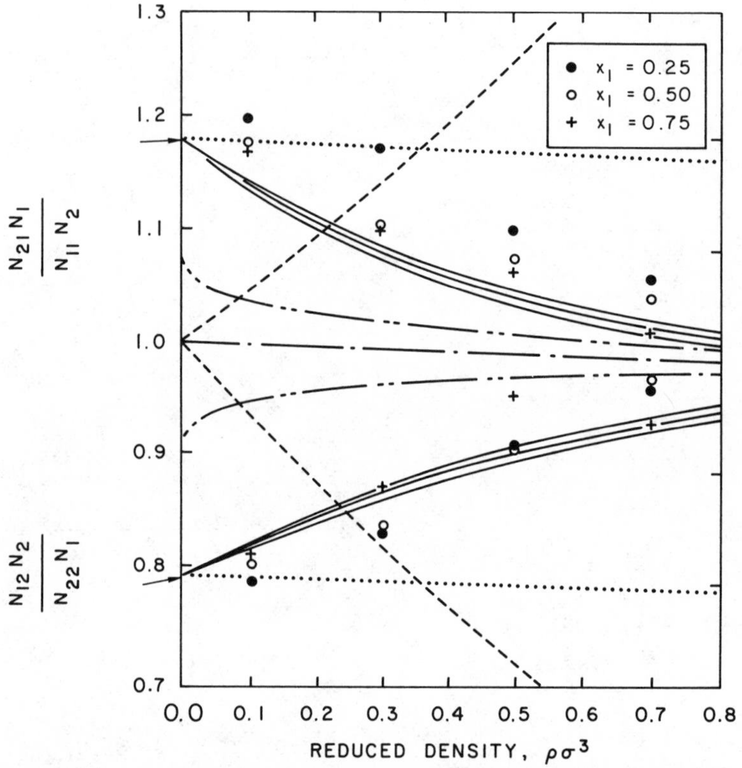

Figure 3. Extent of local composition in square-well mixture for $\varepsilon_{11}/kT = 0.4$, $\varepsilon_{22}/kT = 0.8$, $\sigma_{11} = \sigma_{22}$, $R_1 = R_2 = 1.5$: (●, O and +) Monte Carlo simulation; (····) Wilson (1964); (----) Whiting and Prausnitz (ref. 18); (—— -——) Hu et al. (ref. 19); (——) Lee, Sandler and Patel (ref. 21); (————) random mixing. (→ shows the theoretical low density limit)

overpredict local composition effects at all other densities. random mixing at low density, and strong local composition effects at high density. As can be seen, this is completely opposite from the computer simulation results which are in agreement with both the exact low density result and the Weeks et al. [20] conclusion that attractive forces are of little importance in determining local composition at high density. Also, the Whiting-Prausnitz model shows no composition dependence. Since the underlying local composition mixing rule is incorrect, it is not surprising that in application, the density dependent mixing rule of Equation (53) is found to be of little utility. Recently, Hu et al. [19] have proposed the local composition model of Equation (54); the results of this model also appear in Figure 3. There we see that for equal-size molecules, the Hu et al. model predicts local composition effects which qualitatively have the correct density dependence, but are too small at all densities. The model of Hu et al., like the other models considered previously, shows no composition dependence in the ratios we have plotted.

Also shown in Figure 3 are the predictions for the local composition ratios of Equation (20) from a model we have proposed recently [21]. This model has the correct low density and high density limits, a slight composition dependence, and is in better agreement with simulation results than the other models considered here. At present we are considering the extension of this model to molecules of unequal size, and this work will be presented separately [22]. Also, we consider elsewhere [23] the composition variation of the total coordination number in mixtures of unequal size molecules.

Conclusions

First, we have shown the utility of the generalized van der Waals partition function in that it allows us to interrelate molecular level assumptions to applied thermodynamic models. This was used, in Section II, to ascertain the coordination number models used in a number of equations of state. These coordination number models were tested against our Monte Carlo simulation data for a square-well fluid, and none were found to be satisfactory. A new coordination number model was proposed, and this was found to lead to an equation of state, with no adjustable parameters, which is more accurate for the square-well fluid than multiconstant equations currently in use.

In Section III, the generalized van der Waals partition function for mixtures was used to identify the molecular-level local composition and other assumptions imbedded in commonly used equation of state mixing rules and activity coefficient models. We then compared these assumptions for local composition effects with the results of our own Monte Carlo simulation studies for mixtures of square-well molecules in Section IV. There we found that the models currently in use do not properly account for nonrandom mixing due to attractive energy effects. This suggests that better local composition models are needed. Once they are obtained, the generalized van der Waals partition function will again be useful in developing the macroscopic thermodynamic models which results from these molecular level assumptions.

Acknowledgments

This work was supported, in part, with funds provided by a National Science Foundation Grant (No. CPE-8316913) to the University of Delaware.

Literature Cited

1. Sandler, S. I. Fluid Phase Eq. 1985, 19, 233.
2. Alder, B. J.; Young, D. A.; Mark, M. A. J. Chem. Phys. 1972, 56, 3013.
3. Redlich, O.; Kwong, J. N. S. Chem. Rev. 1949, 44, 233.
4. Peng, D. Y.; Robinson, D. B. I&EC Fund. 1976, 15, 59.
5. Beret, S.; Prausnitz, J. M. AIChE J. 1975, 21, 1123.
6. Donohue, M. D.; Prausnitz, J. M. AIChE J. 1978, 24, 849.
7. Lee, K.-H.; Lombardo, M.; Sandler, S. I. Fluid Phase Eq. 1985, 21, 177.
8. Carnahan, N. F.; Starling, K. E. J. Chem. Phys. 1969, 51, 635.
9. Aim, K.; Nezbeda, I. Fluid Phase Eq. 1983, 12, 235.
10. Ponce, L.; Renon, J. J. Chem. Phys. 1976, 64, 638.
11. Flory, P. J. "Principles of Polymer Chemistry," Ithaca, N.Y., Cornell University Press 1953.
12. Staverman, A. J. Rec. Trav. Chim. Pays-bas 1950, 69, 163.
13. Hildebrand, J. H.; Prausnitz, J. M.; Scott, R. L. "Regular and Related Solutions," Van Nostrand Reinhold Co. 1970.
14. Gierycz, P.; Nakanishi, K. Fluid Phase Eq. 1984, 16, 225.
15. Toukubo, K.; Nakanishi, K. J. Chem. Phys. 1976, 65, 1937.
16. Hoheisel, C.; Deiters, U. Molec. Phys. 1979, 37, 95.
17. See, for example, Hirschfelder, J. O.; Curtiss, C. F.; Bird, R. B. "Molecular Theories of Gases and Liquids," J. Wiley & Sons, Inc., N.Y., 1953.
18. Whiting, W. B.; Prausnitz, J. M. Fluid Phase Equilibria 1982, 9, 119.
19. Hu, Y.; Ludecke, D.; Prausnitz, J. M. Fluid Phase Eq. 1984, 17, 217.
20. Weeks, J. D.; Chandler, D.; Anderson, H. C. J. Chem. Phys. 1971, 54, 5237.
21. Lee, K.-H.; Sandler, S. I.; Patel, N. C. Fluid Phase Equilibria, accepted for publication.
22. Lee, K.-H.; Sandler, S. I. Fluid Phase Eq., to be submitted.
23. Lee, K.-H.; Sandler, S. I.; Monson, P. A. Ninth Thermophysical Properties Meeting, Boulder, CO., June 1985.

RECEIVED November 5, 1985

9

Local Structure of Fluids Containing Short-Chain Molecules via Monte Carlo Simulation

K. G. Honnell and C. K. Hall[1]

Department of Chemical Engineering, Princeton University, Princeton, NJ 08544

Monte Carlo simulation has been used to investigate local structure of model systems of butane and octane at volume fractions ranging from 0.001 to 0.3. At low and moderate densities the bead-bead intermolecular radial distribution function, g(r), near contact was found to be significantly less than one, indicating the presence of an excluded volume surrounding each bead. The value of g(r) near contact increased with increased density, and decreased with increased chain length.

In recent years there has been increased demand for general equations of state for fluids and fluid mixtures of interest to the natural gas and petroleum industries. The techniques of statistical mechanics offer an attractive route to the development of such equations of state since they are based upon fundamental physical considerations. An historical survey of the scientific literature shows however that up until the 1970's most research in statistical mechanics was limited to the development of theories for highly idealized monatomic fluids containing spherically symmetric molecules, e.g., argon, which were relatively uninteresting from an industrial point of view. In the mid 1970's however, research efforts began to focus on more industrially relevant compounds including rigid asymmetric molecules, e.g., N_2, rigid polar molecules, e.g., H_2O, and very long and very flexible molecules, e.g., polymers. The perhaps more difficult problem of intermediate length molecules such as hydrocarbons, which are long and flexible compared to diatomics but short and stiff compared to polymers, also began to receive some attention, largely because such fluids are of great industrial interest.

The primary difficulty in applying statistical mechanics to systems of short chains lies in rigorously accounting for their asymmetric structure and their flexibility. The two major analytical

[1]Current address: Department of Chemical Engineering, North Carolina State University, Raleigh, NC 27695-7905.

0097-6156/86/0300-0201$06.00/0

approaches taken to this problem, the perturbed hard chain theories (1) and the lattice fluid theories (2), of necessity must resort to relatively crude approximations in order to account for molecular asymmetry and flexibility. In contrast, computer simulation is capable of dealing explicitly with these aspects of molecular shape. Thus, computer simulation presents an attractive way to directly examine the cause-effect relationship between the molecular structure of short flexible chains and the macroscopic or thermodynamic behavior of these systems. In addition, the simulated data can be used as a reference against which analytical theories may be tested.

Most of the early simulation work on many-chain systems was done on a lattice (3). The first off-lattice Monte Carlo calculations appear to be those of Curro (4) who simulated chains of fifteen and twenty beads in an effort to compute the bead-bead radial distribution function and the resulting equation of state. His molecular model consisted of fixed bond lengths and angles, a rotational isomeric state approximation and a hard-sphere intermolecular potential. Curro had difficulty simulating at reduced densities ρ above 0.25 (where ρ = volume occupied by chains/volume of the system). Bishop, et al. (5) also used a Monte Carlo technique to calculate the center of mass-center of mass radial distribution function for flexible multichain systems but their resulting curves are hard to interpret.

The molecular dynamics simulation technique has also been used to examine more sophisticated molecular models. Ryckaert and Bellemans (6) studied static and dynamic properties of butane and decane, using a Lennard-Jones intermolecular potential, a continuous torsional rotation potential, and fixed bond angles and bond lengths. Weber (7) also investigated butane and octane using a Lennard-Jones potential coupled to a continuous torsional rotation potential, but he relaxed the constraints of constant bond angles and bond lengths. The results of Ryckaert and Bellemans and of Weber are reasonable for most quantities measured with the exception of the pressure, which is quite poorly predicted.

In this paper we report preliminary Monte Carlo results on the local structure of model systems of butane, octane and dodecane. We have focused our attention on the intermolecular radial distribution function, since this may be used in predicting the pressure. A future paper will report more extensive simulation results for the distribution function as well as results for the pressure.

The Model

A simplified skeletal alkane model was constructed consisting of an n bead chain (where n is the number of carbons) with bond angles, θ, and bond lengths, ℓ, fixed at 109°28' and 1.53Å, respectively. The diameter of each bead (methyl group), σ, was also fixed. Torsional rotations through the angle ϕ_i were modelled using the rotational isomeric state potential, which allows for just three discrete rotational states -- trans, gauche$^+$, and gauche$^-$. The potentials employed were similar to those of Curro (4). The energy of the gauche states was set at 700 cal/mol above that of the trans.

Neighboring g^+/g^- combinations along chain backbones were given an additional energy of 2300 cal/mol, to account for the so-called pentane effect arising from steric interactions between beads separated by four bonds (6). Intermolecular interactions and intramolecular interactions between beads separated by more than four bonds were accounted for by a hard sphere potential.

The molecular model chosen for the Monte Carlo studies is not substance specific but is instead highly idealized. The goal has been to simplify the potential to the point where it may be treated using some analytical statistical mechanical technique but not beyond the point where those essential features of the potential responsible for the fluid's thermodynamic behavior are obscured. In this respect, these studies can serve as a bridge between developing analytical theories and existing experimental and simulated data.

Monte Carlo Simulations

Simulations are performed within the general framework of the algorithm developed by Metropolis, et al. (9). The Monte Carlo program randomly generates a large number of trial configurations for a system of twenty-seven model molecules in a box of size appropriate to give the desired density. Periodic boundary conditions were employed to minimize wall effects. The program begins with an initial configuration at a specified temperature and density. Subsequent configurations are generated by moving one molecule at a time. After allowing the system to relax from its initial state, the properties of each configuration are recorded and at the end of the simulation are averaged to yield equilibrium properties. Initial configurations for the low density simulations were obtained by placing the molecules on a lattice with all torsional angles in the trans position. Higher density initial configurations were obtained by compressing equilibrated low density configurations from previous runs. Starting with a low density system, trial moves were performed until the molecule closest to the wall of the box was displaced toward the center. At this point the wall of the box was moved in toward the center resulting in a smaller container. This shrinking process was repeated until the desired density was reached.

The program uses four different methods to move the molecules: rigid translation of the center of mass, rigid rotation about a central bead, torsional rotation of an end bead, and a reptation technique known as the slithering snake (10). In the slithering snake moves, a new torsional angle is chosen for the chain, but rather than simply rotating into this state, the "tail" of the chain is detached and reattached at the other end, creating a new torsional angle in the process. The net effect is a snake-like wiggling motion of the molecules along their axis throughout the box.

After each attempted move, the new position of the chain is checked against the positions of the other molecules to insure that there is no overlap. In addition, if a torsional rotation or slither results in an increase in the intramolecular energy of the trial chain, the move is accepted with a probability P equal to the Boltzman factor, $P = \exp(-\Delta E/kT)$, where ΔE is the change in the intramolecular energy caused by the move. Moves which result in a decrease in the intramolecular energy of the chain but do not result in overlap are always accepted. For the rigid translational and rotational moves, the maximum displacement parameters were adjusted so that, on average, half of the trial moves were accepted.

Butane, octane and dodecane were simulated at a temperature of 700K and at dimensionless densities ranging from 0.001 to 0.30. The relatively high temperature was chosen to encourage torsional rotation to gauche states and does not influence intermolecular interactions. The dimensionless density is defined to be the volume fraction of the box occupied by molecules and is given by

$$\rho = \frac{Nn\pi\sigma^3}{6V}$$

where N is the number of molecules, n is the number of beads per chain, σ is the hard sphere diameter, and V is the volume. The hard sphere diameter, σ, was set equal to the bond length for these studies. While a more realistic value of σ would be larger, this "pearl necklace" model allows for a number of simplifications in theoretical treatments that larger diameter beads do not (11).

For butane, two hundred thousand configurations were generated at each density, requiring approximately 45 minutes of CPU time each on the Princeton University IBM 3081. The computer time required to adequately generate and sample enough configurations to get reliable statistics increases with increasing chain length and/or density. For example, the generation of 75,000 configurations for dodecane required one hour and fifteen minutes. Equilibration generally required between twenty and forty thousand configurations.

In order to test the reliability of the program, several runs were repeated using different starting configurations and strings of random numbers. Approximate error estimates were obtained by partitioning each simulation into five to ten blocks in which subaverages of $g_{inter}(r)$ were calculated. These subaverages were treated as independent samples contributing to the overall $g_{inter}(r)$. Analysis of the standard deviation of the subaverages about the overall average indicates that the results are accurate to $\sim$ 4%.

The structure of the simulated fluid was examined by calculating the bead-bead intermolecular radial distribution function. The bead-bead intermolecular radial distribution function, $g_{inter}(r)$ is proportional to the probability of finding two beads (methyl groups) on different chains separated by a distance r. Thus, $g_{inter}(r)$ is a measure of the local density of the fluid. A value of $g_{inter}(r) > 1$ implies that the local density is greater than the bulk density and conversely, when $g_{inter}(r) < 1$, the local density is less than the bulk density. By monitoring the local density of the system of chains throughout the course of the simulation, $g_{inter}(r)$ can be found from

$$g_{inter}(r) = \frac{V}{Nn} \frac{\langle M \rangle}{4\pi r^2 \Delta r}$$

where $\langle M \rangle$ is the average number of beads located at a distance between r and r + Δr from a neighboring bead on a different chain. Correlations were monitored out to a maximum distance of one half the box length in order to avoid the periodicity introduced at larger separations by the boundary conditions. In addition to describing the local structure of the fluid, $g_{inter}(r)$ can be used to calculate the pressure and is the key to investigating other model potentials through perturbation expansions.

Results

The results for our Monte Carlo simulations are illustrated in Figs. 1 through 9. Figure 1 shows the intermolecular radial distribution function for butane at reduced densities of 0.01, 0.10 and 0.30 as a function of r/σ where σ is the bead diameter. Also shown for comparison in the inset is the radial distribution function for a monatomic hard sphere fluid which illustrates the familiar peak and valley structure, reflecting the shells of nearest neighbors, followed by a gradual loss of correlations as $g_{inter}(r)$ decays to 1. This behavior is characteristic of all monatomic fluids. What is striking about the lower-density butane results is that $g_{inter}(r)$ at contact is significantly less than 1, indicating a relative absence of neighboring beads in the immediate vicinity of a central bead. The origin of this "correlation hole" can be seen by constructing a physical picture of what may be happening on a molecular level.

A comparison of Figures 2a and 2b provides an explanation for the correlation hole. Figure 2a shows a cluster of monatomic molecules surrounding a central (shaded) monatomic molecule. The monatomic molecules are able to approach each other with relative ease resulting in the familiar shells of neighboring beads, first nearest neighbors, second nearest neighbors, and so on, shown in the inset on Fig. 1. In the case of chain molecules shown in Fig. 2b, neighboring beads on the same chain as the central (shaded) bead are not counted in computing intermolecular correlations. The presence of these adjoining beads, in turn, leads to a shielding effect which makes it more difficult for surrounding chains to approach the central bead. The net effect of this shielding is the creation of an excluded volume surrounding each bead, the so called correlation hole.

If this exclusion effect is indeed the correct explanation for the correlation hole, then one might expect that the value of $g_{inter}(r)$ at contact (i.e. at $r = \sigma$) would increase with density as more and more chains are packed together. This trend is observed in Figure 1; at a reduced density of 0.3, $g_{inter}(r)$ has an intercept greater than 1, and the curve begins to take on a shape reminiscent of a monatomic system.

Another interesting feature of the local structure is illustrated in Figure 3 which show cusps in the $g(r)$ curve for butane at $\sigma = 0.25$ occurring at distances $r/\sigma = 2$ and $r/\sigma = 2.666$. These cusps may be explained by recalling that discontinuities in the intermolecular potential result in corresponding discontinuities in the in the pair correlation function. For example, in the monatomic hard-sphere model, the discontinuity in the potential at $r = \sigma$ results in a discontinuity in $g(r)$ at $r = \sigma$. Likewise, in the monatomic square well model, discontinuities in $g(r)$ occur at $r = \sigma$ and at $r = k\sigma$, $k\sigma$ being the width of the well (12). For chains however, the contact between a bead such as the one labeled 1 in Figure 2b and the central bead can become an effective contact between the central

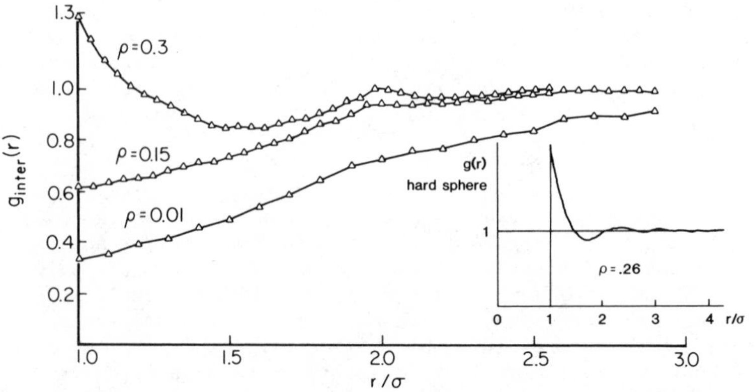

Fig. 1. Comparison of $g_{inter}(r)$ for butane at ρ = 0.01, 0.15 and 0.3 obtained from Monte Carlo simulations at 700K. The inset shows the radial distribution for a monatomic hard sphere system.

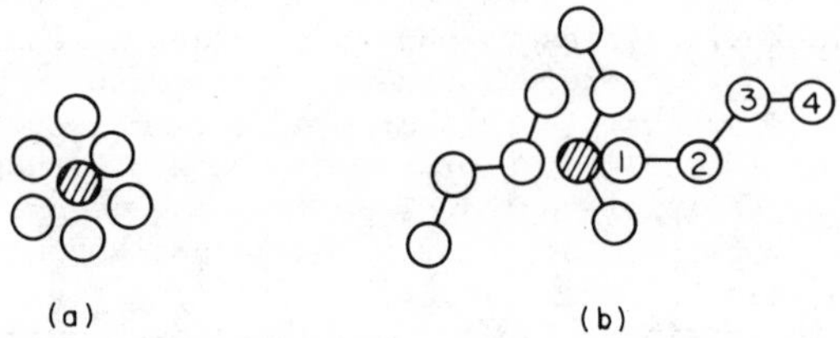

Fig. 2. Local correlations for: (a) fluid of hard sphere monatomics, (b) fluid of hard bead chains.

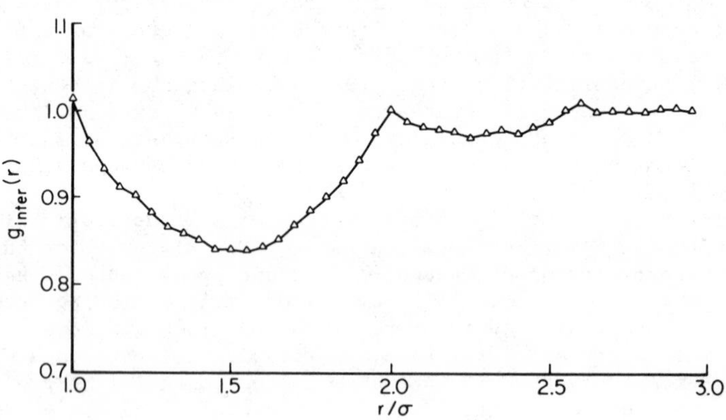

Fig. 3. Intermolecular radial distribution function for butane at ρ = 0.25 predicted by Monte Carlo simulations.

bead and beads 2, 3 or 4 since the rigid bond angles and lengths will not allow beads 2, 3 or 4 to approach the central bead any closer. Discontinuities in the pair potential at $r > \sigma$ will occur whenever beads 2, 3, or 4 become colinear with the central bead and bead 1 (as is the case for bead 2 in Fig. 2b). These effective contact distances are reflected in the cusps which appear in the angle-averaged $g_{inter}(r)$ at $r/\sigma = 2$ for bead 2 and $r/\sigma = 2.66$ for bead 3. The existence of these cusps has been predicted by Hsu, Pratt, and Chandler using RISM (13).

Figure 4 illustrates the effect of chain length on $g_{inter}(r)$. Results are shown for butane, octane and dodecane at a reduced density of 0.15. Although the overall shapes of the curves are the same, the size of the correlation hole is seen to increase with chain length. Longer chains, with more torsional rotation angles, can coil around themselves to a greater extent than shorter molecules leading to an increased degree of shielding and a larger correlation hole.

The effect of temperature on local structure is illustrated in Fig. 5, where $g_{inter}(r)$ is plotted for octane at a dimensionless density of 0.2 and temperatures of 300K and 700K. As might be expected with a hard sphere potential, decreasing the temperature had very little effect on the local structure of the fluid. Thus, the simulated results at 700K are probably indicative of the local structure over a fairly broad temperature range. It is interesting to note, however, that the $g_{inter}(r)$ at 300K is always greater than at 700K. At 300K the chains are more likely to be in a trans conformation, making it easier for neighboring chains to approach each other. One would therefore expect that the magnitude of the temperature effect, like the size of the correlation hole, should increase with chain length.

In addition to the simulations described previously in which the bead diameter was set equal to the bond length, several studies were performed to try to determine the effect of varying the bead diameter while maintaining a constant value for the bond length. In order to relate to more realistic models of alkanes, a value of $\sigma/\ell_o = 2.48$ was considered. This corresponds to a bond length of $\ell = 1.53$Å and a hard sphere diameter of 3.7889Å. This hard sphere diameter was chosen using a method developed by Verlet and Weiss (14) to represent an effective hard sphere diameter for a Lennard Jones system at 298K.

Figure 6 compares $g_{inter}(r)$ vs. r for $\sigma/\ell = 1$ and $\sigma/\ell = 2.48$ for octane at a temperature of 298K and a molecular number density N/V of 0.00333/Å^3 = 0.631 g/cc. We now refer to the number density rather than the reduced density because, in the larger diameter model, the amount of overlap between beads separated by three or four bonds varies with the isomeric state. Consequently the volume of the chain changes with chain conformation. For a given number density, the bead diameter to bond length ratio is seen to have a large effect on fluid structure. The bulkier molecules occupy much more of the available volume than do their skinny counterparts, leading to a tighter packing of chains. This results in a larger value of $g_{inter}(r)$ at contact, a smaller correlation hole, and a more structured appearance of $g_{inter}(r)$ due to the presence of neighboring beads. At $\sigma/\ell = 1$ there is considerably more void space in the box and chains tend to remain relatively far apart because of steric interactions. The resulting $g_{inter}(r)$ is characteristic of a low

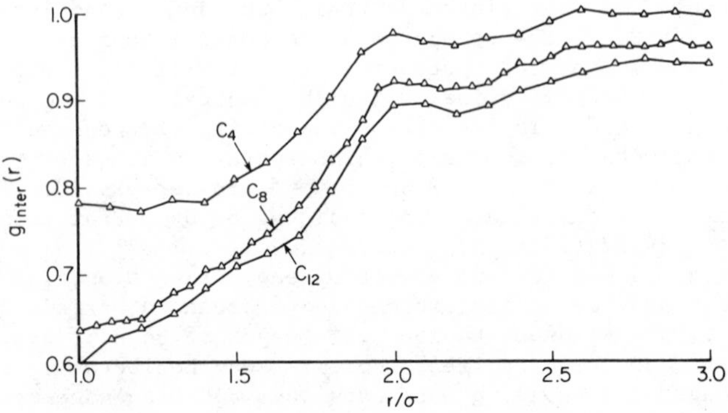

Fig. 4. Comparison of intermolecular radial distributions for butane, octane and dodecane at ρ = 0.15 predicted by Monte Carlo simulations.

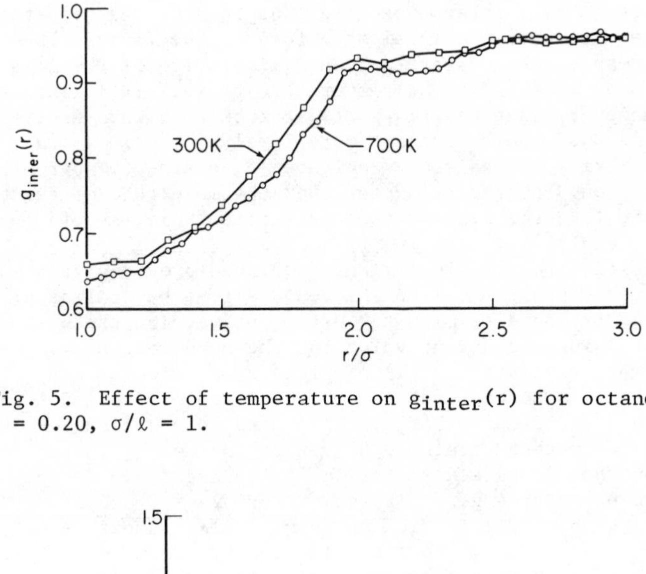

Fig. 5. Effect of temperature on $g_{inter}(r)$ for octane at ρ = 0.20, σ/ℓ = 1.

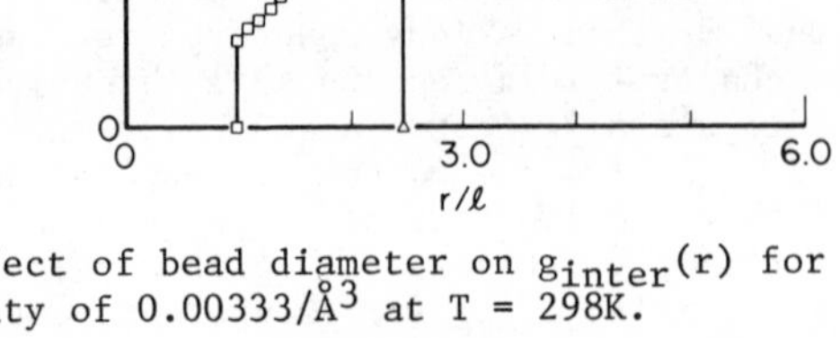

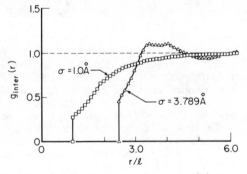

Fig. 6. Effect of bead diameter on $g_{inter}(r)$ for octane at a number density of 0.00333/Å^3 at T = 298K.

density system, with only a small change in slope at a distance $r = \sigma + \ell$. These results indicate that thermodynamic properties of such model systems, particularly the pressure, will be very sensitive to the size of the hard sphere diameter employed.

In Fig. 7, $g_{inter}(r)$ for butane at a temperature of 298K, reduced density of 0.25 and $\sigma/\ell = 1$ is compared with the predictions of RISM, a computationally convenient integral theory for local structure (15). At values of $r/\sigma > 1.3$, good agreement between theory and simulation is obtained, including the prediction of the cusps at $r/\sigma = 2$ and $r/\sigma = 2.67$. Near contact however, RISM predicts a tighter clustering of chains than is observed in our simulations. Since for hard chain systems the pressure is related to the value of the distribution function at contact, these two curves will lead to different perdictions for the pressure. Similar discrepancies near contact between RISM predictions and simulation results have been observed in systems of dimers and trimers (16). In Figure 8, RISM and Monte Carlo results are compared for octane at a temperature of 298K, reduced density of .05, and σ/ℓ ratio of one. Both simulation and theory predict a sizable correlation hole although the overall agreement is not as good as it was for butane. Again the discrepancy between the two curves is largest near contact.

Further insights into the local structure of short chain fluids can be gained by examining the individual correlations between beads at specific locations along the chain backbone. Figure 9 shows g(1,1), g(1,2), and g(4,4) for octane at $N/V = 0.0033/Å^3$ and $\sigma/\ell = 2.48$, where the pairs of numbers refer to the locations of the two beads along the chain backbone, relative to the ends of the chains. For example, g(1,1) represents the distribution function for end beads on different chains, while g(4,4) refers to correlations between center beads on different chains. See Figure 10. Because the chains can be numbered from either end, beads 1 and 8, 2 and 7, 3 and 6, and 4 and 5 are equivalent, resulting in only ten distinct bead-bead pairs comprising $g_{inter}(r)$.

The distribution function g(1,1) has its greatest value at contact, as it is relatively easy for two end beads to approach each other. However, at larger distances g(1,1) drops below one, since if two "heads" of chains are adjacent, their "tails" must be relatively far apart. The (1,2) correlation has its maximum at a distance of a diameter plus a bond length, which corresponds to the distance that would be between beads 1 and 2 if the two heads of the chain were in contact. Near contact, g(1,2) is less than one, since it is more difficult for a bead in the second position to approach another chain end without steric interference from its bonded neighbors. At larger distances g(1,2) begins to resemble g(1,1) since, if beads 1 and 2 are in relative proximity, beads 7 and 8 are likely to be far apart. Steric hindrance plays the biggest role in the (4,4) correlation. Near contact g(4,4) is much less than one, implying that the centers of chains rarely come into contact. At larger values of r however, g(4,4) has a relatively broad peak. This is because the distances between equivalent pairs (4,4), (4,5), (5,4) and (5,5) are likely to be about the same, unlike the pairs contributing to g(1,1) and g(1,2). The behaviors of the seven remaining correlation functions lie in between that of g(1,2) and

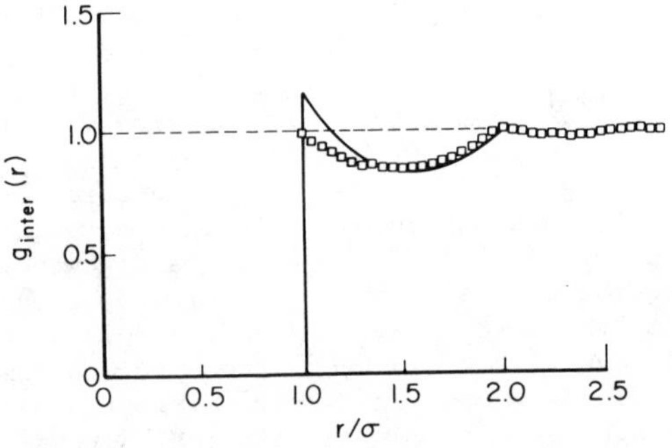

Fig. 7. Comparison of RISM and Monte Carlo predictions of $g_{inter}(r)$ for butane at $\rho = 0.25$, $\sigma/\ell = 1$, T = 298K. Solid curve is RISM, boxes are simulation results. The two curves coincide for $r/\sigma > 2$.

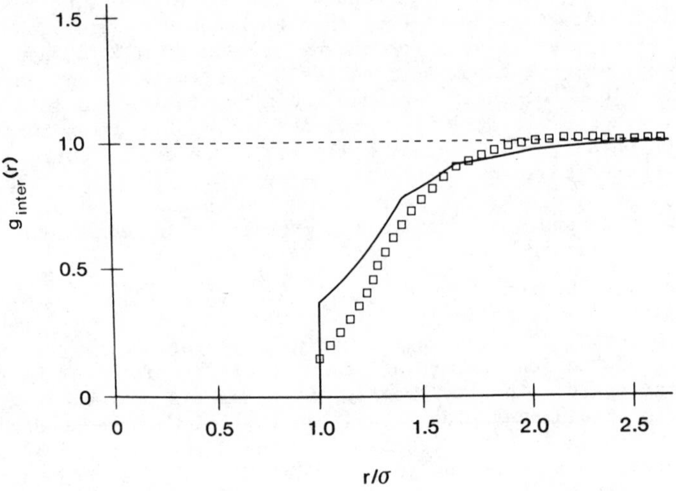

Fig. 8. Comparison of RISM and Monte Carlo predictions for octane at $\rho = .05$, $\sigma/\ell = 1$, T = 298K. Solid curve is RISM, boxes are simulation results.

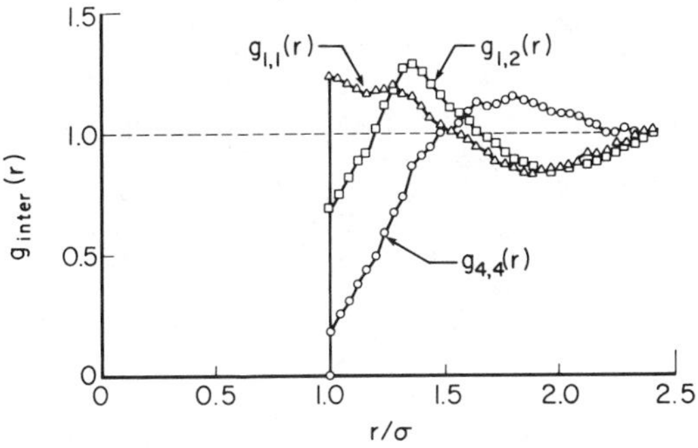

Fig. 9. Comparison of g(1,1), g(1,2), and g(4,4) bead correlations for octane at a number density of 0.00333(1/Å^3), $\sigma/\ell = 2.48$, T = 298K.

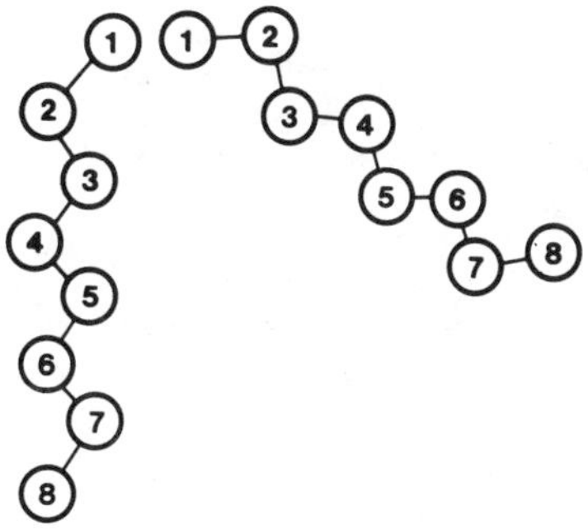

Fig. 10. Location of individual beads along chain backbone.

g(4,4). Each has a contact value less than one, which, when superimposed, leads to the correlation hole seen in $g_{inter}(r)$.

Conclusions and Discussion

Monte Carlo simulation has been used to obtain the intermolecular radial distribution function for systems of interacting short chain molecules. At low and moderate densities, the value of $g_{inter}(r)$ near contact is significantly less than one, indicating the presence of an excluded volume surrounding each bead. The existence of this "correlation hole" appears to stem from a shielding effect associated with neighboring beads along the same chain backbone. An examination of the individual bead-bead correlations comprising $g_{inter}(r)$ indicates that interior beads along the chains are much more heavily shielded than beads near the chain ends.

The size of the correlation hole is seen to increase with increasing chain length and decrease with increasing density. One would expect the correlation hole to be quite small for dimers and trimers. Local structure is strongly dependent on the hard sphere diameter. Preliminary comparisons with RISM calculations show overall good agreement except near contact.

Calculations are currently underway to more thoroughly investigate the effect of bead diameter on local structure as well as to more closely examine the individual end-end, end-middle, and middle-middle correlations which comprise $g_{inter}(r)$. Simulations are being performed at higher densities and at longer chain lengths in order to more fully investigate the effect of density and chain length on local structure. We are also trying to develop simple correlations which summarize the effect of varying molecular parameters on the radial distribution function.

Finally, we should point out that due to our relatively small sample size and relatively short runs the results presented in this paper should be considered suggestive rather than conclusive. More extensive simulations will be reported in a future paper.

Acknowledgments

The authors would like to thank the Gas Research Institute (Grant No. 5082-260-0724) and the Petroleum Research Fund (Grant No. 14851-AC5) for supporting this research. Ms. D. Chirichilla is gratefully acknowledged for her help in performing the RISM calculations. The authors also wish to thank Dr. L. Pratt, E. Helfand, and J. Curro for helpful discussions and suggestions.

Literature Cited

1. Donohue, M. O.; Prausnitz, J. M. AIChE J. 1978, 24, 849.
2. Sanchez, I.; Lacombe, R.; J. Chem. Phys. 1976, 80, 2352, Lacombe, R.; Sanchez, I., J. Chem. Phys. 1976, 80, 2568; Simha, R., Macromolecules, 1977, 10, 1025.
3. DeVos, E.; Bellemans, A., Macromolecules 1975, 8, 651; ibid, 1974, 7, 812.
4. Curro, J. G., J. Chem. Phys. 1976, 64, 2496.

5. Bishop, M.; Ceperley, D.; Frisch, H. L.; Kalos, M. H., J. Chem. Phys. 1980, 72, 3228.
6. Ryckaert, J.-P.; Bellemans, A., Chem. Phys. Lett. 1975, 30, 123; Faraday Soc. Disc 1978, 66, 95.
7. Weber, T. A., J. Chem. Phys. 1979, 69, 2347; 1979, 70, 4277.
8. Flory, P. J. "Statistical Mechanics of Chain Molecules", John Wiley & Sons, 1969, pp. 55-56.
9. Metropolis, N.; Metropolis, A.; Rosenblush, M.; Teller, A.; Teller, E.; J. Chem. Phys. 1953, 21, 1087.
10. Wall, F. T.; Mandel, P., J. Chem. Phys. 1975, 63, 4593.
11. Curro, J. G.; Blatz, P. J.; Pings, C. J., J. Chem. Phys. 1969, 50, 2199.
12. Nelson, P. A. Ph.D. Thesis, Princeton University, New Jersey, 1967.
13. Hsu, C. S.; Pratt, L. R.; Chandler, D., J. Chem. Phys. 1978, 68, 4213.
14. Verlet, M.; Weis, J., Phys. Rev. A 1972, 5, 939.
15. Ladanyi, B. M.; Chandler, D., J. Chem. Phys. 1975, 62, 4308.
16. Gray, C. G.; Gubbins, K. E. "Theory of Molecular Fluids"; Oxford: Clarendon Press, 1984, Vol. I, pp. 400-403.

RECEIVED November 18, 1985

10

Local Composition of Square-Well Molecules by Monte Carlo Simulation

R. J. Lee and K. C. Chao

School of Chemical Engineering, Purdue University, West Lafayette, IN 47907

Local composition in mixtures is determined for square-well molecules by means of Monte Carlo simulation at 29 states from dilute gas to dense liquid at a wide range of temperature for divergent energies of interactions and varying compositions. The results are compared with models in the literature, and a new local composition model is proposed.

Local composition about a molecule in a mixture can differ from the total bulk composition as a result of divergent energies of interaction and molecular size differences. The difference between local composition and bulk composition reflects molecular segregation and preferential orientation. Debye and Hückel (1) developed their ionic solution theory by recognizing the local composition of ions. Wilson (2) postulated the dependence of partial entropy on local composition. Whiting and Prausnitz (3) based their mixture thermodynamics on local composition. The occurrence of local composition is fundamental to thermodynamics and transport properties of mixtures.

Local composition cannot be determined by experiments in the laboratory, but can be simulated by computer calculations. Nakanishi and co-workers (4-7) and Hoheisel and Kohler (8) obtained local composition in mixtures of Lennard-Jones molecules by means of molecular dynamics simulation. In this work we determine local composition of square-well molecules by means of Monte Carlo simulation. Square-well molecules offer the special advantage of an unambiguously defined neighborhood in which local composition occurs. We carry out calculations to states previously unexplored by extending to wide ranges of density, temperature, bulk composition, and energies of interaction.

0097-6156/86/0300-0214$06.00/0

Model and Method

The system investigated in this work is a binary mixture in which molecules interact via the square-well potential,

$$\phi_{ij}(r) = \begin{cases} \infty & \text{as } r \leq \sigma_{ij} \\ -\epsilon_{ij} & \text{as } \sigma_{ij} < r < K\sigma_{ij} \\ 0 & \text{as } K\sigma_{ij} \leq r \end{cases} \tag{1}$$

where ϕ_{ij} is the potential energy between molecules i and j at center-to-center distance r; ϵ is the well depth; σ is the diameter and K is the energy well width factor. In this work $K = 1.5$, and all σ_{ij} are equal; the investigation is directed at energy effects.

The number of molecules j at a distance r to r + dr from a molecule i is given by

$$dz_{ji} = 4\pi\rho_j g_{ji}(r) r^2 dr \tag{2}$$

where ρ_j is the number density of molecules j and g is the radial distribution function. The local composition of j at r from i is therefore

$$x_{ji}(r) = \frac{\rho_j g_{ji}(r)}{\sum_k \rho_k g_{ki}(r)} \tag{3}$$

where x denotes a mole fraction.

We have obtained the total number of molecules j in the energy well of i,

$$z_{ji} = 4\pi\rho_j \int_0^{K\sigma} g_{ji}(r) r^2 dr \tag{4}$$

from which we have evaluated the average local composition

$$x_{ji} = \frac{\rho_j \int_0^{K\sigma} g_{ji}(r) r^2 dr}{\sum_k \rho_k \int_0^{K\sigma} g_{ki}(r) r^2 dr} \tag{5}$$

In this work we report the average local composition as the local composition for brevity and in agreement with the usage of previous investigators (4,8). We have also obtained g_{ji} (r) and x_{ji} (r) which we will report elsewhere.

The upper limit of integration of Equations 4 and 5 is defined without ambiguity for the square-well molecules, while there is an element of arbitrariness in setting the upper limit for continuous potentials such as the Lennard-Jones.

We simulated Equations 2 through 5 with the canonical ensemble method. The procedure of Metropolis *et al.* (9) was applied to 108 molecules in a central cell with periodic boundary conditions and following the minimum distance convention. The procedure was initiated by randomly displacing molecules from a regular face - centered cubic lattice. About 1.0×10^5 initial configurations were discarded for dilute gases. Up to 1.7×10^6 initial configurations were discarded for high density liquids. The total Markov chain varied from 1.2×10^6 configurations for a dilute gas to 4.5×10^6 for a dense liquid. About 10 minutes of computing on a CYBER 205 was required to generate 1 million configurations.

From the Monte Carlo simulation we obtain the radial distribution function,

$$g_{ji}(r) = \frac{3\langle N_{ji,\lambda}\rangle}{4\pi N_i \rho_j (r_\lambda^3 - r_{\lambda-1}^3)} \tag{6}$$

with

$$r_\lambda = r_{\lambda-1} + \delta r$$

$$r = \frac{1}{2}(r_\lambda + r_{\lambda-1})$$

where $\langle N_{ji,\lambda}\rangle$ is the ensemble average number of molecular pairs j i within the λth distance interval from $r_{\lambda-1}$ to r_λ. Equation 5 becomes

$$x_{ji} = \frac{\langle N_{ji}(K\sigma_{ji})\rangle}{\sum_k \langle N_{ki}(K\sigma_{ki})\rangle} \tag{7}$$

where $\langle N_{ji}(K\sigma_{ji})\rangle$ is the ensemble average number of molecular pairs j i at distances of separation up to $K\sigma_{ji}$ and is obtained by summing the $\langle N_{ji,\lambda}\rangle$.

Results of Monte Carlo Simulation

Pure square-well fluid pressure, internal energy, and radial distribution function were generated by our computer programs at a large number of states and compared with previous results by Alder et al. (10), Henderson et al. (11) and Smith et al. (12). By setting K = 1 for some molecules in our computer programs we simulated mixtures of hard sphere and square-well molecules and compared the results with those of Alder et al. (13). Comparison of our results (14) with these previous studies enabled us to check out our programs.

Square-well mixtures have not been simulated before. We compared our results with Lennard-Jones mixtures obtained by Nakanishi et al. (6) at two comparable states. Table I shows that comparable local compositions were obtained.

Our simulated local compositions x_{11}, x_{22} and number of nearest neighbors z_1, z_2 in square-well fluids are presented in Table II, where z_i is given by $\sum_j z_{ji}$. Four variables — density, temperature, interaction factors, and composition — are varied successively to generate the results of Table II. Density in reduced form ρ^* ($= \rho\sigma^3$) varies from 0.01 to 0.8; reduced temperature T^* ($= kT/\epsilon_{11}$) from 1.0 to 4.0; composition from 0.25 to 0.75 mole fraction; unlike interaction factor E_1 ($= \epsilon_{12}/\epsilon_{11}$) from 1.0 to 10.0 and alike interaction factor E_2 ($= \epsilon_{22}/\epsilon_{11}$) from 2.0 to 10.0. The interaction factor ranges are sufficiently wide to include non-polar (except the quantum gases), associating, and hydrogen bonding energies.

The 29 states in Table II are made up of four sections in each of which one variable is systematically varied while all others are held constant. In section 1 consisting of states 1 to 9, density is the variable. At the lowest density of state 1 which is a dilute gas, there is the greatest departure of local composition from the bulk. As density increases, local composition departs less and less from the bulk and tends to approach it as a limit.

In section 2, consisting of states 10 to 15, temperature is the variable. Two sub sections of three states each show the effect of temperature at a different density. The effect of increasing temperature is to homogenize local composition toward the bulk. This is the most expected part of our results.

In section 3, consisting of states 16 to 22, the interaction factors are varied. In all these states $E_2 \geq$ 2.0, and x_{22} is always greater than x_{11}. The tendency to form alike pairs of greater energy is favored. For Lorentz-Berthelot mixtures with $\epsilon_{12} = (\epsilon_{11}\epsilon_{22})^{1/2}$ both x_{11} and x_{22} increase with increasing alike interaction factor E_2. The states with a constant E_2 (= 2.0) show that x_{11} and x_{22} decrease as the unlike interaction factor E_1 increases.

In section 4, consisting of states 23 to 29, bulk composition is changed. Local composition is shown to be strongly dependent on bulk composition. Deviation of local composition from bulk composition is predominantly on the part of the dilute component.

In addition to the variation of one single variable within a section of Table II, the effect of simultaneous variation of two variables can be ascertained upon suitably scanning the table entries in different sections. Higher effects can be similarly discerned. The entire simulated results should be considered in the development of a new model.

Comparison with Previous Models

Wilson (2), Gierycz and Nakanishi (GN) (7), and Hu, Lüdecke, and Prausnitz (HLP) (15) have proposed models of local composition. The models are examined with our simulated results.

Wilson expressed local composition by

$$x_{ji} = \frac{x_j \exp(-\lambda_{ji}/kT)}{\sum_k x_k \exp(-\lambda_{ki}/kT)} \tag{8}$$

Gierycz and Nakanishi's local composition is given by

$$\frac{x_{ji}}{x_{ii}} = \frac{x_j A_i B_{ji} \exp(-\alpha G_{ji}/kT)}{x_i \exp(-\alpha G_{ii}/kT)} \tag{9}$$

$$A_i = (\sigma_{11}\sigma_{22} + \sigma_{ii}\sigma_{22})/2\sigma_{12}^2 \tag{10}$$

$$B_{ji} = \exp[\alpha \sqrt{} G_{11}G_{22}(1 - G_{11}G_{22}/G_{ji}^2)/kT] \tag{11}$$

$$\alpha = 0.4 \tag{12}$$

Hu, Lüdecke, and Prausnitz's model is

$$x_{21} = \frac{x_1 z_1 + x_2 z_2 - [(x_1 z_1 + x_2 z_2)^2 - 4x_1 x_2 z_1 z_2 \tau_{21}]^{1/2}}{2x_1 z_1 \tau_{21}} \tag{13}$$

$$z_i = \sum_j \frac{4}{3}\pi(r_{ji}''^{3} - \sigma_{ji}^3)\rho_j \exp[-\alpha\epsilon_{ji}(r_{ji}')/kT)] \tag{14}$$

$$\tau_{21} = 1 - \exp(\Omega/kT)\frac{x_{11,hS}x_{22,hS}}{x_{12,hS}x_{21,hS}} \tag{15}$$

$$\Omega = \epsilon_{21}(r'_{21}) + \epsilon_{12}(r'_{12}) - \epsilon_{11}(r'_{11}) - \epsilon_{22}(r'_{22}) \tag{16}$$

$$x_{ji,hS} = x_j\sigma_{ji}^3/\sum_k x_k\sigma_{ki}^3 \tag{17}$$

$$\alpha = 0.60 - 0.58\rho^{*0.1856} \tag{18}$$

Figure 1 shows the substantial local composition variation with density of states 1 to 9 of Table II. Both the Wilson and the GN models are independent of density and completely miss this variation. The Wilson model exactly represents the low density limiting behavior of LC according to Chao and Leet (16). The limiting behavior is confirmed by the simulated data.

Figure 2 shows the variation with temperature of our simulated data and the models at two liquid-like densities. For the relatively small energy difference shown in Figure 2A the HLP model agrees better with the simulated data than the other two models. With larger E_1 ($= \epsilon_{12}/\epsilon_{11}$) shown in Figure 2B the HLP model tends to be low for both x_{11} and x_{22} and the deviation becomes larger at lower temperature. The GN model gives good results for x_{22} but low x_{11}. Wilson's model departs widely from data for these high density conditions.

Figure 3 shows the effect of unlike energy factor E_1 while ρ^*, T^*, and E_2 are kept constant. For low E_1 the HLP equation stays closer to the data than both the Wilson and the GN equations. All equations become worse with increasing E_1, apparently due to the Boltzmann exponential factor lending too much weight. The best results of the GN equation are attributable to the use of a factor of $\alpha = 0.4$ to reduce the weight of the Boltzmann term.

Figure 4 shows the effect of bulk composition while ρ^*, T^*, E_1, and E_2 are kept constant. All three models are in reasonable agreement with the data in part A of Figure 4 where the interaction factor values are not large. The HLP equation appears particularly good. However, the equations fail at larger values of the interaction factors as shown in Figure 4B.

Table I
Local Compositions in
Square-Well and Lennard-Jones Fluids

	SW	LJ	SW	LJ
T^*	1.5	1.44	1.5	1.43
ρ^*	0.7	0.716	0.8	0.75
E_1	1.414	1.414	2.0	2.0
E_2	2.0	2.0	2.0	2.0
x_1	0.5	0.5	0.5	0.5
x_{11}	0.491	0.492	0.438	0.449
x_{22}	0.518	0.518	0.458	0.465

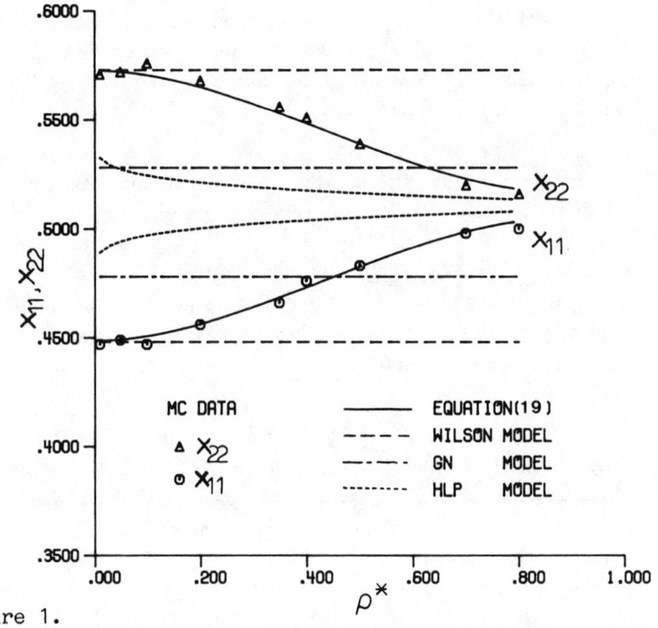

Figure 1.

Local Composition as a Function of Density and Comparison of MC Data with Four Models at $T^* = 2.0$, $E_1 = 1.414$, $E_2 = 2.0$, and $x_1 = 0.5$.

Table II
Simulated Number of Nearest Neighbors and Local Composition in Square Well Mixtures

State	T^*	ρ^*	E_1	E_2	x_1	z_1	z_2	x_{11}	x_{22}
1	2.0	0.01	1.414	2.0	0.5	0.18	0.23	0.447	0.571
2		0.05				0.89	1.14	0.449	0.572
3		0.1				1.74	2.28	0.447	0.576
4		0.2				3.27	4.13	0.456	0.568
5		0.35				5.24	6.26	0.466	0.553
6		0.4				5.93	6.92	0.476	0.552
7		0.5				7.28	8.18	0.482	0.539
8		0.7				10.40	10.86	0.499	0.520
9		0.8				11.78	12.17	0.500	0.516
10	1.5	0.7	1.414	2.0	0.5	10.62	11.23	0.491	0.518
11	3.0					10.20	10.52	0.500	0.516
12	4.0					10.09	10.31	0.497	0.508
13	1.0	0.8	2.0	2.0	0.5	12.60	13.28	0.420	0.450
14	1.5					12.19	12.64	0.438	0.458
15	4.0					11.56	11.77	0.470	0.480
16	2.0	0.8	1.0	2.0	0.5	11.65	12.10	0.579	0.595
17			2.0	2.0		11.97	12.30	0.448	0.462
18			2.0	4.0		11.75	12.95	0.563	0.603
19			3.0	10.0		11.48	14.58	0.607	0.691
20			5.0	2.0		12.71	13.02	0.363	0.378
21			10.0	2.0		13.39	13.83	0.315	0.337
22			10.0	10.0		12.72	15.48	0.364	0.477
23	2.0	0.7	10.0	2.0	0.25	14.05	11.59	0.074	0.626
24					0.333	13.30	11.88	0.145	0.521
25					0.50	12.38	12.73	0.298	0.318
26					0.67	11.48	13.83	0.498	0.167
27					0.75	11.11	14.25	0.620	0.111
28	2.0	0.8	2.0	2.0	0.25	12.10	12.25	0.189	0.733
29					0.75	11.84	12.38	0.726	0.216

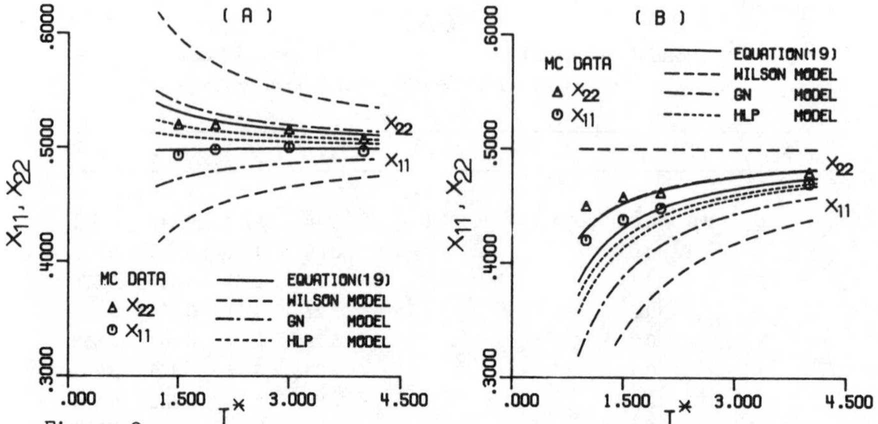

Figure 2.
Local Composition as a Function of Temperature and Comparison of MC Data with Four Models. (A): $\rho^* = 0.7$, $E_1 = 1.414$, $E_2 = 2.0$, and $x_1 = 0.5$. (B): $\rho^* = 08$, $E_1 = 2.0$, $E_2 = 2.0$, and $x_1 = 0.5$.

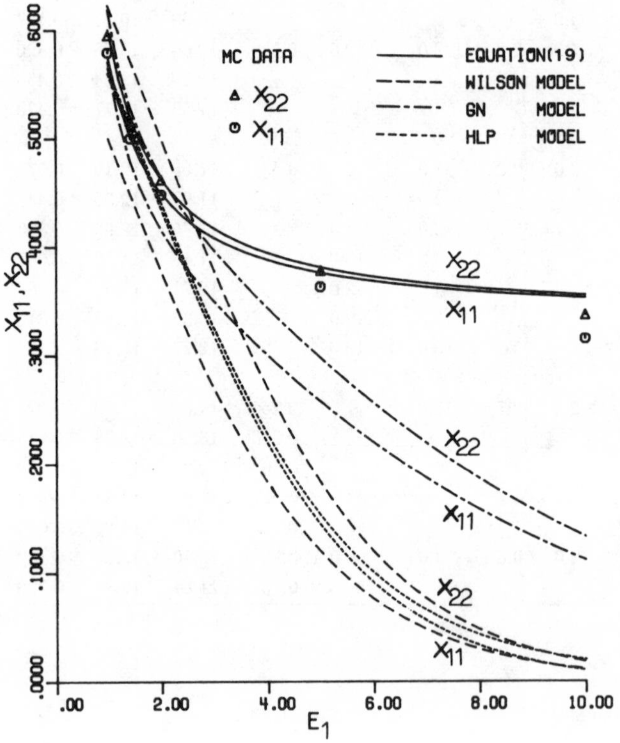

Figure 3.
Local Composition Changing with E_1 for Four Models at $\rho^* = 0.8$, $T^* = 2.0$, $E_2 = 2.0$ and $x_1 = 0.5$.

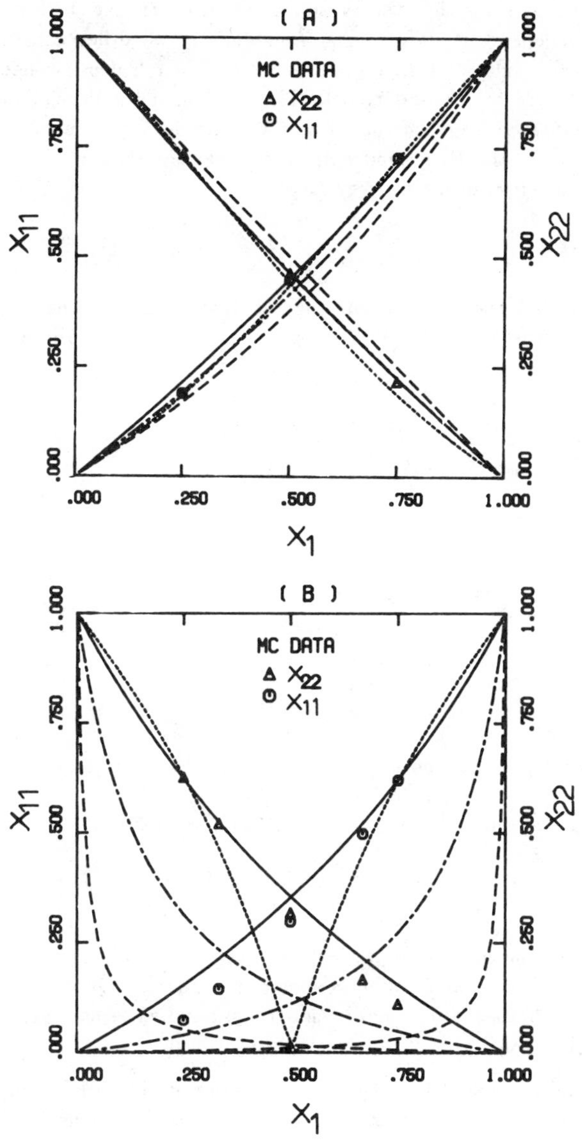

Figure 4.

Local Composition Changing with Bulk Composition for Four Models. (A): $\rho^* = 0.8$, $T^* = 2.0$, $E_1 = 2.0$, and $E_2 = 2.0$ (B): $\rho^* = 0.7$, $T^* = 2.0$, $E_1 = 10.$, and $E_2 = 2.0$.

—— Equation(19); — — — — Wilson Model; — - — - GN Model; ------ HLP Model.

The comparisons show that the Wilson model is correct at the low density limit. Being density independent, it diverges from data at all other densities. The GN model gives good results at a liquid-like density. Though density independent it is much improved over the Wilson model. The HLP model is density dependent and displays the qualitative trend of the data at varying density, temperature, and energies. Quantitatively the HLP model diverges from the data in density effect, and over estimates the effect of high energy factors.

A New Model

The comparisons of the last section suggest the need for a new model for the representation of the MC data. We propose a new local composition model as follows,

$$x_{ji} = \frac{x_j \exp[\gamma_{ji}(\epsilon_{ji} - \epsilon_{ji}^{o})/kT]}{\sum_k x_k \exp[\gamma_{ki}(\epsilon_{ki} - \epsilon_{ki}^{o})/kT]} \tag{19}$$

where

$$\epsilon_{ji}^{o} = \alpha\epsilon_{ii} + (1 - \alpha)\epsilon_{jj} \tag{20}$$

$$\gamma_{ji} = (\epsilon_{ii}\,\epsilon_{jj})^{1/2}/\epsilon_{ji} \tag{21}$$

and α is density dependent and is given by

$$\alpha = 1 + 0.1057\rho^{*} - 2.1694\rho^{*2} + 1.7164\rho^{3} \tag{22}$$

For Lorentz-Berthelot mixtures $\gamma_{ji} = 1$, Equation 19 reduces to the Wilson model as $\rho \rightarrow 0$ and $\alpha \rightarrow 1$.

Comparison of the new model with data is reported in Figures 1 to 4. The density dependence of local composition is shown in Figure 1. Quantitative representation of the data in the entire composition range studied is achieved with the new model and the low density limit is correctly approached.

The temperature dependence of local composition is shown in Figure 2 where the new model agrees with data better than the other models. Figure 3 shows that the new model brings about a substantial improvement at high values of E_1. In Figure 4A all models show good agreement with data at changing bulk composition when the energy differences are not excessive.

Figure 4B shows a case of large energy difference where the new model holds out better than the others. The weight of the Boltzmann factor is reduced in the new model to achieve the improved result but the reduction tends to be on the high side.

Conclusion

1. Local composition in square-well fluid mixtures has been simulated with the canonical ensemble Monte Carlo method.
2. Local composition is found to depend on the state of the system and local composition values are determined for 29 states at various density, temperature, interaction energy, and bulk composition.
3. The predictions of three models are tested with the simulated results. The Wilson and GN models do not show any density dependence. The HLP model does, but underpredicts the dependence.
4. A new model is proposed for improved representation of local composition.

Acknowledgment

Funds for this research were provided by the National Science Foundation through grant CBT-8516449.

Glossary of Symbols

E_1	$= \epsilon_{12}/\epsilon_{11}$, unlike interaction factor
E_2	$= \epsilon_{22}/\epsilon_{11}$, alike interaction factor
G_{ji}	= energy parameter in GN model
g	= radial distribution function
K	= energy well width factor
k	= Boltzmann's constant
$\langle N_{ji,\lambda} \rangle$	= ensemble average number of molecular pairs j i within the λth distance interval
r	= intermolecular separation
r', r''	= diameter of the first and second shells respectively
T	= temperature
T^*	$= kT/\epsilon_{11}$, reduced temperature
x_i	= bulk composition of molecule i
x_{ji}	= local composition of molecule j around molecule i
z_i	= coordination number of molecule i
z_{ji}	= number of molecules j around a central molecule i
α	= parameter in GN and HLP model

γ = energy parameter in the new model

ϵ = well depth of square-well potential

λ_{ji} = energy parameter in Wilson's equation

ρ_j = number concentration of molecule j

ρ^* = $\rho\sigma^3$, reduced density

σ = hard-core diameter

ϕ_{ij} = potential energy between molecules i and j

Literature Cited

1. Debye, P.; Hückel E. Physik Z. 1923, 24, 185, 334.
2. Wilson, G.M. J. Am. Chem. Soc. 1964, 86, 127.
3. Whiting, W.B.; Prausnitz, J.M. Fluid Phase Equil. 1982, 9, 119.
4. Nakanishi, K.; Toukubo, K. J. Chem. Phys. 1979, 70, 5648.
5. Nakanishi, K.; Toukubo, K. Polish J. Chem. 1980, 54, 2019.
6. Nakanishi, K.; Tanaka, H. Fluid Phase Equil. 1983, 13, 371.
7. Gierycz, P.; Nakanishi, K. Fluid Phase Equil. 1984, 16, 255.
8. Hoheisel, C.; Kohler, F. Fluid Phase Equil. 1984, 16, 13.
9. Metropolis, N.; Rosenbluth, A.W.; Rosenbluth, M.N.; Teller, A.H. J. Chem. Phys. 1953, 21, 1087.
10. Alder, B.J.; Young, D.A.; Mark, M.A. J. Chem. Phys. 1972, 56, 3013.
11. Henderson, D.; Madden, W.G.; Fitts, D.D. J. Chem. Phys. 1976, 64, 5026.
12. Smith, W.R.; Henderson, D.; Tago, Y. J. Chem. Phys. 1977, 67, 5308.
13. Alder, B.J.; Alley, W.E.; Rigby, M. Physica 1974, 73, 143.
14. Lee, R.J. M.S. Thesis, Purdue University, 1985.
15. Hu, Y.; Lüdecke, D.; Prausnitz, J.M. Fluid Phase Equil. 1984, 17, 217.
16. Chao, K.C.; Leet, W.A. Fluid Phase Equil. 1983, 11, 201.

RECEIVED November 5, 1985

11

Equations of State for Nonspherical Molecules Based on the Distribution Function Theories

S. B. Kanchanakpan[1], L. L. Lee[1], and Chorng H. Twu[2]

[1] School of Chemical Engineering and Materials Science, University of Oklahoma, Norman, OK 73019
[2] Simulation Science, Inc., Fullerton, CA 92633

Based on the distribution function theories of the liquid state, we are able to derive an expression for the contribution from the attractive energy to the pressure of anisotropic fluids. We have adopted the nonspherical square-well potential with Gaussian overlap to model moderately anisotropic molecules. The repulsive pressure is shown to be given essentially by the underlying hard core repulsion and is represented, for example, by the Nezbeda equation of state for hard convex bodies. It is found that the background correlation function, y(12), plays a major role in the determination of attractive pressures. We exhibit the cluster series for this correlation function and derive therefrom a resummation formula for the pressure. To obtain adequate description of real fluid properties, a two-step form of the square-well potential as proposed by Kreglewski is adopted. The final equation consists of three terms, clearly separated into the hard core, repulsive and attractive parts. It is used to correlate the P-v-T and thermal behavior of some 69 substances, including hydrocarbons, ketones, alcohols, amines and polar solvents. For pressures up to 69 MPa, the errors in density and vapor pressure predictions are, with few exceptions, within 1%. Comparison with similar equations of state, such as the Peng-Robinson and Mohanty-Davis equations, shows that the present equation is uniformly superior.

The use of hard convex molecules as a reference fluid for real fluids has received much attention lately, especially in perturbation theory formulations (1-4) dealing with nonspherical molecules. This is due to the recognition that in liquids the structure is essentially determined by the repulsive part of the molecular interaction potential. On the other hand, the attractive interaction makes a major

0097-6156/86/0300-0227$06.75/0

contribution to the energy of the liquid (Note: the energy derived from the hard convex bodies, HCB, is zero, and is therefore inadequate for thermal properties.) In order to achieve a complete description of the thermodynamics of the liquid state, both contributions, attractive as well as repulsive, should be considered.

Beret and Prausnitz (5) proposed the perturbed hard chain (PHC) equation of state for long chain hydrocarbons. It contained a modified hard sphere term of repulsion and a square well term for attractive pressure. It was applied to hydrocarbons such as methane, ethane, n-decane and eicosane as well as molecules involving polar forces, such as carbon monoxide and water. Later the approach has been extended (Donohue and Prausnitz (6) 1978, Gmehling and Prausnitz (7) 1979) to mixtures and highly polar substances such as methanol, ethanol, acetone and acetic acid by using a chemical theory. The BACK (Boublik-Alder-Chen-Kreglewski) equation proposed by Chen and Kreglewski (8) and Simnick, Lin and Chao (9) was based on the Boublik (10) equation of hard convex bodies. It has been successfully applied to fluids such as methane, neopentane and hydrogen sulfide. In these formulations the attractive contribution to the pressure was expressed in a 24-parameter temperature-density double power series originally given by Alder, Young and Mark (11). This attractive term was later modified to a 21-parameter or a 10-parameter series. These series, however, are cumbersome to use and do not reveal the underlying physical basis.

We propose here a theoretical formulation of the attractive contributions to the equations of state based on the distribution function theories of liquids and to derive equations that are applicable to moderately anisotropic fluids. First we use the Gaussian overlap potential of Berne and Pechukas (12) to represent anisotropic forces. We note that a continuous potential (see Figure 1) can be approximated by an n-step potential. In the limit $n \rightarrow \infty$, the original potential is recovered. The validity of this representation is closely associated with the Riemann integration theory in real analysis. Secondly, we apply the cluster theories of correlation functions for the derivation of the attractive pressure. The approach is therefore different from the perturbation approach. These developments will be presented in Sections 2 and 3.

Earlier studies on rigid nonspherical molecules were based on the so-called scaled particle theory (13) (SPT). This theory made use of the fact that the chemical potential of an N-body system is related to the energy required to create a cavity in the fluid in order to accommodate an additional particle. As developed by Reiss et al., (13) SPT contains certain approximations. Thus the results of the theory are not exact. Gibbons (14,15) first applied this theory to hard convex bodies and obtained an equation of state of the form

$$P/\rho kT = 1/(1-y) + 3\alpha y/(1-y)^2 + 3\alpha^2 y^2/(1-y)^3 \quad (1)$$

where y is the packing fraction, $y=\rho b$, b is the volume of a single hard convex molecule. α is a ratio of geometries of the HCB,

$$\alpha = \bar{r}s/3b \quad (2)$$

where $\bar{r}$ is the mean radius of curvature, and s the surface area of the HCB molecule. For different shapes, the mean radius of curvature is given by different geometric formulas. As an example, for spherocylinders, the geometric factors are

$$\bar{r} = R + L/4, \quad s = 4\pi R^2 + 2\pi RL, \text{ and } b = (4/3)\pi R^3 + \pi R^2 L \tag{3}$$

where R is the radius of the hemispheres and L the bond length between the centers of the hemispheres. For spheres, $\bar{r}$ = R, the radius of the hard sphere (HS).

Equation 1 is accurate at low densities but deteriorates at high densities. Thus it must be improved. Boublik (10,16) proposed a modified form of (1) for prolate spherocylinders that reduces to the Carnahan-Starling (17) equation at α=1:

$$P/\rho kT = 1/(1-y) + 3\alpha y/(1-y)^2 + 3\alpha^2 y^2/(1-y)^3 - \alpha^2 y^3/(1-y)^3 \tag{4}$$

Nezbeda (18) upon considering the virial coefficients, proposed an alternative resummation formula

$$P/\rho kT = [\, 1+(3\alpha-2)y+(\alpha^2+\alpha-1)y^2-\alpha(5\alpha-4)y^3]/(1-y)^3 \tag{5}$$

which, in comparison with simulation results for spherocylinders, is more accurate than Equation 4, especially at high densities.

These equations have been developed for the specialized geometry of spherocylinders. In later studies (Nezbeda and Boublik (19, 20) on fused diatomic hard spheres, it was found that Equation 4 remains applicable if one substitutes for the dumb-bell an equivalent spherocylinder that has the "neck" filled in. Thus the fused spheres can also be described by the equations developed for hard convex bodies. A recent study [Wojcik and Gubbins (21)] showed that Equation 4 is also applicable to mixtures of hard dumb-bells. Nezbeda and Boublik (22) (see also Nezbeda, Smith and Boublik (23), Nezbeda, Pavlicek and Labik (24), Boublik (25)) classified hard convex bodies into three major types (i) linear molecules (e.g. prolate spherocylinders, linear fused hard spheres and diamonds), (ii) disk-like molecules (e.g. oblate spherocylinders) and (iii) cubes. They found that Equation 4, derived for prolate spherocylinders, is accurate for linear molecules, and reasonable for disk molecules, but is less satisfactory for cube-like molecule. We shall adopt the Nezbeda Equation 5 here to describe the harsh repulsive forces in real molecules not of the cubic shape.

Theoretical Developments

We formulate our approach in terms of a square-well (SW) potential. The SW potential has been extensively studied (for a review, see Luks and Kozak (26) and references contained therein). This is a simple potential embodying the essential features of the interaction forces in real molecules, i.e., the excluded volume and attractive forces. For nonspherical molecules, the orientational variations of the pair interaction should also be accounted for. One class of angle-dependent potentials that can be generated from simple spherical ones is the so-called Gaussian overlap model of Corner (27) and

Berne and Pechukas (12). They built the molecular anisotropy into the potentials by incorporating angle dependence into the potential parameters. Therefore a Gaussian overlap model for the square-well (GOSW) potential is, (See Figure 2)

$$u(12) = \begin{cases} \infty, & r < d(\omega_1\omega_2) \\ -\varepsilon(\omega_1\omega_2), & d(\omega_1\omega_2) < r < \lambda(\omega_1\omega_2) \\ 0, & r > \lambda(\omega_1\omega_2) \end{cases} \tag{6}$$

where the potential parameters, d, λ and ε are functions of relative orientation, i.e. the Euler angles ω_1 and ω_2. We note that although Equation 6 is a simple model, it can be generalized to more elaborate potentials by using multistep square wells. For example, the spherical Lennard-Jones potential depicted in Figure 1 can be approximated by an n-step spherical square well potential, particularly as $n \to \infty$. Thus by increasing the number of steps, one can approximate a variety of smooth potentials. The strengths of interaction can also be approximated by the heights of the flights (which yield impulsive forces at the distances of discontinuities). The same procedure can be applied to angle dependent smooth potentials when the same angle dependence is reproduced by the Gaussian overlap parameters. These multistep SW have been applied to simulate hydrogen bonding in water (Dahl and Andersen (28)) and vibrating dumb-bells (Chapela and Martinez-Casas (29)). Kincaid, Stell and Goldmark (30) have shown that with a positive shoulder at λ_1, while setting all attractive energies, $\varepsilon_i=0$, the fluid can support a critical point (see also Kreglewski (31)). For the purpose of this study, we consider a two-step SW potential consisting of a positive shoulder and a negative well (Figure 3).

$$u(12) = \begin{cases} \infty, & r < d(\omega_1\omega_2) \\ \varepsilon_1(\omega_1\omega_2), & d(\omega_1\omega_2) < r < \lambda_1(\omega_1\omega_2) \\ -\varepsilon_2(\omega_1\omega_2), & \lambda_1(\omega_1\omega_2) < r < \lambda_2(\omega_1\omega_2) \\ 0, & r > \lambda_2(\omega_1\omega_2) \end{cases} \tag{7}$$

The virial equation for anisotropic molecules in statistical mechanics is given by [Hansen and MacDonald (32)].

$$P/\rho kT = 1 - \rho/6 \int dr\ 4\pi r^2 \langle r\ (\partial\beta u(12)/\partial r) g(r\omega_1\omega_2)\rangle_{12} \tag{8}$$

where the angular brackets <.> indicate the angle average

$$\langle\ldots\rangle_{12} = 1/(4\pi)^2 \int d\omega_1\ d\omega_2\ (\ldots) \tag{9}$$

$$= (4\pi)^{-2} \int d\theta_1\ \sin\theta_1\ d\phi_1\ d\theta_2\ \sin\theta_2\ d\phi_2\ (\ldots)$$

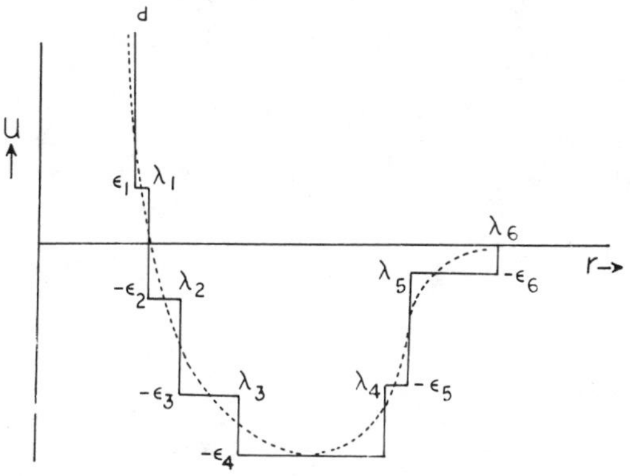

Figure 1. Approximation of the Lennard-Jones potential by a six-step square-well potential. Note that there are three repulsive shoulders and three attractive wells following thereafter. The repulsive step acts effectively as a temperature dependent hard core dimension commonly used in perturbation theories for soft repulsion.

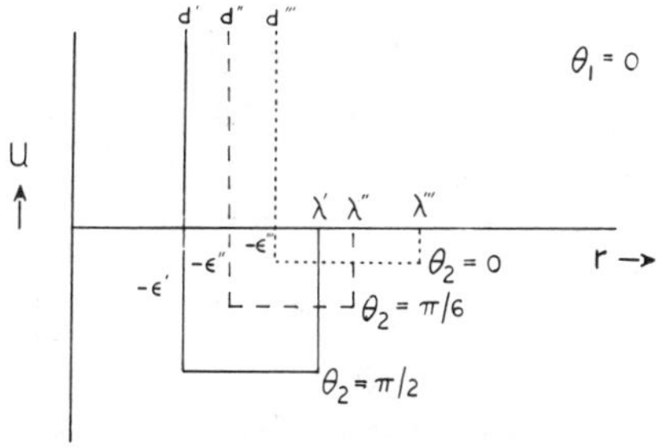

Figure 2. Schematic drawing of the Gaussin overlap anisotropic square-well potential, $u(r,\theta_1,\theta_2,\phi_{12})$, at three values of θ_2 (0, $\pi/6$ and $\pi/2$). θ_1 is kept at 0. The dependence on the azimuthal angle, ϕ_{12}, is not shown. The parameters, d, ε, and λ are the Kihara hard core dimension, the attractive energy and the range of attractive interaction, respectively. The primed quantities are values at different angles.

For nonlinear molecules, one should also integrate over the rotational angles and normalize by $(8\pi^2)^2$. Note that the gradient of the pair potential can be written in terms of the Boltzmann factor, $e(12) = \exp[-\beta u(12)]$, as

$$\partial\beta u(12)/\partial r = -(\partial e(12)/\partial r)\exp[\beta u(12)] \tag{10}$$

However, due to the form of the two-step SW potential, the derivative $\partial e(12)/\partial r$ is a sum of three Dirac deltas, one arising from the hard core repulsion at $d(\omega_1\omega_2)$, the others from the steps at $\lambda_1(\omega_1\omega_2)$ and $\lambda_2(\omega_1\omega_2)$

$$\partial e(12)/\partial r = \delta(r,d)\exp(-\beta\varepsilon_1) + \delta(r,\lambda_1)[\exp(\beta\varepsilon_2)- \exp(-\beta\varepsilon_1)] - \delta(r,\lambda_2)[\exp(\beta\varepsilon_2)-1] \tag{11}$$

Substitution of Equation 10 and Equation 11 in Equation 8 and interchange of r-integration with angle averaging gives

$$\begin{aligned} P/\rho kT = 1 &+ (4\pi/6)\rho \langle d^3 e^{-\beta\varepsilon 1} y_{SW}(d,\omega 1,\omega 2)\rangle_{12} \\ &+ (4\pi/6)\rho \langle \lambda_1{}^3(e^{\beta\varepsilon 2}-e^{-\beta\varepsilon 1}) y_{SW}(\lambda 1,\omega 1,\omega 2)\rangle_{12} \\ &- (4\pi/6)\rho \langle \lambda_2{}^3(e^{\beta\varepsilon 2}-1) y_{SW}(\lambda 2,\omega 1,\omega 2)\rangle_{12} \end{aligned} \tag{12}$$

where y_{SW} is the background distribution function for the square-well potential defined as

$$y_{SW}(12) = g(12)e^{\beta u(12)} \tag{13}$$

The second term is the repulsive contribution at the contact of hard cores. We note that for SW, the contact value, $y_{SW}(d)$, is not the same as that for hard spheres. In the following we shall analyze the simulation data for spherical SW to determine the relation between y_{SW} and y_{HS}. The third term is due to the repulsive shoulder at $\lambda_1(\omega_1\omega_2)$. This term actually plays the role of the temperature-dependent hard core dimension often used in conventional perturbation theories for soft potentials. The fourth term is the attractive contribution. It can be treated as a mean field value in case of continuous potentials. To gain a better understanding, we analyze simulation data on spherical square wells next.

Contact Values of Correlation Functions for the SW Potential

Henderson et al. (33,34) have carried out computer simulations for the spherical one-step square well potential with different widths, $\lambda^* = \lambda/d = 1.125 - 2.0$, over wide ranges of state conditions. The potential well is a negative step, $-\varepsilon$, beginning at d and vanishing at λ. Table 1 shows the contact values of the background correlation functions for SW and HS, respectively. The hard spheres are taken to have the same collision diameter, d, as that of the SW (i.e., at $\beta\varepsilon = 0$, the SW reduces to the HS potential). We make the following observations :

(i) The data clearly show that the Weeks-Chandler-Andersen zeroth order approximation

$$y_{SW}(r) \cong y_{HS}(r) \tag{14}$$

is inadequate. In fact, it is true only in the limit of high temperatures (or low densities).
(ii) For a given density, the simple empirical formula

$$y_{SW}(d) \cong y_{HS}(d)\exp(\zeta\beta\varepsilon) \tag{15}$$

is found to apply with surprising accuracy for wide ranges of well widths, (λ^* from 1.125 to 2.0, see Table 1). The index ζ has the values of -1.18 at $\rho^* = \rho d^3$ =0.8, -1.16 at ρ^* =0.6 and -1.1 at ρ^* =0.4 (for the well width λ^* = 1.5). Since at low densities, both y_{SW} and y_{HS} approach 1, ζ must approach zero at zero density. Figure 4 shows the variations of the index, ζ, with density. The contact value of $y_{SW}(d)$ (Note that the y function is continuous at d) is also dependent on the well width, λ. We have displayed the ζ values for cases λ^* = 1.375, 1.5, 1.625 in Figure 4. The three curves are fairly consistent in their density dependence, ζ is around -1.0 at medium densities (i.e., near ρ^* = 0.5).
(iii) The values of $y_{SW}(\lambda)$ at the attractive wall can be represented by

$$y_{SW}(\lambda) \cong y_{HS}(\lambda)\exp(\xi\beta\varepsilon) \tag{16}$$

The index ξ has the values -0.6 at ρ^* = 0.8, -0.525 at ρ^* = 0.6 and -0.535 at ρ^* = 0.5 (for λ/d = 1.5). Similar results are observed for other well widths (see Table 1). The value of ξ is around -0.5 at high densities. Again, as $\rho^* \to 0$, ξ must be zero. Thus ξ is a function of ρ^*.

Equation 12 reduces, for an angle-independent one-step square-well potential, to

$$P/\rho kT = 1 + (4\pi/6)\rho d^3 y_{SW}(d)e^{\beta e} - (4\pi/6)\rho\lambda^3(e^{\beta\varepsilon}-1)\ y_{SW}(\lambda) \tag{17}$$

The above observations can be utilized to transcribe the pressure equation (for one-step SW) into

$$P/\rho kT = 1 + (4\pi/6)\rho d^3 y_{HS}(d)e^{(1+\zeta)\beta\varepsilon} - (4\pi/6)\rho\lambda^3(e^{\beta\varepsilon}-1)\ y_{HS}(\lambda)e^{\xi\beta\varepsilon} \tag{18}$$

We identify the second term as related to the hard sphere pressure, written as

$$(4\pi/6)\rho d^3 y_{HS}(d) = Z_{HS} - 1 = Z'_{HS} \tag{19}$$

where Z'_{HS} is the configurational part of the compressibility factor. To carry out the calculation, it is necessary to find an expression for the value of $y_{HS}(\lambda)$ at the attractive wall. From distribution

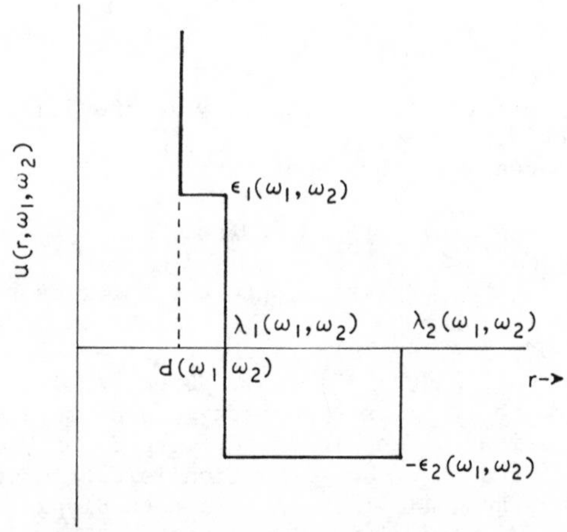

Figure 3. The two-step square-well potential with angle dependent parameters. This potential is used in the present correlation.

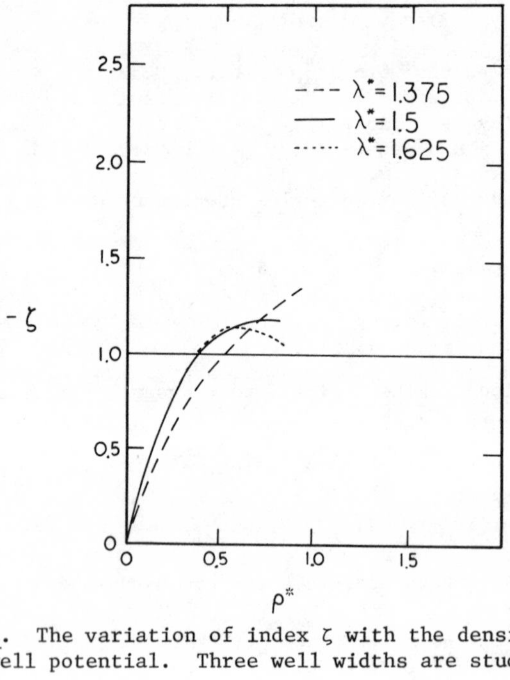

Figure 4. The variation of index ζ with the density for one-step square-well potential. Three well widths are studied.

function theories, we know that the background correlation function
has the following cluster representation (35)

$$y_{HS}(\lambda) = e^{\,\triangle\!\!\!\triangle \,+\, \square \,+\, \triangle\!\!\!\square \,+\, \triangle\!\!\!\square \,+\, \frac{1}{2}\,\square\!\!\!\triangle \,+\, \frac{1}{2}\,\square\!\!\!\triangle} + O(\rho^3) \tag{20}$$

The cluster integral $\triangle\!\!\!\triangle$ for hard spheres is known (36)

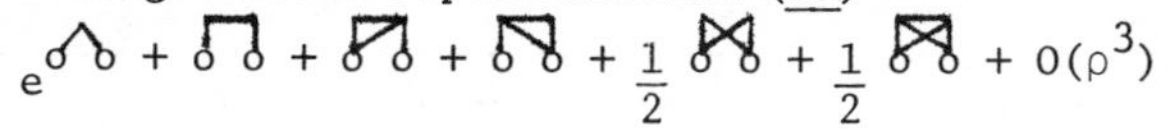

$$\triangle\!\!\!\triangle = (\pi/12)\rho^* \,(\lambda^{*3}-12\lambda^* +16) \tag{21}$$

where $\rho^* = \rho d^3$. For $\lambda^* = 1.5$, $\triangle\!\!\!\triangle = 0.36\ \rho^*$. Higher density terms
of the cluster integrals have also been published (37). However,
for simplicity we propose to approximate Equation 20 by

$$y_{HS}(h) = e^{\,\triangle\!\!\!\triangle}\,[1+f] = e^{\,\triangle\!\!\!\triangle} + e^{\,\triangle\!\!\!\triangle}f \tag{22}$$

where f is a function of density. We next approximate f by

$$e^{\,\triangle\!\!\!\triangle}f = a_o\,\rho^*\,e^{-a''\rho^{*2}} \tag{23}$$

Obviously, other forms of density dependence could be introduced.
We used simulation data (33) to determine the constants, a_o and a''
and found

$$a_o = -0.34378 \text{ and } a'' = 1.2284 \tag{24}$$

for the case $\lambda^* = 1.5$. Thus the complete pressure equation can be
written as

$$P/\rho kT = 1 + Z'_{HS}e^{(1+\zeta)\beta\varepsilon} \tag{25}$$

$$- (4\pi/6)\rho\lambda^3(e^{\beta\varepsilon}-1)\,[e^{0.36\rho^*} + a_o\,\rho^*\,e^{a''\rho^{*2}}]\,e^{\xi\beta\varepsilon}$$

This equation is used to calculate the pressure for SW fluid at
$\rho^* = 0.8$ and for temperatures $\beta\varepsilon = 0.25$ to 1.75 at well width $\lambda =$
1.5d. The results are compared with simulation data in Table II. The
agreement is quite satisfactory over the range of conditions compared. For other densities, similar results are obtained. We therefore adopt the form Equation 25 as one of our working equations for
the equation of state studies.

Generalization of Equation 25 to angle-dependent GOSW potential
is straightforward. For example, the one-step form (with negative ε)
of Equation 12 is

$$P/\rho kT = 1 + (4\pi/6)\rho \,<d^3\, e^{\beta\varepsilon}\, y_{SW}(d,\omega_1,\omega_2)>_{12} \tag{26}$$

$$- (4\pi/6)\rho\, <\lambda^3(e^{\beta\varepsilon}-1)y_{SW}(\lambda,\omega_1,\omega_2)>_{12}$$

By mean value theorem, we may extract out the angle averages

$$P/\rho kT = 1 + (4\pi/6)\rho\bar{d}^3\,e^{\beta\bar{\varepsilon}}\,y_{SW}(\bar{d}) \tag{27}$$

$$- (4\pi/6)\rho\bar{\lambda}^3(e^{\beta\varepsilon}-1)y_{SW}(\bar{\lambda})$$

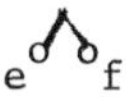

Table I. The Contact Values of Correlation Functions for the One-Step Spherical Square-Well Potential

$\lambda^* = 1.5$				
$\beta\varepsilon$	$y_{SW}(d^+)$ (MC)	$y_{HS}(d)e^{\zeta\beta\varepsilon}$	$y_{SW}(\lambda)$ (MC)	$y_{HS}(\lambda)e^{\xi\beta\varepsilon}$
$\rho d^3 = 0.5$		$\zeta = -1.1$		$\xi = -0.535$
0.0	2.12	2.12	0.98	0.98
0.25	1.62	1.61	0.864	0.857
0.50	1.225	1.223	0.77	0.75
0.75	0.95	0.93	0.675	0.66
1.00	0.72	0.71	0.574	0.574
$\rho d^3 = 0.6$		$\zeta = -1.16$		$\xi = -.525$
0.0	2.56	2.56	0.92	0.92
0.25	1.92	1.91	0.802	0.807
0.50	1.41	1.43	0.716	0.708
0.75	1.02	1.07	0.638	0.621
1.00	0.80	0.80	0.544	0.544
$\rho d^3 = 0.8$		$\zeta = -1.18$		$\xi = -0.6$
0.0	3.97	3.97	0.73	0.73
0.25	3.01	2.96	0.63	0.63
0.50	2.21	2.20	0.55	0.54
0.75	1.67	1.64	0.487	0.47
1.00	1.20	1.22	0.423	0.40
1.25	0.90	0.908	0.35	0.35
1.50	0.72	0.68	0.315	0.3
1.75	0.58	0.50	0.27	0.26
$\lambda^* = 1.625$				
$\rho d^3 = 0.4$		$\zeta = -1.0$		
0.0	1.812#	1.812	--	--
0.333	1.30	1.3		
0.5	1.10	1.1		
0.80	0.85	0.81		
$\rho d^3 = 0.6$		$\zeta = -1.15$		$\xi = -0.538$
0.0	2.56	2.56	--	(0.861)*
0.333	1.78	1.74	0.72	0.72
0.50	1.44	1.44	0.67	0.66
0.80	1.02	1.02	0.56	0.56
$\rho d^3 = 0.8$		$\zeta = -0.96$		$\xi = -0.478$
0.0	3.97	3.97	--	(0.727)
0.333	2.77	2.88	0.62	0.62
0.50	2.58	2.46	0.58	0.57
0.80	1.86	1.84	0.51	0.50
1.25	1.178	1.195	0.40	0.40

$\beta\varepsilon$	$y_{SW}(d^+)$ (MC)	$y_{HS}(d)e^{\zeta\beta\varepsilon}$	$y_{SW}(\lambda)$ (MC)	$y_{HS}(\lambda)e^{\xi\beta\varepsilon}$
$\lambda^* = 1.375$				
$\rho d^3 = 0.8$		$\zeta = -1.19$		$\xi = -0.58$
0.00	3.97	3.97	--	(0.942)
0.667	1.72	1.795	0.65	0.64
1.00	1.21	1.21	0.52	0.527
1.25	0.82	0.897	0.46	0.46
$\lambda^* = 1.125$				
$\rho d^3 = 0.8$		$\zeta = -0.90$		$\xi = -0.684$
0.0	3.97	3.97	--	(2.26)
1.0	1.58	1.61	1.14	1.138
1.5	1.02	1.029	0.79	0.81
2.0	0.66	0.656	0.58	0.575
$\lambda^* = 2.00$				
$\rho d^3 = 0.8$		$\zeta = -0.67$		$\xi = -0.495$
0.0	3.97	3.97	--	(1.2)
0.1667	3.65	3.55	1.11	1.105
0.333	3.17	3.175	1.02	1.017
0.5	2.95	2.84	0.92	0.937

*Values in parentheses are estimated.

Table II. The Pressure Prediction for One-Step Square-Well Potential

$\lambda^* = 1.5$

$\rho d^3 = 0.8$

$\beta\varepsilon$	Z (MC)	Z From Equation 25	$Z_2^{\#}$ (From second virial) /attractive
0.25	6.47	6.34	-1.61
0.50	5.08	5.09	-3.67
0.75	3.84	3.86	-6.32
1.00	2.34	2.65	-9.71
1.25	1.35	1.44	-14.08
1.50	0.18	0.22	-19.69
1.75	-0.65	-1.0	-26.88

*We have used $\zeta = -1.18$ and $\xi = -0.6$ for the calculations.

#Z_2 is the part of second virial coefficient due to attractive contribution.

i.e. $Z_2 = -(4\pi/6)\rho\lambda^3(e^{\beta\varepsilon} - 1)$

This results in temperature (and possibly density) dependent mean value parameters $\bar{d}=\bar{d}(T)$, $\bar{\varepsilon}=\bar{\varepsilon}(T)$ and $\bar{\lambda}=\bar{\lambda}(T)$. We approximate the repulsive pressure by an equivalent hard convex body pressure according to the relations Equation 15 and Equation 19 and the attractive pressure according to Equation 25

$$P/\rho kT = 1+Z'_{HCB}e^{(1+\zeta)\beta\varepsilon} - (4\pi/6)\rho\lambda^3(e^{\beta\varepsilon}-1)\ [e^{b'\rho^*\beta\varepsilon} + b_o\ \rho^*\ e^{b''\rho^{*2}\beta\varepsilon}] \tag{28}$$

Note the replacement of the hard sphere Z_{HS} by the hard convex body Z_{HCB}. Z_{HCB} is set to the Nezbeda form, Equation 5. The term involving b_o, b' and b'' is a representation of the temperature and density dependence of $y_{SW}(\lambda)$. Generalization to the two-step SW potential mentioned above is less obvious. One should use simulation data (these being unavailable at the present time) to verify the correct parametrization. If we insist on the same functional form as Equation 28 but treat the parameters as empirical constants, we obtain

$$\begin{aligned} Z = 1 + Z'_{HCB}e^{-(1+\zeta)\beta\varepsilon_1} &+\rho V_1(e^{\beta\varepsilon_2} - e^{-\beta\varepsilon_1})[e^{-d_1 y/T^*} + d_o y\ e^{-d_2 y^2/T^*}] \\ &-\rho V_2(e^{\beta\varepsilon_2} - 1)[e^{-c_1 y/T^*} + c_o y\ e^{-c_2 y^2/T^*}] \end{aligned} \tag{29}$$

where V_1 and V_2 are, respectively, the repulsive volume of the shoulder at λ_1 and attractive volume of the well at λ_2. The constants c_i and d_i, i=0,1,2, are introduced to account for the state dependence of the correlation functions y_{SW} at the locations of discontinuities. This form of functional dependence is arbitrary. The form is suggested by Equation 25. However, it mimics the cluster form Equation 20. T* is a reduce temperature, to be specified later. We note the similarity of Equation 29 with those derived by Reijnhart (38,39) and Henderson and Chen (40) from different methods. Due to the temperature and density dependence of the mean value parameters $\bar{\varepsilon}_i(T)$, $\bar{\lambda}_i(T)$ and $y_{SW}(\bar{\lambda}_i)$, we have combined the density and temperature terms in the exponentials. Equation 29 gives the pressure expression for the two-step Gaussian overlap square well potential. In the following we shall discuss its applications.

The P-v-T Behavior of Real Fluids

We have applied Equation 29 and its modifications to the calculation of the pressures and enthalpies of real fluids, such as hydrocarbons (e.g. methane, ethane, n-decane, and eicosane), nonhydrocarbons (e.g. carbon dioxide, and hydrogen sulfide), alcohols (1-butanol and phenol), ketones (acetone and 2-butanone), amines (methylamine and diethylamine), ethers (dimethylether and diphenylether) and polar solvents (e.g. tetrahydrofuran). Some 69 substances have been investigated.

The average conformation of a polyatomic molecule in the fluid state is characterized by a geometric ratio, α. We do not assign a specific "shape" to the molecules since conformational changes (e.g. bond rotation and solvent forces) constantly "deform" the molecule. α should reflect the averaged effects of such changes. Other molecular parameters for use in Equation 29 are b, the volume of a single molecule, ε_1 and ε_2, the energies of interaction, λ_1 and λ_2, the widths of potential walls, and the constants c_i and d_i (i=0,1,2) specifying the temperature and density dependence. However, in practice, it soon becomes apparent that the repulsive second virial, $(1+3\alpha)b$, given by Z_{HCB}, is inaccurate, as pointed out by Kohler and Haar (41). To obtain an adequate description of the second virial coefficient of gases, we modify Equation 29 to give

$$Z = 1 + Z'_{HCB}e^{-(1+\zeta)\beta\varepsilon_1} - (1+3\alpha)ye^{-(1+\zeta)\beta\varepsilon_1} + yB_r^*e^{-\beta\varepsilon_1} + yV_1^*(e^{\beta\varepsilon_2} - e^{-\beta\varepsilon_1})[e^{-d_1y/T^*} + d_oy\, e^{-d_2y^2/T^*}] - yV_2^*(e^{\beta\varepsilon_2} - 1)[e^{-c_1y/T^*} + c_oy\, e^{-c_2y^2/T^*}] \tag{30}$$

i.e. we have taken out the HCB second virial, $(1+3\alpha)b\exp[-(1+\zeta)\beta\varepsilon_1]$, and substituted for it the repulsive term

$$B_r\exp(-\beta\varepsilon_1) \tag{31}$$

B_r being an effective repulsive second virial coefficient. (Note, $B_r^* = Br/b$, $V_1^* = V_1/b$, and $V_2^* = V_2/b$) Equation 30 automatically reduces to a second virial expression, i.e. $P/kT = \rho + B_2\rho^2$ with

$$B_2 = B_r\exp[-\beta\varepsilon_1] + V_1\ [\exp(\beta\varepsilon_2) - \exp(-\beta\varepsilon_1)] - V_2[\exp(\beta\varepsilon_2 - 1)] \tag{32}$$

At the moment, we have thirteen parameters to use in the equation of state. This number of parameters is considered excessive in practical calculations, especially when some of the parameters (e.g. the c_i's and d_i's) are empirical and have no physical meaning. They could have been evaluated if simulation data for the 2-step SW fluid were available. In the absence of information we embark on a reduction of parameters by setting

$$\varepsilon_1 = 0 \tag{33}$$

$$d_i = c_i, \qquad i = 0,1,2 \tag{34}$$

$$T^* = kT/\varepsilon_2 \tag{35}$$

We have also found from fitting the data that ε_2/k is related to the critical temperature by

$$\varepsilon_2/k = T_c/2.18601 \tag{36}$$

and

$$V_1 = 1.4\ b \text{ (Kreglewski 1984)} \tag{37}$$

We define an attractive volume (between λ_1 and λ_2), V_a as

$$V_a = V_2 - V_1 \tag{38}$$

or

$$V_2 = 1.4b + V_a \tag{39}$$

The above prescription has reduced the number of parameters from 13 to 7, i.e., B_r, V_a, c_i (i=0,1,2), α and b. Therefore the final working equation is given by

$$Z = (1-y)^{-3}[1+(3\alpha-2)y+(\alpha^2+\alpha-1)y^2-\alpha(5\alpha-4)y^3] \tag{40}$$

$$-(1+3\alpha)y + \rho B_r$$

$$-\rho V_a (e^{1/T*}-1)[e^{-c_1 y/T*} + c_o y e^{-c_2 y^2/T*}]$$

where $y=b\rho$, V_a is given by Equation 38 and T* by Equation 35 and Equation 36.

There is no compelling reason for the choices made in Equation 33 to Equation 39 except that we have achieved economy of parameters. Future research shall address such questions in more detail, preferably based on detailed simulation data.

We first used the P-v-T data of paraffin hydrocarbons from C_1 to C_{20} (methane to eicosane) to determine the parameters. (N.B. We have excluded the critical region in the fit.) As the critical behavior is nonanalytical, we do not expect Equation 40 to apply. Normally, we are ± 2°C away from the critical temperature. The results are listed in Table III. These values are empirically determined. However, they are not entirely arbitrary. The molecular volumes, b, thus determined are directly proportional to the critical and/or van der Waals volumes (Bondi (42)) for hydrocarbons. (See Figure 5). The energy parameter, ε/k, is proportional to the critical temperature as mentioned above. ($\varepsilon = \varepsilon_2$).

The agreement in density predictions for twenty hydrocarbons is around 1%, and in vapor pressures, for the most part, less than 1%. We are also able to predict the configurational enthalpy to within 1 cal/g. (See Table V). This equation is particularly effective in the liquid state, e.g., for liquid ethane and n-pentane up to 69 MPa. In a number of perturbation formulations, the contribution from the attractive second virial coefficient overwhelmed other contributions at low temperatures and liquid densities, thus requiring a high order density correction (up to the sixth virial in, e.g., Bienkowski and Chao 1975). As an example, Table II exhibits the values of the attractive second virial in SW fluid. It reaches -26.9 at $\beta\varepsilon = 1.75$. Use of the exponential form Equation 25 has moderated such influences and obviated the necessity of retaining explicitly high order density terms. The same behavior is obtained by Equation 40.

We have made calculations with other equations of state, such as Peng-Robinson (43) (PR), Mohanty-Davis (44) (MD), De Santis, Gironi and Marrelli (45) (DGM) and the BACK (Simnick et al. (9,46,47,48)) equation. The results based on the same data sets for density and vapor pressure are compared in Table IV from methane to n-decane. Equation 40 is uniformly better than all the equations compared. PR equation is reasonable for vapor pressure calculations (e.g., 1.5%

Table III. The Molecular Parameters* to Be Used in the Equation of State, Equation 40

	α	b, cc/gmol	B_r/b	V_a/b	c_o	c_1	c_2	D/b^2, K
Methane	1.02008	16.325	4.4186	20.036	3.4908	11.652	4.0623	0.0
Ethane	1.10299	23.629	5.4571	24.213	3.5874	14.840	4.7173	0.0
Propane	1.19523	31.795	6.3561	24.277	3.6030	11.715	4.3511	0.0
n-Butane	1.33117	39.831	6.4804	25.995	3.6669	12.950	4.3736	0.0
n-Pentane	1.37924	49.330	6.8402	27.438	3.6114	13.377	4.1813	0.0
n-Hexane	1.46968	56.797	7.3511	28.842	3.7117	8.2301	3.6129	0.0
n-Heptane	1.56411	66.381	7.3663	29.549	3.6766	13.768	3.9829	0.0
n-Octane	1.62814	74.956	8.3639	32.183	3.7150	14.547	4.3176	0.0
n-Nonane	1.73361	84.341	8.3276	33.405	3.6863	15.303	4.1272	0.0
n-Decane	1.79693	94.133	8.6238	34.592	3.6627	15.471	4.0298	0.0
n-Undecane	1.88289	101.72	8.7832	35.242	3.7154	15.539	3.9781	0.0
n-dodecane	1.96885	113.43	9.2483	37.098	3.5195	15.069	3.6788	0.0
n-Tridecane	2.05458	123.38	9.3391	37.462	3.6206	15.123	3.7317	0.0
n-Tetradecane	2.14076	132.02	10.044	40.291	3.6004	16.189	3.9976	0.0
n-Pentadecane	2.21226	141.11	10.651	42.723	3.5717	16.733	4.1281	0.0
n-Hexadecane	2.29972	150.65	11.270	45.207	3.5873	17.477	4.3009	0.0
n-Heptadecane	2.38626	160.94	11.201	44.931	3.5585	16.590	4.0536	0.0
n-Octadecane	2.46985	169.83	11.844	47.512	3.5761	17.528	4.3169	0.0
n-Nonadecane	2.57054	180.02	11.983	48.067	3.5600	16.961	4.1916	0.0
Eicosane	2.65651	193.74	12.457	49.967	3.4807	16.808	4.1385	0.0
Ethylene	1.05120	21.170	5.3742	22.722	3.3543	11.930	4.0591	0.0
Propylene	1.40083	27.870	5.9988	26.160	3.7196	14.069	4.8769	0.0
Isobutane	1.33986	40.237	6.1462	27.186	3.6441	15.352	4.6742	0.0
Isopentane	1.33484	48.689	6.9562	27.182	3.5997	13.145	4.3044	0.0
Benzene	2.05852	36.117	6.3696	26.729	4.6586	13.991	5.7104	0.0
Toluene	1.95468	44.494	8.0161	32.300	4.0585	15.230	5.7809	0.0
H S	1.36413	14.722	5.9335	25.050	3.6959	12.920	4.9114	0.0
Nitrogen	1.10870	14.762	5.2785	21.174	3.4955	13.016	3.7681	0.0
o-Xylene	1.38407	57.6142	7.8234	28.4512	3.6678	12.6131	4.2050	0.0
Bicyclohexyl	1.89254	97.3920	8.7008	33.7047	3.3930	12.396	3.8834	0.0
Fluorine	1.66415	83.0885	12.4446	39.2337	4.0983	16.972	6.1894	0.0
Diphenylmethane	1.82590	83.7528	8.6256	33.5330	3.6182	13.566	4.0255	0.0
Dichloro-difluoromethane	1.69801	30.9599	7.0824	28.4097	4.0143	13.830	5.6238	0.0
Carbon dioxide	2.20215	11.0023	4.8590	20.8229	4.8934	14.153	9.5355	10784.0
Ammonia	1.49024	10.7183	6.0547	28.7928	3.8722	17.299	4.8739	-85.944
Methyl fluoride	1.93017	17.9824	7.8860	31.3100	3.9492	15.114	6.0694	-48.39
Butanol	1.13685	47.9597	6.0482	15.6550	3.2727	1.0914	2.4495	1429.4
Phenol	1.51368	41.6151	8.0264	18.4726	3.1471	4.5436	3.1105	4996.2
p-Cresol	1.1934	57.2452	6.2618	32.1420	2.7713	17.998	2.8097	5869.6

Continued on next page

Table III. Continued

	α	b, cc/gmol	B_r/b	V_a/b	c_o	c_1	c_2	D/b^2, K
o-Cresol	1.01932	57.2771	7.5929	30.4572	2.8199	14.249	2.7697	2181.6
m-Cresol	1.45342	52.8778	7.5816	30.4120	3.3792	14.595	3.7238	8209.
2,3-Xylenol	1.51523	50.5423	7.5123	30.1339	4.0656	19.978	5.9300	19978.2
Acetone	2.41148	27.3953	8.6534	35.7104	4.3692	16.110	7.2340	1008.9
2-Butanone	1.30404	32.2047	5.7769	23.1723	4.7482	16.2711	5.3465	5249.2
2-Pentanone	1.34478	49.5422	8.4338	26.3432	3.4796	9.8353	3.7188	107.4
Dimethyl ether	1.87662	24.0541	7.0549	28.2993	4.2381	13.7299	5.7141	268.2
Methylethyl ether	1.20617	36.8069	5.4309	23.0285	3.4387	12.7078	3.2635	3045.3
Diethyl ether	1.86761	40.9721	6.5979	30.7533	3.9140	16.7066	5.2008	1138.7
Ethylpropyl ether	1.93416	47.4490	7.7616	32.3502	4.1240	16.6954	6.5101	2680.5
Diphenyl ether	1.54833	86.2730	4.9072	30.2940	3.4141	20.4675	3.1343	255.94
Acetic acid	2.77183	22.4067	10.8006	44.8423	4.3707	18.4607	8.1122	-798.44
Methylamine	1.08011	23.5104	5.8656	23.5286	3.4105	13.0372	2.8898	917.3
Dimethylamine	1.64228	29.5128	6.9561	27.7942	3.6289	12.4523	3.5060	150.6
Ethylamine	1.44901	28.1500	6.9604	27.8841	3.6215	13.4346	3.8865	177.4
Diethylamine	1.76156	45.4039	5.1196	29.1512	3.7969	19.1798	4.1423	490.4
Aniline	1.00785	49.2220	8.6346	26.4511	3.1887	9.2086	2.8025	-2274.8
Pyridine	2.21624	35.3380	7.1343	31.8251	4.1165	14.8389	4.7775	-5068.1
4-Methyl pyridine	1.35605	57.4625	9.0103	30.3674	3.1100	10.8422	3.7898	633.56
i-Quinoline	1.65480	56.8107	8.2691	27.9398	4.0630	11.4332	4.7670	-3145.9
Carbazol	2.03042	99.2549	8.9344	35.8262	3.6023	14.6698	4.1678	144.4
Acridine	1.85980	77.4694	12.2423	35.3462	4.0133	10.1169	4.6394	-15394.1
Formamide	1.08191	20.4789	0.8686	22.2851	4.2058	16.0239	3.2574	-13245.1
Ethyl mercaptan	2.16017	26.2166	8.1989	30.7596	4.5221	13.4265	7.6142	410.9
Dimethyl sulfide	1.85038	27.5070	6.5736	29.3502	4.1849	16.0913	6.8876	2469.5
Tetrahydro thiophane	1.28571	43.4649	5.9428	27.3692	3.4524	15.0371	3.4758	-4452.9
Thianaphthene	1.38088	54.8143	7.8006	28.3144	4.0240	13.9649	4.9626	1430.3
Dibenzo-thiophene	1.04294	34.3903	7.3413	26.9661	3.7304	10.8115	4.5357	-28998.1
Tetrahydrofuran	1.56977	68.9898	20.0423	49.7294	4.3846	19.4136	8.6362	-8.722
Dibenzofuran	2.40006	66.1668	9.3165	32.3696	4.3589	9.3444	5.3389	-17853.6

*The parameter, ε, is related to the critical temperature by $\varepsilon/k = T_c/2.18601$

Note that the reduced temperature is calculated as $T^* = kT/\bar{\varepsilon}$

where $\bar{\varepsilon}/k = \varepsilon/k + \frac{D}{b^2 T}$.

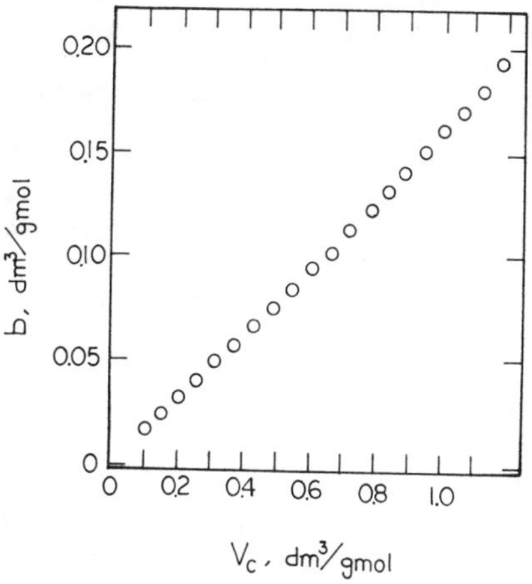

Figure 5. Variations of the hard core volume, b, with the critical volume V_c, for n-paraffins.

for methane and 3% for ethane). However, Equation 40 gives about half the deviations for all hydrocarbons studied (1% for methane and 0.7% for ethane). The BACK equation performs well for lower hydrocarbons (e.g. from methane to n-hexane), but deteriorates for higher hydrocarbons (e.g., to 14% in deviations for density and 6.7% for vapor pressure of n-decane). The vapor pressures predicted by BACK is of the order of 4% as compared to 1% by Equation 40. We note that the PR equation used is in its generalized form, which has some advantages in application.

In contrast to previous equations, such as PHC and BACK, the present equation is based on distribution function theory. Thus further improvements can be made by resummation of cluster integrals and/or inclusion of integral equation relations. The contributions from individual terms can also be checked against simulation data.

The same equation is applied to other nonpolar substances including branched chain hydrocarbons, unsaturated hydercarbons, and ring compounds. The parameters are presented in Table III and the results are summarized in Table V. The overall deviations for 607 data points of density, 435 points of vapor pressure and 147 points of enthalpy are 0.7%, 0.6% and 3 J/g, respectively. We have shown some sample compounds here, the detailed comparison can be found in a thesis (49).

In order to apply the equation to polar fluids, it is necessary to use a mean potential form of the energy parameter. It is given by

$$\bar{\varepsilon}/k = \varepsilon/k + D/(Tb^2) \qquad (41)$$

where one additional parameter, D, is introduced. This formulation is designed to introduce the temperature dependence in the mean value parameter $\bar{\varepsilon}(T)$. It accounts in part for the dipolar moment contributions in polar fluids. The temperature is now reduced according to $T^* = kT/\bar{\varepsilon}$. For example, for ammonia $D/b^2 = -85.94$ K. Some thirty-six amines, ketones, alcohols and ethers have been chosen for correlation. The results are given in Tables III and V. For over a thousand density data and a thousand vapor pressure data, the overall deviation is 1.2% and 0.4%, respectively. Further comparison is made with the Peng-Robinson and BACK equation in Table VI. The results again show that Equation 40 is more accurate based on the same set of experimental data.

We have shown here that, by empirically fitting the two-step square-well equation to experimental data, we are able to reproduce quantitatively the P-v-T information. The given formula can be generalized to highly polar molecules according to known statistical mechanics. To treat strong molecular anisotropy, it will be necessary to let b and λ depend on temperature. We also anticipate further developments of the present formulation to mixtures in the future.

Table IV. Comparison of Density (ρ) and Vapor Pressure (VP) Calculations Obtained from the Peng-Robinson (PR), Mohanty-Davis (MD), De Santis-Gironi-Marrelli (DGM), BACK Equations of State and Equation 40

	Type	No. Data	Temp. Range, K	Press. Range, MPa	Error AAD% Equation 40	PR	MD	DGM	BACK
Methane	ρ	39	115-623	0.9-14.3	0.93	5.4	3.3	8.7	0.4
	VP	32	112-190.6	0.1-4.6	0.99	1.5	2.3	--	1.07
Ethane	ρ	46	133-427.6	0.9-68.9	0.95	5.6	4.5	32.0	2.0
	VP	46	139-305	0.003-4.9	0.62	3.0	5.8	--	4.6
Propane	ρ	134	90-398	0.1-73	0.82	4.3	9.9	7.2	1.1
	VP	55	144-494	0.00001-4.26	0.79	3.1	7.7	--	4.4
n-Butane	ρ	40	144-494	0.1-48.2	0.56	3.7	2.8	5.9	1.2
	VP	52	194-411	0.001-3	0.80	4.11	6.12	--	5.1
n-Pentane	ρ	39	166.5-511	0.1-68.9	1.18	3.3	2.7	9.1	1.9
	VP	45	223-470	0.001-3.4	0.67	1.35	6.11	--	3.9
n-Hexane	ρ	41	177.6-433	0.1-27.4	0.34	1.6	7.0	3.6	1.0
	VP	53	220-508	0.0001-3	0.42	4.92	5.28	--	3.6
n-Heptane	ρ	41	205-511	0.1-21.2	0.77	1.5	3.5	4.6	2.6
	VP	40	231-531.6	0.00006-2.4	0.77	1.1	3.85	--	3.4
n-Octane	ρ	48	216.5-538.7	0.1-1.6	0.72	4.1	5.9	3.1	6.0
	VP	56	230-566.5	0.00001-2.4	0.8	3.73	--	--	3.6
n-Nonane	ρ	43	253-573	0-50	0.67	3.39	--	--	--
	VP	18	245.7-452.6	0.00001-0.2	0.05	4.05	2.15	--	--
n-Decane	ρ	32	311-511	1.4-41.3	0.47	4.7	8.7	--	14.3
	VP	18	350-477	0.003-0.2	0.04	2.09	8.75	--	6.7

Table V. P-v-T and Thermal Properties Calculated by Equation 40 for Some Selected Compounds

	Type	No. Data	Temp. Range, K	Press. Range, MPa	Error, AAD*
n-Hexadecane	ρ	9	463-543	0.007-0.07	0.17%
	VP	10	463-5553	0.007-0.09	0.17%
Eicosane	ρ	25	373-573	5.0-50.0	0.98%
	VP	15	473-623	0.002-0.11	0.42%
Ethylene	ρ	40	116.5-400	0.1-13.8	0.91%
	VP	35	133-412	0.69-5.1	0.94%
	H	34	188.7-400	0.7-1.4	4.7 J/g
Benzene	ρ	102	280-545	0.005-4.0	0.74%
	VP	67	280-562.6	0.005-4.9	0.35%
Carbon dioxide	ρ	39	243-413.2	1.5-38	0.39%
	VP	33	216.5-302.6	0.5-7.1	0.81%
	H	39	243-413.2	3-50.6	4.1 J/g
Ammonia	ρ	300	238-598	0.09-81.4	2.37%
	VP	172	221-403.6	0.04-11	0.08%
Phenol	ρ	14	323-673	0.0003-5.1	0.78%
	VP	15	323-694	0.0003-6.1	1.33%
Acetone	ρ	28	273.2-503	0.1-4.5	0.62%
	VP	20	329-508	0.1-4.8	0.24%
Acetic acid	ρ	21	392-583	0.1-5	0.52%
	VP	23	392-595	0.1-5.8	0.36%
Diethyl amine	ρ	17	329-483	0.1-3	0.50%
	VP	18	329-493	0.1-3.5	0.80%
Aniline	ρ	29	273.2-673	0-4	0.44%
	VP	34	273.2-648.6	0-3	0.76%
Pyridine	ρ	26	253-613	0.0001-5.3	0.76%
	VP	25	253-613	0.0001-5.2	0.88%
Tetrahydro-furan	ρ	10	253-333	0.002-0.08	0.27%
	VP	25	253-540	0.002-5.2	0.32%

*AAD is defined as

$$(1/n) \sum_{i=1}^{n} [V_{calc,i} - V_{expt,i}]/V_{expt,i}$$

For density (ρ) and vapor pressure (VP) predictions, the percent (x100%) AAD is used.
For enthalpy (H) predictions, the absolute value in Joule/g is used.
The sources of experimental data used here are listed in reference 49.

Table VI. Comparison of Equation 40 with the BACK and Peng-Robinson Equations of State

Component	Property	Number of Points	AAD		
			Equation 40	BACK	PR
Ethylene	RHO	40	0.91		3.19
	VP	35	0.94		5.22
	ENTH	34	2.00		2.77
Propylene	RHO	57	1.28		2.31
	VP	28	0.80		3.72
i-Butane	RHO	116	0.79		3.77
	VP	82	0.75		2.39
	ENTH	73	1.34		1.67
i-Pentane	RHO	38	0.84		3.01
	VP	22	0.29		0.90
Benzene	RHO	102	0.74	1.79	3.35
	VP	67	0.35	3.16	1.21
Toluene	RHO	13	0.31	1.82	0.54
	VP	33	0.89	5.24	2.31
o-Xylene	RHO	59	0.37		1.17
	VP	43	0.25		5.27
Dichloro difluoro methane	RHO	182	0.47		1.69
	VP	40	0.53		2.50
	ENTH	40	0.26		1.36
Carbon dioxide	RHO	39	0.39		1.89
	VP	33	0.81		0.61
	ENTH	39	1.76		3.09
Nitrogen	RHO	38	0.79	0.30	4.32
	VP	19	0.69	0.33	3.26
	ENTH	77	1.33	2.36	1.21
Hydrogen sulfide	RHO	40	0.53	0.74	3.69
	VP	24	0.39	2.12	4.23
Methyl fluoride	RHO	131	0.74		4.14
	VP	28	0.54		18.1
Phenol	RHO	14	0.78		10.5
	VP	15	1.33		6.11
Acetone	RHO	46	0.62		13.1
	VP	20	0.24		0.99
Dimethyl ether	RHO	34	2.90		3.77
	VP	19	0.26		3.46
Methyl ethyl ether	RHO	28	3.25		6.51
	VP	18	0.23		2.27
Diethyl ether	RHO	43	1.85		5.11
	VP	23	0.35		0.95
Aniline	RHO	29	0.44		2.98
	VP	34	0.76		8.09

Literature Cited

1. Kohler, F.; Quirke, N. J. Chem. Phys. 1979, 71, 4128.
2. Boublik, T. Fluid Phase Equil. 1979, 3, 85.
3. Tildesley, D. Mol. Phys. 1980, 41, 341.
4. Chung, T. H.; Khan, M. M.; Lee, L. L.; Starling, K. E. Fluid Phase Equil. 1984, 17, 351.
5. Beret, S.; Prausnitz, J. M. AIChE J. 1975, 21, 1123.
6. Donohue, M. D.; Prausnitz, J. M. AIChE J. 1978, 24, 849.
7. Gmehling, J.; Liu, D. D.; Prausnitz, J. M. Chem. Eng. Sci. 1979, 34, 951.
8. Chen, S. S.; Kreglewski, A. Ber. Bunsengesellschaft Phys. Chem. 1977, 81, 1048.
9. Simnick, J. J.; Lin, H. M.; Chao, K. C. Adv. Chem. Series 1979, 182, 209.
10. Boublik, T. J. Chem. Phys. 1975, 63, 4084.
11. Alder, B. J.; Young, D. A.; Mark, M. A. J. Chem. Phys. 1972, 56, 3013.
12. Berne, B. J.; Pechukas, P. J. Chem. Phys. 1972, 56, 4213.
13. Reiss, H.; Frisch, J. L.; Lebowitz, J. L. J. Chem. Phys. 1959, 31, 369.
14. Gibbons, R. M. Mol. Phys. 1969, 17, 81.
15. Gibbons, R. M. Mol. Phys. 1970, 18, 809.
16. Boublik, T. Mol. Phys. 1974, 27, 1415.
17. Carnahan, N. F.; Starling, K. E. J. Chem. Phys. 1969, 51, 635.
18. Nezbeda, I. Chem. Phys. Lett. 1976, 41, 55.
19. Nezbeda, I.; Boublik, T. Chem. Phys. Lett. 1977a, 46, 315.
20. Neabeda, I; Boublik, T. Czech. J. Phys. 1977b, B27, 1071.
21. Wojcik, M.; Gubbins, K. E. Mol. Phys. 1983, 49, 1401.
22. Nezbeda, I.; Boublik, T. Mol. Phys. 1984, 51, 1443.
23. Nezbeda, I.; Smith, W. B.; Boublik, T. Mol. Phys. 1979, 37, 985.
24. Nezbeda, I.; Pavlicek, J.; Labik, S. Coll. Czech. Chem. Commun. 1979, 44, 3555.
25. Boublik, T. Equation of State of Nonspherical Hard Body Systems in "Molecular-Based Study of Fluids"; Haile, J. M.; Mansoori, G. A., Eds.; ADVANCES IN CHEMISTRY SERIES No. 204, American Chemical Society, Washington, D.C., 1983.
26. Luks, K. D.; Kozak, J. J. Adv. Chem. Phys. 1978, 37, 139.
27. Corner, J. Proc. Roy. Soc. (London) 1948, A192, 275.
28. Dahl, L. W.; Andersen, H. C. J. Chem. Phys. 1983, 78, 1980.
29. Chapela, G. A.; Matinez-Casas, S. E. Mol. Phys. 1983, 50, 129.
30. Kincaid, J. M.; Stell, G; Goldmark, E. J. Chem. Phys. 1976, 65, 2176.
31. Kreglewski, A. "Equilibrium Properties of Fluids and Fluid Mixtures", Texas A&M University Press, College Station, Texas, 1984.
32. Hansen, J.-P.; McDonald, I. R. "Theory of Simple Liquids", Academic, New York, 1976.
33. Henderson, D.; Madden, W. G.; Fitts, D. D. J. Chem. Phys. 1976, 64, 5026.
34. Henderson, D.; Scalise, O. H.; Smith, W. R. J. Chem. Phys. 1980, 72, 2431.
35. Watts, R. O.; McGee, I. J. "Liquid State Chemical Physics", Wiley, New York, 1976.

36. Kirkwood, J. G. J. Chem. Phys. 1935, 3, 300.
37. Nijboer, B. R. A.; van Hove, L. Phys. Rev. 1952, 85, 777.
38. Reijnhart, R. Physica 1976, 83A, 533-547.
39. Peters, C. J.; van der Kooi, J. J.; Reijnhart, R.; Diepen, G. A. M. Physica 1979, 98A, 245.
40. Henderson, D; Chen, M. J. Chem. Phys. 1975, 16, 2042.
41. Kohler, F; Haar, L. J. Chem. Phys. 1981, 75, 388.
42. Bondi, A. Physical Properties of Molecular Crystals, Liquids and Glasses (Wiley, New York) 1968.
43. Peng, D. Y.; Robinson, D. B. Ind. Eng. Chem. Fundam. 1976, 15, 59.
44. Mohanty, K. D.; Davis, H. T. AIChE J. 1979, 25, 701.
45. DeSantis, R,; Gironi, F.; Marrelli, L. Ind. Eng. Chem. Fundamen. 1976, 15, 183.
46. Kreglewski, A; Chen, S. S. J. De Chim. Phys. 1978, 75, 347.
47. Bienkowski, D. R.; Chao, K. C. J. Chem. Phys. 1975a, 62, 615.
48. Bienkowski, D. R.; Chao, K. C. Fourth International Conference of Chemical Thermodynamics, Montpelier, 1975b, p. 28.
49. Kanchanakpan, S. B. Ph.D. Thesis, "Molecular Theoretical Equation of State for Applications to Selected Pure Liquids and Dense Fluids of Natural Gas, Petroleum Hydrocarbons and Coal Chemicals", University of Oklahoma, Norman, Oklahoma, 1984.

RECEIVED November 8, 1985

12

Application of a New Local Composition Model in the Solution Thermodynamics of Polar and Nonpolar Fluids

M. H. Li, F. T. H. Chung[1], C.-K. So[2], L. L. Lee, and K. E. Starling

School of Chemical Engineering and Materials Science, University of Oklahoma, Norman, OK 73019

A local composition model in solution thermodynamics is developed for the calculation of vapor-liquid equilibria of mixtures of molecules vastly different in size, polarity and strength of interaction. The concepts of nearest neighbor numbers, coordination shell, and pair interaction energies are interpreted in terms of modern liquid theory. For broad ranged applications, an accurate equation of state is introduced in the spirit of the mean density approximation by differentiation of the Helmholtz free energy. Calculations of the vapor-liquid equilibria of 83 binary and ternary systems, including nonpolar hydrocarbons, hydrogen-bonding alcohols, water, ammonia, and carbon dioxide show good agreement with experimental data.

Since its introduction, the local composition model (LCM) for the excess free energy has been used in solution thermodynamics for the calculation of vapor-liquid equilibria of highly nonideal mixtures. The systems included polar fluids (such as alcohols, water and acetone) (1-3). The formulation was in a form suitable for calculation of activity coefficients. However, the original method was not useful for the calculation of densities. It also did not apply to the near critical region. On the other hand, mixture

[1]Current address: NIPER, PO Box 2128, Bartlesville, OK 74005
[2]Current address: Aspen Tech, 251 Vassar St., Cambridge, MA 02139

0097-6156/86/0300-0250$08.75/0

properties have been predicted by using equations of state with mixing rules for the equation parameters (4). The predicted properties included density, enthalpy and vapor-liquid equilibrium (VLE).

The overlapping applications of these methods have coexisted for many years while the two approaches remained separate. Only recently have there been efforts to assimilate the useful features of the local composition activity coefficient model into equations of state(5). The theoretical groundwork connecting both approaches was laid by Lee, et al. (6) where a molecular theory was established for the LCM. In this work, we demonstrate the feasibility of the approach by calculating for the vapor-liquid equilibria as well as the densities of highly nonideal mixtures. We find that local composition mixing rules perform better than conformal solution mixing rules for most of the systems studied.

Theory

New Free Energy Model. The conventional practice in solution thermodynamics is to employ equations of state to describe the gaseous state of mixtures while using the so-called activity coefficient models (e.g., van Laar, Margules, Redlich-Kister) to describe the liquid mixtures. We attempt a synthesis here by combining the activity coefficient model with the equation of state. The basic approach is outlined below, and discussions that follow will give the details.

Outline of Theoretical Steps

1. Local composition expression of energy, (by integration: $A/NkT = \int d\beta\ (U/N)$) — U ↓
2. Helmholtz free energy (H.F.E.) (by the relation: $P = -\partial A/\partial V|_T$) — A ↓
3. The pressure — P

We first give a statistical mechanical definition of the local compositions. For simplicity, we consider a binary mixture of spherical molecules of types A and B. The number of neighboring B molecules surrounding a central A molecule is given in terms of the radial distribution function, $g_{BA}(r)$,

$$n_{BA}(L) = \rho_B \int_0^L dr\ 4\pi r^2\ g_{BA}(r) \tag{1}$$

where L is the range (radius) of the coordination. Similarly, the number of neighbors A surrounding the center A is

$$n_{AA}(L) = \rho_A \int_0^L dr\ 4\pi r^2\ g_{AA}(r) \tag{2}$$

We call n_{AA} and n_{BA} the nearest neighbor numbers. The coordination number of the center A is then the sum of its A neighbors and B neighbors,

$$z_A = n_{AA} + n_{BA} \tag{3}$$

As a consequence of definitions Equations 1-3, the 'local compositions' of B molecules and A molecules surrounding a center of type A are

$$x_{BA} = n_{BA}/z_A = n_{BA}/(n_{AA} + n_{BA}) \tag{4}$$

and

$$x_{AA} = n_{AA}/z_A = n_{AA}/(n_{AA} + n_{BA}) \tag{5}$$

Substituting the integral expressions into Equations 4 and 5, we have

$$x_{BA} = \rho_B \int_0^{L_{BA}} dr\ 4\pi r^2\ g_{BA}(r) / \left[\rho_A \int_0^{L_{AA}} dr\ 4\pi r^2 g_{AA}(r) + \rho_B \int_0^{L_{BA}} dr\ 4\pi r^2 g_{BA}(r)\right] \tag{6}$$

$$x_{AA} = \rho_A \int_0^{L_{AA}} dr\ 4\pi r^2 g_{AA}(r) / \left[\rho_A \int_0^{L_{AA}} dr\ 4\pi r^2 g_{AA}(r) + \rho_B \int_0^{L_{BA}} dr\ 4\pi r^2 g_{BA}(r)\right] \tag{7}$$

If we divide through by $\int dr\ 4\pi r^2 g_{AA}(r)$

$$x_{BA} = x_B \Lambda_{BA}/(x_A + x_B \Lambda_{BA}) \tag{8}$$

and

$$x_{AA} = x_A/(x_A + x_B \Lambda_{BA}) \tag{9}$$

where

$$\Lambda_{BA} = \int_0^{L_{BA}} dr\ 4\pi r^2 g_{BA}(r) / \int_0^{L_{AA}} dr\ 4\pi r^2 g_{AA}(r) \tag{10}$$

Similar definitions can be made for x_{BB} and x_{AB}. It is interesting to note that Equations 8 and 9 are of the same form as given by Wilson (1). The analogy can be carried further by noting the mean value theorem in integral calculus,

$$\int_0^{L_{BA}} dr\ 4\pi r^2 g_{BA}(r) = \int_0^{L_{BA}} dr\ 4\pi r^2 \exp\left[-\beta W_{BA}(r)\right]$$
$$= V_{BA} \exp\left[-\beta \bar{W}_{BA}\right] \qquad (11)$$

where $\bar{W}_{BA} = W_{BA}(r_0)$ is the potential of mean force (7) evaluated at some mean value, r_0, and V_{BA} is the spherical volume, $4\pi L_{BA}^3/3$. Similarly,

$$\int_0^{L_{AA}} dr\ 4\pi r^2 g_{AA}(r) = V_{AA} \exp\left[-\beta \bar{W}_{AA}\right] \qquad (12)$$

Thus

$$\Lambda_{BA} = (V_{BA}/V_{AA}) \exp\left[-\beta(\bar{W}_{BA} - \bar{W}_{AA})\right] \qquad (13)$$

in analogy to Wilson's $\Lambda_{ji} = (v_j/v_i)\exp\left[(g_{ii} - g_{ji})/kT\right]$. The energy equation in statistical mechanics is given by (6)

$$U' = \left[N_A(N_A - 1)/2V\right] \int dr\ g_{AA}(r) u_{AA}(r)$$
$$+ (N_A N_B/2V) \int dr\ g_{AB}(r)\ u_{AB}(r)$$
$$+ (N_A N_B/2V) \int dr\ g_{BA}(r)\ u_{BA}(r)$$
$$+ \left[N_B(N_B - 1)/2V\right] \int dr\ g_{BB}(r)\ u_{BB}(r) \qquad (14)$$

By mean value theorem,

$$U' = (\bar{u}_{AA}/2)\rho_A^2 V \int dr\ g_{AA}(r)$$
$$+ (\bar{u}_{BA}/2)\rho_B \rho_A V \int dr\ g_{BA}(r)$$
$$+ (\bar{u}_{AB}/2)\rho_A \rho_B V \int dr\ g_{AB}(r)$$
$$+ (\bar{u}_{BB}/2)\ \rho_B^2 V \int dr\ g_{BB}(r) \qquad (15)$$

where U' is the configurational first neighbor internal energy and $u_{ij}(r)$ is the pair potential between species i and j. For angle-dependent and multi-body potentials, an effective pair potential can be used. Using Equations 1-5,

$$U'/N = (z_A/2)\ x_A x_{AA}\ \bar{u}_{AA} + (z_A/2)\ x_A x_{BA}\ \bar{u}_{BA}$$
$$+ (z_B/2)\ x_B x_{BB}\ \bar{u}_{BB} + (z_B/2)\ x_B x_{AB}\ \bar{u}_{AB} \qquad (16)$$

The key element of Wilson's original formulation was his free energy expression. Here we discuss the form due to Whiting and Prausnitz (5):

$$-\frac{a A'}{NkT} = x_A\ \ln(x_A \exp\left[-\beta \bar{W}_{AA}\right] + x_B F_{BA} \exp\left[-\beta \bar{W}_{BA}\right])$$

$$+ x_B \ln(x_B \exp[-\beta\overline{W}_{BB}] + x_A F_{AB} \exp[-\beta\overline{W}_{AB}]) \tag{17}$$

At low densities it can be shown that α is related to the coordination number, z.

$$\alpha = 2/z \tag{18}$$

Since in a liquid, the coordination number, z_A, of molecules A is not necessarily the same as the coordination number, z_B, of molecules B, we can generalize the Whiting-Prausnitz (WP) formula by introducing two α parameters,

$$-\frac{A'}{NkT} = (x_A/\alpha_A)\ln\{x_A \exp[-\beta\overline{W}_{AA}] + x_B F_{BA} \exp[-\beta\overline{W}_{BA}]\} + (x_B/\alpha_B)\ln\{x_B \exp[-\beta\overline{W}_{BB}] + x_A F_{AB} \exp[-\beta\overline{W}_{AB}]\} \tag{19}$$

Under weak conditions (6), one can show that Equation 19 is consistent with Equation 16 through the Gibbs-Helmholtz relation. This equation satisfies the pure fluid limits of

$$A_A^0/NkT = (1/\alpha_A)\ \overline{W}_{AA}^0/kT \tag{20}$$

Equation 19 serves as a free energy model for future calculations. Next we shall look at the mixing rules.

N-Fluid Theories and Pressures. When an equation of state developed for a pure substance is extended to mixtures, one of the questions is the composition dependence of the new equation. This dependence is usually incorporated through mixing rules applied to the state variables (dependent, independent or both) and/or the parameters of the equation. A fundamental understanding of composition dependence can be gained through studying some model mixtures in statistical mechanics.

The virial pressure equation for a binary mixture of N_A molecules of type A and N_B molecules of type B is given in terms of the pair potentials and pair correlation functions (pcf) as

$$P/(\rho kT) = 1 - \beta\rho/6 \times \sum_{ij}\sum x_i x_j \int d\underline{r}\, r(\partial u_{ij}/\partial r_{ij})\, g_{ij}(r;T,\rho,\underline{x}) \quad i,j = A,B \tag{21}$$

This equation serves as our starting point of

examination of the mixing rules. We note that the composition dependence has an 'explicit' part, under the double summation (underlined by ⟷) and an 'implicit' part, contained in the functional dependence of the pair correlation functions, g_{ij}. Attempts have been made to approximate Equation 21 by the so-called van der Waals n-fluid theories (i.e. one-fluid, two-fluid and three-fluid theories) where the composition dependence is simplified and the pcf's are evaluated at reduced states characterized by the corresponding energy and size parameters. In the following we shall examine first the effects of molecular sizes on mixing rules in the absence of attractive energies (e.g. in hard sphere mixtures), and then of attractive energies in addition to the size differences (e.g. in mixtures of Lennard-Jones molecules). The formulation due to Henderson and Leonard (9) for multifluid theories is summarized below.

<u>Hard Sphere Mixtures</u>. For hard sphere mixtures, the difference in species is manifested in size. We shall denote the big spheres as those with diameter d_{AA} and small spheres with diameter d_{BB}. The cross interaction is characterized by the diameter d_{AB} $(=(d_{AA} + d_{BB})/2)$. The virial equation reduces in the hard sphere case to

$$P/(\rho kT) = 1 + (4\pi/6)\sum_{ij}\sum x_i x_j \rho d_{ij}^3 \; g_{ij}(d_{ij}^+;\rho,\underset{\sim}{x}) \qquad i,j=A,B \tag{22}$$

where $g_{ij}(d_{ij}^+)$ is the contact value of the ij pcf. Equation 22 can be contrasted with the pure hard sphere pressure equation

$$P/(\rho kT) = 1 + (4\pi/6)\rho d^3 g_o(d^+;\rho) \tag{23}$$

where g_o denotes the pure hard sphere pcf. The question posed in the n-fluid theories is the relation between the mixture pcf, $g_{ij}(r)$, and the pure $g_o(r)$. As far as the pressure is concerned, it is their contact values that are important. For example in the three-fluid theory (vdw3) for binary mixtures, one assumes

$$g_{AA}(d_{AA}^+;\rho,\underset{\sim}{x}) \cong g_o(d_{AA}^+;\rho d_{AA}^3) \tag{24}$$

$$g_{BB}(d_{BB}^+;\rho,\underset{\sim}{x}) \cong g_o(d_{BB}^+;\rho d_{BB}^3) \tag{25}$$

and

$$g_{AB}(d_{AB}^+;\rho,\underset{\sim}{x}) \cong g_o(d_{AB}^+;\rho d_{AB}^3) \tag{26}$$

In the two fluid theory, the following assumptions are made,

$$g_{AA}(d_{AA}^+;\rho,\underset{\sim}{x}) \cong g_o(d_{xA}^+;\rho d_{xA}^3) \tag{27}$$

$$g_{BB}(d_{BB}^{+};\rho,x) \cong g_o(d_{xB}^{+};\rho d_{xB}^{3}) \quad (28)$$

and

$$g_{AB}(d_{AB}^{+};\rho,x) \cong [g_{AA} + g_{BB}]/2 \quad (29)$$

where

$$d_{xA}^{3} = \sum_j x_j d_{jA}^{3} \quad (30)$$

and

$$d_{xB}^{3} = \sum_j x_j d_{jB}^{3} \quad (31)$$

Finally, the one-fluid theory is simply given by

$$g_{ij}(d_{ij}^{+};\rho,x) \cong g_o(d_x^{+};\rho d_x^{3}) \quad (32)$$

where

$$d_x^{3} = \sum_i\sum_j x_i x_j d_{ij}^{3} \quad (33)$$

We can examine the validity of all three models by comparing the pressure values obtained above with known simulation results for hard spheres. For example, the contact values are accurately given by the Carnahan-Starling (C-S) formula,

$$g_o(d^{+}) = \frac{4-2y}{4(1-y)^3} \quad (34)$$

where $y=\frac{\pi}{6}\rho d^3$. The results are listed in Table I.

We choose three typical mixtures for discussion: (i) low density with diameter ratio (d_{AA}/d_{BB}=) 1.9; (ii) high density with diameter ratio 1.1 and (iii) high density with a high diameter ratio of 3.3. At low densities (ρd_{AA}^3 = 0.3036 and x_A = 0.5) the compressibility from the C-S equation is 1.406. All three theories give comparable results with the vdw1 giving the best prediction. At higher density, (ρd_{AA}^3 = 1.0825) with diameter ratio d_{AA}/d_{BB} = 1.111, vdw1 gives a value for Z of 11.993 as compared to the Monte Carlo value of 12.3. vdw2 gives 12.48 and vdw3 gives 13.035. Since the diameters are very close, we can expect all three theories to be fairly accurate. (In the limit $d_{AA}=d_{BB}$, all three theories are exact). At a higher diameter ratio d_{AA}/d_{BB} = 3.333, corresponding to a molecular volume ratio of ~37., MC gives $\beta P/\rho$ = 8.814, while vdw1 gives 6.027. Vdw2 and vdw3 theories are totally inadequate at this condition, being 15.49 and 5208, respectively. The vdw3 theory fails to give a reasonable value due to the magnitude of the contact value $g_{AA}(d_{AA}^{+})$=5453. This is caused by the high value of the reduced density ρd_{AA}^3=1.8225. Namely, the vdw3

Table I. The Compressibility Factor, $\beta P/\rho$, Comparison of the 1-Fluid, 2-Fluid and 3-Fluid Theories with Exact Results for Hard Sphere Mixtures of Species A and B

	d_{AA}/d_{BB} [+]	ρd_{AA}^3	x_A	MC	Vdw1	Vdw2	Vdw3
			Compressibility Factor: $\beta P/\rho$				
i	1.111	1.0825	0.01	7.5571	7.5553	7.5661	7.5770
			0.25	9.4378	9.3894	9.6602	9.9620
			0.50	12.300*	11.993	12.480	13.035
			0.75	15.762	15.657	16.160	16.709
			0.99	20.736	20.728	20.765	20.801
ii	1.111	1.1266	0.01	8.3720	8.3698	8.3837	8.3977
			0.25	10.577*	10.552	10.905	11.304
			0.50	13.835	13.715	14.363	15.113
			0.75	18.417	18.280	18.964	19.718
			0.99	24.797	24.786	24.837	24.888
iii	1.905	0.3036	0.01	1.1018	1.1017	1.1020	1.1023
			0.25	1.2265	1.2238	1.2324	1.2432
			0.50	1.4061	1.3988	1.4170	1.4396
			0.75	1.6518	1.6420	1.6628	1.6865
			0.99	1.9687	1.9678	1.9694	1.9710
iv	3.333	1.8225	0.01	1.1328	1.1320	1.1435	3.2353
			0.25	2.3269	2.1117	3.2066	1303.5
			0.50	8.8140*	6.0273	15.494	5208.6

+d_{AA} = diameter of large spheres and d_{BB} = diameter of small spheres. $d_{AB} = (d_{AA} + d_{BB})/2$

*Monte Carlo data of Lee and Levesque (<u>10</u>); others calculated from the formulation of Mansoori, et al. (<u>11</u>)

Vdw1, Vdw2 and Vdw3 are the van der Waals one-fluid, two-fluid and three-fluid theories.

theory presupposes in one of its terms a hypothetical pure fluid composed entirely of large spheres. This picture is physically unsound due to the fact that half of the molecules in the mixture are small spheres which, being interspersed amongst the large ones, leave a larger portion of 'empty space' or 'free volume' accessible to other molecules. Consequently, the total pressure of vdw3 is considerably higher than that given by the real fluid. The above comparison clearly demonstrates that the vdw1 theory gives the best results for high densities and large size ratios. The two other theories (vdw2 and vdw3) are inadequate by comparison.

Mixtures of LJ Molecules. Recently, Hoheisel and Lucas (12) have made extensive studies of binary mixtures of Lennard-Jones molecules with different size ratios (up to 2) and energy ratios (up to 5). They compared the pcf obtained from vdw1 and the mean density approximation (MDA) with simulation data. Their results showed that both theories are reasonable; however, the MDA theory gives better J-integrals. The mean density approximation was proposed by Mansoori and Leland (13)

$$\sigma_x^3 = \sum_{ij}\sum x_i x_j \sigma_{ij}^3 \tag{35}$$

and

$$g_{ij}(r) \cong g_o\ (r/\sigma_x; kT/\varepsilon_{ij}, \rho\sigma_x^3) \tag{36}$$

while the vdw1 in this case is given by Equation 35 plus

$$\sigma_x^3 \varepsilon_x = \sum_{ij}\sum x_i x_j \sigma_{ij}^3 \varepsilon_{ij} \tag{37}$$

and

$$g_{ij}(r) = g_o\ (r/\sigma_x; kT/\varepsilon_x, \rho\sigma_x^3) \tag{38}$$

Thus the MDA differs from the vdw1 in that individual energy parameters, ε_{ij}, are used in characterizing the temperature. The two theories coincide when $\varepsilon_{AA} = \varepsilon_{BB}$. The virial pressure for Lennard-Jones molecules can be written as

$$\beta P/\rho = 1 - 16\pi \sum_{ij}\sum x_i x_j (\beta\varepsilon_{ij})(\rho\sigma_{ij}^3)\left[J_{ij}^{(6)} - 2J_{ij}^{(12)}\right] \tag{39}$$

where the J-integral is defined as

$$J^{(n)} = \int dr^*\ r^{*2} g(r^*)/r^{*n} \tag{40}$$

where r* is the reduced distance. The results are presented in Table II.

Table II. The Compressibility Factor, $\beta P/\rho$, for Equimolar Mixtures of Lennard-Jones Molecules of Species A and B

$\rho^* = \rho\sigma_x^3 = 0.8$

COMPRESSIBILITY FACTOR: $\beta P/\rho$

T,K	$\varepsilon_{BB}/\varepsilon_{AA}$	σ_{BB}/σ_{AA}	MC*	MDA**	Vdw1	Vdw2	Vdw3
270	1.50	1.00	3.1563	3.1346	3.1680		
270	3.00	1.00	3.1483	3.1431	3.2772		
270	5.00	1.00	2.8555	2.9402	3.1211		
270	1.00	1.15	3.1702	3.0711	3.0711		
270	1.50	1.15	3.1414	3.0697	3.0474		
270	3.00	1.15	3.0417	3.0196	3.0760		
270	5.00	1.15	2.9185	2.8014	2.8909		
270	1.00	1.30	3.1425	3.1167	3.1167		
200	1.50	1.30	4.0571	3.9323	3.9531	4.4724	4.9951
200	3.00	1.30	3.9158	3.9533	4.2409		
200	5.00	1.30	3.6924	3.6644	3.7173		
200	1.00	1.65	4.1416	3.9395	3.9395		
200	1.50	1.65	4.3014	3.9544	3.9879		
200	3.00	1.65	4.3526	3.9320	4.0978		
200	5.00	1.65	4.1050	3.6296	3.6015		
200	1.00	2.00	4.3519	4.1583	4.1583		
200	1.50	2.00	4.6298	3.9738	4.0277		
200	3.00	2.00	4.8513	3.9341	3.9303		
200	5.00	2.00	4.6499	3.6627	3.5321		
200	0.33	1.30	3.5412	3.5952	3.6183		
200	0.20	1.30	3.3245	3.4565	3.4941		
34	1.50	1.30	1.0974	0.9648	1.1034		

* Monte Carlo data of Hoheisel and Lucas (12)

** MDA=Mean Density Approximation

Vdw1, Vdw2 and Vdw3 are the van der Waals one-fluid, two-fluid and three-fluid theories

It shows that for small size ratios, (σ_{BB}/σ_{AA} < 1.3), both MDA and vdw1 give reasonable results ($\pm$ 0.7%). As σ_{BB}/σ_{AA} reaches 1.65, vdw1 deteriorates at large $\varepsilon_{BB}/\varepsilon_{AA}$ (-12% at $\varepsilon_{BB}/\varepsilon_{AA}$ = 5). The size disparity coupled with large energy difference proves to be beyond the scope of both theories. For example at σ_{BB}/σ_{AA} = 2.00 and $\varepsilon_{BB}/\varepsilon_{AA}$ = 3.00, vdw1 gives an error of -19%, while MDA is similarly -19% in error. There is little to choose between MDA and vdw1 in this case.

Our conclusions are in agreement with recent studies on the Lennard-Jones mixtures. In the following we shall consider e.g. the MDA approximation for use in our equation of state.

Mixture Equation of State

Given an equation of state for a pure fluid, the

following procedure can be used to build an equation for mixtures.

We first derive the Helmholtz free energy, A, from the given P-V-T relation. By identifying the potential of mean force as the Helmholtz free energy, we can immediately write down the free energy for mixtures according to Equation 19, i.e.

$$-A'/(NkT) = (x_A/\alpha_A)\ln\left[x_A\exp(-\alpha_A\hat{A}'_{AA}/kT) + x_B F_{BA}\exp(-\alpha_A\hat{A}'_{BA}/kT)\right] + (x_B/\alpha_B)\ln\left[x_B\exp(-\alpha_B\hat{A}'_{BB}/kT) + x_A F_{AB}\exp(-\alpha_B\hat{A}'_{AB}/kT)\right] \quad (41)$$

In this equation the approximation is made that for mixtures the potentials of mean force are set proportional to the component Helmholtz free energies (see Hill(14)). To obtain the pressure, we differentiate A' with respect to volume, according to

$$P = -\partial A/\partial V \mid_T \quad (42)$$

thus obtaining

$$P = x_A(x_{AA}P_{AA} + x_{BA}P_{BA}) + x_B(x_{BB}P_{BB} + x_{AB}P_{AB}) \quad (43)$$

where

$$P_{ij} = -\partial\hat{A}_{ij}/\partial V \mid_T \quad (44)$$

Thus, the mixture equation of state, Equation 43, depends on the composition in terms of the local compositions. We note that Equation 43 satisfies the explicit composition dependence of Equation 21 (N.B. the local compositions, x_{ij}, contains the mole fraction, x_i). So far the development has been general. To make practical calculations, we must adopt a function form for the specific Helmholtz free energies.

Mixture Parameters. $\hat{A}'_{ji}$ and P_{ji} in Equations 41-44, being properties of molecules with j-i interactions, are calculated from an equation of state for pure substances developed by Chung, et al. (15). The equation of state has been tested for normal paraffins (C_2-nC_{20}), ring compounds, and polar compounds (H_2S, acetone, etc.), and associating compounds (NH_3, water, and alcohols, etc.). The pure fluid equation of state is accurate enough for engineering design calculations.

Reduced Temperature and Density. For the equation of state used here, $\hat{A}'_{ij}$ and P_{ij} are functions of the reduced temperature, T^*_{ij}, reduced density, ρ^*_{ij}, and structure parameter, λ_{ij} (see (15)). The relations used for T^*_{ij} and ρ^*_{ij} are

$$T^*_{ij} = kT/\varepsilon_{ij} \quad (45)$$

$$\rho^*_{ij} = \rho V^*_x \quad (46)$$

where

$$V^*_x = \sum_m\sum_n x_m x_n V^*_{mn} \quad (47)$$

The relation for T^*_{ij} corresponds to the MDA model, while the relation for ρ^*_{ij} corresponds to a one fluid model. Combination of Equations 45-47 with Equation 41 gives a mean potential LCM model for fluid mixtures.

__The Combining Rules.__ The characterization parameters λ_{ij}, V^*_{ij}, and ε_{ij} are calculated by the following combining rules:

$$\lambda_{ij} = \frac{\lambda_{ii} + \lambda_{jj}}{2} \quad (48)$$

$$V^*_{ij} = \xi^3 (V^*_{ii} V^*_{jj})^{1/2} \quad (49)$$

$$\varepsilon_{ij}{}^o = \zeta(\varepsilon_{ii}^o \varepsilon_{jj}^o)^{1/2} \quad (50)$$

$$D_{ij} = \frac{D_{ii} + D_{jj}}{2} \quad (51)$$

$$\varepsilon_{ij}/k = \varepsilon^o_{ij}/k + \frac{D_{ij}}{T} \quad (52)$$

where the subscripts ii and jj refer to the pure component parameters and ξ and ζ are binary interaction parameters, BIPs.

Results and Discussion

The values of α_i in Equation 41 are related to the coordination numbers. We have chosen $\alpha' = \alpha_A = \alpha_B = 0.5$ in this work, i.e., the value recommended by Whiting and Prausnitz (5). The quantity F_{ij} is the ratio of V_{ij} to V_{jj}, where V_{jj} is the first-neighbor volume. To account for V_{ij} ($i \neq j$), a binary interaction parameter δ is introduced, where $V_{ij} = \delta^3 (V_{ii} V_{jj})^{1/2}$. Furthermore, the first-neighbor volume, V_{ii}, is approximated as proportional to the molar covolume V^*_{ii}, thus

$$F_{ij} = \delta^3 \left(\frac{V^*_{ii}}{V^*_{jj}}\right)^{1/2} \quad (53)$$

In the local composition model used here, three binary interaction parameters (ξ, ζ, and δ) are introduced. They are determined by fitting the mixture equation of state to mixture experimental data.

For purpose of comparison, vapor-liquid equilibria calculations were also performed using conformal solution mixing rules, which correspond to the vdw1 model. The following conformal solution mixing rules are adopted to obtain mixture characterization parameters

$$V_x^* = \sum_i \sum_j x_i x_j V_{ij}^* \tag{54}$$

$$\varepsilon_x V_x^* = \sum_i \sum_j x_i x_j \varepsilon_{ij} V_{ij}^* \tag{55}$$

$$\lambda_x V_x^* = \sum_i \sum_j x_i x_j \lambda_{ij} V_{ij}^* \tag{56}$$

The subscript x represents the mixture characterization parameters. The combining rules used to calculate the cross interaction parameters are the same as Equations 48-52. Hereafter, we refer to the two mixing rules as LCM (local composition mixing rules) and CSM (conformal solution mixing rules), respectively.

__Binary Systems__. We have applied the above equation of state to the calculation of mixture densities and vapor-liquid equilibria of binary systems. In these calculations, the same combining rules, Equations 48-52, with two binary interaction parameters, ξ and ζ, were used in both LCM and CSM. For the LCM, a third parameter, δ, was introduced to calculate the value of Fij, i.e., from Equation 53. We have increased the number of binary interaction parameters in the CSM to three and then to four (for D_{ij} and λ_{ij}). When tested for the methanol-carbon dioxide system, no appreciable improvement was obtained. Thus we retain only two binary parameters for CSM. For systems with both vapor-liquid equilibrium and mixture density data, the optimal values of the binary interaction parameters for each system were determined by using both data simultaneously in multiproperty regression analyses. The experimental vapor-liquid equilibrium and density ranges studied are given in Tables III and IV.

For mixture density calculations, five systems were selected for study. The results of mixture density calculations using both LCM and CSM are given in Table IV. The two mixing rules yield similar results except for the system acetone-water. The LCM gives better results (1.7%) than the CSM (7.9%) for forty-one density data points in the acetone-water system. The mixture acetone-water is a strongly nonideal system; its

Table III. Vapor-Liquid Equilibrium Calculations for Binary Systems

System	Reference	No. of Data Points	Temp. Range K	Pres. Range bar	Mixing* Rules	Binary Interaction Parameters ξ	ζ	δ	A.A.D.% + K_1	K_2
Methane-Ethane	(17)	19	192-200	2.7-48	LCM	1.0581	0.9268	1.0344	1.5	1.6
Methane-Propane	(18)	54	172-214	1.9-65	LCM	1.1609	0.8326	1.1152	2.0	9.8
Methane-n-Butane	(19)	27	278-378	11-131	LCM	1.2428	0.7874	1.1479	5.3	8.4
Methane-n-Pentane	(20)	21	224-273	7-124	LCM	1.1267	0.7976	1.1073	6.1	34.
Methane-n-Hexane	(21)	8	344	35-198	LCM	1.2203	0.8247	1.1926	1.2	20.
Methane-n-Heptane	(22)	33	222-255	7-155	LCM	0.8519	0.5295	0.4705	9.6	89.
Methane-n-Octane	(23)	28	298-423	20-71	LCM	1.2147	0.4415	0.7908	1.7	27.
Methane-n-Nonane	(24)	63	323-423	10-319	LCM	1.2833	0.3423	0.7801	6.5	66.
Methane-n-Decane	(25)	12	543-583	30-126	LCM	1.0884	0.5675	0.6281	4.0	2.4
					CSM	1.0741	1.1136	--	4.9	4.3
Ethane-Propane	(26)	11	233	1.4-7	LCM	1.0144	0.9904	0.9983	1.2	2.5
Ethane-n-Butane	(27)	19	338-394	32-55	LCM	1.0047	0.9843	1.0283	1.7	2.3
					CSM	1.0006	0.9992	--	1.3	2.2
Propane-n-Butane	(28)	19	363-393	21-38	LCM	0.9718	1.0074	1.0554	1.7	1.7
Benzene-n-Hexane	(29)	18	298-328	.15-.63	LCM	1.0085	0.9730	1.0897	0.5	1.2
Benzene-n-Heptane	(30)	8	334-347	0.53	LCM	0.9399	0.9738	0.6139	0.8	1.6

Continued on next page

Table III Continued

System	Reference	No. of Data Points	Temp. Range K	Pres. Range bar	Mixing Rules	Binary Interaction Parameters ξ	ζ	δ	A.A.D.% K_1	K_2
Hydrogen-Methane	(31)	41	163-183	22-160	LCM	1.1492	1.1462	1.2528	14.	5.2
Hydrogen Ethane	(32)	21	158-200	7-138	LCM	1.1420	0.8926	1.2572	23.	8.5
Hydrogen Propane	(33)	6	173-348	17-207	LCM	1.1970	1.3977	1.3250	3.4	4.7
Nitrogen-Methane	(34)	88	122-183	2.8-50	LCM	1.0364	0.9310	1.0296	1.6	2.3
Nitrogen-Ethane	(35)	11	150	3.4-89	LCM	1.0860	0.8356	1.2095	2.8	15.
Nitrogen-Propane	(36)	18	222-273	14-138	LCM	1.1730	0.8422	1.2053	16.	9.5
Nitrogen-Hydrogen	(37)	11	90-95	8.4-46	LCM	1.1376	0.9611	1.2797	3.8	2.2
H_2S-Methane	(38)	41	278-344	14 121	LCM	1.0747	0.8501	1.1650	3.6	3.0
H_2S-Ethane	(39)	45	200-283	0.6-30.5	LCM	1.0698	0.8486	1.0012	8.7	3.5
H_2S-Propane	(40)	35	218-344	1.4-28	LCM	1.0508	0.8599	1.0130	6.7	4.3
H_2S-n-Heptane	(41)	28	311-478	6.1-96	LCM	1.1232	0.6975	0.8873	6.0	9.8
H_2S-n-Nonane	(42)	15	311-478	1.4 28	LCM	1.2537	0.5928	0.8962	2.5	9.2
CO-Methane	(43)	30	123-178	3.6 47	LCM	0.9779	0.9707	1.0032	4.4	2.7
CO-Ethane	(44)	18	223-273	8.6-117	LCM	1.0658	0.8809	1.1196	6.0	9.7
CO-Propane	(44)	16	273-323	13-138	LCM	1.2022	0.8552	1.2255	19.	11.

System	Reference	No. of Data Points	Temp. Range K	Pres. Range bar	Mixing Rules	Binary Interaction Parameters ξ	ζ	δ	A.A.D.% K_1	K_2
Nitrogen-CO_2	(45)	19	233-273	18-139	LCM	1.1837	0.8310	1.2120	13.	4.5
Nitrogen-H_2S	(46)	24	300-344	35-206	LCM	1.1679	0.8154	1.2988	3.3	12.
Nitrogen-CO	(47)	8	84	1.4-2.0	LCM	0.7606	1.1330	0.8331	2.1	5.5
CO_2-H_2S	(48)	37	233-361	6.9-83	LCM	1.0019	0.9186	1.0847	1.7	5.2
CO_2-Hydrogen	(49)	38	220-290	11-203	LCM	1.2744	1.0091	1.3834	20.	5.9
CO_2-CO	(50)	34	223-283	8.3-131	LCM	1.1236	0.8460	1.2171	9.7	5.1
H_2S-CO	(51)	37	233-293	3.5-237	LCM	1.0019	0.9186	1.0847	29.	7.1
Hydrogen-CO	(52)	19	85	5.2-211	LCM	0.7362	1.1018	1.0833	16.	7.8
CO_2-Methane	(53)	20	210-219	6-64	LCM	1.0429	0.8549	1.1320	3.2	3.5
					CSM	0.9867	0.9718	--	5.0	3.7
CO_2-Ethane	(54)	55	223-293	6-63	LCM	1.1312	0.8602	1.2824	1.9	2.0
CO_2-Propane	(55)	9	233-273	3.4-27	LCM	1.1081	0.8647	1.0564	4.2	4.8
CO_2-n-Pentane	(56)	48	278-378	2.3-96	LCM	1.1930	0.6349	0.9125	5.6	11.
CO_2-n-Hexane	(57)	10	298	4-51	LCM	1.1736	0.7819	1.1824	4.3	12.
					CSM	1.0430	0.8723	--	8.9	15.
CO_2-n-Heptane	(58)	62	311-477	1.9-133	LCM	1.2521	0.6703	0.9675	9.1	16.
CO_2-n-Decane	(59)	16	462-583	14-51	LCM	1.2433	0.7511	1.1255	3.7	2.6
					CSM	1.0539	0.9082	--	8.6	5.4

Continued on next page

Table III Continued

System	Reference	No. of Data Points	Temp. Range K	Pres. Range bar	Mixing Rules	Binary Interaction Parameters ξ	ζ	δ	A.A.D.% K_1	K_2
CO_2-	(59)	16	462-	19-	LCM	1.3504	0.7200	1.1274	10.	9.8
n-Hexadecane			663	50	CSM	1.0799	0.9051	--	13.	8.9
CO_2-	(57)	17	298-	8-	LCM	1.2044	0.7870	1.1369	1.6	13.
Benzene			313	76	CSM	1.0268	0.9405	--	7.7	15.
Acetone-	(60)	8	298	4.7-	LCM	0.8993	0.8914	1.0497	13.	9.9
Ethane				39	CSM	0.9588	0.9644	--	38.	22.
Acetone-	(61)	8	350	3.4-	LCM	1.1227	0.9218	1.0981	5.8	4.3
Propane				27						
Acetone-	(62)	11	304-	1.01	LCM	1.1459	0.8048	0.9510	7.8	3.4
n-Pentane			322							
Acetone-	(63)	8	338	.76-	LCM	1.2544	0.6982	0.8997	2.8	5.0
n-Heptane				1.4	CSM	0.9480	0.8919	--	6.5	7.5
Acetone-	(64)	11	298	.15-	LCM	0.9677	0.9820	0.9886	2.1	1.6
Benzene				0.3	CSM	0.9888	0.9767	--	2.1	1.9
Methanol-	(60)	5	298	11-	LCM	0.8746	0.8752	0.9739	2.2	19.
Ethane				41						
Methanol-	(65)	21	303-	1.00	LCM	0.8623	0.8302	0.7345	8.1	26.
n-Pentane			335							
Methanol-	(66)	18	323-	1.01	LCM	0.9356	0.8325	0.7982	23.	19.
n-Hexane			336							
Methanol-	(67)	8	332-	1.01	LCM	0.9761	0.7873	0.7879	20.	17.
n-Heptane			344							
Methanol-	(68)	18	331-	1.01	LCM	1.0170	0.8308	0.8198	6.9*	5.5
Benzene			349							
Ethanol-	(69)	16	400-	6-	LCM	1.1148	0.8659	1.0288	3.7	4.9
Propane			424	54						
Ethanol-	(70)	17	328	.46-	LCM	0.9687	0.8252	0.8331	11.*	2.8
n-Hexane				.9	CSM	0.9089	0.9497	--	13.*	3.5

System	Reference	No. of Data Points	Temp. Range K	Pres. Range bar	Mixing Rules	Binary Interaction Parameters ξ	ζ	δ	A.A.D.% K_1	K_2
Ethanol-n-Heptane	(71)	10	327-336	0.53	LCM	0.8511	0.8095	0.6861	16.	29.
Ethanol-n-Decane	(72)	11	353-433	1.01	LCM	1.2490	0.7011	0.9643	12.	4.3
Ethanol-Benzene	(73)	9	298	.11-.16	LCM	1.0848	0.9669	0.9175	7.8	2.9
1-Propanol-n-Hexane	(74)	8	339-362	1.01	LCM	0.9090	0.9305	0.9614	6.6	11.
1-Propanol-n-Decane	(75)	10	348-371	0.39	LCM	1.2883	0.7440	0.9935	5.4	9.0
					CSM	1.1232	0.8840	--	6.9	11.
1-Propanol-Benzene	(76)	50	350-370	1.01	LCM	1.1644	0.8373	0.9327	7.0	7.8
Water-Methane	(77-78)	36	411-444	13-689	LCM	0.9955	0.6796	1.2024	2.7	9.2
Water-Ethane	(79-80)	36	411-444	13-689	LCM	1.0353	0.6728	1.2554	3.0	9.8
Water-Ethylene	(81)	12	378-411	13-344	LCM	1.0737	0.7249	1.2488	4.2	6.4
Acetone-Methanol	(82)	12	329-338	1.01	LCM	0.8864	0.9438	0.8794	1.4	2.0
Acetone-Ethanol	(82)	9	330-350	1.01	LCM	0.8696	0.9892	0.9352	3.7	3.8
Acetone-Water	(83)	25	473	16-30	LCM	1.0527	0.8972	1.1327	2.8*	1.3
					CSM	0.9777	1.0017	--	9.2*	2.4
Methanol-CO_2	(84)	13	298	2-60	LCM	1.0535	0.8479	1.0728	4.2	2.5
					CSM	0.9711	1.1170	--	29.	28.
Methanol-Ethanol	(82)	12	338-350	1.01	LCM	0.8140	1.0302	1.1264	1.3	9.7

Continued on next page

Table III Continued

System	Reference	No. of Data Points	Temp. Range K	Pres. Range bar	Mixing Rules	Binary Interaction Parameters ξ	ζ	δ	A.A.D.% K_1	K_2
Methanol-1-Propanol	(85)	9	323	.16-.51	LCM	0.9884	0.9646	0.4838	2.2	4.7
Methanol-Water	(83)	9	423	5.3-13	LCM	1.0549	0.9494	1.0824	7.9	3.8
					CSM	1.0675	0.9315	--	12.	4.1
Ethanol-1-Propanol	(86)	9	353-367	1.01	LCM	1.1224	0.9492	0.9364	2.9	3.1
Ethanol-Water	(87)	12	351-369	1.01	LCM	0.8718	1.0036	1.0025	3.0	2.9
CO_2-Water	(88)	9	548	99-690	LCM	1.0315	0.9381	1.2034	9.6	6.2
H_2S-Water	(89)	11	444	13-170	LCM	0.8134	0.9177	1.0359	4.7	3.7
Ammonia-Water	(90)	20	283-443	1.-10.1	LCM	1.0233	1.0828	0.9392	5.7	11.
Acetone-CO_2	(84)	11	298	4-55	LCM	1.2054	0.8265	1.0986	9.0	2.6
					CSM	1.0268	0.9644	--	20.	6.2

* LCM denotes local composition mixing rules.
CSM denotes conformal solution mixing rules.

\+ $\text{A.A.D.\%} = 1/NP \sum_{1}^{NP} |(Kexp-Kcal)/Kexp| x100$, where NP = number of data points.

★ Results do not cover the entire composition range.

components differ appreciably in both size and strength of interactions (polar and associating effects).

Table IV. Density Calculations Using Local Composition and Conformal Solution Mixing Rules

System	Ref.	No of Data Points	T oK	P bar	Mixing Rules	Density A.A.D.%
Methane-n-decane	(91)	10	310-511	137-620	LCM	1.04
					CSM	1.07
Ethane-n-butane	(92)	49	273-413	5.1-43	LCM	1.96
					CSM	2.15
Methanol-benzene	(93)	27	293-313	1.013	LCM	1.69
Acetone-water	(94-95)	41	293-353	1.013	LCM	1.70
					CSM	7.97
Methanol-water	(96)	50	298-323	1.013	LCM	1.71
					CSM	3.22

For vapor-liquid equilibrium calculations, sixteen binary systems, ranging from nonpolar-nonpolar, nonpolar-polar, to polar and associating fluids, were chosen for comparison. A summary of the results is given in Table III. The overall average absolute percentage deviation (A.A.D.%) of the equilibrium K-value (K1) for these fifteen systems are 5% for the LCM and 10% for the CSM, respectively. For nonpolar systems ethane-n-butane and methane-n-decane, both the LCM and CSM fit the vapor-liquid equilibria data well. The LCM fits the data better than the CSM at the temperature of 583 K for the system methane-n-decane (see Figure 1). For nonpolar-polar systems, CO_2-methane, CO_2-n-hexane, CO_2-n-decane, CO_2-n-hexadecane, CO_2-benzene, acetone-ethane, acetone-n-heptane, acetone-benzene, ethanol-n-hexane and 1-propanol-n-decane, the LCM yields better results than the CSM (see Table III). For systems of CO_2-n-hexane and CO_2-benzene, the CSM gives poor results in liquid composition while the LCM yields improved results (see Figure 2). For the CO_2-n-hexadecane system, the ratio of the covolume parameters, V*, is 1 : 7.6. The A.A.D.% for the equilibrium K-values (K1) calculated from the LCM is 10% and from the CSM is 13% for temperatures ranging from 462 K to 663 K (see Table III). The acetone-n-heptane system forms a maximum pressure azeotrope at 338.15 K. The CSM gives poor results near the azeotrope while the LCM gives reasonable results (see Figure 3). For polar and associating fluids, acetone-water, methanol-CO_2, methanol-water and acetone-CO_2, CSM gives poor results for the VLE calculations. The improvement in the fit of the VLE data by the LCM is dramatic (see Table III).

Overall, for nonpolar mixtures such as ethane-n-

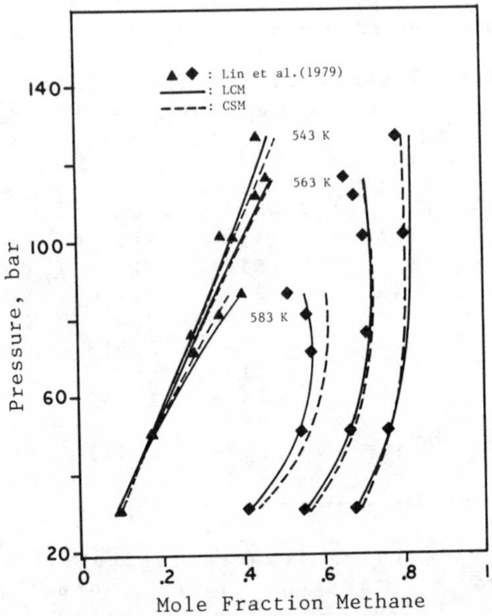

Figure 1. Experimental and calculated vapor-liquid equilibria for the system methane-n-decane.

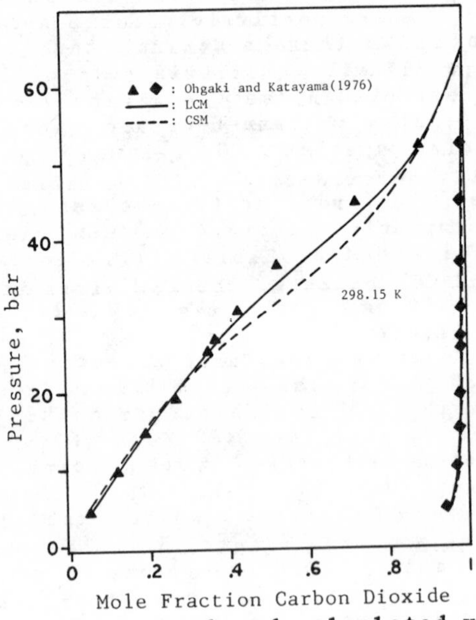

Figure 2. Experimental and calculated vapor-liquid equilibria for the system carbon dioxide-n-hexane at 298.15 K.

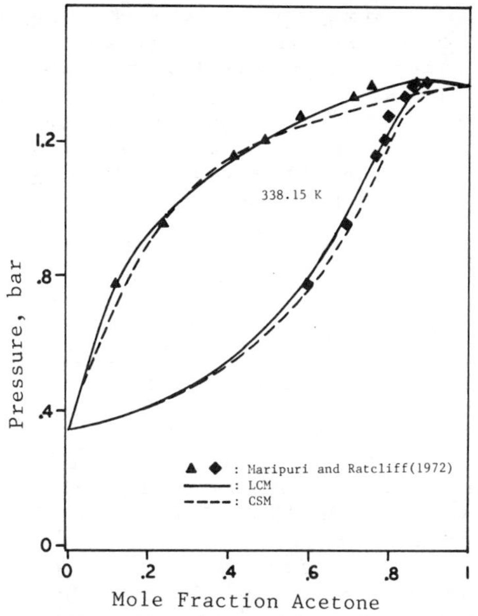

Figure 3. Experimental and calculated vapor-liquid equilibrium for the system acetone-n-heptane.

butane and methane-n-decane, both LCM and CSM yield similar results. For strongly nonideal systems such as methanol-CO_2, acetone-water, n-hexane-ethanol and CO_2-n-hexane, the LCM is clearly superior to the CSM.

The applicability of the LCM has also been extensively tested for 1.) natural gas systems, including normal paraffins up to n-decane, nitrogen, carbon dioxide, and hydrogen sulfide, and 2.) polar systems, including water, methanol, ethanol, 1-propanol, and acetone. The summary of the results for vapor-liquid equlibrium calculations is given in Table III.

We have covered many mixtures with methane (all together 15 systems) i.e. from ethane to n-decane and from CO_2 to water. The accuracy of the calculations for these systems are shown to be satisfactory.

The systems hydrocarbons-alcohols are highly nonideal and are therefore difficult to fit. The ethanol-n-heptane system forms minimum boiling temperature azeotrope at 0.533 bar as shown in Figure 4. The LCM describes this system reasonably well over the entire concentration range. The LCM also gives satisfactory VLE results for other minimum boiling temperature azeotropes such as methanol-n-pentane, methanol-n-hexane, methanol-n-heptane, methanol-benzene and benzene-1-propanol (see Table III). It is demonstrated that the composition dependence of the LCM is very effective for VLE calculations for strongly nonideal systems, even near the region of the azeotrope.

The mixtures of methane-water, ethane-water and ethylene-water are of practical interest. The LCM gives adequate VLE results for methane-water system for pressures up to 700 bar (see Figure 5.) Even though methanol and 1-propanol are both polar and associating substances, the methanol-1-propanol mixture is not a strongly nonideal system. As shown in Figure 6, the LCM yields satisfactory VLE results for this system.

Ternary Systems. Ternary and multicomponent vapor-liquid equilibria are obviously more important in industry than binary systems. Though the amount of binary vapor-liquid equilibria data available in the literature is large, highly accurate data for multicomponent systems are rather scarce. Thus, it is of great importance that mixing rules can be applied to multicomponent systems using only binary parameters. The ternary systems methane-ethane-propane, acetone-methanol-ethanol and n-hexane-ethanol-benzene were selected for test. The results for these three ternary systems are given in Table V. The LCM predicts the mole fractions in both the liquid and the vapor phases reasonably well for these three ternary systems.

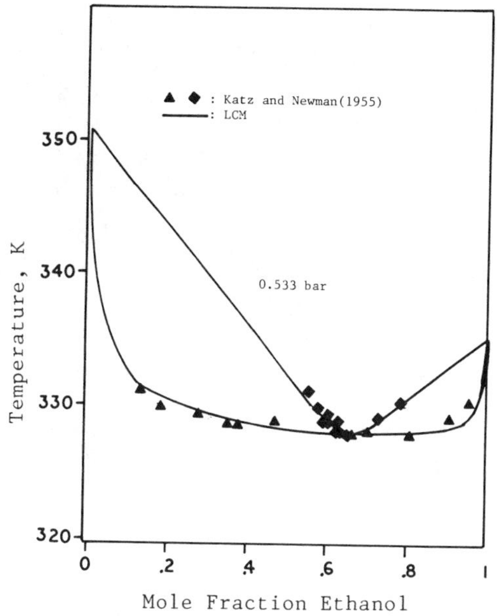

Figure 4. Experimental and calculated vapor-liquid equilibria for the system ethanol-n-heptane at 0.533 bar.

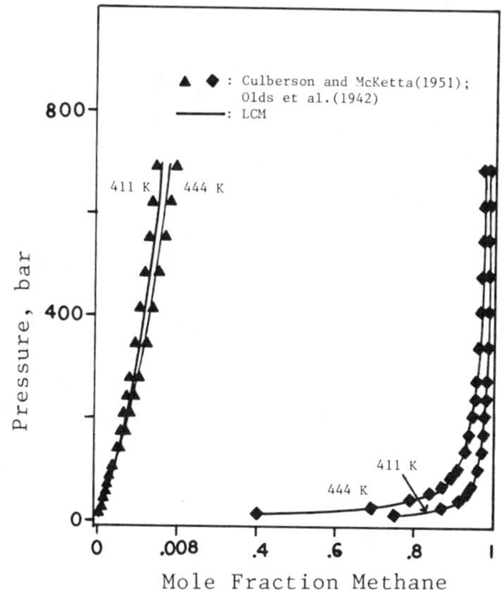

Figure 5. Experimental and calculated vapor-liquid equilibria for the system methane-water.

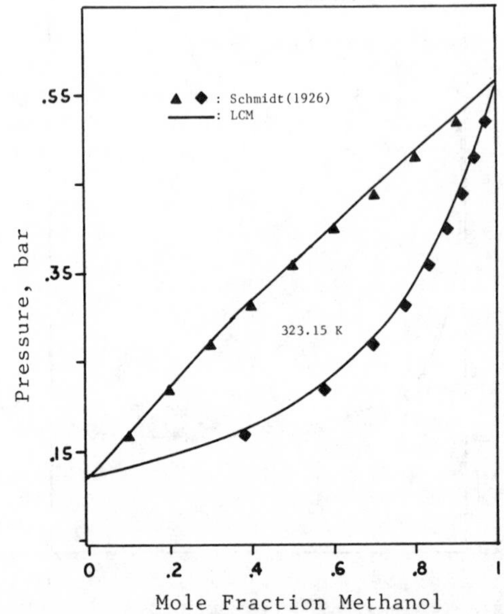

Figure 6. Experimental and calculated vapor-liquid equilibria for the system methanol-1-propanol at 323.15K.

Table V. Vapor-liquid equilibria predictions using local composition mixing rules for three ternary systems

System	Reference	No. of data points	T (K)	P (bar)	K_1	K_2	K_3
					A.A.D %		
Methane(1)-Ethane(2)-Propane(3)	(97)	33	158-214	2.2-55	3.08	6.83	11.2
Acetone(1)-Methanol(2)-Ethanol(3)	(82)	81	329-348	1.0130	8.04	8.45	10.3
n-Hexane(1)-Ethanol(2)-Benzene(3)	(29)	43	328.15	.54-.88	7.75	9.79	7.72

Conclusions

A local composition model has been developed to describe the composition dependence of mixture thermodynamic properties. The method employs a recently proposed equation of state. A multifluid model is adopted for the reduced temperature and one fluid model adopted for the reduced density.

For strongly nonideal mixtures, such as methanol-CO_2, acetone-water, n-hexane-ethanol, and CO_2-benzene, the LCM describes vapor-liquid equilibria behavior accurately while the conformal solution model, which is a one fluid model, does not work well for these mixtures.

It should be noted that the present local composition mixing rules can be used to extend virtually any corresponding-states type equation of state to mixtures.

In summary, the present formulation has the following useful features:

1). The method works well for natural gas systems.
2). The method works for mixtures in which components differ appreciably in size.
3). The method works well for a wide variety of fluid mixtures, ranging from nonpolar to highly polar fluids.
4). The method can be applied to extend any corresponding-states type equation of state to mixtures.

Acknowledgments

We thank the Gas Research Institute for their support that made this research possible.

Legend of Symbols

A	Helmholtz free energy
$\hat{A}'_{ij}$	configurational Helmholtz free energy of molecules with i-j interactions
d_{ij}	hard sphere diameter
D	characterization parameter for polar and associating effects
F_{ij}	coordination volume ratio
$g_{ij}(r)$	radial distribution function between ij pairs
k	Boltzmann's constant
K_i	equilibrium constant
L_{ij}	range (distance) of first coordination shell for i molecules surrounding the center j molecule
n_{ij}	nearest neighbor number of i molecules in the first coordination shell surrounding a central j molecule
N	total number of molecules in the system
N_{ij}	global number of nearest neighbor pairs between i and j molecules

P	pressure
r	intermolecular distance measured from center to center
R	gas constant
T	temperature
U	internal energy
$\bar{u}_{ij}$	average pair intermolecular energy
V	volume
V*	molecular volume parameter
V_{ij}	coordination volume
$\bar{W}_{ij}$	potential of mean force between species i and j molecules
x_i	mole fraction of component i
x_{ij}	local composition of i molecules around a j molecule
z_i	coordination number of first neighbors
Z	compressibility factor

<u>Greek Letters</u>

α'	relates to the coordination number, equal to 0.5 in this work
β	1/kT
δ	binary interaction parameter
ε	energy parameter
ε^o	energy parameter for nonpolar contribution
ζ	binary interaction parameter
λ	structure parameter
ρ	density
ξ	binary interaction parameter

<u>Literature Cited</u>

1. Wilson, G.M. <u>J. Am. Chem. Soc.</u>, 1964, 86, 127.
2. Orye, R.V. and Prausnitz, J.M. <u>Ind. Eng. Chem.</u>, 1965, 57, 18.
3. Abrams, D.S. and Prausnitz, J.M. <u>AIChE J.</u>, 1975, 21, 116.
4. Benedict, M., Webb, G.B. and Rubin, L.C. <u>J. Chem. Phys.</u>, 1940, 8, 334.
5. Whiting, W.B. and Prausnitz, J.M. <u>Fluid Phase Equilibria</u>, 1982, 9, 119.
6. Lee, L.L., Chung, T.H. and Starling, K.E. <u>Fluid Phase Equilibria</u>, 1983, 12, 105.
7. Kirkwood, J.G. <u>J. Chem. Phys.</u>, 1935, 3, 300.
8. Starling, K.E., Lee, L.L., Chung, T.H. and Li, M.H., 'Equation of state study for polar fluids,' 1983, third annual report to the Gas Research Institute, University of Oklahoma, Norman, OK.
9. Henderson, D. and Leonard, P.J., 'Liquid Mixtures' in Physical Chemistry: An Advanced Treatise, Volume VIIIB (Academic Press, New York 1971), pp. 413-510.
10. Lee, L.L. and Levesque, D., <u>Mol. Phys.</u>, 1973, 26, 1351.

11. Mansoori, G.A., Carnahan, N.F., Starling, K.E., and Leland, T.W., J. Chem. Phys., 1971, 54, 1523.
12. Hoheisel, C. and Lucas, K. Mol. Phys., 1984, 53(1), 51.
13. Mansoori, G.A. and Leland, T.W. J. Chem. Soc. Faraday Trans. II, 1972, 68, 320.
14. Hill, T.L., 1956, Statistical Mechanics. McGraw-Hill, New York.
15. Chung, T.H., Khan, M.M., Lee, L.L. and Starling, K.E. Fluid Phase Equilibria, 1984, 17, 351.
16. Guggenheim, E.A. 'Mixtures', 1952, Oxford University Press, London.
17. Wichterle, I. and Kobayashi, R.K. J. Chem. Eng. Data, 1972, 17(1), 9.
18. Wichterle, I. and Kobayashi, R.K. J. Chem. Eng. Data, 1972, 17(1), 4.
19. Roberts, L.R., Wang, R.H., Azarnoosh, A., and McKetta, J.J. J. Chem. Eng. Data, 1962, 7(4), 484.
20. Chu, T.-C., Chen, R.J.J., Chappelear, P.S., and Kobayashi, R.K. J. Chem. Eng. Data, 1976, 21(1), 41.
21. Poston, R.S. and McKetta, J.J. J. Chem. Eng. Data, 1966, 11(3), 362.
22. Chang, H.L., Hurt, L.J., and Kobayashi, R.K. AIChE J., 1966, 12(6), 1212.
23. Kohn, J.P. and Bradish, W.F., J. Chem. Eng. Data, 1964, 9, 15.
24. Shipman, L.M. and Kohn, J.P. J. Chem. Eng. Data, 1966, 11(2), 176.
25. Lin, H.M., Sebastian, H.M., Simnick, J.J., and Chao, K.C. J. Chem. Eng. Data, 1979, 24(2), 146.
26. Haines, M.S. Master Thesis, University of Oklahoma, 1967.
27. Mehra, V.S. and Thodos, G. J. Chem. Eng. Data, 1965, 10(4), 307.
28. Kay, W.B. J. Chem. Eng. Data, 1970, 15(1), 48.
29. Ho, J.C.K. and Lu, B.C.Y. J. Chem. Eng. Data, 1963, 8(4), 549.
30. Nielsen, R.L. and Weber, J.H. J. Chem. Eng. Data, 1959, 4(2), 145.
31. Hong, J.H. and Kobayashi, R.K. J. Chem. Eng. Data, 1981, 26(2), 127.
32. Cohen, A.E., Hopkins, H.G., and Koppany, C.R. Chem. Eng. Prog. Symp. Ser., 1967, 63-81, 10.
33. Trust, D.B. and Kurata, F. AIChE J., 1971, 17(1), 86.
34. Stryjek, R., Chappelear, P.S., and Kobayashi, R. J. Chem. Eng. Data, 1974, 19(4), 334.
35. Stryjek, R., Chappelear, P.S., and Kobayashi, R. J. Chem. Eng. Data, 1974, 19(4), 340.
36. Schindler, D.L., Swift, G.W., and Kurata, F. Hydrocarbon Processing, 1966, 45(11), 205.
37. Maimoni, A. AIChE J., 1961, 7, 371.

38. Reamer, H.H., Sage, B.H., and Lacey, W.N. Ind. Eng. Chem., 1951, 43(4), 976.
39. Robinson, D.B., Kabra, H., Krishnan, T., and Miranda, R.D. GPA Report RR-15, 1975.
40. Brewer, J., Rodewald, N. and Kurata, F. AIChE J., 1961, 7(1), 13.
41. Ng, H.-J., Kabra, H., Robinson, D.B., and Kubota, H. J. Chem. Eng. Data, 1980, 25(1), 51.
42. Eakin, B.E. and DeVaney, W.E. AIChE Symposium Series, 1974, No. 140, Vol. 70, 80.
43. Christiansen, L.J., Fredenslund, A. and Mollerup, J. Cryogenics, 1973, 13(7), 405.
44. Trust, D.B. and Kurata, F. AIChE J., 1971, 17-2, 415.
45. Zenner, G.H. and Dana, L.I. Chem. Eng. Prog. Symp. Ser., 1963, 59, 36.
46. Robinson, D.B. and Besserer, G.J. GPA Report RR-7, 1972.
47. Sprow, F.B. and Prausnitz, J.M. AIChE J., 1966, 12(4), 780.
48. Sobocinski, D.P. and Kurata, F. AIChE J., 1959, 5(4), 545.
49. Spano, J.O., Heck, C.K., and Barrick, P.L. J. Chem. Eng. Data, 1968, 13(2), 168.
50. Christiansen, L.J., Fredenslund, A., and Gardner, N. Adv. Cryog. Eng., 1974, 19, 309.
51. Fredenslund, A. and Mollerup, J. J. Chem. Thermo., 1975, 7, 677.
52. Tsang, C.Y. and Streett, W.B. Fluid Phase Equilibria, 1981, 6, 261.
53. Mraw, S.C., Hwang, S.C. and Kobayashi, R. J. Chem. Eng. Data, 1978, 23(2), 135.
54. Fredenslund, A. and Mollerup, J. J. Chem. Soc. Faraday Trans. I, 1974, 70(9), 1653.
55. Akers, W.W., Kelley, R.E., and Lipscomb, T.G. Ind. Eng. Chem., 1954, 46(12), 2535.
56. Besserer, G.J. and Robinson, D.B. J. Chem. Eng. Data, 1973, 18(4), 416.
57. Ohgaki, K. and Katayama, T. J. Chem. Eng. Data, 1976, 21, 53.
58. Kalra, H., Kubota, H., Robinson, D.B., and Ng, H.-J. J. Chem. Eng. Data, 1978, 23(4), 317.
59. Sebastian, H.M., Simnick, J.J., Lin, H.M., Chao, K.C. J. Chem. Eng. Data, 1980, 25(2), 138.
60. Ohgaki, K., Sano, F. and Katayama, T J. Chem. Eng. Data, 1976, 21(1), 55.
61. Gomez-Nieto, M. and Thodos, G. Chem. Eng. Sci., 1978, 33, 1589.
62. Lo, T.C., Bieber, H.H., and Karr, A.E. J. Chem. Eng. Data, 1962, 7(3), 327.
63. Maripuri, V.O. and Ratcliff, G.A. J. Chem. Eng. Data, 1972, 17(3), 366.
64. Tasić, A., Djordjević, B., Grozdanić, D., Afgan, N., and Malić, D. Chem. Eng. Sci., 1978, 33, 189.

65. Tenn, F.G. and Missen, R.W. Can. J. Chem. Eng., 1963, 41, 12.
66. Raal, J.D., Code, R.K., and Best, D.A. J. Chem. Eng. Data, 1972, 17, 211.
67. Benedict, M., Johnson, C.A., Solomon, E., and Rubin, L.C. Trans. AIChE, 1945, 41, 371.
68. Nagata, I. J. Chem. Eng. Data, 1969, 14, 418.
69. Gómez-Nieto, M. and Thodos, G. AIChE J., 1968, 24(4), 672.
70. Ho, J.C.K. and Lu, B.C.-Y. J. Chem. Eng. Data, 1963, 8(4), 549.
71. Katz, K. and Newman, M. Ind. Eng. Chem., 1955, 48(1), 137.
72. Ellis, S.R.M. and Spurr, M.J. Brit. Chem. Eng., 1961, 6, 92.
73. Smith, V.C. and Robinson, R.L. J. Chem. Eng. Data, 1970, 15(3), 391.
74. Prabhu, P.S. and Van Winkle, M. J. Chem. Eng. Data, 1963, 8, 210.
75. Ellis, S.R.M., McDermott, C., Williams, J.C.L. Proc. of the Intern. Symp. on Dist., 1960, Inst. Chem. Engrs., London.
76. Ccon, J., Tojo, G., Bao, M., and Arce, A. An. Real. Soc. Espan. De Fis.Y Quim., 1973, 69, 1177.
77. Culberson, O.L. and McKetta, J.J., Jr. Petro. Trans., AIME, 1951, 192, 223.
78. Olds, R.H., Sage, B.H., and Lacey, W.N. Ind. Eng. Chem., 1942, 34(10), 1223.
79. Culberson, O.L. and McKetta, J.J., Jr. Petro. Trans., AIME, 1950, 189, 319.
80. Reamer, H.H., Olds, R.H., Sage, B.H., and Lacey, W.N. Ind. Eng. Chem., 1943, 35(7), 790.
81. Anthony, R.G. and McKetta, J.J. J. Chem. Eng. Data, 1967, 12(1), 17.
82. Amer, H.H., Paxton, R.R., and Van Winkle, M. Ind. Eng. Chem., 1956, 48(1), 142.
83. Griswold, J. and Wong, S.Y. Chem. Eng. Prog. Sympos. Series, 1952, 48(3), 18.
84. Katayama, T. Ohgaki, K., Maekawa, G., Goto, M., and Nagano, T. J. Chem. Eng. Japan, 1975, 8(2), 89.
85. Schmidt, G.C. Z. Phys. Chem., 1926, 121, 221.
86. Ochi, K. and Kojima, K. Kagaku Kogaku, 1969, 33, 352.
87. Paul, R.N. J. Chem. Eng. Data, 1976, 21(2), 165.
88. Takenouchi, S. and Kennedy, G.C. Am. J. of Sci., 1961, 262, 1055.
89. Selleck, F.T., Carmichael, L.T., and Sage, B.H. Ind. Eng. Chem., 1952, 44, 2219.
90. Clifford, I.L. and Hunter, E. J. Phys. Chem., 1933, 37, 101.
91. Reamer, H.H., Olds, R.H., Sage, B.H., and Lacey, W.N. Ind. Eng. Chem., 1942, 34(12), 1526.
92. Kay, W.B. Ind. Eng. Chem., 1940, 32(3), 353.
93. Sumer, K.M. and Thompson, A.R. J. Chem. Eng. Data, 1967, 12(4), 489.

94. Noda, K., Ohashi, M., and Ishida, K. J. Chem. Eng. Data, 1982, 27, 326.
95. Thomas, K.T., McAllister, R.A. AIChE J., 1957, 3(2), 161.
96. Mikhail, S.Z. and Kimel, W.R. J. Chem. Eng. Data, 1961, 6(4), 533.
97. Wichterle, I. and Kobayashi, R. 'Low Temperature Vapor-Liquid Equilibria in the Methane-Ethane-Propane Ternary and Associated Binary Methane Systems with Special Consideration of the Equilibria in the Vicinity of the Critical Temperature of Methane', 1970, Monogram (Rice University, Houston, Texas).

RECEIVED November 5, 1985

13

Equation of State of Ionic Fluids

Douglas Henderson[1], Lesser Blum[2], and Alessandro Tani[3]

[1]**IBM Research Laboratory, San Jose, CA 95193**
[2]**Department of Physics, University of Puerto Rico, Rio Piedras, PR 00931**
[3]**Instituto di Chimica Fisica, Università di Pisa, via Risorgimento 35, 56100 Pisa, Italy**

An ionic fluid is modelled as a mixture of dipolar hard spheres (the solvent) and charged hard spheres (the ions). The free energy is expanded in a power series in the inverse temperature. Some of the terms in this expansion are infinite. However, these terms can be resummed to give a finite result. This expansion reduces to previously known results for pure dipolar hard spheres and pure charged hard spheres. As is the case for the pure dipolar and charged hard sphere systems, the convergence of the expansion can be enhanced by means of a Padé approximant. Simple expressions for the integrals appearing in the expansion are given.

Most theoretical studies of the equation of state of ionic fluids have been based upon the primitive model in which the solvent is modelled as a dielectric continuum and the ions are modelled as charged hard spheres. It is clearly desirable to model the solvent as a collection of molecules. A simple model of the solvent would be the dipolar hard sphere system. Although deficient as a model of water, the dipolar hard sphere system may be a fairly reasonable model for many organic solvents. Even for water, the dipolar hard sphere model is a first step and is clearly an improvement over a dielectric continuum.

Using this model of the solvent, an ionic fluid can be modelled as a mixture of dipolar hard spheres and charged hard spheres. Such a model would include the pure solvent and the molten salt as limiting cases and could, in principle, be applied to a continuously miscible system.

This model was first discussed by Blum and colleagues[1] and Adelman and Deutch[2] who solved analytically the mean spherical approximation (MSA) for the restricted model in which all the spheres have the same diameter. Extensions and generalizations have also been given.[3-5] Patey and colleagues[5,6] have considered simplified versions of the hypernetted chain (HNC) approximation for this fluid. There is also the modified Poisson-Boltzmann (MPB) approximation considered by Outhwaite[7] and the study of Adelman and Chen.[8]

Unfortunately, there are few exact results or computer simulations which can be used to determine the accuracy of the theory. There is the simulation of Patey and Valleau[9] for very dilute solutions. Very recently, the one-dimensional mixture of charged and dipolar hard spheres was solved exactly by Vericat and Blum.[10] In the high temperature limit, the one-dimensional MSA is recovered.

0097–6156/86/0300–0281$06.00/0

The solution of the MSA is analytic but implicit. Explicit results can be obtained only by some expansion. The MPB results are resonably simple but are restricted to small dielectric constants or low densities. The HNC results are numerical. Clearly, there is a need for a simple explicit theory.

Perturbation theory[11] is a very promising candidate for such a simple explicit theory. In perturbation theory, the free energy is expanded in powers of the inverse temperature using a hard sphere fluid as a reference system. This theory, which we have reviewed in previous ACS symposia,[12] is a very satisfactory theory of simple fluids, such as the inert gases and the pure dipolar and pure charged hard sphere fluids. Prior to this work, it has not been applied to a mixture of dipolar and charged hard spheres. However, it should be a good approximation for this system also.

In this publication, we apply perturbation theory to this mixture. For simplicity, we restrict ourselves to the case where the dipolar and charged hard spheres have an equal diameter, σ. The general case of differing diameters would be rather complex. However, we give some plausible expressions for the case where the dipolar diameter σ_d differs from the ion diameter σ_i.

Perturbation Expansion

Using perturbation theory,[11-13] the Helmboltz free energy can be expanded in powers of $\beta=1/kT$, where k is Boltzmann's constant and T is the temperature. To third order, the result is

$$\frac{\beta(A-A_0)}{N} = -\frac{1}{4}\rho\beta^2\sum_{ij} x_i x_j \int g_{ij}^0(12)\langle u_{ij}^2(12)\rangle d\underset{\sim}{r}_2$$

$$-\frac{1}{6}\rho^2\beta^3\sum_{ijk} x_i x_j x_k \int g_{ijk}^0(123)\langle u_{ij}(12)u_{jk}(23)u_{ik}(13)\rangle d\underset{\sim}{r}_2 d\underset{\sim}{r}_3 , \quad (1)$$

where A_0 is the free energy of the reference hard sphere system and *includes the entropy of mixing term*, $\Sigma x_i \ell n x_i$, $\rho=N/V$, the number of hard spheres per unit volume, $x_i=N_i/N$, the fraction of hard spheres of species i, $g_{ij}^0(12)$ and $g_{ijk}^0(123)$ are the reference pair and triplet correlation functions for a pair or triplet of species i and j or i, j, and k, respectively, and the terms in angular brackets are orientationally averaged pair interactions, *i.e.*,

$$\langle u_{ij}^2(12)\rangle = \int u_{ij}^2(12)d\Omega_1,d\Omega_2/\int d\Omega_1 d\Omega_2 \quad (2)$$

$$\langle u_{ij}(12)u_{jk}(23)u_{ik}(13)\rangle = \int u_{ij}(12)u_{jk}(23)u_{ik}(13)d\Omega_1 d\Omega_2 d\Omega_3/\int d\Omega_1 d\Omega_2 d\Omega_3 . \quad (3)$$

In principle, other terms are present in Eq. (1). However, for the system considered here, they vanish either because for the dipolar hard spheres

$$\langle u_{ij}(12)\rangle = \langle u_{ij}^3(12)\rangle = 0 \quad (4)$$

or because of the charge neutrality of the charged hard spheres,

$$\sum_j x_i z_i = 0 \quad (5)$$

where z_i is the valence (including sign) of the ion of species i.

For simplicity, we assume that the salt is binary and symmetric and that all the hard spheres have the same diameter, σ. Thus, $x_1=x_2=x/2$, $x_3=1-x$ and $u_{ij}(r)=\infty$ for $r<\sigma$. For $r>\sigma$,

$$u_{11}(r) = u_{22}(r) = \frac{z^2e^2}{r} = -u_{12}(r) \quad (6)$$

$$u_{13}(r) = -\frac{ze\mu}{r^2}(\hat{r}\cdot\hat{\mu}_1) \quad (7)$$

$$u_{23}(r) = \frac{ze\mu}{r^2}(\hat{r}\cdot\hat{\mu}_2) \quad (8)$$

and

$$u_{33}(r) = -\frac{\mu^2}{r^3}D(12) \quad (9)$$

where species 1 and 2 are the charged hard spheres and species 3 is a dipolar hard sphere. The quantities e and μ are the electronic charge and the magnitude of the dipole moment of the dipolar hard spheres, respectively, and $z=|z_i|$. The caret above a vector indicates that it is a unit vector. Finally,

$$D(12) = 3(\hat{\mu}_1\cdot\hat{r}_{12})(\hat{\mu}_2\cdot\hat{r}_{12}) - (\hat{\mu}_1\cdot\hat{\mu}_2) \quad (10)$$

Thus,

$$<D^2(12)> = 2/3 \quad (11)$$

and

$$<(r\cdot\mu_1)^2> = \frac{1}{3} \quad (12)$$

The triplet orientation averages have been worked out by Rasaiah and Stell[14] and are

$$<\quad>_{ccd} = \frac{\cos\theta_3}{3} \quad (13)$$

$$<\quad>_{cdd} = \frac{2}{9}\{\cos(\theta_2-\theta_3) + \cos\theta_2\cos\theta_3\} \quad (14)$$

and

$$<\quad>_{ddd} = \frac{1+3\cos\theta_1\cos\theta_2\cos\theta_3}{9} \quad (15)$$

where θ_3 is the interior angle at the vertex 3 in a triangle formed by spheres 1, 2, 3, *etc.*, and ccd, cdd, and ddd mean that the three members of the triplet are a charge-charge-dipole, a charge-dipole-dipole, and a dipole-dipole-dipole, respectively. The subscript ccc will be used to denote three charges. Defining

$$\kappa_0^2 = 4\pi\beta z^2e^2\rho x \quad (16)$$

and

$$y = \frac{4\pi}{9}\beta\mu^2\rho(1-x) \quad (17)$$

Eq. (1) becomes

$$\frac{\beta(A-A_0)}{N} = -\frac{\kappa_0^4}{16\pi\rho}\int_\sigma^\infty g_0(r)dr - \frac{3\kappa_0^2 y}{8\pi\rho}\int_\sigma^\infty \frac{g_0(r)dr}{r^2}$$

$$-\frac{27y^2}{8\pi\rho}\int_\sigma^\infty \frac{g_0(r)dr}{r^4}$$

$$+\frac{\kappa_0^6}{6(4\pi)^3\rho}\int_{r_{ij}\geq\sigma} \frac{g_0(123)d\underset{\sim}{r}_2 d\underset{\sim}{r}_3}{r_{12}r_{13}r_{23}}$$

$$+\frac{3\kappa_0^4 y}{2(4\pi)^3\rho}\int_{r_{ij}\geq\sigma} \frac{g_0(123)\cos\theta_3}{r_{12}r_{13}^2 r_{23}^2} d\underset{\sim}{r}_2 d\underset{\sim}{r}_3$$

$$+\frac{9\kappa_0^2 y^2}{(4\pi)^3\rho}\int_{r_{ij}\geq\sigma} \frac{g_0(123)[\cos(\theta_2-\theta_3)+\cos\theta_2\cos\theta_3]}{r_{12}^2 r_{13}^2 r_{23}^3} d\underset{\sim}{r}_2 d\underset{\sim}{r}_3$$

$$+\frac{27y^3}{2(4\pi)^3\rho}\int_{r_{ij}\geq\sigma} \frac{g_0(123)[1+3\cos\theta_1\cos\theta_3\cos\theta_3]}{r_{12}^3 r_{13}^3 r_{23}^3} d\underset{\sim}{r}_2 d\underset{\sim}{r}_3 \qquad (18)$$

where $g_0(r)$ and $g_0(123)$ are the pair and triplet correlation functions of a pure fluid of hard spheres of diameter σ.

The first of the pair integrals and the first and second of the triplet integrals diverge. The divergence of the pair integrals arises because of the asymptotic value of unity for $g_0(r)$ at large r. The divergence of the second triplet integral also arises because $g_0(123)$ tends to unity when all three hard spheres are far apart. The divergence of the first triplet integral arises because of the asymptotic form $g_0(123)$ where even two spheres are far apart.

Let us consider first the divergence arising from the asymptotic character of the g_0's when all spheres are far apart. Thus,

$$\frac{\beta(A-A_0)}{N} = -\frac{\kappa_0^4}{16\pi\rho}\int_0^\infty dr + \frac{\kappa_0^6}{6(4\pi)^3\rho}\int_{r_{ij}\geq 0} \frac{d\underset{\sim}{r}_2 d\underset{\sim}{r}_3}{r_{12}r_{13}r_{23}}$$

$$+\frac{3\kappa_0^4 y}{2(4\pi)^3\rho}\int_{r_{ij}\geq 0} \frac{\cos\theta_3}{r_{12}r_{13}^2 r_{23}^2} d\underset{\sim}{r}_2 d\underset{\sim}{r}_3$$

$$-\frac{\kappa_0^4}{16\pi\rho}\int_0^\infty h_0(r)dr$$

$$+\frac{\kappa_0^6}{6(4\pi)^3\rho}\int_{r_{ij}\geq 0} \frac{g_0(123)-1}{r_{12}r_{13}r_{23}} d\underset{\sim}{r}_2 d\underset{\sim}{r}_3$$

$$+\frac{3\kappa_0^4 y}{2(4\pi)^3\rho}\int_{r_{ij}\geq 0} \frac{\cos\theta_3}{r_{12}r_{13}^2 r_{23}^2}[g_0(123)-1]d\underset{\sim}{r}_2 d\underset{\sim}{r}_3 + \cdots \qquad (19)$$

where $h_0(r)=g_0(r)-1$. The missing integrals in Eq. (19) indicated by "+···" are the same as in Eq. (18).

The first three integrals can be summed. It is convenient to consider the energy rather than the free energy. Thus, considering only the first three terms of Eq. (19) and differentiating with respect to β,

$$\frac{\beta(U-U_0)}{N} = -\frac{\kappa_0^4}{8\pi\rho}\int_0^\infty dr + \frac{\kappa_0^6}{2(4\pi)^3\rho}\int_{r_{ij}\geq 0}\frac{d\underset{\sim}{r}_2 d\underset{\sim}{r}_3}{r_{12}r_{13}r_{23}}$$

$$+\frac{9\kappa_0^4 y}{2(4\pi)^3\rho}\int_{r_{ij}\geq 0}\frac{\cos\theta_3}{r_{12}r_{13}^2r_{23}^2}d\underset{\sim}{r}_2 d\underset{\sim}{r}_3 + \cdots \tag{20}$$

$$= -\frac{\kappa_0^4}{8\pi\rho}\int_{r_{ij}\geq 0}\left\{\frac{1}{r_{12}} - \frac{\kappa_0^2}{4\pi}\int\frac{d\underset{\sim}{r}_3}{r_{13}r_{23}}\right.$$

$$\left. - \frac{9y}{4\pi}\int\frac{\cos\theta_3}{r_{13}^2 r_{23}^2}d\underset{\sim}{r}_3\right\}r_{12}dr_{12} + \cdots \tag{21}$$

Now

$$\int_{r_{ij}\geq 0}\frac{\cos\theta_3}{r_{13}^2 r_{23}^2}d\underset{\sim}{r}_3 = \frac{4\pi}{r_{12}} \tag{22}$$

Therefore

$$\frac{\beta(U-U_0)}{N} = \frac{\kappa_0^4}{8\pi\rho}\int_{r_{ij}\geq 0}\left\{\frac{1}{r_{12}}(1-9y) - \frac{\kappa_0^2}{4\pi}\int\frac{d\underset{\sim}{r}_3}{r_{13}r_{23}}\right\}r_{12}dr_{12} + \cdots$$

$$= -\frac{\kappa_0^4}{8\pi\rho}(1-9y)\int_{r_{ij}\geq 0}\left\{\frac{1}{r_{12}}\right. \tag{23}$$

$$\left. - \frac{\left[\kappa_0\left(1-\frac{3}{2}y\right)\right]^2}{4\pi}\int\frac{d\underset{\sim}{r}_3}{r_{13}r_{23}}\right\}r_{12}dr_{12} + \cdots$$

The first and second forms of Eq. (23) are equivalent to order β^3.

Defining the ring or chain sum

$$\mathscr{C}(r_{12}) = \frac{1}{r_{12}} - \frac{\left[\kappa_0\left(1-\frac{3}{2}y\right)\right]^2}{4\pi}\int\frac{d\underset{\sim}{r}_3}{r_{13}r_{23}} + \cdots \tag{24}$$

and taking the Fourier transform we have

$$\tilde{\mathscr{C}}(k) = \frac{4\pi}{k^2}\left\{1 - \frac{\left[\kappa_0\left(1-\frac{3}{2}y\right)\right]^2}{k^2} + \cdots\right\}$$

$$= 4\pi\frac{1}{k^2 + \left[\kappa_0\left(1-\frac{3}{2}y\right)\right]^2} \tag{25}$$

Hence,

$$\mathscr{C}(r) = \frac{\exp\left\{-\kappa_0\left(1-\frac{3}{2}y\right)r\right\}}{r} \tag{26}$$

Thus,

$$\frac{\beta(U-U_0)}{N} = -\frac{\kappa_0^4}{8\pi\rho}(1-9y)\int_0^\infty \exp\left\{-\kappa_0\left(1-\frac{3}{2}y\right)r\right\}dr$$

$$= -\frac{\kappa_0^3}{8\pi\rho}\left(1-\frac{15}{2}y\right)+\cdots \tag{27}$$

and

$$\frac{\beta(A-A_0)}{N} = -\frac{\kappa_0^3}{12\pi\rho}\left(1-\frac{9}{2}y\right)+\cdots \tag{28}$$

The expansion of the dielectric constant is just

$$\varepsilon = 1+3y+\cdots \tag{29}$$

Therefore,

$$\kappa = \frac{\kappa_0}{\sqrt{\varepsilon}} = \kappa_0\left(1-\frac{3}{2}y\right)+\cdots \tag{30}$$

and Eq. (28) can be rewritten as

$$\frac{\beta(A-A_0)}{N} = -\frac{\kappa^3}{12\pi\rho}+\cdots \tag{31}$$

so that the Debye-Hückel result is obtained. Resummation has transformed the series from an expansion in powers of β to an expansion in power of $\beta^{1/2}$.

Returning to Eqs. (18) and (19)

$$\frac{\beta(A-A_0)}{N} = -\frac{\kappa_0^3}{12\pi\rho}\left(1-\frac{9}{2}y\right)-\frac{\kappa_0^4}{16\pi\rho}\int_0^\infty h_0(r)dr$$

$$+\frac{\kappa_0^6}{6(4\pi)^3\rho}\int_{r_{ij}\geq 0}\frac{g_0(123)-1}{r_{12}r_{13}r_{23}}d\underset{\sim}{r}_2 d\underset{\sim}{r}_3$$

$$+\frac{3\kappa_0^4 y}{2(4\pi)^3\rho}\int_{r_{ij}\geq 0}\frac{\cos\theta_3}{r_{12}r_{13}^2r_{23}^2}[g_0(123)-1]d\underset{\sim}{r}_2 d\underset{\sim}{r}_3+\cdots \tag{32}$$

The three body integrals in Eq. (32) can be simplified by introducing

$$h_0(123) = g_0(123)-1-h_0(12)-h_0(13)-h_0(23) \tag{33}$$

since

$$\int_{r_{ij}\geq 0}\frac{h_0(12)}{r_{12}r_{13}r_{23}}d\underset{\sim}{r}_2 d\underset{\sim}{r}_3 = 8\pi^2\left\{2\int_0^\infty rh_0(r)dr\int_0^\infty ds - \int_0^\infty r^2h_0(r)dr\right\} \tag{34}$$

$$\int_{r_{ij}\geq 0}\frac{h_0(12)\cos\theta_3}{r_{12}r_{13}^2r_{23}^2}d\underset{\sim}{r}_2 d\underset{\sim}{r}_3 = 16\pi^2\int_0^\infty h_0(r)dr \tag{35}$$

and

$$\int_{r_{ij}\geq 0} \frac{h_0(13)\cos\theta_3}{r_{12}r_{13}^2r_{23}^2}\, d\underset{\sim}{r}_2 d\underset{\sim}{r}_3 = 8\pi^2 \int_0^\infty h_0(r)dr \tag{36}$$

The first integral on the RHS of Eq. (34) contains a divergent term. However, it is resummable. In analogy to Eqs. (24)-(27), the divergent integral is of order κ_0^{-1}. Unfortunately, it is difficult to perform the resummations with only two terms (of order β^2 and β^3) in the expansion. As a result, we adopt an *ad hoc* procedure. We assume that a truncated perturbation expansion correctly identifies the terms which appear but does not give the correct values for some of the coefficients because of resummed contributions from higher order terms.

For example, comparison with the Stell-Lebowitz[15] expansion for pure charged hard spheres,

$$\frac{\beta(A-A_0)}{N} = -\frac{\kappa_0^3}{12\pi\rho} - \frac{\kappa_0^4}{16\pi\rho}\int_0^\infty h_0(r)dr + \frac{\kappa_0^5}{8\pi\rho}\int_0^\infty r h_0(r)dr$$

$$-\frac{\kappa_0^6}{8\pi\rho}\left[\int_0^\infty r^2 h_0(r)dr - \frac{1}{6}\int_0^\infty dr_{12}\int_0^\infty dr_{13}\int_{|r_{12}-r_{13}|}^{r_{12}+r_{13}} h_0(123)dr_{23}\right] + \cdots \tag{37}$$

shows that third order perturbation theory correctly identifies all the terms appearing in the expansion but does not give the coefficients correctly for the term of order κ_0^5 or the pair integral in the κ_0^6 term. Note that the pair term of order $\kappa_0^4(\beta^2)$ and the triplet term of order $\kappa_0^6(\beta^3)$ are given correctly.

The procedure we adopt here is to choose the coefficients of the pair integrals of order κ_0^5 and κ_0^6 from the Stell-Lebowitz series and the pair terms of order $\kappa_0^2 y^2$ and $\kappa_0^4 y$ from the expansion of the MSA free energy for the charged hard sphere/dipolar hard sphere mixture in powers of κ_0. When the g_0's are set equal to their low density limit of unity for nonoverlapping spheres, the perturbation expansion must give the corresponding MSA result.

We emphasize that perturbation theory, if taken to high enough order, would give the correct coefficients for all the various integrals. Our *ad hoc* procedure is adopted only to simplify the derivation of the final expression for the free energy.

Numerical Results for Perturbation Theory Integrals

To evaluate the various integrals appearing in perturbation, we assume that the hard sphere $g_0(r)$ is given adequately by the Percus-Yevick (PY) theory.[16] Two integrals can be obtained immediately and are

$$I'_{cc} = \frac{1}{\sigma^2}\int_0^\infty r h_0(r)dr = -\frac{10-2\eta+\eta^2}{20(1+2\eta)} \tag{38}$$

and

$$I''_{cc} = \frac{1}{\sigma^3}\int_0^\infty r^2 h_0(r)dr = \frac{(\eta-4)(\eta^2+2)}{24(1+2\eta)} \tag{39}$$

where $\eta=\pi\rho\sigma^3/6$. The other pair integrals can be obtained numerically from G(s), the Laplace transform of $g_0(r)$. Thus,

$$\int_0^\infty h_0(r)dr = \int_0^\infty \left[G(s) - \frac{1}{s^2}\right]ds \tag{40}$$

and

$$\int_0^\infty \frac{g(r)}{r^n}\,dr = \frac{1}{n!}\int_0^\infty s^n G(s)ds, \quad n\geq 2 \tag{41}$$

where

$$G(s) = \frac{sL(s)}{12\eta[L(s) + e^s S(s)]} \tag{42}$$

$$L(s) = 12\eta[1 + 2\eta + s(1 + \eta/2)] \tag{43}$$

$$S(s) = -12\eta(1 + 2\eta) + 18\eta^2 s + 6\eta(1 - \eta)s^2 + (1 - \eta)^2 s^3 \tag{44}$$

We have calculated these three integrals and have found them to be well represented by

$$I_{cc} = \frac{1}{\sigma}\int_0^\infty h_0(r)dr = -\frac{1 + 0.97743\rho\sigma^3 + 0.05257\rho^2\sigma^6}{1 + 1.43613\rho\sigma^3 + 0.41580\rho^2\sigma^6} \tag{45}$$

$$I_{cd} = \sigma\int_0^\infty \frac{g_0(r)}{r^2}\,dr = \frac{1 + 0.79576\rho\sigma^3 + 0.104556\rho^2\sigma^6}{1 + 0.486704\rho\sigma^3 - 0.0222903\rho^2\sigma^6} \tag{46}$$

and

$$I_{dd} = \sigma^3\int_\sigma^\infty \frac{g_0(r)}{r^4}\,dr = \frac{1 + 0.18158\rho\sigma^3 - 0.11467\rho^2\sigma^6}{3(1 - 0.49303\rho\sigma^3 + 0.06293\rho^2\sigma^6)} \tag{47}$$

We have also calculated the three body integrals both by MC simulation and by numerical integration using the superposition approximation

$$g_0(123) = g_0(12)g_0(13)g_0(23) \tag{48}$$

or equivalently,

$$h_0(123) = h_0(12)h_0(13)h_0(23) + h_0(12)h_0(13) + h_0(13)h_0(23) + h_0(13)h_0(23) \tag{49}$$

with $g_0(r)$ or $h_0(r)$ given by the PY approximation, and have found that the relevent three body integrals are well represented by

$$I_{ccc} = \frac{1}{\sigma^3}\int^\infty dr_{12}\int_0^\infty dr_{13}\int_{|r_{12}-r_{13}|}^{r_{12}+r_{13}} h_0(123)dr_{23}$$

$$= \frac{3(1 - 1.05560\rho\sigma^3 + 0.26591\rho^2\sigma^6)}{2(1 + 0.53892\rho\sigma^3 - 0.94236\rho^2\sigma^2)} \tag{50}$$

$$I_{ccd} = \frac{1}{\sigma}\int_0^\infty dr_{12}\int_0^\infty \frac{dr_{13}}{r_{13}}\int_{|r_{12}-r_{13}|}^{r_{12}+r_{13}} h_0(123)\,\frac{\cos\theta_3}{r_{23}}\,dr_{23}$$

$$= \frac{11(1 + 2.25642\rho\sigma^3 + 0.05679\rho^2\sigma^6)}{6(1 + 2.64178\rho\sigma^3 + 0.79783\rho^2\sigma^6)} \tag{51}$$

$$I_{cdd} = \sigma \int_0^\infty \frac{dr_{12}}{r_{12}} \int_0^\infty \frac{dr_{13}}{r_{13}} \int_{|r_{12}-r_{13}|}^{r_{12}+r_{13}} g_0(123) \frac{\cos(\theta_2 - \theta_3) + \cos\theta_2\cos\theta_3}{r_{23}^2} dr_{23}$$

$$= 0.94685 \frac{1 + 2.97323\rho\sigma^3 + 3.11931\rho^2\sigma^6}{(1 + 2.70186\rho\sigma^3 + 1.22989\rho^2\sigma^6)} \quad (52)$$

and

$$I_{ddd} = \sigma^3 \int_0^\infty \frac{dr_{12}}{r_{12}^2} \int_0^\infty \frac{dr_{13}}{r_{13}^2} \int_{|r_{12}-r_{13}|}^{r_{12}+r_{13}} g_0(123) \frac{1 + 3\cos\theta_1\cos\theta_2\cos\theta_3}{r_{23}^2} dr_{23}$$

$$= \frac{5(1 + 1.12754\rho\sigma^3 + 0.56192\rho^2\sigma^6)}{24(1 - 0.05495\rho\sigma^3 + 0.13332\rho^2\sigma^6)} \quad (53)$$

Previous experience[17,18] with the evaluation of similar triplet integrals indicates that the use of the PY g_0's and the superposition approximation does not introduce appreciable error.

The leading term in Eqs. (50) and (53) of 3/2 and 5/24, respectively, were obtained previously by Larsen *et al.*[19] and Rushbrooke *et al.*[20] and have been reobtained by ourselves. The leading term of 11/6 in Eq. (51) is new. Unfortunately, we have not yet obtained an analytic result for the leading term in Eq. (52).

Although none of these integrals is independent of ρ, the variation of the integrals I_{cc}, I'_{cc}, I''_{cc}, I_{cd}, I_{dd}, I_{ccc}, I_{ccd}, I_{cdd}, and I_{ddd} is small compared to that of the prefactor ρ for the pair integrals or ρ^2 for the triplet integrals. Because of this, it may not be too bad an approximation to assume these integrals to be a constant, provided the density does not vary overly much. Thus, at low densities,

$$I_{cc} \quad -1 \quad (54a)$$

$$I'_{cc} \quad -1/2 \quad (54b)$$

$$I''_{cc} \quad -1/3 \quad (54c)$$

$$I_{cd} \quad 1 \quad (54d)$$

$$I_{dd} \quad 1/3 \quad (54e)$$

and

$$I_{ccc} \quad 3/2 \quad (55a)$$

$$I_{ccd} \quad 11/6 \quad (55b)$$

$$I_{cdd} \quad 0.94685 \quad (55c)$$

and

$$I_{ddd} \quad 5/24 \quad (55d)$$

while at high densities

$$I_{cc} \quad -0.75 \tag{56a}$$

$$I'_{cc} \quad -0.25 \tag{56b}$$

$$I''_{cc} \quad -0.10 \tag{56c}$$

$$I_{cd} \quad 1.24 \tag{56d}$$

$$I_{dd} \quad 0.55 \tag{56e}$$

and

$$I_{ccc} \quad 0.59 \tag{57a}$$

$$I_{ccd} \quad 1.44 \tag{57b}$$

$$I_{cdd} \quad 1.29 \tag{57c}$$

and

$$I_{ddd} \quad 0.45 \tag{57d}$$

Expansion of MSA Results

According to Blum *et al.*,[1] in the MSA for the charged hard sphere/dipolar hard sphere model for equal size spheres,

$$\frac{\beta(U-U_0)}{N} = \frac{\kappa_0^2 b_0 - 2(3\kappa_0 y)^{1/2} b_1 - 6yb_2}{4\pi\rho} \tag{58}$$

where b_0, b_1, and b_2 are parameters whose expansions in powers of y and κ_0 are

$$b_0 = \frac{\kappa}{2}\,(1 - \kappa + \frac{1}{4}\,y\kappa\sigma) + \cdots \tag{59}$$

$$b_1 = \frac{1}{2}\,\kappa(3y)^{1/2}\beta_6\beta_{12}\left(1 - \kappa\sigma\,\frac{\beta_{12}}{\beta_3}\right) + \cdots \tag{60}$$

$$b_2 = \frac{3y}{2} - \frac{45}{64}\,y^2 - \frac{13}{384}\,y\kappa_0^2\sigma^2 + \cdots \tag{61}$$

and

$$\beta_3 = 1 + b_2/3 \tag{62}$$

$$\beta_6 = 1 - b_2/6 \tag{63}$$

$$\beta_{12} = 1 + b_2/12 \tag{64}$$

In Eqs. (59)-(61), $\kappa = \kappa_0/\sqrt{\varepsilon}$. At $\kappa_0=0$, the parameter b_2 is related to the dielectric constant by

$$\varepsilon = \frac{\beta_{12}^4 \beta_3^2}{\beta_6^6} \tag{65}$$

If κ is expanded as

$$\kappa = \kappa_0\left(1 - \frac{3}{2}\,y + \cdots\right) \tag{66}$$

Equations (58)-(65) may be combined to give the expansion

$$\frac{\beta(U - U_0)}{N} = \frac{1}{4\pi\rho}\left\{-\frac{\kappa_0^3}{2}\left(1 - \frac{15}{2}\,y\right) + \frac{\kappa_0^4\sigma}{2}\left(1 - \frac{13}{4}\,y\right) - \frac{3\kappa_0^2 y}{\sigma}\left(1 - \frac{39}{16}\,y\right) - \frac{9y^2}{\sigma^3}\left(1 - \frac{15}{32}\,y\right)\right\} + \cdots \tag{67}$$

which may be integrated to give

$$\frac{\beta(A - A_0)}{N} = -\frac{\kappa_0^3}{12\pi\rho}\left(1 - \frac{9}{2}\,y\right) + \frac{\kappa_0^4\sigma}{16\pi\rho}\left(1 - \frac{13}{6}\,y\right) - \frac{3\kappa_0^2 y}{8\pi\rho\sigma}\left(1 - \frac{13}{8}\,y\right) - \frac{9}{8\pi\rho^2\sigma}\,y^2\left(1 - \frac{5}{16}\,y\right) + \cdots \tag{68}$$

Equation (68) bears a close resemblance to perturbation theory. It lacks the terms in κ_0^5 and κ_0^6. This is not surprising since Eq. (68) was constructed from a low concentration (κ_0 small) expansion and is not a complete β expansion.

If desired, Eq. (68) can be rewritten using $\varepsilon - 1 = 3y(1 + y) + \cdots$ to obtain

$$\frac{\beta(A - A_0)}{N} = -\frac{\kappa^3}{12\pi\rho} + \frac{\kappa^4\sigma}{16\pi\rho}\left[1 + \frac{23}{18}(\varepsilon - 1)\right] - \frac{\kappa^2(\varepsilon - 1)}{8\pi\rho\sigma}\left[1 + \frac{3}{8}\,y\right] - \frac{9}{8\pi\rho\sigma^3}\,y^2\left[1 - \frac{5}{16}\,y\right] + \cdots \tag{69}$$

If we recognize that $1 + 3y/8$ is just the expansion of $2/(1 + \lambda^{-1})$, where $\lambda = \beta_3/\beta_6$, then Eq. (69) becomes

$$\frac{\beta(A - A_0)}{N} = -\frac{\kappa^3}{12\pi\rho} + \frac{\kappa^4\sigma}{16\pi\rho}\left[1 + \frac{23}{18}(\varepsilon - 1)\right] - \frac{\kappa^2(\varepsilon - 1)}{4\pi\rho\sigma(1 + \lambda^{-1})} - \frac{9}{8\pi\rho\sigma^3}\,y^2\left[1 - \frac{5}{16}\,y\right]\cdots \tag{70}$$

When written in the above form, the third term gives exactly the free energy of solvation as calculated by Garisto *et al.*[6] from the MSA. We refrain from expressing y in the last term of Eq. (70) in terms of ε because the free energy of the pure dipolar fluid is conventionally written in terms of a series in y rather than in ε.

Free Energy of an Ionic Fluid

Our procedure is to combine Eqs. (18), (32), (34), (35), and (36) with the coefficients of the integrals in Eqs. (34)-(36) chosen so as to reproduce Eq. (37) when y=0 and Eq. (68) when the integrals I_{cc}, I'_{cc}, I''_{cc}, I_{cd}, I_{dd}, I_{ccc}, I_{ccd}, I_{cdd}, and I_{ddd} have the values given in Eqs. (54) and (55). While *ad hoc*, this procedure is suggested by the form of perturbation theory and reduces to all previously known expansions in the appropriate limits.

An extension to the case where $\sigma_i \neq \sigma_d$ is desirable. To do this without further approximation would require calculating the integral I_{cc}, *etc.*, for every concentration. This would be acceptable for the pair integrals but would be far too time consuming for the triplet integrals. Instead, we assume that Eqs. (38)-(39), (45)-(47), (50)-(53) can be used with $\rho\sigma^3$ replaced by $\rho\langle\sigma^3\rangle$, where

$$\langle\sigma^3\rangle = x\sigma_i^3 + (1-x)\sigma_d^3 \tag{71}$$

This type of averaging has proven useful in theories of nonelectrolyte solutions[21] and should be useful for this sytem provided σ_i and σ_d are not too different. Thus, for $\sigma_i \neq \sigma_d$,

$$\begin{aligned}\frac{\beta(A-A_0)}{N} = &-\xi^{3/2}\frac{\kappa_0^3}{12\pi\rho} - \xi^2\left\{\frac{\kappa_0^4\sigma_i}{16\pi\rho} I_{cc} + \frac{3\kappa_0^2 y}{8\pi\rho\sigma_{id}} I_{cd} + \frac{27y^2}{8\pi\rho\sigma_d^3} I_{dd}\right\} \\ &+ \xi^{5/2}\left\{\frac{\kappa_0^5\sigma_i^2}{8\pi\rho} I'_{cc} + \frac{3\kappa_0^3 y}{8\pi\rho}\right\} \\ &+ \xi^3\left\{-\frac{\kappa_0^6\sigma_i^3}{8\pi\rho}\left(I''_{cc} - \frac{1}{6} I_{ccc}\right)\right. \\ &+ \frac{\kappa_0^4 y\sigma_i}{16\pi\rho}\left[\left(6 + \frac{5}{3}\left[\frac{\sigma_d}{\sigma_i}\right]\right) I_{cc} + 3I_{ccd}\left(\frac{\sigma_d}{\sigma_i}\right)\right] \\ &+ \frac{3\kappa_0^2 y^2}{8\pi\rho\sigma_{id}}\left[\left(2 - 3.21555\left[\frac{\sigma_d}{\sigma_{id}}\right]\right) I_{cd} + 3I_{cdd}\left(\frac{\sigma_d}{\sigma_{id}}\right)\right] \\ &\left. + \frac{27y^3}{16\pi\rho\sigma_d^3} I_{ddd}\right\} + \cdots\end{aligned} \tag{72}$$

where $\xi=1$ is a parameter whose power denotes the order in β of the term which follows that power of ξ and

$$\sigma_{id} = \frac{\sigma_i + \sigma_d}{2} \tag{73}$$

Equation (72) reduces to the Stell-Lebowitz series for charged hard spheres in a vacuum when y=0 (or $\varepsilon=1$) and reduces to the perturbation series of Rushbrooke *et al.*[20] for dipolar hard spheres when $\kappa_0=0$. When $\sigma_i=\sigma_d$, Eq. (72) reduces to the MSA expansion, Eq. (68), plus the terms in κ_0^5 and κ_0^6, if Eqs. (54) and (55) are used for the integrals.

When $\sigma_i \neq \sigma_d$, the MSA expansion corresponding to Eq. (68) has a complex dependence on σ_d and σ_i. However, the form of Eq. (72) is suggested by that result. The combination $y\sigma_d$ ensures that the free energy, in excess of the free energy of the solvent and the free energy of solvation, reduces to the Stell-Lebowitz expansion with

ε expanded in powers of y for charged hard spheres in a dielectric continuum when σ_d tends to zero. The free energy of solvation becomes the Born result for the free energy of solvation of a charged hard sphere in a dielectric continuum.

It might be preferable to eliminate y from part of Eq. (72) by using

$$\varepsilon = 1 + 3y + \cdots \tag{74}$$

$$\frac{\varepsilon - 1}{\varepsilon} = 3y(1 - 2y) + \cdots \tag{75}$$

and

$$\left(1 + \frac{\sigma_d}{\lambda\sigma_i}\right)^{-1} = \frac{\sigma_i}{2\sigma_{id}}\left[1 + \frac{3}{8}\,y\left(\frac{\sigma_d}{\sigma_{id}}\right)\right] + \cdots \tag{76}$$

Hence,

$$\begin{aligned}\frac{\beta(A - A_0)}{N} = &- \xi^{3/2}\frac{\kappa^3}{12\pi\rho} - \xi^2\left\{\frac{\kappa^4\sigma_i}{16\pi\rho}I_{cc} + \frac{\kappa^2}{4\pi\rho\sigma_i}\left(\frac{\varepsilon - 1}{1 + \frac{\sigma_d}{\lambda\sigma_i}}\right)I_{cd}\right.\\ &\left. + \frac{27y^2}{8\pi\rho\sigma_d^3}I_{dd}\right\} + \xi^{5/2}\frac{\kappa^5\sigma_i^2}{8\pi\rho}I'_{cc}\\ &+ \xi^3\left\{-\frac{\kappa^6\sigma_i^3}{8\pi\rho}\left(I''_{cc} - \frac{1}{6}I_{ccc}\right) + \frac{\kappa^4\sigma_d}{16\pi\rho}(\varepsilon - 1)\left(\frac{5}{9}I_{cc} + I_{ccd}\right)\right.\\ &\left. - \frac{\kappa^2\sigma_d}{6\pi\rho\sigma_i^2}\left(\frac{\varepsilon - 1}{1 + \frac{\sigma_d}{\lambda\sigma_i}}\right)^2(2.84055I_{cd} - 3I_{cdd}) + \frac{27y^3}{16\pi\rho\sigma_d^3}I_{ddd}\right\}\end{aligned} \tag{77}$$

The factors of y are retained in the coefficients of I_{dd} and I_{ddd} so that the series reduces to that of Rushbrooke *et al.*[20] when $\kappa=0$.

For Eq. (77) to be useful, the dielectric constant, ε, must be calculated as a function of concentration. The fact that this information is not necessary for Eq. (72) may be an advantage for this equation. To the order in which we are working, perturbation theory yields the result

$$\varepsilon = 1 + 3y + \cdots \tag{29}$$

For the pure dipolar hard sphere fluid

$$\varepsilon = 1 + 3y + 3y^2 + 3y^3\left(\frac{9I_{dd\Delta}}{16\pi^2} - 1\right) \tag{78}$$

where $I_{dd\Delta}$ is an integral calculated by Tani *et al.*,[18] seems to be a good approximation. Comparison of Eqs. (29) and (78) suggests that the same series might be used for the dielectric constant of the mixture. In this approximation, ε would decrease with increasing concentration as y decreased because 1−x is decreasing. Such a decrease does not seem large enough compared with experiment.[22,23] Empirically,

$$\epsilon = 1 + 3y \frac{1 + y + y^2\left(\frac{9I_{dd\Delta}}{16\pi^2} - 1\right)}{1 + c\kappa^2\sigma_i^2} \tag{79}$$

where c is an empirically adjusted parameter and $I_{dd\Delta}$ is assumed to be the function of Tani *et al.* with $\rho\sigma^3$ replaced by $\rho<\sigma^3>$ if $\sigma_i \neq \sigma_d$, seems reasonable. For an aqueous solution of NaCl, c~0.1 gives fairly good agreement with experiment.

For pure charged spheres and pure dipolar hard spheres, the studies of Larsen *et al.*[19] and Rushbrooke *et al.*,[20] respectively, suggest that Eqs. (72) or (77) will converge slowly for large κ or y but that good results can be obtained from a Padé approximant. Thus, we write the series

$$\frac{\beta(A - A_0)}{N} = \xi^{3/2}A_{3/2} + \xi^2 A_2 + \xi^{5/2}A_{5/2} + \xi^3 A_3 + \cdots \tag{80}$$

in the form

$$\frac{\beta(A - A_0)}{N} = \xi^{3/2} \frac{n_1 + \xi^{1/2}n_2}{1 + d_1\xi^{1/2} + d_2\xi} \tag{81}$$

where the coefficients n, n_2, d_1, and d_2 are chosen to reproduce $A_{3/2}$, A_2, $A_{5/2}$, and A_3. Hence,

$$n_1 = A_{3/2}, \tag{82}$$

$$n_2 = \frac{A_{3/2}^2 A_3 - 2A_{3/2}A_2A_{5/2} + A_2^3}{A_2^2 - A_{3/2}A_{5/2}} \tag{83}$$

$$d_1 = \frac{A_{3/2}A_3 - A_2A_{5/2}}{A_2^2 - A_{3/2}A_{5/2}} \tag{84}$$

and

$$d_2 = -\frac{A_2A_3 - A_{5/2}^2}{A_2^2 - A_{3/2}A_{5/2}} \tag{85}$$

For y=0, all the A_n are, in general, nonvanishing and Eq. (81) is the Padé approximant of Larsen *et al.* For $\kappa_0=0$, $A_{3/2}=A_{5/2}=0$ so that

$$n_1 = 0 \tag{86}$$

$$n_2 = A_2 \tag{87}$$

$$d_1 = 0 \tag{88}$$

$$d_2 = -\frac{A_3}{A_2} \tag{89}$$

and Eq. (81) is the Padé approximant of Rushbrooke *et al.* Thus, Eq. (80) has the appropriate limiting forms.

There is a little ambiguity as to whether the terms $\kappa^2(\varepsilon-1)$, $\kappa^4(\varepsilon-1)$, and $\kappa^2(\varepsilon-1)^2$ should be considered to be of order ξ^2, ξ^3, and ξ^3, respectively, as in Eq. (77), or of order ξ, ξ^2, and ξ, respectively. The question is purely academic in Eq. (79) since $\xi=1$. However, it is a question of importance when a Padé is formed. For the moment, we leave Eq. (77) in the above form. However, experience may cause us to rethink this issue.

Summary

We have given two promising forms for the free energy of an ionic fluid. These two expressions are derived from perturbation theory with plausibility arguments used to determine the coefficients of some of third order terms. These expressions reduce to the previously known perturbation expression for a pure fluid of charged hard spheres when dipole moment or the diameter of the dipolar hard spheres vanishes and when the concentration of dipoles is zero. They also reduce to the previously known perturbation expression for a pure fluid of dipolar hard spheres when the charge or concentration of the ions is zero.

The expressions developed in this paper should be of value in electrochemical studies, including corrosion, and in studies of biological systems. The perturbation expansion is straightforward but slowly convergent for most practical applications. Unfortunately, there is a great deal of flexibility in the summation procedures used to obtain practical expressions. Equations (80)-(85) give just one possibility. Many more are possible and equally plausible. Before a final choice is made, application to experimental data will be required. Unfortunately, simulated data are not available. Simulation studies of ion-dipole mixtures are clearly required. We hope that the chemical engineering community will join us in generating such data and in making calculations based on the expressions presented here and in comparing the results with experimental and simulated data.

Literature Cited

1. Blum, L. Chem. Phys. Lett. 1974, 26, 200; J. Chem. Phys. 1974, 61, 2129; J. Stat. Phys. 1978, 18, 451; Vericat, F.; Blum, L. J. Stat. Phys. 1980, 22, 593; Peréz-Hernández, W.; Blum, L. ibid. 1981, 24, 451.
2. Adelman, S. A.; Deutch, J. M. J. Chem. Phys. 1974, 60, 3935.
3. Vericat, F.; Blum, L. Mol. Phys. 1982, 45, 1067.
4. Chan., D. Y. C.; Mitchell, D. J.; Ninham, B. W.; Pailthorpe, B. J. Chem. Phys. 1978, 69, 691; 1979, 70, 1578; Chan, D. Y. C.; Mitchell, D. J.; Ninham, B. W. ibid. 1979, 70, 2946.
5. Garisto, F.; Kusalik, P. G.; Patey, G. N. J. Chem. Phys. 1983, 79, 6294.
6. Levesque, D.; Weis, J. J.; Patey, G. N. Phys. Lett. 1978, 66A, 115; J. Chem. Phys. 1980, 72, 1887; Patey, G. N.; Carnie, S. L. ibid., 1983, 78, 5183; Kusalik, P. G.; Patey, G. N. J. Chem. Phys. 1983, 79, 4468.
7. Outhwaite, C. W. Mol. Phys. 1976, 31, 1345; 1977, 33, 1229.
8. Adelman, S. A.; Chen, J.-H. J. Chem. Phys. 1979, 70, 4291.
9. Patey, G. N.; Valleau, J. P. J. Chem. Phys. 1975, 63, 2334.
10. Vericat, F.; Blum, L. J. Chem. Phys. 1985, 82, 1492.
11. Barker, J. A.; Henderson, D. J. Chem. Phys. 1967, 47, 2856; 1967, 47, 4714; Rev. Mod. Phys. 1976, 48, 587.
12. Henderson, D. Adv. in Chem. 1979, 182, 1; 1983, 204, 47.
13. Henderson, D.; Barojas, J.; Blum, L. Revista Mexicana de Fisica 1984, 30, 139.
14. Rasaiah, J. C.; Stell, G. Chem. Phys. Lett. 1974, 25, 519.
15. Stell, G.; Lebowitz, J. L. J. Chem. Phys. 1968, 49, 3796.

16. Percus, J. K.; Yevick, G. J. Phys. Rev. 1958, 110, 1.
17. Barker, J. A.; Henderson, D.; Smith, W. R. Mol. Phys. 1969, 17, 579.
18. Tani, A.; Henderson, D.; Barker, J. A.; Hecht, C. E. Mol. Phys. 1983, 48, 863.
19. Larsen, B.; Rasaiah, J. C.; Stell, G. Mol. Phys. 1977, 33, 987.
20. Rushbrooke, G. S.; Stell, G.; Hoye, J. S. Mol. Phys. 1973, 26, 1199.
21. Henderson, D.; Leonard, P. J. In "Physical Chemistry – An Advanced Treatise"; Eyring, H.; Henderson, D.; Jost, P. W., Eds.; Academic Press: New York, 1971; Vol. 8B, Chap. 7.
22. Halsted, J. B.; Ritson, D. M.; Collie, C. H. J. Chem. Phys. 1948, 16, 1.
23. Hubbard, J. B.; Onsager, L.; van Beek, W. M.; Mandel, M. Proc. Natl. Acad. Sci., U.S.A., 1977, 74, 401.

RECEIVED November 8, 1985

14

Thermodynamics of Multipolar Molecules

The Perturbed-Anisotropic-Chain Theory

P. Vimalchand, Marc D. Donohue, and Ilga Celmins

Department of Chemical Engineering, The Johns Hopkins University, Baltimore, MD 21218

The Perturbed-Hard-Chain theory (PHCT) is modified to treat rigorously dipolar and quadrupolar fluids and their mixtures. The multi-polar interactions, which are treated explicitly, are calculated by extending the perturbation expansion of Gubbins and Twu to chain-like molecules. Moreover, the square-well potential used to characterize the spherically symmetric interactions in the PHCT has been replaced by a soft-core (Lennard-Jones) potential. Theoretical calculations and data reduction on a number of pure fluids and mixtures indicate that the above changes result in physically more meaningful pure component parameters, and the properties of even highly non-ideal mixtures can be predicted accurately without the use of a binary interaction parameter.

Although considerable phase equilibrium data are available for hydrocarbon systems of low molecular weight, data for high molecular weight hydrocarbons, especially aromatic hydrocarbons, are scarce. Unfortunately, none of the theories that have been developed for the properties of lower molecular weight aliphatic compounds encountered in natural gas and oil are accurate for heavy hydrocarbons such as coal derivatives. They require large values of an adjustable binary interaction parameter, and therefore, the predictive ability of these theories becomes poor for multicomponent mixtures and for systems for which there are no experimental data.

There are two main reasons why these theories fail for high molecular weight multipolar compounds. First, these theories are valid only for compounds with molecular weight below 150 because they ignore the effect of rotational and vibrational motions on thermodynamic properties. While this deficiency was overcome, in part, by the polymer theories of Flory (1) and Prigogine (2), their theories are valid only at high density. The second deficiency in all the theories that were developed for oil and gas-refining operations concerns the nature of intermolecular forces. All these theories tacitly assume that molecules interact with London dispersion forces (also referred to as van der Waals forces). Since coal and its derivatives are generally high molecular

0097-6156/86/0300-0297$06.00/0

weight (150 to 1000) compounds that have numerous benzene rings and functional groups which have strong anisotropic (dipolar and quadrupolar) potential functions, any theory for coal compounds must reflect this difference. These two deficiencies are taken into account in developing the Perturbed-Anisotropic-Chain theory (PACT) which enables calculation of thermodynamic properties for a wide variety of pure fluids and mixtures.

The PACT equation is based on the Perturbed-Soft-Chain theory (PSCT) and on improvements to Pople's perturbation expansion (3) by Gubbins and Twu (4). The PSCT equation is essentially the Perturbed-Hard-Chain theory (PHCT) of Donohue and Prausnitz (5), but the potential energy function used to characterize the interaction between the molecules is different. The original PHCT equation uses a square-well intermolecular potential, while in the PSCT equation, interactions are calculated with the Lennard-Jones potential. Multipolar interactions in the PACT are treated by combining the perturbation expansion of Gubbins and Twu for anisotropic molecules with lattice theory for chain-like molecules.

The PHCT and PSCT equations have been shown to predict accurately the properties of pure fluids and mixtures of non-polar molecules of varying size and shape [Beret and Prausnitz (6); Donohue and Prausnitz (5); Kaul et al. (7) and Morris (8)], including polymeric systems [Beret and Prausnitz (6) and Liu and Prausnitz (9)]. For mixtures containing quadrupolar molecules (such as carbon dioxide - ethylene system; carbon dioxide - ethane system; benzene - 1-methyl naphthalene system), the PACT equation predicts properties better than either the PHCT or PSCT equations [Vimalchand and Donohue (10)]. In this paper, we extend the PACT to include dipolar interactions. The theoretical results show that the PACT equation also can be used for predicting accurately the properties of highly non-ideal mixtures (without the use of a binary interaction parameter) containing strongly dipolar molecules such as acetone.

The Perturbed–Anisotropic–Chain Theory

The canonical ensemble partition function used in the Perturbed-Anisotropic-Chain theory is of form:

$$Q(N,V,T) = \frac{1}{N!}\left(\frac{V}{\Lambda^3}\right)^N \left(\frac{V_f}{V}\right)^{Nc} \left(\exp\frac{-\phi^{iso}}{2ckT}\right)^{Nc} \left(\exp\frac{-\phi^{ani}}{2ckT}\right)^{Nc} \quad (1)$$

where N is the number of molecules at temperature T and volume V; Λ is the thermal deBroglie wavelength, and k is Boltzmann's constant. The partition function is based on the generalized van der Waals theory where molecular translational motions are governed by intermolecular attractions and repulsions. The molecular repulsions are written in terms of free volume, V_f, which is the volume available to the center of mass of a single molecule as it moves about the system holding the positions of all other molecules fixed. The molecular attractions are generalized in terms of a potential field, $\phi/2$, which is the intermolecular potential energy of one molecule due to the presence of all other molecules. For non-central force molecules (with dipolar or quadrupolar forces), the intermolecular potential energy function can be written as a sum of isotropic and anisotropic interactions. The isotropic pair potential, which depends only on the distance between the molecules, is taken as an unweighted average over all orientations. Summing the intermolecular interactions, $\phi/2$ is given by the sum of isotropic and anisotropic interactions of one molecule with all other molecules in the system, $\phi^{iso}/2$ and $\phi^{ani}/2$, respectively.

The rotational and vibrational motions, which are strongly density-dependent, are affected by intermolecular interactions. To account for deviations in equation of state due to rotational and vibrational motions, a parameter c is defined such that $3c$ is the total number of density-dependent degrees of freedom. This includes the three density-dependent translational degrees of freedom. Further, the non-idealities in the equation of state (i.e., due to attractive and repulsive forces) caused by each of the $3c-3$ density-dependent rotational and vibrational degrees is approximated as equivalent to the non-idealities caused by each of translational degree of freedom. The inclusion of the rotational and vibrational motions in an approximate way through the parameter c improves the simultaneous prediction of various thermodynamic properties.

The partition function in equation 1 approaches the correct ideal-gas limit as density approaches zero and a Prigogine- and Flory-type partition function at liquid densities. Moreover, this partition function is valid for both non-polar and anisotropic molecules. For non-polar molecules, with $\phi^{ani}=0$, the partition function in equation 1 reduces to a form analogous to that used in the PHCT.

The isotropic interactions, ϕ^{iso}, are calculated using a Lennard-Jones potential energy function, and following Gubbins and Twu (4), the isotropic dipole-induced dipole interactions are calculated assuming an average polarizability for the molecule. In this work, for anisotropic interactions, ϕ^{ani}, we are considering only systems in which the dipolar forces are predominant, and all other anisotropic interactions are assumed to be negligible.

The PACT equation of state, obtained by differentiating the partition function in equation 1, is given by

$$P = \frac{RT}{v}\left(1 + c\, z^{rep} + c\, z^{iso} + c\, z^{ani}\right) \tag{2}$$

In the PACT repulsions due to hard-chains are calculated using the parameter c and the equation of Carnahan-Starling (11) for hard-sphere molecules. The attractive Lennard-Jones isotropic interactions are calculated using the perturbation expansion of Barker and Henderson (12). Higher-order terms in the perturbation expansion are accounted for by using a Pade′ approximation for Helmholtz free energy. The perturbation expansion results for spherical molecules are extended to chain-like molecules with the following reduced quantities:

$$\tilde{T} = \frac{T}{T^*} = \frac{ckT}{\epsilon q} \tag{3}$$

and

$$\tilde{v}_d = \frac{v}{\tilde{v}_d^*} = \frac{v}{N_A r d^3/\sqrt{2}} \tag{4}$$

where N_A is Avagadro's number. The equation of state for pure non-polar molecules contains three characteristic parameters. Besides c, the parameters are molecular soft-core size, v^* (or the related hard-core size, v_d^*), and characteristic temperature, T^*. Values for these three parameters are obtained from data reduction, using experimental vapor pressure and liquid-density data. The other parameters ϵ/k, q, r, and d are only necessary for mixture calculations. The Parameter q is proportional to the molecular surface area, ϵ is the energy per unit external surface area of a molecule and r is the number of segments of diameter d (or σ) in a chain-like molecule. For pure fluids, the parameters ϵ and q and parameters r and σ^3 always appear as a product. For mixtures, σ and ϵ are determined by correlating v^* and cT^* for a large number of similar fluids.

In PACT, both the anisotropic multipolar interactions and isotropic dipole-induced dipole interactions are calculated using the perturbation expansion of Gubbins and Twu (4) assuming the molecules to be effectively linear. The anisotropic interactions are calculated by treating these forces as a perturbation over isotropic molecules. Higher-order terms in the perturbation expansion are accounted for by using a Pade′ approximation for Helmholtz free energy. Gubbins and Twu obtained the isotropic dipole-induced dipole interactions by treating these forces as a perturbation over Lennard-Jones molecules and truncating the series after the first perturbation term. Their results for small molecules are extended to chain-like molecules by use of reduced quantities in equations 3 and 4 and by defining the following characteristic reduced temperatures. For dipolar interactions,

$$\tilde{T}_{\mu} = \frac{T}{T^*_{\mu}} = \frac{ckT}{\epsilon_{\mu}\, q} \tag{5}$$

and for isotropic dipole-induced dipole interactions

$$\tilde{T}_{\alpha\mu} = \frac{T}{T^*_{\alpha\mu}} = \frac{v^* ckT}{\alpha\, \epsilon_{\mu} q} \tag{6}$$

where ϵ_{μ} characterizes the segment-segment dipolar interactions and α is the average polarizability. For pure fluids, parameters ϵ_{μ} and q always appear as a product which can be evaluated from known values of dipole moment, μ, and using the relation

$$\epsilon_{\mu}\, q = \frac{\mu^2}{\sqrt{2}\, v^*/N_A} \tag{7}$$

For mixtures, the segmental anisotropic dipolar interaction energy can be obtained from q $[=cT^*/(\epsilon/k)]$ and equation 7. Complete equations for the configurational Helmholtz energy and the equation of state involving dipolar and isotropic dipole-induced dipole interactions are given by Vimalchand (13).

Mixtures. The pure-component partition function is extended to mixtures using a one-fluid approximation, however, without the usual random-mixing assumption. Following Donohue and Prausnitz, the mixing rules for both the isotropic and anisotropic terms are derived using a Lattice theory model. The pure component partition function given by equation 1 is extended to mixtures of chain molecules satisfying the following conditions: (i) For mixtures of both small and large (polymeric) molecules, mixture properties should be based on surface and volume fractions rather than on mole fractions. (ii) Both the isotropic and anisotropic part of the second virial coefficient should have a quadratic dependence on mole fraction. (iii) The pressure and other excess thermodynamic properties must remain finite when the chain length becomes infinitely large. (iv) The athermal entropy of mixing must reduce to the well-known Flory-Huggins entropy of mixing when the reduced volume of the pure components and solution are identical. (v) For correct prediction of Henry's constant, the residual chemical potential should be proportional to the product $q_i r_j$ rather than $q_i r_i$. (vi) For molecules with significantly different intermolecular potential energies, nonrandom-mixing due to molecular clustering becomes important. For mixtures of spherical molecules, the mixing rules should agree with the nonrandom-mixing theory of Henderson (14).

Details of the derivation of mixing rules satisfying the above conditions are given by Vimalchand (13). Complete expression for the configurational Helmholtz free energy is given in the Appendix.

Results and Discussion

Pure Fluids. The Perturbed-Anisotropic-Chain theory has been tested with a wide variety of fluids in which molecules interact with substantial quadrupolar [Vimalchand and Donohue (10)] and dipolar forces. The properties of pure fluids were predicted with parameters which are independent of temperature and pressure. The pure-component partition function was fitted to available experimental data and then pure-component parameters were correlated to ensure reliability of data and data reduction.

Pure-component parameters have been obtained for eleven fluids with appreciable dipole moments and these are tabulated in Table I.

Table I. Pure-Component Parameters

	T^* K	$100v^*$ $\frac{lit}{mol}$	c	$\frac{\epsilon}{k}$	T^*_μ K
Dipolar Fluids					
Carbon monoxide	101.9	2.0274	1.0275	105	2.5
Diethyl ether	297.2	5.9382	1.8054	115	42.8
Chloroform	351.9	4.6651	1.6158	140	49.5
Methyl acetate	324.0	4.5508	1.8314	145	105.7
Dimethyl ether	278.6	3.6650	1.3214	115	106.0
Hydrogen sulfide	271.9	2.1695	1.1181	150	127.2
Sulfur dioxide	279.1	2.4765	1.5037	105	214.7
Methyl iodide	417.3	3.8332	1.0002	160	216.4
Methyl chloride	280.5	3.2858	1.1194	178	315.7
Acetone	325.3	4.4500	1.3768	145	411.9
Acetonitrile	272.1	3.4482	1.0013	180	1094.3
Nonpolar fluids					
Ethane	225.1	3.1947	1.1636	105	
Pentane	311.4	6.2948	1.6501	105	
Isopentane	305.8	6.4009	1.6078	105	
Hexane	326.4	7.3984	1.8224	105	
Cyclohexane	371.3	6.3798	1.5507	118	
Carbon disulfide	439.9	3.7143	1.0045	142	
Carbon tetrachloride	377.8	5.7982	1.4883	118	

Parameters for eight non-polar fluids are also given in Table I. For dipolar fluids, the product $\epsilon_\mu q$ was determined using the dipole moments reported by McCellan (15). The remaining three pure-component parameters (T^*, v^*, and c) for each compound were found by fitting the partition function to experimental liquid-density and vapor pressure data. For mixture calculations, either parameter ϵ or q need to be determined independently. The segmental energy parameter, ϵ, was determined by examining the predicted binary mixture data where one of the components is a C_5 or higher alkane. Following Kaul et al. (7), ϵ for normal and branched alkanes was assigned a value of 105 K.

With the parameters given in Table I, errors in calculated vapor pressure and liquid-density are typically within 2% over a wide range of temperature and pressure. Figure 1 shows the experimental and calculated vapor pressures for hydrogen sulfide, sulfur dioxide and acetone from triple point to critical point. The average error is less than 2%. Liquid densities were also calculated over a wide range of temperature and presure with similar accuracy. Also, the saturated liquid and vapor volumes usually were predicted within 2% error (up to 0.95 of the critical temperature) as shown in Figure 2 for hexane and for the dipolar fluids, sulfur dioxide, acetone and methyl acetate.

Since both acetone and acetonitrile have large dipole moments, we included the isotropic dipole-induced dipole interactions assuming an average polarizability [Landolt-Bornstein (15)] for these molecules. The parameters obtained by fitting the experimental vapor pressure and liquid-density data are given in Table II. The average errors in vapor pressure and liquid-density are similar to those reported above. The inclusion of isotropic induction forces yields a characteristic energy parameter T^* of 303.4 K compared to $T^* = 325.3$ K obtained without induction forces while the other parameters (v^* and c) are nearly identical in both cases. Similarly for acetonitrile T^* decreases to 226.4 K from a high value of 272.1 obtained without induction forces. The values of characteristic dispersive energy ($=cT^*$) of acetone and acetonitrile, when induction forces are included may be the more realistic and as a result parameters for these fluids can be expected to correlate well with other fluids.

Mixtures. Fluid-mixture properties are calculated with pure-component parameters and a single binary interaction parameter defined by

$$\epsilon_{ij} = \sqrt{\epsilon_{ii}\,\epsilon_{jj}}\,(1 - k_{ij}) \tag{8}$$

Experimental K-factor (ratio of vapor phase mole fraction to liquid phase mole fraction) data are used to determine the binary interaction parameter, k_{ij}, which is independent of temperature, density, and composition.

Table II. Parameters including average polarizability for compounds with large dipole moments

Dipolar fluids	T^* K	$100v^*\ \frac{lit}{mol}$	c	$\frac{\epsilon}{k}$	T^*_μ K	$\tilde{\alpha}$
Acetone	303.4	4.5515	1.3628	145	406.8	0.0592
Acetonitrile	226.4	3.7124	1.0000	180	1017.9	0.0514

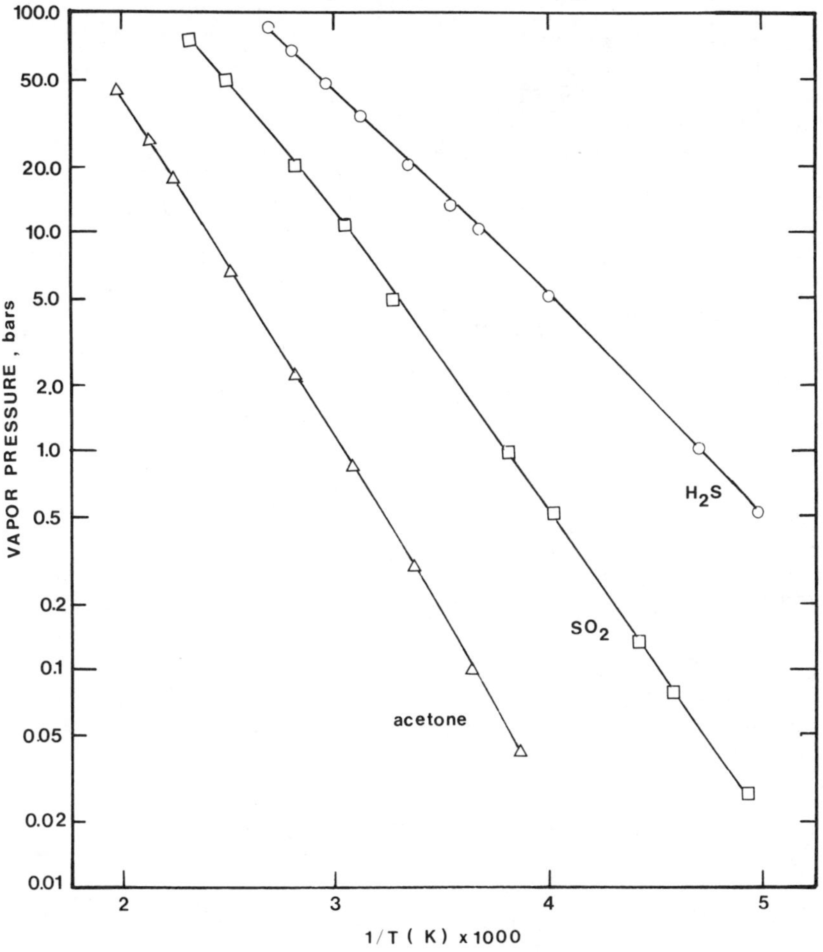

Figure 1. Comparison of experimental and predicted vapor pressures of dipolar fluids from their triple point to critical point.

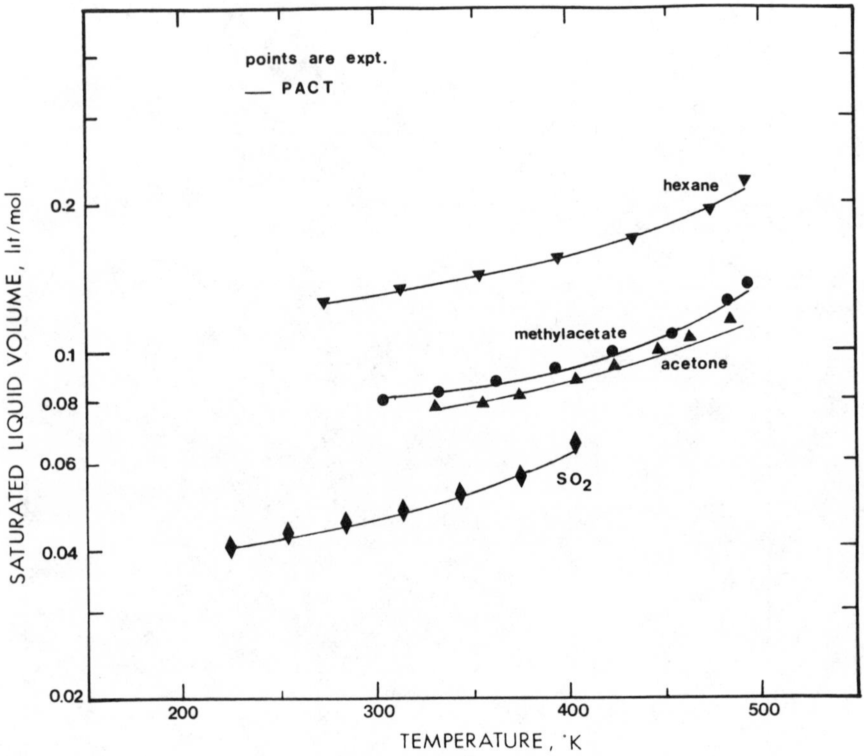

Figure 2. Comparison of calculated and experimental saturated liquid molar volumes (up to 0.95 of the critical temperature).

The binary interaction parameter corrects for inadequacies in theory and in mixing rules. To predict properties of multicomponent mixtures with ease and reasonable accuracy, it is desirable to develop an equation to predict mixture properties without the use of k_{ij} or at least, with small values of k_{ij}. The PHCT, compared to many other equations of state, predicts accurately the properties of non-polar mixtures with small values of k_{ij} (less than 0.05). The PHCT, however, still requires large values of k_{ij} for fluid mixtures containing a dipolar and a quadrupolar substance. Consideration of dipolar and quadrupolar forces in the PACT allows fairly accurate prediction of mixture properties from pure-component parameters alone. Usually, the PACT requires small values of k_{ij} to predict mixture properties with accuracy.

Binary interaction parameters for 9 dipolar fluid mixtures and 13 nonpolar - dipolar fluid mixtures are given in Table III. The inclusion of isotropic dipole-induced dipole interaction does not change k_{ij} significantly and therefore it is neglected except for the system acetone - pentane and acetone - cyclohexane. For these two systems, the acetone parameters given in Table II were used.

Like many widely used equations of state [such as the Peng - Robinson equation (17)], both the PHCT and PACT have (three) adjustable parameters for pure-components which are fitted to experimental data. As a result, each fits the pure-component properties rather well. However, a much more stringent test of the theory is the prediction of mixture properties from pure-component parameters alone. Such a prediction is made in Figures 3 to 6. The K-factors calculated from the PACT with $k_{ij} = 0$ are compared with experimental values for mixtures containing dipolar fluids. Also for comparison, K-factors calculated with $k_{ij} = 0$ using the PHCT and the Peng - Robinson equation of state (PR) are shown.

The inclusion of dipolar interactions improves the prediction of K-factors significantly for the systems sulfur dioxide - acetone (Figure 3), hydrogen sulfide - pentane (Figure 4), and methyl iodide - acetone (Figure 5). Sulfur dioxide, acetone, hydrogen sulfide, and methyl iodide are all dipolar fluids. Acetone forms a weak complex with sulfur dioxide and yet the PACT gives a good prediction of K-factors. In the hydrogen sulfide - pentane system, where molecular sizes differ considerably, both the PACT and the PHCT predict better K-factors than the Peng - Robinson equation. In all these figures, it can be seen that the PACT predicts the experimental data much more closely than either the PHCT or the Peng - Robinson equation.

The y-x phase diagram for the systems and acetone - cyclohexane (Figure 6) show the effect of including, explicitly, the dipole interactions of acetone, rather than the use of the equivalent attractive dispersion interaction as was done in the PHCT. The PACT predicts the azeotrope and closely follows the experimental data, however, the PHCT fails to predict the azeotrope and in addition, poorly follows the experimental values. This may be explained as follows: In the alkane-rich liquid phase each acetone molecule is surrounded by alkane molecules and therefore there are no acetone-acetone (dipole-dipole) interactions. However, when the dipole interaction between pure acetone molecules is empirically replaced by an equivalent dispersion interaction (as was done in the PHCT), the equation predicts an erroneously large attractive energy between acetone and the surrounding alkane molecules. This additional attractive energy incorrectly lowers the predicted mole fraction of acetone in the vapor phase as shown in the Figure 6. Similar arguments indicate that the PHCT when compared to the PACT would predict an increase in mole fraction of acetone in the vapor phase in equilibrium with an acetone-rich liquid phase where alkane molecules are surrounded by acetone molecules.

Mixture predictions by the PACT with pure-component parameters alone are quantitatively correct but usually not within experimental error. However, mixture properties usually can be fit with small errors using small values of a binary interaction parameter given in Table III. Phase equilibria of even complex hydrogen-bonding systems like chloroform - acetone and chloroform - diethyl ether are predicted well with small values of interaction parameters.

Conclusion. A new theoretical equation of state, the Perturbed-Anisotropic-Chain theory has been developed. This equation takes into account the effects of differences in molecular size, shape, and intermolecular forces including anisotropic dipolar and quadrupolar forces. While prediction of properties of pure fluids is no better for the PACT than many other equations of state, the prediction of mixture properties, especially highly non-ideal mixtures is improved significantly. The PACT accurately predicts mixture properties from pure-component parameters alone or with small values of a binary interaction parameter.

Table III. Binary Interaction Parameters

Binary mixtures	$100k_{ij}$
Dipolar fluid mixtures	
Sulfur dioxide - Acetone	-0.75
Sulfur dioxide - Methyl acetate	-1.50
Sulfur dioxide - Chloroform	1.74
Acetone - Methyl acetate	-1.24
Acetone - Diethyl ether	-1.00
Acetone - Chloroform	-5.00
Acetone - Methyl iodide	0.96
Methyl acetate - Chloroform	-2.55
Diethyl ether - Chloroform	-4.18
Non– polar and Dipolar fluid mixtures	
Carbon monoxide - Ethane	0.20
Ethane - Hydrogen sulfide	3.40
Ethane - Methyl acetate	3.78
Ethane - Diethyl ether	0.82
Hydrogen sulfide - Pentane	2.90
Acetone - Pentane	-0.62
Acetone - Carbon disulfide	1.60
Acetone - Cyclohexane	-0.11
Acetone - Carbon tetrachloride	0.55
Isopentane - Diethyl ether	0.60
Hexane - Chloroform	0.50
Chloroform - Carbon tetrachloride	0.21
Methyl iodide - Carbon tetrachloride	0.96

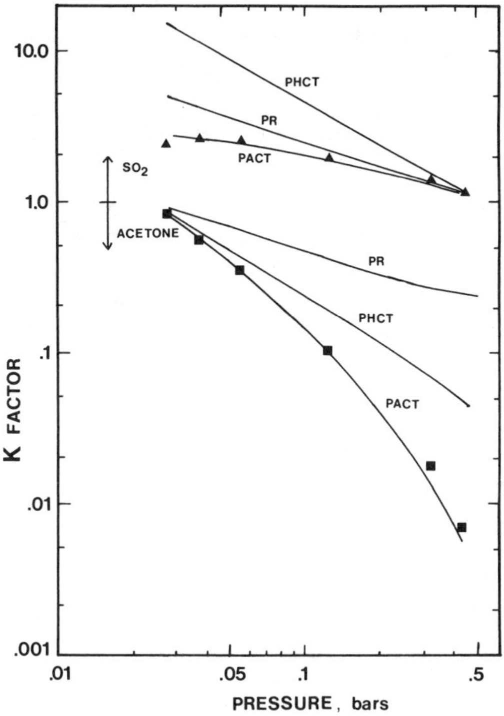

Figure 3. Comparison of experimental K-factors for sulfur dioxide - acetone system and calculated values (with $k_{ij} = 0$).

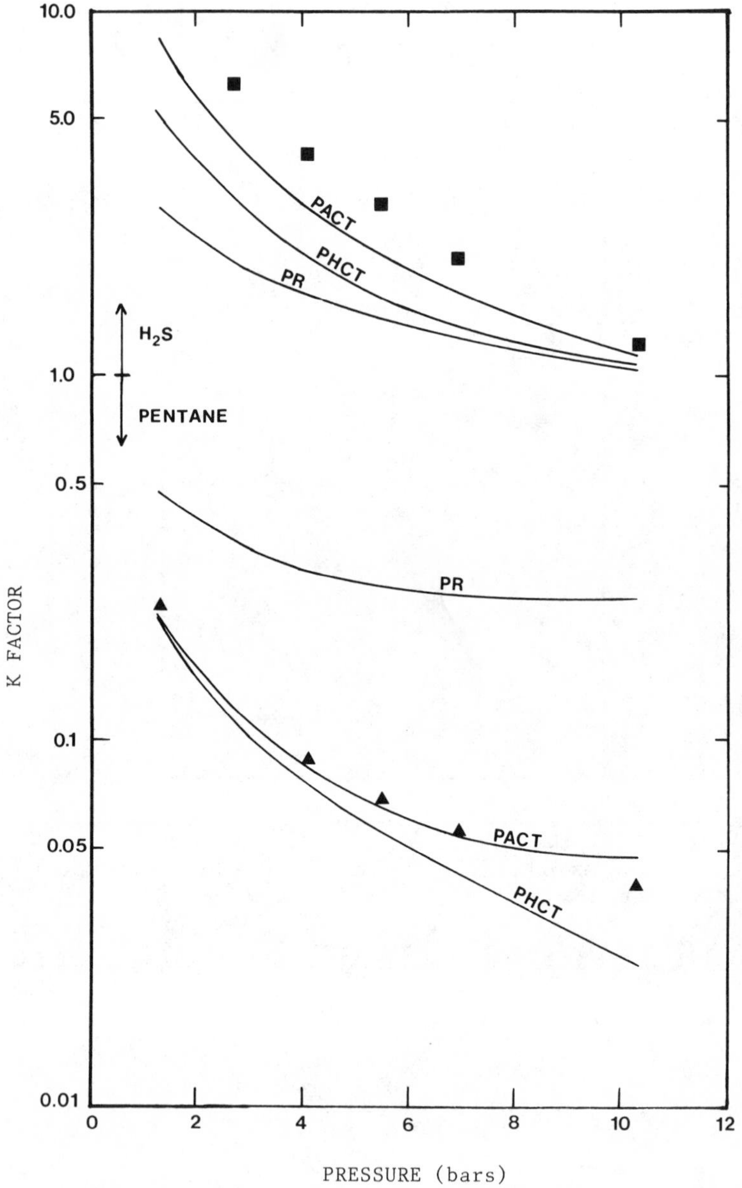

Figure 4. Comparison of experimental K-factors for hydrogen sulfide - pentane system and calculated values (with $k_{ij} = 0$).

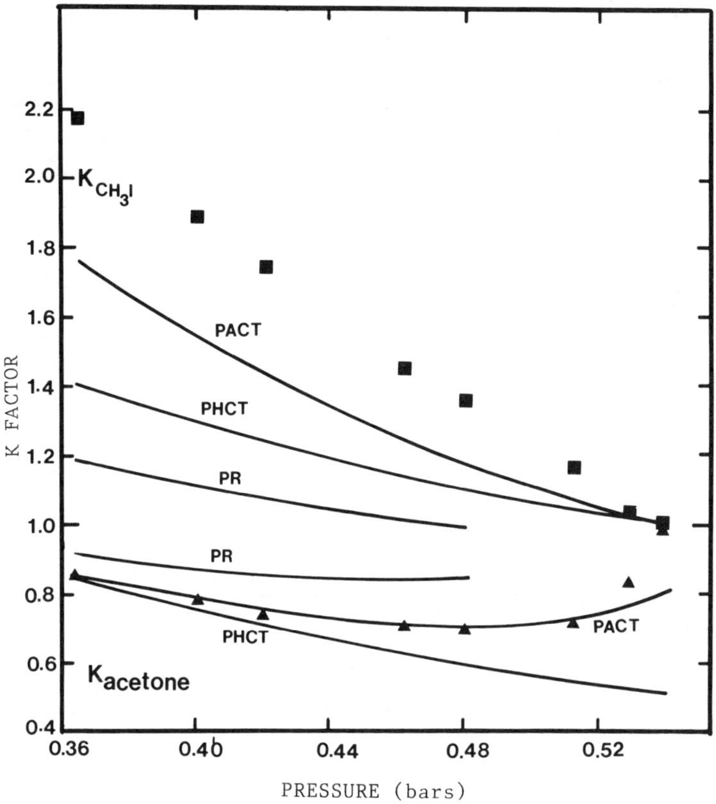

Figure 5. Comparison of experimental K-factors for methyl iodide - acetone system and calculated values (with $k_{ij} = 0$).

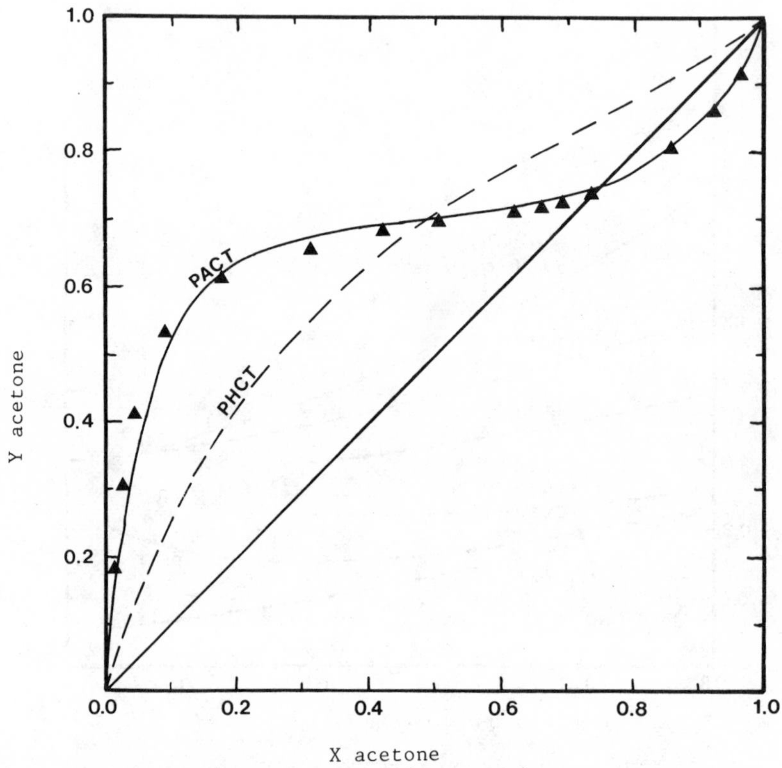

Figure 6. Comparison of experimental y - x phase equilibrium data of acetone in acetone - cyclohexane system and calculated values (with $k_{ij} = 0$).

Because of this success, the PACT may prove very valuable in predicting, to a fair degree of accuracy, mixture properties of systems containing intermediate molecular weight compounds for which no experimental data exist. We are currently extending the PACT for systems involving a dipolar and a quadrupolar molecule as well as for complex systems involving polymeric and hydrogen-bonding molecules.

Acknowledgments

Support of this research by the Chemical Sciences division of the Office of Basic Energy Sciences, U. S. Department of Energy under contract number DE-AC02-81ER10982-A004 is gratefully acknowledged.

Appendix

A summary of various terms in the Helmholtz free energy expression used for calculating the mixture properties with the Perturbed-Anisotropic-Chain theory is given below.

$$A - A^{IG}(T, V) = A^{rep} + A^{LJ} + A^{\mu ind \mu} + A^{\mu\mu}$$

Repulsions:

$$\frac{A^{rep}}{NkT} = <c> \frac{4\left(\dfrac{\tau<v_d^*>}{v}\right) - 3\left(\dfrac{\tau<v_d^*>}{v}\right)^2}{\left(1 - \dfrac{\tau<v_d^*>}{v}\right)^2}$$

where $< \cdots >$ represents a mixture property, $\tau = 0.7405$ and

$$<c> = \sum_i x_i c_i \; ; \qquad <v_d^*> = \sum_i x_i \frac{r_i d_{ii}^3}{\sqrt{2}}$$

The ratio of segmental diameters, d (hard-core) and σ (soft-core), is evaluated as a function of temperature and fitted to a polynomial in reduced temperature [Vimalchand and Donohue (10)].

Lennard-Jones Attractions:

$$\frac{A^{LJ}}{NkT} = \frac{A_1^{LJ}}{NkT}\left[1 - \frac{A_2^{LJ}/NkT}{A_1^{LJ}/NkT}\right]^{-1}$$

where

$$\frac{A_1^{LJ}}{NkT} = \frac{1}{T}\sum_m \frac{A_{1m}<cT^* v_d^*><v_d^*>^{m-1}}{v^m}$$

$$\frac{A_2^{LJ}}{NkT} = \frac{1}{T^2}\sum_m \frac{C_{1m}<cT^{*2} v_d^*><v_d^*>^{m-1}}{2\, v^m} +$$

$$\frac{C_{2m}<cT^* v_d^*><T^*>_L <v_d^*>^m}{v^{m+1}} + \frac{C_{3m}<cT^* v_d^*><T^*>^{(2)}<v_d^*>^{m+1}}{2\, v^{m+2}}$$

The results obtained for pure-component are generalized to multicomponent mixtures by extending the van der Waals one-fluid theory to segmental interactions. For example,

$$<cT^{*n} v_d^*> = \frac{1}{\sqrt{2}} \sum_{ij} x_i x_j \; c_i \left(\frac{\epsilon_{ij} q_i}{c_i k} \right)^n r_j d_{ji}^3$$

The one-fluid theory is applied differently to each term of the perturbation expansion. This eliminates the approximation that the mixture is completely random. Details of the derivation, expressions for $<T^*>_L$ and $<T^*>^{(2)}$, and the constants A_{1m} and C_{nm} are given by Donohue and Prausnitz (5) and Vimalchand and Donohue (10).

Dipole-Induced Dipole Interactions:

$$\frac{A^{\mu\, ind\, \mu}}{NkT} = -8.886 \; J_{\mu\mu}(\tilde{v}, \tilde{T}) \; \frac{<cT^*_{\alpha\mu} v^*>}{v \; T}$$

where

$$<cT^*_{\alpha\mu} v^*> = \sum_{ij} x_i x_j \; c_i \left(\frac{\alpha_j}{r_j \sigma_{ji}^3} \frac{\epsilon_{\mu_{ii}} q_i}{c_i \; k} + \frac{\alpha_i}{r_i \sigma_{ij}^3} \frac{\epsilon_{\mu_{jj}} q_j}{c_j \; k} \right) r_j \sigma_{ji}^3$$

Dipolar interactions:

$$\frac{A^{\mu\mu}}{NkT} = \frac{A_2^{\mu\mu}}{NkT} \left[1 - \frac{A_3^{\mu\mu} / NkT}{A_2^{\mu\mu} / NkT} \right]^{-1}$$

where

$$\frac{A_2^{\mu\mu}}{NkT} = -2.962 \; J_{\mu\mu}(\tilde{v}, \tilde{T}) \; \frac{<cT_\mu^{*2} v^*>}{v \; T^2}$$

$$\frac{A_3^{\mu\mu\mu}}{NkT} = 43.596 \; K_{\mu\mu\mu}(\tilde{v}, \tilde{T}) \; \frac{<cT_\mu^{*(3)} v^{*(2)}>}{v^2 \; T^3}$$

with

$$\tilde{v} = \frac{v}{<v^*>} = \frac{v}{\sum_i x_i \dfrac{r_i \sigma_{ii}^3}{\sqrt{2}}}$$

$$\sigma_{ij} = \frac{\sigma_{ii} + \sigma_{jj}}{2}$$

and

$$\tilde{T} = \frac{T}{<T^*>} = \frac{T}{\sum_{ij} x_i x_j \dfrac{\epsilon_{ij} q_i}{c_i k}}$$

Expressions for $< cT_{\mu}^{*2} v^{*} >$ and $< cT_{\mu}^{*(3)} v^{*(2)} >$ can be obtained from the mixing rules given by Vimalchand and Donohue (10) for quadrupolar interactions by replacing T_Q^* and ϵ_Q with T_μ^* and ϵ_μ respectively. The cross-term, $\epsilon_{\mu_{ij}}$, is given by

$$\epsilon_{\mu_{ij}} = \sqrt{\epsilon_{\mu_{ii}} \epsilon_{\mu_{jj}}}$$

The terms $J_{\mu\mu}$ and $K_{\mu\mu\mu}$ involve integrals over the radial distribution function for Lennard-Jones molecules. Gubbins and Twu (4) evaluated the integrals as a function of reduced volume and reduced temperature.

Literature Cited

1. Flory, P. J. J.Am.Chem.Soc. 1965, 87, 1833.
2. Prigogine, I. "The Molecular Theory of Solutions", North-Holland, Amsterdam, 1957.
3. Pople, J. A. Proc.Roy.Soc. 1954, A221, 498.
4. Gubbins, K. E.; Twu, C. H. Chem.Eng.Sci. 1978, 33, 863; *ibid.*, 1978, 33, 879.
5. Donohue, M. D.; Prausnitz, J. M. AIChEJ. 1978, 24, 849.
6. Beret, S.; Prausnitz, J. M. AIChEJ. 1975, 21, 1123; Macromolecules 1975, 8, 878.
7. Kaul, B.; Donohue, M. D.; Prausnitz, J. M. FluidPhaseEquilib. 1980, 4, 171.
8. Morris, W. O. M.S. Thesis, The Johns Hopkins University, Baltimore, 1985.
9. Liu, D. D.; Prausnitz, J. M. Macromolecules 1979, 12, 454; J.Appl.Polym.Sci. 1979, 24, 725; Ind.Eng.Chem.,Prod.Des.Dev. 1980, 19, 205.
10. Vimalchand, P.; Donohue, M. D. Ind.Eng.Chem.Fundam. 1985, 24, 246.
11. Carnahan, N. F.; Starling, K. E. AIChEJ. 1972, 18, 1184.
12. Barker, J. A.; Henderson, D. J.Chem.Phys. 1967, 47, 4714.
13. Vimalchand, P. Ph.D. Thesis, The Johns Hopkins University, Baltimore, 1985.
14. Henderson, D. J.Chem.Phys. 1974, 61, 926.
15. McCellan, A. L. "Tables of Experimental Dipole Moments", W. H. Freeman and Co., London, 1963.
16. Landolt-Bornstein, "Zahlenwerte und Funktionen", Springer, 6th ed., Vol. 1, part 3, p. 510, Berlin, 1951.
17. Peng, D.; Robinson, D. B. Ind.Eng.Chem.Fundam. 1976, 15, 59.

RECEIVED December 2, 1985

15

Mixing Rules for Cubic Equations of State

G. Ali Mansoori

Department of Chemical Engineering, University of Illinois, Chicago, IL 60680

Through the application of conformal solution theory of statistical mechanics a coherent theory for the development of mixing rules is produced. This theory allows us to use different approximations for the mixture radial distribution functions for derivation of a variety of sets of conformal solution mixing rules some of which are density and temperature dependent. The resulting mixing rules are applied to the van der Waals, Redlich-Kwong, and Peng-Robinson equations of state as the three representative cubic equations of state.

There exists a wealth of information in the literature about cubic equations of state applicable to varieties of fluids of chemical and engineering interest. Although cubic equations of state are generally empirical modifications of the van der Waals equation of state, they have found widespread applications in process design calculations because of their simplicity. Extension of their applicability to mixtures is generally acieved by introduction of mixing rules for their parameters. Mixing rules are expressions relating parameters of a mixture equation of state to pure fluid parameters through, usually, some composition dependent expressions. Except for the van der Waals equation of state the mixing rules for cubic equations of state are empirical expressions. In the present report we introduce a statistical mechanical conformal solution technique through which we can derive varieties of sets of mixing rules applicable to cubic equations of state. This pressure, energy, and compressibility equations of statistical mechanics. In Part II of the present report we introduce the conformal solution theory of polar fluid mixtures (1) and its relationship to the idea of mixing rules. In Part III we introduce the concept of the conformal

0097-6156/86/0300-0314$06.00/0

solution mixing rules and we produce different sets of mixing rules based on different approximations for the mixture radial distribution functions. In Part IV we review the existing forms of the cubic equations of state for mixtures and the deficiencies of their mixing rules and combining rules. Finally, in Part IV we introduce guidelines for the use of conformal solution mixing rules and combining rules in equations of state and we demonstrate application of such mixing rules and combining rules for three representative cubic equations of state.

II. Conformal Solution Theory of Mixtures

Conformal solutions refer to substances whose intermolecular potential energy function, ϕ_{ij}, are related to each other and to those of a reference fluid, usually designated by sub-script (oo), according to (1,2)

$$\phi_{ij} = f_{ij}\phi_{oo}(r/h_{ij}^{1/3}) \tag{1}$$

For substances whose intermolecular potential energy function can be represented by an equation of the form

$$\phi_{ij} = E_{ij}[(L_{ij}/r)^n - (L_{ij}/r)^m] \tag{2}$$

and for which exponents m and n are the same as for the reference substance, conformal parameters f_{ij} and h_{ij} will be defined by the following relations with respect to the intermolecular potential energy parameters E_{ij} and L_{ij}:

$$f_{ij} = E_{ij}/E_{oo}, \qquad h_{ij} = (L_{ij}/L_{oo})^3 \tag{3}$$

Thus the configurational thermodynamic properties of a pure substance of type (a) are related to those of the reference substance according to the following relations:

$$F_a(V, T) = f_{aa}F_o(V/h_{aa}, T/f_{aa}) - NkT\ell nh_{aa} \tag{4}$$

$$P_a(V, T) = (f_{aa}/h_{aa})P_o(V/h_{aa}, T/f_{aa}) \tag{5}$$

$$S_a(V, T) = S_o(V/h_{aa}, T/f_{aa}) + Nk\ell nh_{aa} \tag{6}$$

$$G_a(P, T) = f_{aa}G_o(Ph_{aa}/f_{aa}, T/f_{aa}) - NkT\ell nh_{aa} \tag{7}$$

and

$$H_a(P, S) = f_{aa}H_o(Ph_{aa}/f_{aa}, S_o) \tag{8}$$

where F, P, S, G, and H are the Helmholtz free energy, pressure, entropy, Gibbs free energy, and enthalpy, respectively. According to the above equations, all the thermodynamic properties of substance (a) can be expressed in terms of the properties of a reference pure substance (o) through the conformal parameters f_{aa} and h_{aa}. The conformal solution treatment of fluids composed of polar molecules is more complicated than for non-polar fluids. This is mainly due to electrostatic interactions which cause a departure of the intermolecular potential from spherical symmetry. The electrostatic potential between two otherwise neutral molecules arises from permanent asymmetry in the charge distribution within the molecules. For any pair of localized charge distribution, the mutual electrostatic interaction energy can be written in terms of an infinite series of inverse powers of separation of any two points. For no overlap between the charge distributions the series converges(1). Thus the true pair-potential of polar molecules is orientation-dependent and is the sum of dispersion force as well as electrostatic interactions. In order to extent utility of the above formulation of the conformal solution theory to polar fluids we have proposed the following angle-averaged potential function for polar molecular interactions which represents the first order contribution to the anisotropic forces (1)

$$\phi_{ij}(r,T) = K\epsilon_{ij}[(\sigma_{ij}/r)^n - (\sigma_{ij}/r)^m] - \mu_i^2\mu_j^2/(3kTr^6)$$
$$+ 7\mu_i^4\mu_j^4/[450(kT)^3r^{12}] - (\mu_i^2Q_j^2+\mu_j^2Q_i^2)/(2kTr^8)$$
$$- Q_i^2Q_j^2/(1.4kTr^{10}) - (\alpha_i\mu_j^2+\alpha_j\mu_i^2)/r^6 \quad (9)$$

where $K = [n/(n-m)](n/m)^{m/(n-m)}$, and where μ_i, Q_i, and α_i are the dipole moment, quadrupole moment, and polarizability of molecule i, respectively. For a polar fluid, whose intermolecular potential energy function can be represented by eq.9 the conformal parameters f_{aa} and h_{aa} will have the following forms:

$$f_{aa} = E_{aa}(T,r)/E_{oo}(T,r) \qquad h_{aa} = (L_{aa}(T,r)/L_{oo}(T,r)] \quad (10)$$

where $E_{ij}(T,r) = K\epsilon_{ij}A_{ij}(T,r)[H_{ij}(T,r)]^{n/m}$

$$L_{ij}(T,r) = \sigma_{ij}[H_{ij}(T,r)]^{-1/m}$$

$$H_{ij}(T,r) = [C_{ij}(T,r)/A_{ij}(T,r)]^{m/(n-m)}$$

$$A_{ij}(T,r) = 1 + 7\mu_i^4\mu_j^4/[1800(kT)^3r^{12-m}\sigma_{ij}^nK\epsilon_{ij}]$$

and $$C_{ij}(T,r) = 1 + \mu_i^2\mu_j^2/[12kTr^{6-m}\sigma_{ij}^{m}K\epsilon_{ij}]$$
$$+ (7/20)Q_i^2Q_j^2/[kTr^{10-m}\sigma_{ij}^{6}K\epsilon_{ij}$$
$$+ (\mu_i^2Q_j^2+\mu_j^2Q_i^2)/[8kTr^{8-m}\sigma_{ij}^{m}K\epsilon_{ij}]$$
$$+ (\alpha_i\mu_j^2+\alpha_j\mu_i^2)/[4r^{6-m}\sigma_{ij}^{m}K\epsilon_{ij}]$$

The basic concept of the CST of mixtures is the same as for pure fluids, except that f_{aa} and h_{aa} in eqs.4-8 should be replaced with f_{xx} and h_{xx}, the mixture conformal parameters, as given below

$$f_{xx} = f_{xx}(f_{ij}, h_{ij}, x_i) \qquad h_{xx} = h_{xx}(f_{ij}, h_{ij}, x_i) \tag{11}$$

Eqs.11 are called the conformal solution mixing rules. Functional forms of these mixing rules will be different for different theories of mixtures as it will be demonstrated later in this report. In the formulation of a mixture theory we also need to know the combining rules for unlike-interaction potential parameters which are usually expressed by the following expressions

$$f_{ij}=(1- k_{ij})(f_{ii}\ f_{jj})^{1/2}; \quad h_{ij}=(1- \ell_{ij})[(h_{ii}^{1/3} +h_{jj}^{1/3})/2]^3 \tag{12}$$

where k_{ij} and ℓ_{ij} are adjustable parameters.

III. Statistical Mechanical Theory of Mixing Rules

The most important requirement in the development of the CST of mixtures are mixing rules. In the discussion presented here we have introduced a new technique to re-derive the existing mixing rules and derive a number of new mixing rules some of which are density- and temperature-dependent. According to statistical mechanics the macroscopic thermodynamic properties of a pure fluid are related to its microscopic molecular characteristics by the following three equations (3,4)

$$u = u_{ig} + 2\pi\rho\int_0^\infty \phi(r)g(r)r^2dr \tag{13}$$

$$P = \rho RT + (2/3)\pi\rho\int_0^\infty r\phi'(r)g(r)r^2dr \tag{14}$$

$$\kappa_T = 1/\rho RT - (4\pi/RT)\int_0^\infty [g(r)-1]r^2dr \tag{15}$$

where u is the internal energy, P is the pressure and κ_T is the isothermal compressibility, $\phi(r)$ is the pair intermolecular potential energy function, and g(r) is the radial (or pair) distribution function. Eqs.13-15 are commonly called the energy equation, the virial (or pressure) equation, and the compressibility equation, respectively. For a multicomponent mixture these equations assume the following forms (3-5)

$$u = u_{ig} + 2\pi\rho\sum_i\sum_j x_i x_j \int_0^\infty \phi_{ij}(r) g_{ij}(r) r^2 dr \quad (16)$$

$$P = \rho RT + (2/3)\pi\rho\sum_i\sum_j x_i x_j \int_0^\infty r\phi'_{ij}(r) g_{ij}(r) r^2 dr \quad (17)$$

$$\kappa_T = (1/\rho RT)|B| / \sum_i\sum_j x_i x_j |B|_{ij} \quad (18)$$

In the above equations summations are over all the (c) components of the mixture, x_i and x_j are the mole fractions, and $|B|$ is a cxc determinant with its representative terms in the following form

$$B_{ij} = x_i\delta_{ij} + x_i x_j \rho G_{ij} \qquad G_{ij} = 4\pi\int_0^\infty [g_{ij}(r)-1] r^2 dr$$

where δ_{ij} is the Kroneeker delta, and $|B|_{ij}$ is the cofactor of term B_{ij} in determinant $|B|$. Eqs.13-18 can be used in the manner presented below in order to derive mixing rules based on different mixture theory approximations:

III.1. One-Fluid Theory of Mixing Rules: For the development of one-fluid mixing rules we introduce a pseudo-pure fluid which can represent the configurational properties of a mixture provided that the pseudo-pure fluid and the mixture molecular interactions obey eq.1. By replacing eq.1 in eqs.13, 14, 17, and 18 and then equating configurational internal energy, pressure, and isothermal compressibility of the pseudo-pure fluid and the mixture we will obtain the following equations

$$f_{xx}h_{xx}\int\phi_{oo}(y)g_{oo}(y)y^2dy=\sum_i\sum_j x_i x_j f_{ij} h_{ij}\int\phi_{oo}(y)g_{ij}(y)y^2dy \quad (19)$$

$$f_{xx}h_{xx}\int y\phi'_{oo}(y)g_{oo}(y)y^2dy=\sum_i\sum_j x_i x_j f_{ij} h_{ij}\int y\phi'_{oo}(y)g_{ij}(y)y^2dy \quad (20)$$

$$\{1-4\pi\rho h_{xx}\int[g_{oo}(y)-1]y^2dy\}-1=\sum_i\sum_j x_i x_j |B|_{ij}/|B| \quad (21)$$

It should be pointed out that for the case of the hard-sphere fluid eq.19 vanishes, eq.21 remains the same, while eq.20 reduces to the following form

$$h_{xx}g_{oo}(1)=\sum_i\sum_j x_i x_j h_{ij} g_{ij}(1) \quad (22)$$

Solution of eqs.19–21 should produce the two necessary expressions (mixing rules) relating f_{xx} and h_{xx} of the pseudo-pure fluid to f_{ij} and h_{ij} of components of the mixture. For this purpose we should use an approximation technique relating the radial distribution functions (RDF) in the mixture to the pure reference fluid RDF. However, at a first glance it seems that we have in our hand three equations and two unknowns. As it will be demonstrated below for most of the approximations of the mixture RDFs which are used here these three equations produce two mixing rules. In the previous investigations for the development of mixing rules (5–11) all the investigators have used only eq.19 and/or eq.20. Our studies indicate that while eqs.19 and 20 are essential in the development of mixing rules, eq.21 can add a new dimension which could be significant in the calculation of properties of mixtures. In what follows different approximations will be used for relating g_{ij} to g_{oo} in order to derive different sets of mixing rules.

<u>III.1.i. Random Mixing Approximation (RMA) for Mixture RDFs</u>: In this approximation it is assumed that the non-scaled RDF of all the components of the mixture and the interaction RDFs are identical (5), i.e.

$$g_{11}(r) = g_{22}(r) = \ldots = g_{ij}(r) = \ldots \tag{23}$$

When this approximation is replaced in eqs.19–21, eq.21 will vanish and eq.19 and 20 will produce the following mixing rules

$$\phi_{xx}(r) = \Sigma_i\Sigma_j x_i x_j \phi_{ij}(r) \tag{24}$$

$$\phi'_{xx}(r) = \Sigma_i\Sigma_j x_i x_j \phi'_{ij}(r) \tag{25}$$

For example, in the case of the Lennard-Jones (12-6) intermolecular potential function we will derive the following mixing rules (12) from eqs.13 and 14.

$$f_{xx}h_{xx}^2 = \Sigma_i\Sigma_j x_i x_j f_{ij}h_{ij}^2 \tag{26}$$

$$f_{xx}h_{xx}^4 = \Sigma_i\Sigma_j x_i x_j f_{ij}h_{ij}^4 \tag{27}$$

For a hard-sphere potential we will derive only one mixing rule through the RMA and that is derived by replacing eq.23 in 22. The resulting mixing rule will be

$$h_{xx}^{1/3} = \Sigma_i\Sigma_j x_i x_j h_{ij}^{1/3} \tag{28}$$

III.1.ii. Conformal Solution Approximation (CSA) for Mixture RDFs: This approximation technique seems more logical for use in the development of mixing rules than RMA. According to this approximation the scaled RDFs in a mixture are all identical (5), i.e.

$$g_{11}(y) = g_{22}(y) = \ldots = g_{ij}(y) = \ldots \tag{29}$$

When we use this approximation in eqs.19 and 20 they both produce the same mixing rule which is

$$f_{xx}h_{xx} = \Sigma_i\Sigma_j x_i x_j f_{ij} h_{ij} \tag{30}$$

Now, by replacing eq.29 in 21 an additional mixing rule will be produced which is the following

$$|B^*|/\rho RT\kappa_{Txx} = \Sigma_i\Sigma_j x_i x_j |B^*|_{ij} \tag{31}$$

where $|B^*|_{ij} = x_i[\delta_{ij} + x_j(h_{ij}/h_{xx})(\rho RT\kappa_{Txx}-1)]$. Eq.30 is actually the second van der Waals mixing rule which is well known, but eq.31 is a new mixing rule for h_{xx} which is replacing the first van der Waals mixing rule. This new mixing rule, in principle, is a composition-, temperature-, and density-dependent mixing rule. This is because κ_{Txx} which appears in the right and left hand sides of this equation is generally temperature- and density-dependent. For example, for a binary mixture eq.31 can be written in the following form (5)

$$h_{xx} = \{\Sigma_i\Sigma_j x_i x_j h_{ij} + x_1 x_2(h_{11}h_{22}-h_{12}^2)(\rho RT\kappa_{Txx}-1)\}/\{1+x_1x_2(h_{11}+h_{22}-2h_{12})(\rho RT\kappa_{Txx}-1)\} \tag{31-1}$$

By using the hard-sphere potential (by replacing eq.29 in 22) we will derive the following mixing rule

$$h_{xx} = \Sigma_i\Sigma_j x_i x_j h_{ij} \tag{32}$$

This mixing rule is the first van der Waals mixing rule which, in conjunction with eq.30 is usually used for calculation of mixture thermodynamic properties (7,8,10,11). It should be pointed out that eq.32 constitutes another mixing rule for hard-sphere mixtures. As a result, while the CSA approximation produces two mixing rules for potential functions with two parameters, it also produces two mixing rules for a hard-sphere potential which is a one-parameter potential function.

III.1.iii. Hard-Sphere Expansion (HSE) Approximation for Mixture RDFs: It is demonstrated that the RDF of a pure fluid (x) can be expanded around the hard-sphere (hs) RDF in the form (3)

$$g_{xx}(y) = g^{hs}(y) + (f_{xx}/T_o^*)g_1(y) + (f_{xx}/T_o^*)^2 g_2(y) + \ldots \quad (33)$$

Let us also assume that we could make a similar expansion for RDFs in a mixture around the hard-sphere mixture RDFs as the following

$$g_{ij}(y) = g_{ij}^{hs}(y) + (f_{ij}/T_o^*)g_1(y) + (f_{ij}/T_o^*)^2 g_2(y) + \ldots \quad (34)$$

The justification behind this expansion is given elsewhere (6,9). Now by replacing eqs.33 and 34 in either of eqs.19 or 20 we will be able to derive the following two mixing rules by equating the coefficients of the second and third order inverse temperature terms of the resulting expression.

$$f_{xx}h_{xx} = \Sigma_i\Sigma_j x_i x_j f_{ij} h_{ij} \quad (35)$$

$$f_{xx}^2 h_{xx} = \Sigma_i\Sigma_j x_i x_j f_{ij}^2 h_{ij} \quad (36)$$

These mixing rules are used for calculation of excess properties of a mixture over the hard-sphere mixture (13) at the same thermodynamic conditions (9). Application of the HSE approximation in eq.21 will not produce any additional mixing rule.

III.1.iv. Density Expansion (DEX) Approximation for Mixture RDFs: It has been demonstrated that the RDF of a pure fluid can be expanded around the dilute gas RDF, exp[$-\phi(r)/kT$], in the form (14)

$$g_{xx}(y) = [1 + F_{xx}(y)] \exp[-\phi_{xx}(r)/kT] \quad (37)$$

Let us also assume that we could make a similar expansion for RDFs in a mixture around the dilute gas mixture RDFs as the following

$$g_{ij}(y) = [1 + F_{xx}(y)] \exp[-\phi_{ij}(r)/kT] \quad (38)$$

Now by replacing eqs.37 and 38 in eq.19 and after a number of algebraic manipulations we will derive the following mixing rule

$$f_{xx}h_{xx} = \Sigma_i\Sigma_j x_i x_j f_{ij} h_{ij} \{1 - (f_{ij}/f_{xx} - 1)[u - u_{ig})/kT + T(C_v - C_{vig})/(u - u_{ig})]\} \quad (39)$$

The latter mixing rule can be used, joined with another mixing rule, for calculation of mixture properties. Similar approximations can be used in order to derive other mixing rules from the virial and compressibility equations.

III.2. Multi-Fluid Theory of Mixing Rules: The basic assumption in developing the multi-fluid mixing rules is the same as the one-fluid approach except that in this case we will search for a hypothetical multicomponent ideal mixture which could represent the configurational properties of a multicomponent real mixture, both with the same number of components and at the same thermodynamic conditions. In this case eqs.19-21 will be replaced by the following set of equations

$$f_{xi}h_{xi}\int\phi_{oo}(y)g_{oo}(y)y^2dy=\Sigma_jx_jf_{ij}h_{ij}\int\phi_{oo}(y)g_{ij}(y)y^2dy \quad (40)$$

$$f_{xi}h_{xi}\int y\phi'_{oo}(y)g_{oo}(y)y^2dy=\Sigma_jx_jf_{ij}h_{ij}\int y\phi'_{oo}(y)g_{ij}(y)y^2dy \quad (41)$$

$$\{1-4\pi\rho h_{xi}\int[g_{oo}(y)-1]y^2dy\}-1=\Sigma_jx_j|B|_{ij}/|B| \quad (42)$$

Expressions for B_{ij} and G_{ij} will be the same as in eq.18. In the case of the hard-sphere fluid eq.40 will reduce to the following form

$$h_{xi}g_{oo}{}^{hs}(1)=\Sigma_jx_jh_{ij}g_{ij}{}^{hs}(1), \quad (40-1)$$

eq.41 will vanish and eq.43 will remain the same.

III.2.i. Average Potential Model (APM) for Mixture RDFs: In this approximation it is assumed that (5),

$$g_{ij}(r) = [g_{ii}(r) + g_{jj}(r)]/2 \qquad g_{ii}(r) \neq g_{jj}(r) \quad (43)$$

When this approximation is replaced in eqs.40-42, eq.42 will vanish and eq.40 and 41 will produce the following mixing rules

$$\phi_{xi}(r) = \Sigma_jx_j\phi_{ij}(r) \quad (44)$$

$$\phi'_{xi}(r) = \Sigma_jx_j\phi'_{ij}(r) \quad (45)$$

For example, in the case of the Lennard-Jones (12-6) intermolecular potential function we will derive the following mixing rules(12).

$$f_{xi}h_{xi}{}^2 = \Sigma_jx_jf_{ij}h_{ij}{}^2 \quad (46)$$

$$f_{xi}h_{xi}^4 = \Sigma_j x_j f_{ij} h_{ij}^4 \quad (47)$$

For a hard-sphere potential we will derive only one mixing rule

$$h_{xi}^{1/3} = \Sigma_j x_j h_{ij}^{1/3} \quad (48)$$

<u>III.2.ii. Multi-fluid CSA Approximation for Mixture RDFs</u>: According to this approximation the scaled RDFs in a mixture are related as the following

$$g_{ij}(y) = [g_{ii}(y) + g_{jj}(y)]/2 \qquad g_{ii}(r) \neq g_{jj}(y) \quad (49)$$

When we use this approximation in eqs.40 and 41 they both produce the same mixing rule which is

$$f_{xi}h_{xi} = \Sigma_j x_j f_{ij} h_{ij} \quad (50)$$

Now, by replacing eq.49 in 42 an additional mixing rule will be produced which is the following

$$|B^*|/\rho RT\kappa_{Txi} = \Sigma_j x_j |B^*|_{ij} \quad (51)$$

where

$$|B^*|_{ij} = x_i\{\delta_{ij} + (x_j h_{ij}/2)[(\rho RT\kappa_{Txi}-1)/h_{xi} + (\rho RT\kappa_{Txj}-1)/h_{xj}]\}$$

Eq.50 is actually the second van der Waals multi-fluid mixing rule, but eq.51 is a new mixing rule for h_{xi}. By using the hard-sphere potential (by replacing eq.49 in 40-1) we will derive the following mixing rule

$$h_{xi} = \Sigma_j x_j h_{ij} \quad (52)$$

This mixing rule is the first multi-fluid van der Waals mixing rule which, in conjunction with eq.50 is usually used for calculation of mixture thermodynamic properties. It should be pointed out that eq.52 constitutes another mixing rule for hard-sphere mixtures.

<u>III.2.iii. Multi-Fluid HSE Mixing Rules</u>: In a similar manner as the one-fluid case we can derive the following mixing rules

$$f_{xi}h_{xi} = \Sigma_j x_j f_{ij} h_{ij} \quad (53)$$

$$f_{xi}^2 h_{xi} = \Sigma_j x_j f_{ij}^2 h_{ij} \quad (54)$$

These mixing rules are used for calculation of excess

properties of a mixture over the hard-sphere mixture.

III.2.iv. Multi-Fluid DEX Mixing Rules: In a similar manner as the one-fluid case we can derive the following mixing rule

$$f_{xi}h_{xi} = \sum_j x_j f_{ij} h_{ij} \{1-(f_{ij}/f_{xi}-1)[u_i-u_{ig})/kT$$

$$+T(C_{vi}-C_{vig})/(u_i-u_{ig})]\} \qquad (55)$$

This mixing rule can be used, joined with another mixing rule, for calculation of mixture properties.

IV. Application of Mixing Rules for Cubic Equations of State
In order to apply the varieties of the conformal solution mixing rules which are introduced here for cubic and other equations of state the following considerations should be taken into account:

(i) Conformal solution mixing rules are for the molecular conformal volume parameter, h, and the molecular conformal energy parameter, f.

(ii) Conformal solution mixing rules are applicable for constants of an equation of state only. Before using a set of mixing rules for an equation of state one has to express the parameters of the equation of state with respect to the molecular conformal parameters h and f. This will then make it possible to write the combining rules and mixing rules for the equation of state. In what follows mixing rules and combining rules for three representative cubic equations of state are derived and tabulated.

IV.1. Mixing Rules for the van der Waals Equation of State: The van der Waals equation of state (15) can be written in the following form

$$Z = Pv/RT = v/(v-b) - a/vRT \qquad (56)$$

Parameter b of this equation of state is proportional to molecular volume ($b \propto h$) and parameter a is proportional to (molecular volume)(molecular energy) ($a \propto fh$). Then, in order to apply the mixing rules introduced in this report for the van der Waals equation of state we must replace h with b and f with a/b in all the mixing rules. In Table I mixing rules for the van der Waals equation of state based on different theories of mixtures are reported. The combining rules for a_{ij} and b_{ij} ($i \neq j$) of this equation of state, consistent with eqs.12 will be

Table I: Mixing Rules for the van der Waals Equation of State

One-Fluid Mixing Rules	
RMA Theory	$a = [\sum_i\sum_j x_i x_j a_{ij} b_{ij}]^{3/2} / [\sum_i\sum_j x_i x_j a_{ij} b_{ij}^3]^{1/2}$
	$b = [\sum_i\sum_j x_i x_j a_{ij} b_{ij}^3 / \sum_i\sum_j x_i x_j a_{ij} b_{ij}]^{1/2}$
vdW Theory	$a = \sum_i\sum_j x_i x_j a_{ij}$
	$b = \sum_i\sum_j x_i x_j b_{ij}$
HSE Theory	$a = \sum_i\sum_j x_i x_j a_{ij}$
	$b = [\sum_i\sum_j x_i x_j a_{ij}]^2 / \sum_i\sum_j x_i x_j a_{ij}^2 / b_{ij}$
DEX Theory	$a = [a_{vdW} + (b/vRT) \sum_i\sum_j x_i x_j a_{ij}^2 / b_{ij}] / [1 + a_{vdW}/vRT]$
	$b = \sum_i\sum_j x_i x_j b_{ij}$
CSA Theory	$a = \sum_i\sum_j x_i x_j a_{ij}$
	$1+\Delta_{xx} = \lvert B^*\rvert / \sum_i\sum_j x_i x_j \lvert B^*\rvert_{ij}; \quad B^*_{ij} = x_i(\delta_{ij} + x_j \Delta_{xx} b_{ij}/b)$
Multi-Fluid Mixing Rules	
APM Theory	$a_i = [\sum_j x_j a_{ij} b_{ij}]^{3/2} / [\sum_j x_j a_{ij} b_{ij}^3]^{1/2}$
	$b_i = [\sum_j x_j a_{ij} b_{ij}^3 / \sum_j x_j a_{ij} b_{ij}]^{1/2}$
vdW Theory	$a_i = \sum_j x_j a_{ij}$
	$b_i = \sum_j x_j b_{ij}$
HSE Theory	$a_i = \sum_j x_j a_{ij}$
	$b_i = [\sum_j x_j a_{ij}]^2 / \sum_j x_j a_{ij}^2 / b_{ij}$
DEX Theory	$a_i = [a_{ivdW} + (b_i/vRT) \sum_j x_j a_{ij}^2 / b_{ij}] / [1 + a_{ivdW}/vRT]$
	$b_i = \sum_j x_j b_{ij}$
CSA Theory	$a_i = \sum_j x_j a_{ij}$
	$1+\Delta_{xi} = \lvert B^*\rvert / \sum_j x_j \lvert B^*\rvert_{ij}; \; B^*_{ij} = x_i[\delta_{ij} + x_j b_{ij}(\Delta_{xi}/b_i + \Delta_{xj}/b_j)/2]$

$$\Delta_{xx} = \rho RT\kappa_{Txx} - 1 = [2a(v-b)^2 - RTb(2v-b)]/[RTv^2 - 2a(v-b)^2]$$

$$\Delta_{xi} = \rho RT\kappa_{Txi} - 1 = [2a_i(v-b_i)^2 - RTb_i(2v-b_i)]/[RTv^2 - 2a_i(v-b_i)^2]$$

$$\begin{cases} a_{ij} = (1-k_{ij})b_{ij}(a_{ii}a_{jj}/b_{ii}b_{jj})^{1/2}; \\ b_{ij} = (1-\ell_{ij})[(b_{ii}^{1/3}+b_{jj}^{1/3})/2]^3 \end{cases} \tag{57}$$

IV.2. Mixing Rules for the Redlich-Kwong Equation of State: The Redlich-Kwong equation of state (16) which is an empirical modification of the van der Waals equation can be written in the following form

$$Z = Pv/RT = v/(v-b) - a/[RT^{3/2}(v+b)] \tag{58}$$

Parameter b of this equation of state is proportional to molecular volume ($b \propto h$) and parameter a is proportional to (molecular volume)(molecular energy)$^{3/2}$ or ($a \propto f.h^{3/2}$). Then, in order to apply the mixing rules introduced in this report for the Redlich-Kwong equation of state we must replace h with b and f with $(a/b)^{2/3}$ in all the mixing rules. In Table II mixing rules for the Redlich-Kwong equation of state based on different theories of mixtures are reported. The combining rules for a_{ij} and b_{ij} ($i \neq j$) of this equation of state, consistent with eqs.12 will be the same as eqs.57.

IV.3. Mixing Rules for the Peng-Robinson Equation of State: The Peng-Robinson equation of state (17) which is another empirical modification of the van der Waals equation can be written in the following form

$$Z = Pv/RT = v/(v-b) - a(T)v/\{RT[v(v+b)+b(v-b)]\} \tag{59}$$

Parameter b of this equation of state is a constant which is proportional to the molecular volume ($b \propto h$). However, parameter a of the Peng-Robinson equation of state is not a constant and it is a function of temperature as the following.

$$a(T) = a_c\{1+\theta[1-T/T_c)^{1/2}]\}^2 \tag{60}$$

where

$$a_c = 0.45724R^2T_c^{\,2}/P_c \qquad \theta = 0.37464+1.54226\omega-0.26992\omega^2$$

Table II: Mixing Rules for the Redlich-Kwong Equation of State

One-Fluid Mixing Rules

Theory	Mixing rules
RMA Theory	$a = [\sum_i\sum_j x_i x_j a_{ij}^{2/3} b_{ij}^{4/3}]^{5/2} / \sum_i\sum_j x_i x_j a_{ij}^{2/3} b_{ij}^{10/3}$ $b = [\sum_i\sum_j x_i x_j a_{ij}^{2/3} b_{ij}^{10/3} / \sum_i\sum_j x_i x_j a_{ij}^{2/3} b_{ij}^{4/3}]^{1/2}$
vdW Theory	$a = [\sum_i\sum_j x_i x_j a_{ij}^{2/3} b_{ij}^{1/3}]^{3/2} / [\sum_i\sum_j x_i x_j b_{ij}]^{1/2}$ $b = \sum_i\sum_j x_i x_j b_{ij}$
HSE Theory	$a = [(\sum_i\sum_j x_i x_j a_{ij}^{2/3} b_{ij}^{1/3})(\sum_i\sum_j x_i x_j a_{ij}^{4/3} b_{ij}^{-1/3})]^{1/2}$ $b = [\sum_i\sum_j x_i x_j a_{ij}^{2/3} b_{ij}^{1/3}]^2 / \sum_i\sum_j x_i x_j a_{ij}^{4/3} b_{ij}^{-1/3}$
DEX Theory	$a = \sum_i\sum_j x_i x_j a_{ij} (a/b)^{1/3} \{1-[(a_{ij}/b_{ij})^{2/3}(b/a)^{2/3}-1]\zeta\}$ $b = \sum_i\sum_j x_i x_j b_{ij}$
CSA Theory	$a = [\sum_i\sum_j x_i x_j a_{ij}^{2/3} b_{ij}^{1/3}]^{3/2} / b^{1/2}$ $1+\Delta_{xx} = \lvert B^*\rvert / \sum_i\sum_j x_i x_j \lvert B^*\rvert_{ij}; \quad B^*_{ij} = x_i(\delta_{ij} + x_j \Delta_{xx} b_{ij}/b)$

Multi-Fluid Mixing Rules

Theory	Mixing rules
APM Theory	$a_i = [\sum_j x_j a_{ij}^{2/3} b_{ij}^{4/3}]^{5/2} / \sum_j x_j a_{ij}^{2/3} b_{ij}^{10/3}$ $b_i = [\sum_j x_j a_{ij}^{2/3} b_{ij}^{10/3} / \sum_j x_j a_{ij}^{2/3} b_{ij}^{4/3}]^{1/2}$
vdW Theory	$a_i = [\sum_j x_j a_{ij}^{2/3} b_{ij}^{1/3}]^{3/2} / [\sum_j x_j b_{ij}]^{1/2}$ $b_i = \sum_j x_j b_{ij}$
HSE Theory	$a_i = [(\sum_j x_j a_{ij}^{2/3} b_{ij}^{1/3})(\sum_j x_j a_{ij}^{4/3} b_{ij}^{-1/3})]^{1/2}$ $b_i = [\sum_j x_j a_{ij}^{2/3} b_{ij}^{1/3}]^2 / \sum_j x_j a_{ij}^{4/3} b_{ij}^{-1/3}$
DEX Theory	$a_i = \sum_j x_j a_{ij} (a/b)^{1/3} \{1-[(a_{ij}/b_{ij})^{2/3}(b_i/a_i)^{2/3}-1]\zeta_i\}$ $b_i = \sum_j x_j b_{ij}$
CSA Theory	$a_i = [\sum_j x_j a_{ij}^{2/3} b_{ij}^{1/3}]^{3/2} / b_i^{1/2}$ $1+\Delta_{xi} = \lvert B^*\rvert / \sum_j x_j \lvert B^*\rvert_{ij}; \quad B^*_{ij} = x_i(\delta_{ij} + x_j b_{ij}[\Delta_{xi}/b_i + \Delta_{xj}/b_j])$

$\zeta = (3/2)(a/bR)T^{-3/2} \ln[v/(v+b)] - 1/2;$

$\zeta_i = (3/2)(a_i/b_i R)T^{-3/2} \ln[v/(v+b_i)] - 1/2$

$\Delta_{xx} = \rho RT\kappa_{Txx} - 1 = -1 + RT^{3/2}(v^2-b^2)^2 / [RT\{v(v+b)\}^2 - a(2v+b)(v-b)^2]$

$\Delta_{xi} = \rho RT\kappa_{Txi} - 1 = -1 + RT^{3/2}(v^2-b_i^2)^2 / [RT\{v(v+b_i)\}^2 - a_i(2v+b_i)(v-b_i)^2]$

and T_c and P_c are the critical point temperature and pressure, respectively; and ω is the acentric factor. In order to utilize the statistical mechanical mixing rules for the Peng-Robinson equation of state we must first separate thermodynamic variables from constants of this equation of state. For this purpose we may write this equation of state in the following form

$$Z = Pv/RT = v/(v-b) - [(A/RT+C-2(AC/RT)^{1/2}]/[(v+b)+(b/v)(v-b)] \quad (61)$$

where $A = a_c(1+\Theta)^2$ and $C = a_c\Theta^2/RT_c$. This new form of the Peng-Robinson equation of state indicates that there exist three independent constant parameters in this equation which are A, b, and C. Parameters b and C are proportional to the molecular volume ($b \propto h$ and $C \propto h$) while parameter A is proportional to (molecular volume)(molecular energy) or ($A \propto fh$). Based on different theories of mixtures mixing rules for this new form of the Peng-Robinson equation of state are reported in Table III. The combining rules for the unlike interaction parameters of this equation of state are as the following

$$A_{ij} = (1-k_{ij})b_{ij}(A_{ii}A_{jj}/b_{ii}b_{jj})^{1/2} \quad (62)$$

$$b_{ij} = (1-\ell_{ij})[(b_{ii}{}^{1/3}+b_{jj}{}^{1/3})/2]^3 \quad (63)$$

$$C_{ij} = (1-m_{ij})[(C_{ii}{}^{1/3}+C_{jj}{}^{1/3})/2]^3 \quad (64)$$

Similar procedures to those demonstrated above can be used for derivation of conformal solution mixing rules for other cubic equations of state.

Table III: Mixing Rules for the Peng–Robinson Eq. of State

Theory	Mixing Rules
One-Fluid Mixing Rules	
RMA Theory	$A = [\sum_i\sum_j x_i x_j A_{ij} b_{ij}]^{3/2} / [\sum_i\sum_j x_i x_j A_{ij} b_{ij}^3]^{1/2}$
	$b = [\sum_i\sum_j x_i x_j A_{ij} b_{ij}^3 / \sum_i\sum_j x_i x_j A_{ij} b_{ij}]^{1/2}$
	$C = [\sum_i\sum_j x_i x_j A_{ij} C_{ij}^3 / \sum_i\sum_j x_i x_j A_{ij} C_{ij}]^{1/2}$
vdW Theory	$A = \sum_i\sum_j x_i x_j A_{ij}$
	$b = \sum_i\sum_j x_i x_j b_{ij}$
	$C = \sum_i\sum_j x_i x_j C_{ij}$
HSE Theory	$A = \sum_i\sum_j x_i x_j A_{ij}$
	$b = [\sum_i\sum_j x_i x_j A_{ij}]^2 / \sum_i\sum_j x_i x_j A_{ij}^2 / b_{ij}$
	$C = [\sum_i\sum_j x_i x_j A_{ij}]^2 / \sum_i\sum_j x_i x_j A_{ij}^2 / C_{ij}$
DEX Theory	$A = \sum_i\sum_j x_i x_j A_{ij}\{1-[(A_{ij}/b_{ij})(b/A)-1]\xi\}$
	$b = \sum_i\sum_j x_i x_j b_{ij}$
	$C = \sum_i\sum_j x_i x_j C_{ij}$
CSA Theory	$A = \sum_i\sum_j x_i x_j A_{ij}$
	$1+\Delta_{xx} = \lvert B^*\rvert / \sum_i\sum_j x_i x_j \lvert B^*\rvert_{ij}; \quad B^*_{ij} = x_i(\delta_{ij} + x_j \Delta_{xx} b_{ij}/b);$
	$C = \sum_i\sum_j x_i x_j C_{ij}$
Multi-Fluid Mixing Rules	
APM Theory	$A_i = [\sum_j x_j A_{ij} b_{ij}]^{3/2} / [\sum_j x_j A_{ij} b_{ij}^3]^{1/2}$
	$b_i = [\sum_j x_j A_{ij} b_{ij}^3 / \sum_j x_j A_{ij} b_{ij}]^{1/2}$
	$C_i = [\sum_j x_j A_{ij} C_{ij}^3 / \sum_j x_j A_{ij} C_{ij}]^{1/2}$
vdW Theory	$A_i = \sum_j x_j A_{ij}$
	$b_i = \sum_j x_j b_{ij}$
	$C_i = \sum_j x_j C_{ij}$
HSE Theory	$A_i = \sum_j x_j A_{ij}$
	$b_i = [\sum_j x_j A_{ij}]^2 / \sum_j x_j A_{ij}^2 / b_{ij}$
	$C_i = [\sum_j x_j A_{ij}]^2 / \sum_j x_j A_{ij}^2 / C_{ij}$
DEX Theory	$A_i = \sum_j x_j A_{ij}\{1-[(A_{ij}/b_{ij})(b/A)-1]\xi_i\}$
	$b_i = \sum_j x_j b_{ij}$
	$C_i = \sum_j x_j C_{ij}$
CSA Theory	$A_i = \sum_j x_j A_{ij}$
	$1+\Delta_{xi} = \lvert B^*\rvert / \sum_j x_j \lvert B^*\rvert_{ij}; \quad B^*_{ij} = x_i[\delta_{ij} + (x_j b_{ij}/2)(\Delta_{xi}/b_i + \Delta_{xj}/b_j)]$
	$C_i = \sum_j x_j C_{ij}$

$\Delta_{xx} = \rho RT\kappa_{Txx} - 1 = -1 + RT/\{RTv^2/(v-b)^2 - 2Av^3/(v^2+b^2)^2\}$

$\Delta_{xi} = \rho RT\kappa_{Txi} - 1 = -1 + RT/\{RTv^2/(v-b_i)^2 - 2A_i v^3/(v^2+b_i^2)^2\}$

$\xi = \{[A-\sqrt{(ACRT)}]/(2bRT\sqrt{2})\}\ln[(v+b-b\sqrt{2})/(v+b+b\sqrt{2})] + \sqrt{(ACRT)}/\{2[\sqrt{(ACRT)}-A]\}$

$\xi_i = \{[A_i-\sqrt{(A_iC_iRT)}]/(2b_iRT\sqrt{2})\}\ln[(v+b_i-b_i\sqrt{2})/(v+b_i+b_i\sqrt{2})] + \sqrt{(A_iC_iRT)}/\{2[\sqrt{(A_iC_iRT)}-A_i]\}$

Acknowledgment: The author thanks Professor Carol Hall for her helpful comments and corrections. This research is supported by the U.S. Department of Energy Grant No. DE-FG02-84ER13229.

Literature Cited

1. Massih, A.R.; Mansoori, G.A. Fluid Phase Equilibria 1983, 10, 57.
2. Brown, W.B. Proc. Roy. Soc. London Series A, 1957, 240;Phil. Trans. Roy. Soc. London Series A, 1957, 250.
3. Hill, T.L. "Statistical Mechanics" McGraw-Hill, New York, N.Y. 1956.
4. Kirkwood, J.G.; Buff, F. J. Chem. Phys. 1951, 19, 774.
5. Mansoori, G.A.; Ely, J. F. Fluid Phase Equilibria 1985, 22, 253.
6. Lan, S.S.; Mansoori, G.A. Int. J. Eng. Science 1977, 15, 323.
7. Leach, J.W.; Chappelear, P.S.; Leland, T.W. AIChE J. 1968, 14, 568; Proc. Am. Petrol. Inst. Series III, 1966, 46, 223.
8. Leland, T.W. Adv. Cryogenic Eng. 1976, 21, 466.
9. Mansoori, G.A.; Leland, T.W. J. Chem. Soc., Faraday Trans.II 1972, 68, 320.
10. Mansoori, G.A. J. Chem. Phys. 1972, 57, 198.
11. Rowlinson, J.S.; Swinton, F.L. "Liquids and Liquid Mixtures" 3rd Ed., Butterworths, Wolborn, Mass. 1982.
12. Scott, R.L. J. Chem. Phys. 1956, 25,193.
13. Mansoori, G.A.; Carnahan, N.F.; Starling, K.E.; Leland, T.W. J. Chem. Phys. 1971, 54, 1523.
14. Mansoori, G.A.; Ely, J.F. J. Chem. Phys. 1985, 82, 406.
15. Van der Waals, J.D. "Over de continuiteit van den gasen vloeistoftoestand" Leiden, 1873.
16. Redlich. O.; Kwong, J.N.S. Chem. Rev. 1949, 44, 233.
17. Peng, D.Y.; Robinson, D.B. Ind. Eng. Chem. Fundam. 1976, 15, 59.

RECEIVED November 5, 1985

16

Improved Mixing Rules for One-Fluid Conformal Solution Calculations

James F. Ely

Chemical Engineering Science Division, Center for Chemical Engineering, National Bureau of Standards, Boulder, CO 80303

During the past few years there has been great interest in improving equation of state mixing rules for fluid modeling. In this report new one-fluid mixing rules are proposed which explicitly take size difference effects into account. The resulting rules give the hard sphere mixture compressibility factor exactly. Comparisons of predicted excess properties for Lennard-Jones mixtures of varying size and energy ratios are presented. The results of the new mixing rules are superior to the van der Waals one-fluid model, especially for the excess volume.

During the past few years there has been increased interest the deveopment of improved mixing rules for use with engineering equation of state models of fluids. This activity has been focused in two areas--mixing rules developed for the parameters of simple (cubic) equations of state (1-5) and theoretically derived mixing rules which are used in high accuracy pure fluid equations of state via conformal solution arguments (6-10). Examples of the latter approach are the extended corresponding states and hard sphere expansion models proposed by Leland and coworkers (10-12) while the former includes applications of modified Redlich-Kwong type equations of state (13,14) to complex mixtures.

A somewhat surprising result obtained by comparing these two approaches is that the modified cubic equations of state using van der Waals mixing rules typically give as good if not better phase equilibrium results than the conformal solution model using the same mixing rules (15). This result becomes much more pronounced as the size differences in the mixture become large, much more so than for systems which display large energy differences. In addition to these observations based on data for real fluid mixtures, recent computer simulation studies (16-18) on model Lennard-Jones mixtures have dramatically pointed out the failures of the various conformal solution models as a function of size difference.

In this report, a new set of mixing rules is proposed for use in conformal solution mixture calculations. These new mixing rules are derived by making a new "mean density" approximation and then expanding the radial distribution function of a binary pair about that of a hard sphere. The new rules reduce to the van der Waals one-fluid model in the limit of low density and are exact for the compressibility factor of hard sphere mixtures. Some preliminary results obtained with the mixing rules are compared with computer simmulations for Lennard-Jones mixtures with varying size differences.

Derivation

The derivation of the new mixing rules closely follows the derivation of the hard sphere expansion (HSE) mixing rules presented by Mansoori and Leland (19). One first assumes that the N-body configurational energy of a mixture or pure fluid is pairwise additive and that each pairwise potential can be written as a sum of an effective hard sphere repulsion and long range attraction which is a universal function of reduced intermolecular separation r/σ_{ij}

$$\Phi_{ij}(r) = u_{ij}^{HS}(r) + \varepsilon_{ij}\, u_{ij}^{*}\,(r/\sigma_{ij})$$

It is also generally assumed that the pairwise potential parameters may be temperature and/or density dependent (12).

The next step is to make an approximation concerning the radial distribution functions in the mixture. In the HSE, Mansoori and Leland introduced the mean density approximation (MDA)

$$g_{ij}(r;[\rho_\alpha],T;[\varepsilon_{\alpha\beta}],[\sigma_{\alpha\beta}]) = g_o(r/\sigma_{ij};kT/\varepsilon_{ij},\rho\bar{\sigma}^3)$$

In this equation $[\rho_\alpha]$ denotes the set of molecular number densities in the mixture and $[\varepsilon_{\alpha\beta}]$ and $[\sigma_{\alpha\beta}]$ indicate that in the mixture, the radial distribution function depends upon the complete set of intermoleuclar potential parameters. The subscript o denotes a pure fluid radial distribution function. In words, the mean density approximation says that the distribution of molecules in a mixture as a function of reduced distance and temperature is the same as that in a pure fluid evaluated at an effective or mean reduced number density, $\rho\bar{\sigma}^3$.

The mean density approximation is a great improvement over the more common van der Waals one-fluid approximation which assumes that

$$g_{ij}(r;[\rho_\alpha],T;[\varepsilon_{\alpha\beta}],[\sigma_{\alpha\beta}]) = g_o(r/\sigma_{ij};kT/\bar{\varepsilon},\rho\bar{\sigma}^3)$$

although the mean density approximation and van der Waals (VDW1) approximations are identical in the case of equal energy parameters. The shortcomings of the VDW1 approximation are illustrated in Figure 1 which has been redrawn from (20). This figure illustrates the radial distribution functions for a binary mixture and hypothetical pure fluid (denoted by "o"). Nonetheless, a problem still exists with the MDA in that as the size differences in the mixture increase, the mean density no longer supplies an adequate estimate of the environment encountered by the i-i or j-j pairs, although it almost always provides a good estimate of the i-j distribution (30).

In order to at least partially improve this situation we have developed a simple modification to the MDA given by the following

$$g_{ij}(r;[\rho_\alpha],T;[\varepsilon_{\alpha\beta}],[\sigma_{\alpha\beta}]) = R_{ij}\, g_o(r/\sigma_{ij};kT/\varepsilon_{ij};\rho\bar{\sigma}^3) \qquad (1)$$

where

$$R_{ij} = g_{ij}^{HS}(\sigma_{ij};[\rho_\alpha];[\sigma_{\alpha\beta}])/g_o^{HS}(\bar{\sigma};\rho\bar{\sigma}^3) \qquad (2)$$

We shall refer to this as the "modified" mean density approximation (MMDA). One way of viewing this approximation is is that we not only must scale the effective pure distribution function in the radial direction but also in the probability direction, thereby avoiding the problems demonstrated in Figure 1. To illustrate the effects of this approximation more clearly, Figure 2 compares the MDA and MMDA radial distribution functions to the simulated argon-argon distribution in an argon-krypton mixture reported by Mo, et al. (21). As is shown in this figure, there is a definite improvement at small intermolecular separations even for a system where the size difference is small ($\sigma_2/\sigma_1 = 1.07$). An obvious shortcoming of this approximation is that MMDA $g_{ij}(r)$ does not approach 1 as $r \to \infty$. In practice this is not a problem since we always deal with integrals of $g_{ij}(r)$ and $g_o(r)$ does approach the correct limit. Thus, one should view the MMDA as a scaling of integrals of $g_{ij}(r)$.

Continuing with the derivation of the mixing rules, the pure fluid distribution function appearing in Equation 1 is expanded in a power series in 1/T to obtain

$$g_{ij}(r;[\rho_\alpha],T;[\varepsilon_{\alpha\beta}],[\sigma_{\alpha\beta}]) = R_{ij}[g_o(r/\sigma_{ij};\rho\bar{\sigma}^3) + \sum_{n=1}^{\infty} (\varepsilon_{ij}/kT)^n\, \Psi_n(r/\sigma_{ij};\rho\bar{\sigma}^3)] \qquad (3)$$

A similar expansion for a hypothetical pure fluid with parameters $\bar{\sigma}$ and $\bar{\varepsilon}$ yields

$$g(r;\rho,T;\bar{\varepsilon},\bar{\sigma}) = g_o(r/\bar{\sigma};\rho\bar{\sigma}^3) + \sum_{n=1}^{\infty} (\bar{\varepsilon}/kT)^n\, \Psi_n(r/\bar{\sigma};\rho\bar{\sigma}^3)] \qquad (4)$$

Finally, to complete the derivation, one takes the virial equation (22) for a mixture

$$Z_{mix} - 1 = 2\pi\beta\rho \sum_{i=1}^{N} \sum_{j=1}^{N} x_i x_j \int_o^\infty r^3\, (du_{ij}/dr)\, g_{ij}(r)\, dr$$

and hypothetical pure fluid at the same number density

$$Z_o - 1 = 2\pi\beta\rho \int_o^\infty r^3\, (du_o/dr)\, g_o(r)\, dr$$

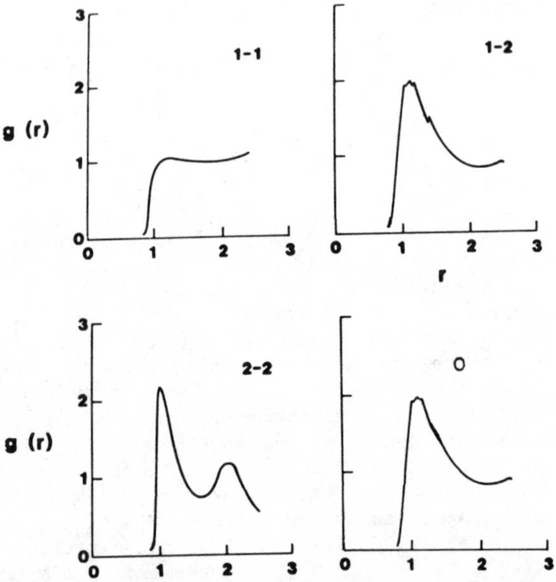

Figure 1. Radial distribution functions for a mixture of soft spheres with a size ratio of approximately two (20). The van der Waals approximation (denoted by o) yields a distribution function which is very reasonable for the 1-2 pair but unsatisfactory for the 1-1 and 2-2 pairs in the mixture.

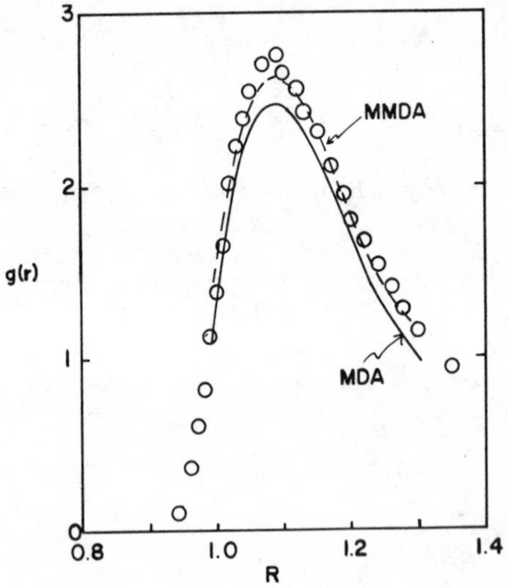

Figure 2. Comparison of the mean density and modified mean density approximations to the radial distribution function for the krypton-krypton pair in an equimolar argon-krypton mixture (21). The open circles are from the computer simulation, solid line is the MDA and dashed line is the MMDA.

substitutes Equations 3 and 4 for the radial distribution functions and subtracts to find

$$\begin{aligned} Z_{mix} - Z_o = & \sum_{i=1}^{N}\sum_{j=1}^{N} [x_i x_j \sigma_{ij}^3 R_{ij} - \bar{\sigma}^3] \; J_o \\ & + \frac{1}{kT}\sum_{i=1}^{N}\sum_{j=1}^{N} [x_i x_j \varepsilon_{ij} \sigma_{ij}^3 R_{ij} - \overline{\varepsilon\sigma^3}] \; J_1 \\ & + \sum_{n=1}^{\infty} \left(\frac{1}{kT}\right)^{n+1} I_n \sum_{i=1}^{N}\sum_{j=1}^{N} [x_i x_j \sigma_{ij}^3 R_{ij} \varepsilon_{ij}^{n+1} - \bar{\sigma}^3 \bar{\varepsilon}^{n+1}] \end{aligned} \tag{5}$$

In this equation J_n and I_n are integrals over the hard sphere distribution function g_o and the Ψ_n, respectively, and the derivative of the intermolecular potential. By setting the zeroth and first order terms of the expansion given in Equation 5 equal to zero, one obtains the desired mixing rules

$$\bar{\sigma}^3 = \sum_{i=1}^{N}\sum_{j=1}^{N} x_i x_j \sigma_{ij}^3 R_{ij} \tag{6}$$

and

$$\overline{\varepsilon\sigma^3} = \sum_{i=1}^{N}\sum_{j=1}^{N} x_i x_j \varepsilon_{ij} \sigma_{ij}^3 R_{ij} \tag{7}$$

where R_{ij} is defined in Equation 2.

Note that as written, $\bar{\sigma}$ appears on both sides of these equations, and that the hard sphere distribution functions which define R_{ij} depend upon density and composition. Also since

$$\lim_{\rho \to o} g_{ij}^{HS} = \lim_{\rho \to o} g_o^{HS} = 1$$

the mixing rules reduce to the common van der Waals one-fluid rules

$$\sigma_{vdw}^3 = \sum_{i=1}^{N}\sum_{j=1}^{N} x_i x_j \sigma_{ij}^3 \tag{8}$$

and

$$\varepsilon_{vdw}\,\sigma_{vdw}^3 = \sum_{i=1}^{N}\sum_{j=1}^{N} x_i x_j \varepsilon_{ij} \sigma_{ij}^3$$

in the limit of low density.

Hard Spheres

The first notable aspect of the new mixing rules (MMDA rules) is that they are exact for the compressibility factor of hard spheres

($\epsilon_{ij} = 0$). This is obvious from the well known relations for the equations of state for pure hard spheres and hard sphere mixtures

$$Z_o^{HS} = 1 + 4\,\eta_o g_o^{HS}(\sigma_o) \tag{9}$$

and

$$Z_{mix}^{HS} = 1 + 4 \sum_{i=1}^{N} \sum_{j=1}^{N} x_i x_j \eta_{ij} g_{ij}^{HS}(\sigma_{ij}) \tag{10}$$

where $\eta_o = \frac{\pi}{6}\rho\,\sigma_o^3$ and $\eta_{ij} = \frac{\pi}{6}\rho\,\sigma_{ij}^3$. Equating 9 and 10 results in exactly Equation 5.

One can solve for $\bar{\sigma}$ for hard spheres algebraically given values for the contact distribution functions. We shall consider two cases, the Percus-Yevick and that given by Carnahan and Starling. From the Wertheim (23) and Thiele (24) solution for the Percus-Yevick equation one has

$$g_{ij}^{HS,PY}(\sigma_{ij}) = \frac{1}{(1 - \eta_m)} + \frac{3}{2}\frac{\chi_m}{(1 - \eta_m)^2}\, d_{ij} \tag{11}$$

and

$$g_o^{HS,PY}(\sigma_o) = \frac{(1 + \eta_o/2)}{(1 - \eta_o)^2} \tag{12}$$

where

$$\eta_m = \frac{\pi}{6}\rho S_3 \,, \quad \chi_m = \frac{\pi}{6}\rho S_2$$

$$S_n = \sum_{i=1}^{N} x_i \sigma_i^n$$

and

$$d_{ij} = 2\sigma_i\sigma_j/(\sigma_i + \sigma_j)$$

η_o is defined above. In the Carnahan-Starling (25) and corresponding MCSL (26) case for mixtures one finds

$$g_{ij}^{HS,MCSL}(\sigma_{ij}) = \frac{1}{(1 - \eta_m)} + \frac{3}{2}\left(\frac{S_2}{S_3}\right)\frac{\eta_m d_{ij}}{(1 - \eta_m)^2} + \frac{3}{4}\left(\frac{S_2}{S_3}\right)^2 \frac{\eta_m^2 d_{ij}^2}{(1 - \eta_m)^3} \tag{13}$$

and

$$g_o^{HS,CS}(\sigma_o) = \frac{(2 - \eta_o)}{2(1 - \eta_o)^3} \tag{14}$$

Solving Equation 5 for $\bar{\eta}$ $(= \frac{\pi}{6} \rho\bar{\sigma}^3)$ using the contact distribution functions given by Equations 11 and 12 one finds

$$\bar{\eta} = - \frac{(1 + 2G) - 1 + 6G}{(1 - 2G)} \tag{15}$$

where

$$G = \sum_{i=1}^{N} \sum_{j=1}^{N} x_i x_j \eta_{ij} g_{ij}^{HS,PY}(\sigma_{ij})$$

Likewise, the solution of Equation 5 for $\bar{\eta}$ using the contact distributions given by Equations 13 and 14 is the real root of

$$\bar{\eta}^3 - \frac{6G' + 1}{G'} \bar{\eta}^2 + \frac{6G' + 1}{2G'} \bar{\eta} - 1 = 0 \tag{16}$$

where G' is evaluated with the MCSL contact distribution function.

To illustrate the differences between the new mixing rules and the VDW-1 rules, Figures 3-4 show the hard sphere results obtained with Equations 6, 13 and 14 relative to that from the VDW-1 mixing rule, Equation 8 at two packing fractions, $\eta = 0.2$ and $\eta = 0.4$. We note that the effect is asymmetric in the concentration, x_2, of the larger component in that small concentrations of large molecules have large effects on the effective mean density. In Figure 5 the same ratio of new one-fluid dimater to the van der Waals result is shown as a function of density (packing fraction) at a constant composition of $x_1 = 0.8$. As one would expect, the difference is most pronounced at the highest densities and increases as the size ratio increases.

In Figure 6, the difference between the VDW Carnahan-Starling compressibility factor is plotted versus the "exact" value given by the MCSL equation. In this case there is no error associated with the new mixing rules. Figure 7, however, plots the error in the Helmholtz free energy from the van der Waals rule and that obtained from Equations 6, 13 and 14. The lines above the zero line represent the VDW-1 case and the lines below the zero line are the new results. The VDW-1 results are uniformly in greater error than the new rules and are also of opposite sign.

Finally, Figures 8 and 9 compare the one-fluid diameters calculated using Equations 11, 12 and 15, i.e., the Percus-Yevick solution for hard spheres to the MCSL results at two compositions. We note that there is negligible difference between the two (maximum being 0.04%) thus one is justified in using the algebraically simpler Percus-Yevick results for practical calculations.

Although the van der Waals diameter is reasonable, Equations 15 and 16 are definite improvements. Also, these results for hard

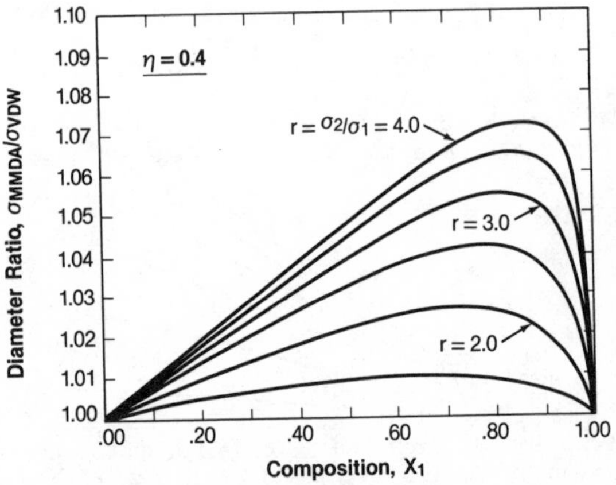

Figure 3. Ratio of the hard sphere diameter resulting from the MMDA to that obtained from the van der Waals-1 model at various size ratios and a fixed packing fraction of 0.4.

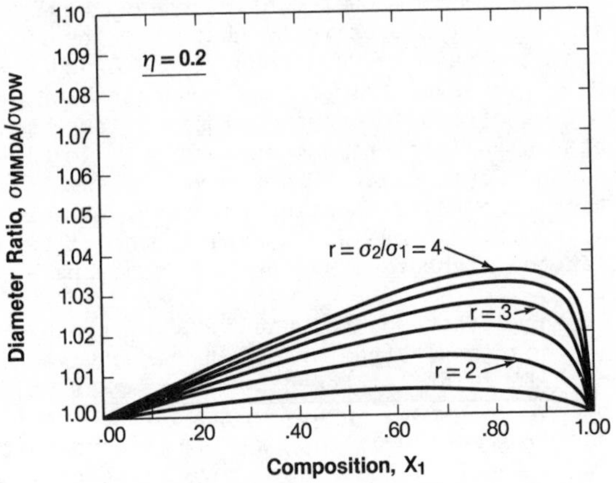

Figure 4. Ratio of the hard sphere diameter resulting from the MMDA to that of the VDW1 model as a function of size ratio and composition at packing fraction of 0.2.

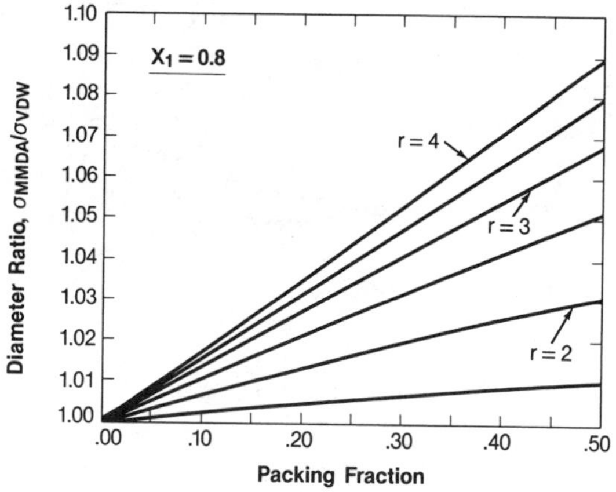

Figure 5. Ratio of the MMDA hard sphere diameter to that of the VDW1 model as a function of packing fraction and size ratio at a fixed composition of $x_1 = 0.8$.

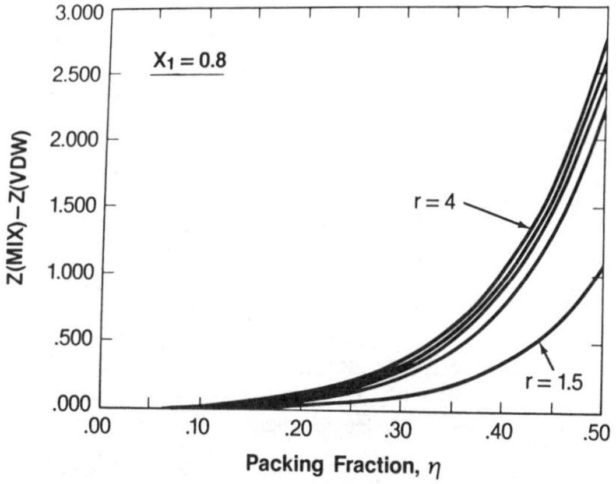

Figure 6. Error in the mixture hard sphere compressibility factor as a function of size ratio and packing fraction at $x_1 = 0.8$. The MMDA is exact for this hard sphere property.

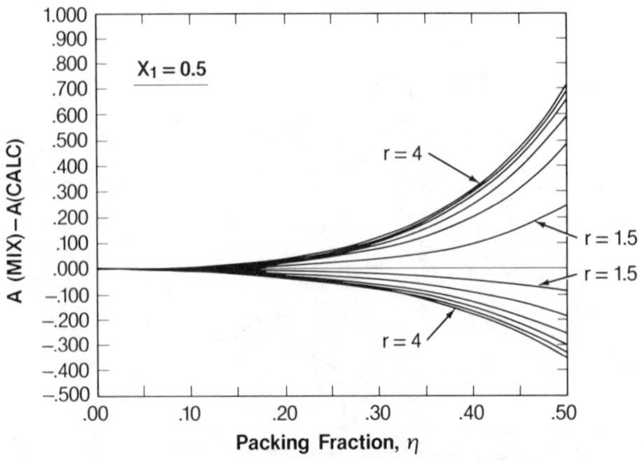

Figure 7. Error in the mixture residual Helmholtz free energy as a function of size ratio and packing fraction at $x_1 = 0.5$. The curves above the zero line are obtained with the VDW1 diameter while those below the zero line were calculated with the MMDA.

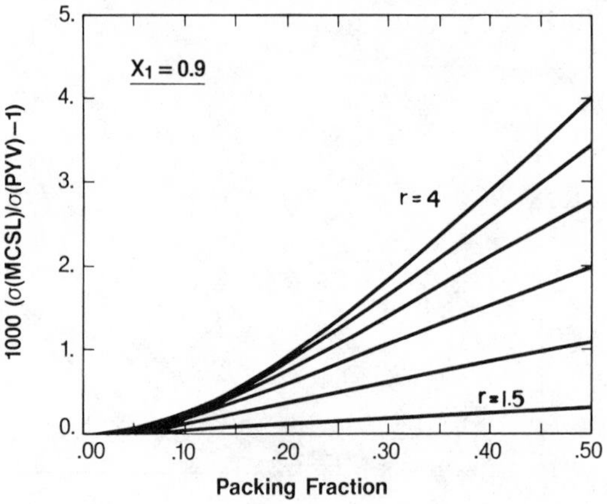

Figure 8. Comparison of the MMDA hard sphere diameter obtained with the MCSL contact hard sphere distribution functions and those obtained with the Percus-Yevick virial equation. Results are shown as a function of size ratio (r) at various packing fractions at a fixed composition of $x_1 = 0.9$.

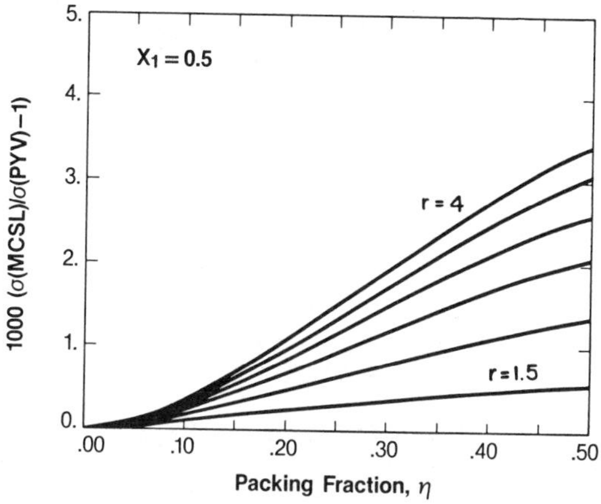

Figure 9. Comparison of the MMDA hard sphere diameter obtained with the MCSL contact hard sphere distribution functions and those obtained with the Percus-Yevick virial equation. Results are shown as a function of size ratio (r) at various packing fractions at a fixed composition of $x_1 = 0.5$.

spheres clearly demonstrate that mixing rules must be density dependent to accurately represent contributions from repulsive forces.

Lennard-Jones Mixtures

In order to test the new mixing rules for mixtures consisting of more realistic molecules, conformal solution calculations using Equations 6, 7, 11 and 12 have been performed using a Lennard-Jones 6-12 reference fluid. The 32 term BWR type equation of state reported by Nicolas, et al. (27) was used to obtain the Lennard-Jones 6-12 properties. In general, the calculation procedure yielded the density, excess volume, excess enthalpy, excess Gibbs energy and excess internal energy. Mixing was performed at constant pressure, i.e.,

$$X^E(T,p,[x_\alpha]) = X^{mix}(T,p,[x_\alpha]) - \sum_{i=1}^{N} x_i X_i(p,T)$$

where X^E is any excess property and X_i denotes a pure fluid value of that property. In all calculations the Lorentz-Berthelot combing rules were used for the unlike pair potential parameters.

Comparisons were made with three different sets of computer simulation results: Singer and Singer's (28) Monte Carlo results and the more recent molecular dynamics simulations of Gupta (18) and Hoheisel, et al. (16). Figures 10-11 compare the excess volumes calculated by the HSE-MDA model (29), VDW-1 and the MMDA presented in this work as a function of size ratio at three different energy ratios. In general, the MDA is slightly better than the MMDA for the excess volume but both the MMDA and MDA are substantial improvements overthe VDW-1 model. Table I compares the predictions for V^E, H^E and G^E to those obtained with the VDW-1 model and the Monte Carlo calculations.

Figure 12 compares the MMDA to the excess volume and Figure 13 compares the excess Gibbs energy to the recent isothermal-isochoric results of Gupta and Hailefor $\sigma_2/\sigma_1 = 2$. Included on these figures are the corresponding results for the VDW-1 model. Figure 13 also shows the calculated Percus-Yevick results reported by Gupta. Table II compares the calculated results to simulations. It is apparent from these comparisons that the MMDA offers a greatly improved prediction of the excess volume for systems where the size differences are large. It is apparent, however, that even though the predictions of the excess Gibbs energy are better than the VDW-1 model, some improvement is still in order.

To explore the effect of large energy and size differences, Table III compares MMDA and VDW-1 predictions to the recent results of Hohesiel, et al. (16). Again, the MMDA offers a substantial improvement over the VDW-1 results, especially for the excess volume.

Summary and Conclusions

We have presented a simple modification to the mean density approximation which improves the results obtained from conformal solution predictions. The model is exact for the compressibility factor of hard sphere mixtures. The results for realistic fluids are not

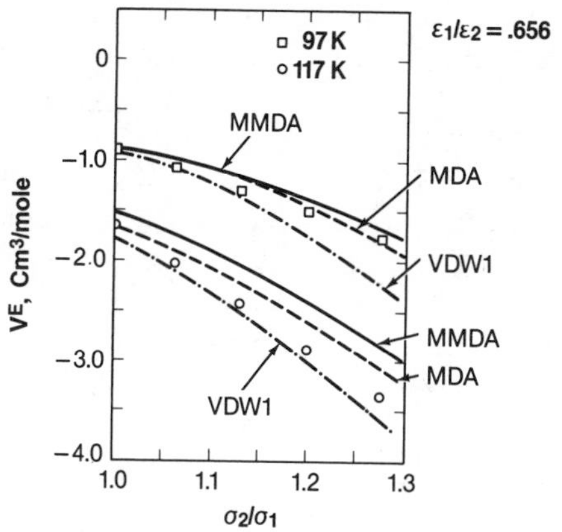

Figure 10. Comparisons of calculated and simulated excess volumes as a function of size ratio. The simulated and VDW1 values were taken from (28) while the MDA values were obtained from (29).

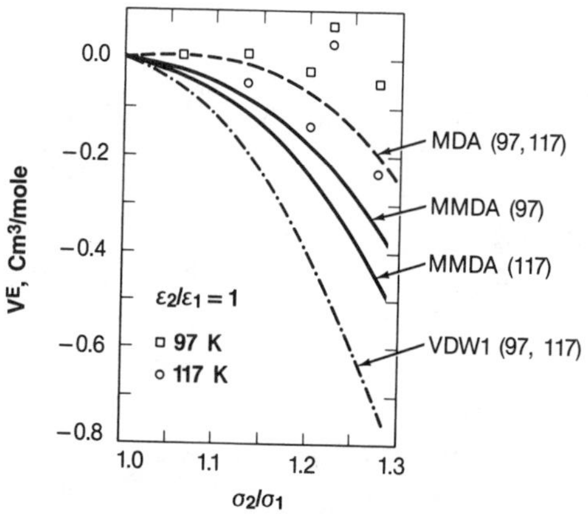

Figure 11. Comparisons of calculated and simulated excess volumes as a function of unity ratio. The simulated and VDW1 values were taken from (28) while the MDA values were obtained from (29).

Table I. Comparison of the MMDA, VDW-1 and Monte Carlo Results (28) for Equimolar Lennard-Jones Mixture Excess Properties.

			V^E		H^E		G^E	
$\varepsilon_2/\varepsilon_1$	σ_2/σ_1	Model	97 K	117 K	97 K	117 K	97 K	117 K
0.656	1.000	MC	-0.89	-1.66	0.15	0.12	0.176	0.159
		MMDA	-0.84	-1.53	0.13	0.09	0.181	0.158
		VDW1	-0.92	-1.77	0.11	0.09	0.186	0.167
	1.0619	MC	-1.08	-2.02	0.24	0.19	0.242	0.208
		MMDA	-0.97	-1.74	0.27	0.21	0.268	0.222
		VDW1	-1.11	-2.08	0.23	0.19	0.263	0.224
	1.1277	MC	-1.29	-2.43	0.34	0.26	0.305	0.250
		MMDA	-1.17	-2.07	0.41	0.32	0.352	0.282
		VDW1	-1.39	-2.48	0.35	0.29	0.334	0.274
	1.1978	MC	-1.50	-2.87	0.45	0.34	0.368	0.287
		MMDA	-1.39	-2.42	0.53	0.42	0.402	0.331
		VDW1	-1.76	-3.00	0.47	0.39	0.402	0.317
	1.2727	MC	-1.76	-3.35	0.58	0.42	0.423	0.319
		MMDA	-1.68	-2.68	0.64	0.49	0.489	0.368
		VDW1	-2.23	-3.61	0.58	0.48	0.454	0.354
1.000	1.000	MC	0.00	0.00	0.00	0.00	0.000	0.000
		MMDA	0.00	0.00	0.00	0.00	0.000	0.000
		VDW1	0.00	0.00	0.00	0.00	0.000	0.000
	1.0619	MC	0.01	-0.02	0.01	0.00	-0.001	-0.003
		MMDA	-0.02	-0.02	0.00	0.00	-0.003	-0.003
		VDW1	-0.05	-0.05	0.00	0.00	-0.003	-0.003
	1.1277	MC	0.02	-0.05	0.03	0.01	-0.005	-0.011
		MMDA	-0.08	-0.11	-0.02	-0.01	-0.013	-0.013
		VDW1	-0.18	-0.20	0.00	0.00	-0.011	-0.011
	1.1978	MC	-0.05	-0.14	0.05	0.02	-0.013	-0.024
		MMDA	-0.20	-0.26	-0.05	-0.03	-0.030	-0.029
		VDW1	-0.41	-0.49	0.00	0.00	-0.024	-0.024
1.000	1.2727	MC	-0.10	-0.24	0.09	0.04	-0.023	-0.044
		MMDA	-0.34	-0.46	-0.09	-0.05	-0.054	-0.052
		VDW1	-0.72	-0.79	0.00	0.00	-0.043	-0.042
1.235	1.000	MC	-0.96	-1.72	0.04	0.08	0.178	0.161
		MMDA	-0.84	-1.55	0.13	0.18	0.181	0.159
		VDW1	-0.92	-1.77	0.11	0.09	0.185	0.167

Table I. Continued.

			V^E		H^E		G^E	
$\varepsilon_2/\varepsilon_1$	σ_2/σ_1	Model	97 K	117 K	97 K	117 K	97 K	117 K
	1.0619	MC	-0.80	-1.43	-0.01	0.00	0.122	0.018
		MMDA	-0.76	-1.36	-0.27	-0.04	0.084	0.086
		VDW1	-0.81	-1.58	0.00	-0.01	0.103	0.105
	1.1277	MC	-0.68	-1.18	-0.10	-0.07	0.040	0.049
		MMDA	-0.69	-1.24	-0.20	-0.17	-0.021	0.016
		VDW1	-0.81	-1.48	-0.12	-0.11	0.014	0.037
	1.1978	MC	-0.55	-0.95	-0.17	-0.14	-0.033	-0.016
		MMDA	-0.69	-1.18	-0.38	-0.31	-0.132	-0.081
		VDW1	-0.90	-1.47	-0.25	-0.21	-0.082	-0.036
	1.2727	MC	-0.45	-0.76	-0.24	-0.20	-0.113	-0.087
		MMDA	-0.72	-1.16	-0.58	-0.46	-0.249	-0.173
		VDW1	-1.06	-1.57	-0.36	-0.31	-0.178	-0.115

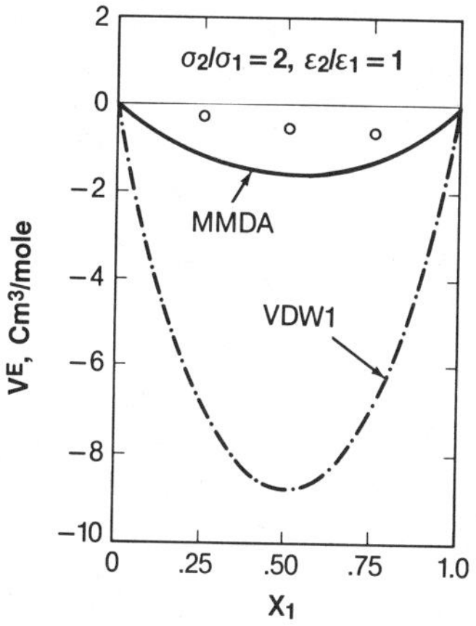

Figure 12. Comparison of calculated and simulated excess volumes for a system of Lennard-Jones molecules as a function of composition. The open circles are the simulated points calculated from (18). Note the substantial improvement shown by the MMDA as compared to the VDW1 model.

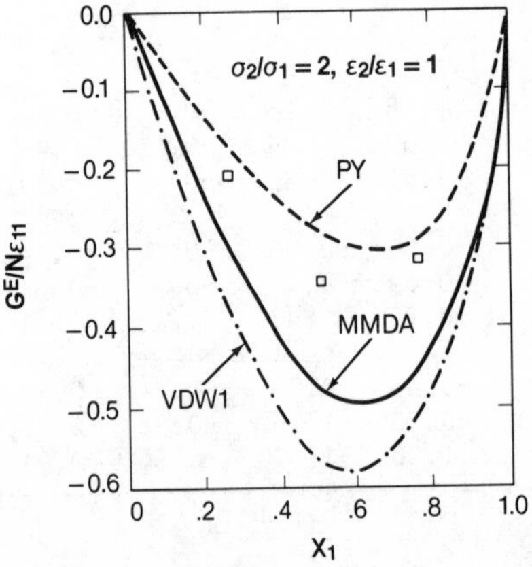

Figure 13. Comparison of simulated and calculated excess Gibbs energy values as a function of composition. Simulated results are from (<u>18</u>).

Table II. Comparison of MMDA and VDW1 Predictions to Simulation Results (<u>18</u>) ($\varepsilon_{22}/\varepsilon_{11}$ = 1 in all cases).

σ_{22}/σ_{11}	x_1	Model	V^E, $cm^3/mole$	G^E/NkT
1.25	0.25	MD	-0.098	-0.053
		MMDA	-0.131	-0.035
		VDW1	-0.373	-0.036
	0.50	MD	-0.145	-0.075
		MMDA	-0.175	-0.057
		VDW1	-0.500	-0.050
	0.75	MD	-0.131	-0.059
		MMDA	-0.133	-0.042
		VDW1	-0.377	-0.040
1.50	0.25	MD	-0.378	-0.070
		MMDA	-0.437	-0.109
		VDW1	-1.542	-0.119
	0.50	MD	-0.388	-0.111
		MMDA	-0.530	-0.166
		VDW1	-2.067	-0.174
	0.75	MD	-0.383	-0.099
		MMDA	-0.405	-0.147
		VDW1	-1.558	-0.147
1.75	0.25	MD	-0.369	-0.133
		MMDA	-0.784	-0.196
		VDW1	-3.629	-0.235
	0.50	MD	-0.813	-0.229
		MMDA	-1.071	-0.311
		VDW1	-4.824	-0.350
	0.75	MD	-0.865	-0.236
		MMDA	-0.817	-0.294
		VDW1	-3.629	-0.305
2.0	0.25	MD	-0.228	-0.207
		MMDA	-1.172	-0.287
		VDW1	-6.672	-0.377
	0.50	MD	-0.480	-0.339
		MMDA	-1.669	-0.468
		VDW1	-8.888	-0.566
	0.75	MD	-0.644	-0.320
		MMDA	-1.265	-0.463
		VDW1	-6.672	-0.504

Table III. Comparisons of MMDA Predictions to Simulated Results for Equimolar Mixtures With Large Size and Energy Differences (16).

σ_2/σ_1	$\varepsilon_2/\varepsilon_1$	T, K	P, Bar	Model	V^E, cm^3/mole	G^E/NkT	H^E/NkT
1.3	1.5	201.4	3563	MD	-0.040	0.041	-0.025
				MMDA	-0.011	-0.029	-0.066
				VDW1	-0.284	-0.088	-0.102
	4.5	200.5	3551	MD	-0.302	-0.140	-0.108
				MMDA	-0.287	-0.117	-0.056
				VDW1	-0.362	-0.112	-0.064
2.0	1.5	198.4	1629	MD	-0.268	-0.564	-0.161
				MMDA	-0.894	-0.586	-0.226
				VDW1	-3.306	-0.540	-0.563
	4.5	199.0	1904	MD	-1.261	-0.409	-0.375
				MMDA	-0.625	-0.485	-1.299
				VDW1	-5.246	-0.890	-1.107

quite as accurate as those obtained from the HSE theory, but are simpler to incorporate since the MMDA gives the optimal effective hard sphere diameter directly. Other similar approximations could be made which might further improve the results of this type of model. To some degree progress in this area has been hampered by the lack of detailed information concerning the radial distribution functions in mixtures. This situation is improving rapidly with the availability of new mixture simulation results and thus one can speculate that improved radial distribution function based conformal solution models will be forthcoming.

Acknowledgments

The author would like to acknowledge Professor J. M. Haile of Clemson University for providing Dr. Gupta's simulation results prior to publication, Ms. Karen Bowie for her assistance in the preparation of the manuscript and the NASA-Lewis Research Center, Grant No. C-60875-D for its financial support.

Literature Cited

1. Vidal, J. Chem. Eng. Sci. 1978, 33, 787-91.
2. Huron, M.; Vidal, J. Fluid Phase Equil. 1979, 3, 255-271.
3. Mathias, P. M.; Copeman, T. W. Fluid Phase Equil. 1983, 13, 91-108.
4. Whiting, W. B.; Prausnitz, J. M. Fluid Phase Equil. 1982, 9, 119-147.
5. Dieters, U. Chem. Eng. Sci. 1981, 36, 1139-51; 1982, 37, 855-61; Fluid Phase Equil. 1983, 13, 109-20.
6. Rowlinson, J. S.; Watson, I. D. Chem. Eng. Sci. 1969, 24, 1156-1574.
7. Leland, T. W.; Chappelear, P. S. Ind. Eng. Chem. 1968, 60, 16-43.
8. Massih, A. R.; Mansoori, G. A. Fluid Phase Equil. 1983, 10, 57-72.
9. Mollerup, J. Adv. Cryo. Eng. 1975, 20, 172-94; 1978, 23, 550-60.
10. Ely, J. F. Proc. 63rd Gas Processors Association Annual Convention, 1984, p. 9.
11. Naumann, K. H.; Leland, T. W. Fluid Phase Equil. 1984, 18, 1-45.
12. Hang, T.; Ely, J. F.; Leland, T. W. Fluid Phase Equil. 1985, submitted for publication.
13. Soave, G. Chem. Eng. Sci. 1978, 33, 787-91.
14. Peng, D. Y.; Robinson, D. B. Ind. Eng. Fundam. 1976, 15, 59-64.
15. Mentzer, R. A.; Greenkorn, R. A.; Chao, K. C. Ind. Eng. Chem. Process Des. Dev. 1981, 20, 240-52.
16. Hoheisel, C.; Dieters, U.; Lucas, K. Mol. Phys. 1983, 49, 159-70.
17. Shing, K. S.; Gubbins, K. E. Mol. Phys. 1982, 46, 1109-1128; 1983, 49, 1121-38.
18. Gupta, S. Ph.D. Dissertation, Clemson University, 1984.
19. Mansoori, G. A.; Leland. T. W. J. Chem. Soc. Faraday Trans. II 1972, 6, 320-44.

20. Hanley, H. J. M.; Evans, D. J. Int. J. Thermophysics 1981, 2, 1-9.
21. Mo, K. C.; Gubbins, K. E.; Jacucci, G.; McDonald, I. R. Mol. Phys. 1974, 27, 1173-83.
22. McQuarrie, D. A. "Statistical Mechanics"; Harper and Row: New York, 1976; Ch. 13.
23. Wertheim, M. S. Phys. Rev. Lett. 1963, 10, 321-4.
24. Thiele, E. J. Chem. Phys. 1963, 39, 474-9.
25. Carnahan, N. F.; Starling, K. E. J. Chem. Phys. 1970, 53, 600-3.
26. Mansoori, G. A.; Carnahan, N. F.; Starling, K. E.; Leland, T. W. J. Chem. Phys. 1970, 54, 1523-5.
27. Nicolas, J. J.; Gubbins, K. E.; Street, W. B.; Tildesley, D. J. Mol. Phys. 1979, 37, 1429-54.
28. Singer, J. V. L.; Singer, K. Mol. Phys. 1972, 24, 357-90.
29. Chang, I. J. M.S. Thesis, Rice University, 1978.
30. Hoheisel, G.; Lucas, K. Mol. Phys. 1984, 53, 51-7.

RECEIVED November 8, 1985

ENGINEERING AND TECHNOLOGY

17

Recent Mixing Rules for Equations of State

An Industrial Perspective

Thomas W. Copeman and Paul M. Mathias

Air Products and Chemicals, Inc., Allentown, PA 18105

The number of applications for equations of state capable of describing phase behavior of mixtures with asymmetric interactions is high - and continuing to grow. Gas-processing applications such as enhanced oil recovery and glycol dehydration involve carbon dioxide, water, hydrogen sulfide, and hydrocarbons at high pressure. Organic-chemical and polymer production involve an increasingly wide variety of conditions, not restricted to low pressures (1). Supercritical extraction is actively being applied to the purification of natural products, which involves mixtures of complex polar compounds near the critical point of a gas like carbon dioxide (2).

Engineering models to describe phase equilibrium can be divided into two broad categories: equations of state and activity coefficient models. Equations of state have been successfully applied to mixtures of nonpolar and slightly polar compounds at all conditions of engineering interest. These models have been used most extensively by the gas processing industry for the design of various processes (see for example Refs. 3 and 4).

Conversely the mathematical flexibility of activity coefficient models has conventionally been considered necessary to describe systems which exhibit high liquid-phase nonideality. These models have been most extensively used by the chemicals and polymer industries for design of various processes. Process design for the production of polyvinyl alcohol (5) is one example. The activity-coefficient approach is adequate at low reduced temperatures where the liquid phase is incompressible and up to moderate pressures. The use of different models for the various phases precludes correct description of mixture critical points and additional problems arise for supercritical components (6).

The equation-of-state approach does not inherently suffer from these limitations and the extension of equations of state to describe asymmetric interactions is currently a highly active area of research. This paper reviews recent developments of mixing rules for equations of state with an engineering-design perspective.

0097–6156/86/0300–0352$06.00/0

Evolution of Equation of State Mixing Rules

Considerable effort has been expended over the past fifty years to develop new equations of state for pure components. Much less attention has been given to improving mixing rules. The one-fluid mixing rules of van der Waals (7) (known hereafter as the vdW-1 mixing rules) are still in widespread use; the equations of state of Soave (8) and Peng and Robinson (9) are two such examples.

It should be noted that the vdW-1 mixing rules are reasonably well based in both theory and the typical excess-function behavior of normal fluids. Leland, Chappelear, and Gamson (10) have shown that these mixing rules arise from very reasonable assumptions for the ji radial distribution functions. (Also see Refs. 11 and 12). Henderson and Leonard (13, 14) have shown that these mixing rules provide good agreement with the quasi-experimental machine-simulation results for hard-sphere and Lennard-Jones (6:12) mixtures. Further, Vidal (15) has demonstrated that the vdW-1 mixing rules predict excess functions very similar to regular solution theory and thus they should provide a good description of most nonpolar mixtures.

An important landmark in the development of industrially significant mixing rules was the work of Stotler and Benedict (16) who suggested the use of binary interaction parameters. They found that the one-fluid mixing rules suggested for the Benedict-Webb-Rubin equation (17, 18) could be markedly improved by the use of small pair-dependent correction terms. Another important effort was the work of Prausnitz and Gunn (19) who noted that the interaction parameter was not a totally empirical correction factor, but was related to the theory of intermolecular forces.

Many efforts to improve the vdW-1 mixing rules attempted to find better forms while retaining the one-fluid concept. Plocker et al. (20), Radosz et al. (21) and Lee et al. (22) varied an exponent in the mixing rules, which can be represented as

$$\varepsilon\sigma^n = \Sigma\Sigma x_i x_j \ \varepsilon_{ji}\sigma_{ji}^n \tag{1}$$

$$\sigma^m = \Sigma\Sigma x_i x_j \ \sigma_{ji}^m \tag{2}$$

where n = m = 3 represents the van der Waals one-fluid mixing rules.

Plocker (20) chose n=0.75 and m=3, while Radosz (21) chose n=-0.25 and m=3. Lee (22) used n=m=4.5. While some degree of success in improving mixture predictions is claimed for each of the methods, their application to highly asymmetric polar-nonpolar systems is limited.

Perturbation theory is useful as a guide to the development of both pure fluid and mixture equations of state for engineering calculations (23). Donohue and Prausnitz (24) developed an equation of state for mixtures of molecules with large size differences based on a perturbation expansion for square-well fluids at low densities and ideas from polymer solution theory. The expansion was then truncated after the fourth attractive term. While a simple quadratic composition dependence was obtained for the first attractive term, each of the higher-order terms yields additional (non-quadratic) mixing rules. Good results were obtained for mixtures containing alkanes (up to C_{30}), aromatics and light inorganic gases.

Wilson (25) has suggested that the liquid-phase excess Gibbs energy data can be used to provide information for improving mixing rules at high densities. Vidal (15), Huron and Vidal (26), Heyen (27), and Won (28) have further proposed non-quadratic mixing rules based on the behavior of liquid-phase activity coefficients. These mixing rules do not reproduce the required quadratic composition dependence of the second virial coefficient, as pointed out by Huron and Vidal (26) and later by Whiting and Prausnitz (29, 30) and Mollerup (31). These initial local-composition equations of state, however, did provide an important advance by demonstrating that complex polar-nonpolar systems could be correlated with equations of state in the same (empirical) manner as with activity-coefficient models. For example, vapor-liquid equilibrium correlations were developed for ethanol-benzene and acetone-water by Heyen (27) and Huron and Vidal (26), respectively.

To overcome the above-mentioned deficiency at low pressure, Whiting and Prausnitz (29, 30), Mollerup (31) and Won (32) proposed density-dependent local-composition mixing rules. A common proposal in all three models is the use of conventional mixing rules for the repulsive term in the equation of state and local-composition mixing rules with density-dependent Boltzmann factors for the attractive term. These equations of state pleasingly resemble the Wilson (33) activity-coefficient model at high densities and conventional equations of state at low densities. Mathias and Copeman (34) however demonstrated that the local-composition Peng-Robinson equation of state derived by Mollerup (31) lacks adequate predictive capability for nonpolar mixtures with even moderate size differences. The local-composition effect is too large, which can result in greatly underpredicted fugacity coefficients (for example, 3-4 orders of magnitude for methane in decane). Sizable interaction parameters along with size parameters (e.g., surface area) are necessary for reasonable predictions of phase equilibrium. The Mollerup (31) Peng-Robinson equation of state, along with the Redlich-Kwong and other analogs, was concluded to be too unreliable for general application to industrial problems.

An important idea has been proposed by Dimitrelis and Prausnitz (35). They propose that local-composition effects do not result when ji interactions are different from ii but rather when ji are different from some "ideal" combinations of ii and jj interactions, referred to as ji°. Dimitrelis and Prausnitz defined ji° as an arithmetic mean in their work. Mathias and Copeman (34) and Mollerup (36), however, chose the geometric mean and developed a density-dependent local-composition Peng-Robinson equation of state that meets an important additional limit: non-conformal effects are not necessarily large (or even non-zero) for an asymmetric system. This feature enables the model to retain the good predictions of the standard Peng-Robinson equation of state for nonpolar systems and facilitates correlation of phase equilibrium of highly non-ideal polar systems. Additional forms of the equation of state were suggested by Mathias and Copeman (34) based on expansion of the Boltzmann factors in the attractive part with truncation after the second term in the series. Reduced computational time along with surprisingly good results for binary hydrocarbon-water liquid-liquid equilibria were shown.

Ludecke and Prausnitz (37) have proposed a similar equation of state based on the van der Waals form. They chose the Mansoori et al. (38) expression for the hard-sphere part and a simple van der Waals expression for the attractive part. For mixtures, the Mansoori et al. expression is used for the hard-sphere contribution, while for the attractive contribution the conventional density-independent quadratic mixing rule is used as a leading term and a density-dependent correction (cubic in mole fraction) is used to describe "non-central" forces. Good results were obtained for vapor-liquid and liquid-liquid equilibria in binary mixtures containing water, phenol, pyridine, methanol and hydrocarbons. The equation of state, however, over-predicts the two-liquid region in the ternary systems studied. The above approach is a simple, yet potentially very effective, way to describe polar, asymmetric systems and should be developed further for industrial applications.

A possible way to avoid violating the quadratic dependence of the second virial coefficient and yet have separate mixing rules for high and low densities has been suggested by Larsen and Prausnitz (39). These researchers have divided the attractive contribution to the Helmholtz energy into a second-virial portion and a dense-fluid portion, thereby allowing an arbitrary mixing rule for the dense-fluid part. Good results have been obtained for the methane-water binary. This work is important since it can be extended to proper identification of various types of contributions at the pure-fluid level, leading to the use of theoretically suggested mixing rules for each contribution.

Li et al. (40) have combined the pure-fluid equation of state developed by Chung et al. (41) with local-composition mixing rules. The local-composition mixing rules along with van der Waals one-fluid mixing rules were evaluated against both vapor-liquid equilibrium and density data for binary systems. The predictive capability of the Li et al. model should be evaluated.

Table I presents a list of systems studied by various investigators using local-composition mixing rules. The majority of studies have not included evaluation of liquid-liquid equilibrium and density predictions.

It appears that improved mixture models will come from the concept of density-dependent mixing rules. However, it is not clear at the present time which of the proposed forms is the "best." Most appear to provide significant improvement over vdW-1 for correlation of the phase equilibrium of binary systems, including complex behavior like liquid-phase immiscibility. However, very few studies have addressed the more difficult problem of multicomponent predictions.

Parallel work in computer simulation and theoretical statistical mechanics is extremely important as it provides a guideline for the development of empirical models.

Approaches to Derive Local Composition Mixing Rules

The early derivations of local-composition mixing rules were based on lattice theories and/or empirical formulations for internal, Helmholtz, or Gibbs energy (33, 42 - 44). The derivation by Kemeny and Rasmussen (44), based on lattice partition functions

Table I. Binary Systems Studies with Equation of State Local-Composition Mixing Rules

Investigators	Year	Systems	Type
Huron and Vidal (25)	1979	1-8	VLE
Heyen (26)	1981	9-11	VLE
Won (27)	1981	12, 13	VLE
Whiting and Prausnitz (29)	1982	14	VLE
Won (31)	1983	5, 15-18	VLE
Mathias and Copeman (33)	1983	1, 10, 11, 119-24	VLE, LLE
Mollerup (35)	1983	1, 5, 20, 25-34	VLE
Ludecke and Prausnitz (36)	1985	20, 22, 23, 25, 35-41	VLE, LLE
Li et al. (39)	1984	1, 5, 6, 14, 16, 19, 20, 25, 29, 31, 42-46	VLE, density

1. acetone-water
2. carbon dioxide-ethane
3. ethane-acetone
4. ethane-methyl acetate
5. methanol-carbon dioxide
6. propane-ethanol
7. methanol-1,2 dichloroethane
8. acetone-cyclohexane
9. ethanol benzene
10. butanol-water
11. carbon dioxide-methane
12. carbon dioxide-napthalene
13. ethylene-napthalene
14. water-methane
15. carbon dioxide-n-octane
16. carbon dioxide-n-decane
17. carbon dioxide-butanol
18. water-carbon dioxide
19. methane-n-decane
20. methanol-benzene
21. isobutylene-methanol
22. benzene-water
23. hexane-water
24. 1-methylnapthalene-water
25. methanol-water
26. ethanol-water
27. 2-propanol-water
28. acetone-carbon dioxide
29. methanol-n-hexane
30. methanol-cyclohexane
31. ethanol-n-hexane
32. ethanol-n-heptane
33. ethanol-cyclohexane
34. ethanol-methylcyclohexane
35. pyridine-water
36. pyridine-benzene
37. water-cyclohexane
38. water-heptane
39. water-octane
40. phenol-water
41. propane-water
42. ethane-butane
43. carbon dioxide-benzene
44. ammonia-water
45. ethanol-n-decane
46. ethane-water

and two-liquid theory, provided a thorough explanation of the assumptions required to obtain the Wilson equation. Following the quasichemical approach, the combinatorial factor in the configurational partition function is expressed in terms of the count of contacts and a correction factor, denoted by h. To meet the proper random-mixture limit, an expression was derived for h using a Taylor series expansion about complete randomness. The lattice theory derivation clearly shows that the Wilson equation is most valid for near-random mixtures of similar sized molecules. Kemeny and Rasmussen also presented an equation for the combinatorial factor which leads to expressions for local area fractions, as used in the successful UNIQUAC activity coefficient model (45). Lattice theories have provided little further progress in developing improved mixing rules since the mid 1970's.

More recently, Lee et al. (46) derived general relations for local compositions in terms of radial distribution functions and related potentials of mean force to the Wilson formulation. This work represents a significant contribution, bridging the gap between conventional local-composition mixing rules and rigorous fluid-phase statistical mechanics. Mansoori and Ely (47) have further derived a unified treatment of the local-composition concept for fluid-phase mixtures. The energy, pressure, and compressibility equations were expressed in terms of local particle numbers. For example, the internal energy is given by

$$E = -NkT/2 \sum_i \sum_j x_i \int_0^\infty n_{ji}(r)\ (dU_{ji}(r)/dr)\ dr \qquad (3)$$

Various approximations for n_{ji} were analyzed by Mansoori and Ely. Their general procedure is based on equating the mixture potential energy function and local particle number to that of a hypothetical pure fluid. An advantage of the Mansoori and Ely approach is that microscopic approximations to the radial distribution functions can be translated into expressions for local particle numbers and mixing rules. The relations, based on the compressibility equation, are novel and merit further evaluation. One probable disadvantage for industrial application is that the mixing-rule expressions are generally not explicit with respect to the mixture parameters, requiring significant additional computational expense.

Seaton and Glandt (48) have explored local compositions for mixtures described by the adhesive intermolecular potential. The unique feature of their work is the unambiguous definition of nearest neighbors since attractive forces are extremely short-ranged and neighboring molecules are in contact. The local particle numbers were shown to be

$$n_{ji} = \pi \rho_i \lambda_{ji} \sigma_{ji}^3 / 3 \qquad (4)$$

where the λ_{ji} are obtained from the solution of three quadratic algebraic equations for a binary mixture. The Wilson equation was compared to the adhesive-potential model and concluded to be inaccurate in describing local compositions for mixtures of unequal-sized molecules. This conclusion is similar to that reached by Gierycz and Nakanishi (49) and suggests that more attention should be given to size differences in local-composition formulations.

Comparison of Local Composition Formulations to Computer Simulation

Nakanishi and co-workers have used molecular dynamics and Monte Carlo simulations to provide useful information for the local structure of various kinds of mixtures (49 - 52). Insight into local fluid structure has been obtained by comparing local compositions calculated from computer simulations to local compositions predicted from semi-empirical models (like Wilson, 25).

The number of particles j around a particle i is determined from the simulation results as

$$n_{ji} = 4\pi N_j/V \int_0^R r^2 g_{ji}(r)dr \tag{5}$$

The local composition of j around i is

$$x_{ji} = n_{ji} \;/\; (\Sigma_k n_{ki}) \tag{6}$$

Nakanishi and co-workers have set the upper limit of integration in Equation (5) to the distance of the first peak in the radial distribution function. Lennard-Jones potential and Lorentz-Berthelot combining rules were chosen for calculations on mixtures with varied size and energy parameters. The Lennard-Jones energy parameters were substituted for the potentials of mean force in the local-composition formulations.

In their most extensive works (49, 52), molecular dynamic calculations were made over a range of overall mole fractions and the local compositions were compared to the Wilson (25) and Renon and Prausnitz (42) formulations. The Wilson equation was found to predict local compositions poorly for the mixtures described above, while the Renon and Prausnitz equation with $\alpha = 0.4$ showed better agreement (α is the nonrandomness parameter). Since the Boltzmann factors of these two models differ only by a factor of 0.4, the Wilson formulation over-predicts the local-composition effect. For mixtures of unequal-sized molecules the Renon-Prausnitz model predictions are less accurate. Firm conclusions cannot be drawn on the ability of these models to correlate the activity-coefficient behavior of real fluids, but these studies do provide a guideline for model development. Gierycz and Nakanishi (49) proposed improvements to the Renon-Prausnitz model based on their comparisons. Evaluation of these improvements using real mixtures of industrial interest is a valuable next-step.

While a number of assumptions were used in Nakanishi's comparisons (setting R to the first peak of the radial distribution function and substituting Lennard-Jones energy parameters for potentials of mean force), this work contributes to the understanding of local liquid structure and local-composition formulations.

Additional work to assess the impact of deficiencies in mixing rules on properties used in process design, such as activity coefficients, is needed. Recently, Hu et al. (53) have compared their equation-of-state predictions to the computer-simulation results of Shing and Gubbins (54). Of particular interest is the direct comparison of residual chemical potentials at low dilution, a stringent test of the equation of state.

van der Waals Partition Function and Square-Well Fluids

In an on-going series of papers, Sandler and co-workers (55-57) have attempted to use a combination of theory and computer simulations to develop fundamentally-based equations of state. They have used the generalized van der Waals partition function to simplify the development of new models and to examine the assumptions in existing models. Further, they have performed Monte Carlo simulations on square-well fluids to develop new models of engineering utility.

The analysis by Sandler and co-workers of "local-compositions" and their effect on equation-of-state mixing rules is of particular importance to the present review. A series of interesting observations and conclusions were presented in their studies.

- The square-well potential is valuable for investigative studies because it is realistic, yet has a well-defined, finite attractive range.
- The Boltzmann factor that is commonly used to model local compositions is only correct at low densities. At high densities the Boltzmann factor overpredicts the local compositions since the fluid structure is predominantly determined by repulsive forces.
- Local compositions are best defined in terms of local particle numbers, which is equal to an integral over the radial distribution function. In the case of infinite range potentials (e.g., Lennard-Jones) it is probably better to model the configurational energy directly.
- The vdW-1 and local-composition models are poorly named since both can be viewed to arise from local-composition assumptions. The key difference between the two is that the first class of models arises from the assumption that the coordination number varies with composition, whereas in the second class the coordination number is fixed.
- The density-dependent local-composition models of Whiting and Prausnitz (29, 30), Mollerup (31) and Won (31) are qualitatively in error since they assume that the effect of the attractive forces is more important at high densities than at low densities.

We present our viewpoint on the statements in the following section:

Analysis

There seems to be on-going confusion about various concepts and conflict regarding results and assumptions among the several approaches to model development.

vdW-1 Vs. Random Mixing. Many researchers assume or imply that the vdW-1 mixing rules result from the assumption that the mixture is random. In terms of distribution function theory, a random mixture is one in which all the ji pair distribution functions are identical. Leland and Chappelear (12) have shown that this assumption leads to a mixing rule that depends on the pair potential. If all pairs in the mixture obey the Lennard-Jones potential, the mixing rule that follows is

$$\varepsilon = \frac{(\sum_i \sum_j x_i\, x_j\, \varepsilon_{ji}\, \sigma_{ji}^{\;6})^2}{(\sum_i \sum_j x_i\, x_j\, \varepsilon_{ji}\, \sigma_{ji}^{\;12})} \tag{7}$$

$$\sigma^3 = \frac{(\sum_i \sum_j x_i\, x_j\, \varepsilon_{ji}\, \sigma_{ji}^{\;12})^{1/2}}{(\sum_i \sum_j x_i\, x_j\, \varepsilon_{ji}\, \sigma_{ji}^{\;6})^{1/2}} \tag{8}$$

The above mixing rules lead to extremely poor predictions even when there are small differences in molecular sizes and energies of attraction.

The vdW-1 mixing rules results from a considerably more appealing assumption. It is assumed that the ij radial distribution function can be approximated as

$$g_{ji}(r, \rho_1, \rho_2, \ldots \rho_n, T) = g^{(0)}\left(\frac{r}{\sigma_{ji}}, \rho\bar{\sigma}^3\right)\left[1 + \left(\frac{\varepsilon_{ji}}{kT}\right)\Psi_1\left(\frac{r}{\sigma_{ji}}, \rho\bar{\sigma}^3\right) + \left(\frac{\varepsilon_{ji}}{kT}\right)^2 \Psi_2\left(\frac{r}{\sigma_{ji}}, \rho\bar{\sigma}^3\right) + \ldots\right] \tag{9}$$

Equation (9) assumes that the ji distribution function at dimensionless distance (r/σ_{ji}) is the same as that of a pure hard-sphere fluid at the same reduced density as that of the mixture with perturbation terms to account for the attractive forces. The power of the model is that the functions $g^{(0)}$ and Ψ need not be known. Now it can be shown that the vdW-1 mixing rules (Equations (1) and (2)) are exact up to order 1/T (12).

The vdW-1 mixing rules are clearly a simplification. Terms of order $(1/T)^2$ and greater are neglected, which should introduce errors at low temperatures. Additionally, the model assumes that $g^{(0)}$ at the point of contact is the same for all ji pairs, which is known to be wrong at high densities. However, the reasonableness and the clarity of the assumptions must also be recognized as we attempt to develop improved mixing rules (58).

The Predictive and Correlative Power of the vdW-1 Mixing Rules. In addition to their foundation in theory, the vdW-1 mixing rules have provided good predictive and correlative capability. We have already mentioned the work of Henderson and Leonard (13, 14) who found that the vdW-1 mixing rules provided good predictions of machine-simulation results for hard-sphere and Lennard-Jones mixtures. We reiterate that comparison with computer simulations is a strong test of theoretical formulations since the parameters are known and cannot be adjusted. However, these comparisons should be made in the dilute region not (as is usual) for equimolar mixtures.

Sandler and co-workers have obtained machine-simulation results for both pure fluids and mixtures interacting with the square-well potentials (56, 57), thus enabling a test of the simple vdW-1 mixing rules. These simulations are for equisized molecules. Therefore the vdW-1 mixing rules reduce to the simple form

$$\varepsilon = \sum_i \sum_j x_i x_j \varepsilon_{ji} \tag{10}$$

The results for the vdW-1 predictions are presented in Figures 1 and 2. Figure 1 shows that in the case where the temperature is relatively high (note that $\varepsilon/kT_c \simeq 0.8$), the vdW-1 mixing rules provide excellent predictions of the configurational energy at all compositions and densities. In the case where the temperature is lower (Figure 2), the vdW-1 mixing rules provide good predictions at high densities, and underpredict the configurational energy at low densities.

A better mixing rule for low-density mixtures is a local-composition model based on the second virial coefficient:

$$E = \sum_i \sum_j x_i x_j E(\varepsilon_{ji}; T,\rho) \tag{11}$$

Equation (11) is exact at zero density (55). The dashed line in Figure 2 shows that the low-density mixing rule provides improved predictions of the mixture data. But the model still underpredicts the data. This is perhaps caused by mixture conditions within the unstable region of the fluid, which is supported by the scatter in the simulation data.

Several investigators have pointed out that the simple vdW-1 mixing rules provide good predictions for systems containing equi-sized molecules with differing energies of interaction (for example, see Fischer, 58). However, the vdW-1 form is inadequate for the more difficult mixtures containing molecules with large size differences (54, 48).

It should be noted that it is not necessary to use the vdW-1 approximation for the repulsive part of an equation of state since theoretically-based, tractable models are available for unequal-sized hard-sphere mixtures (60, 38). Hu et al. (53) have used the Mansoori-Carnahan Starling-Leland equation (38) for the repulsive term and obtained good agreement with computer-simulation data.

Applications to Important Industrial Equations of State. It is useful to apply the radial-distribution-function framework for local compositions (46, 55) to the important industrial equations of state. This allows examination of the local compositions implied by the practically-successful vdW-1 mixing rules and those of the attempts to improve them (29 - 31). Perhaps more important, there are indications that the simplifications inherent in the pure-fluid equations of state must be considered when applying the mixing rules suggested by a rigorous theoretical framework.

Consider the internal energy obtained by applying the vdW-1 mixing rules to the common cubic equations of state

$$E = - F_v \sum_i \sum_j x_i x_j \frac{\partial a_{ji}/T}{\partial\, 1/T} \tag{12}$$

where the function F_v is dependent on the form chosen for the attractive term of the equation of state. For example:

van der Waals: $$F_v = \rho \tag{13}$$

Redlich-Kwong: $$F_v = \frac{1}{b} \ln(1 + \rho b) \tag{14}$$

Peng-Robinson: $$F_v = \frac{1}{2\sqrt{2}b} \ln \left[\frac{1 + (1+\sqrt{2})\, b\rho}{1 + (1-\sqrt{2})\, b\rho} \right] \tag{15}$$

The Redlich-Kwong and Peng-Robinson forms for F_v reduce to the simple linear dependence on density (Equation (13)) in the limit of low density. (See Figure 3.)

Now consider the statistical mechanical expression for the internal energy of a system where the total intermolecular potential is pairwise additive.

$$E = \rho/2 \sum_i \sum_j x_i x_j \int_0^\infty u_{ji}(r) g_{ji}(r)\, 4\pi r^2 dr \tag{16}$$

As $\rho \to 0$, the radial distribution function $g_{ji}(r)$ approaches the Boltzmann factor

$$\lim_{\rho \to 0} g_{ji}(r) = e^{-u_{ji}(r)/RT} \tag{17}$$

If Equation (17) is substituted into Equation (16), it can be shown (46) that the internal energy at low densities is related to the temperature derivative of the second virial coefficient:

$$E = R\rho \sum_i \sum_j x_i x_j \frac{\partial B_{2ji}}{\partial\, 1/T} \tag{18}$$

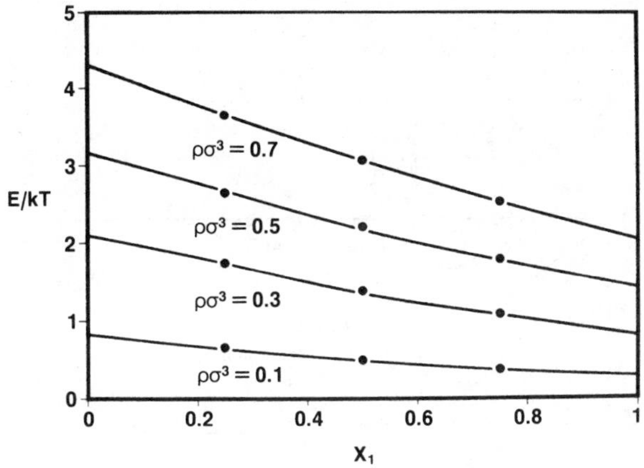

Figure 1. vdW-1 predictions of square-well configuration energies. Case I: $\varepsilon_{11}/kT = 0.4$; $\varepsilon_{22}/kT = 0.8$

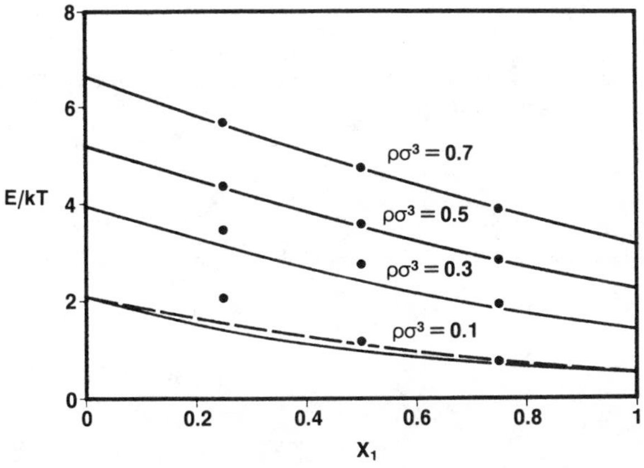

Figure 2. vdW-1 predictions of square-well configuration energies. Case II: $\varepsilon_{11}/kT = 0.6$; $\varepsilon_{22}/kT = 1.2$

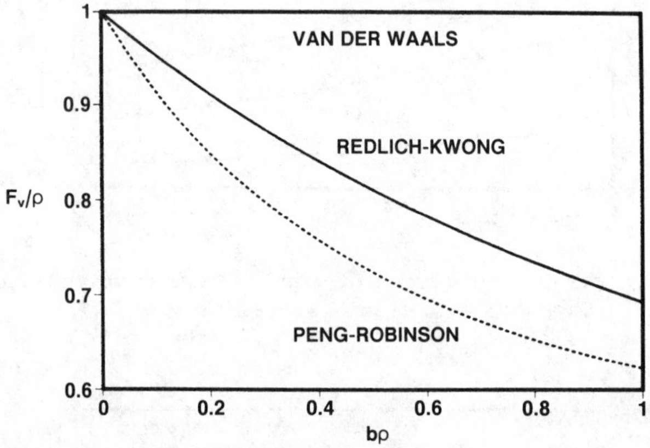

Figure 3. Density dependence of attractive term - Simple equations of state

Comparing Equation (18) with the low-density limit of Equation (12) ($F_v \simeq \rho$), we obtain an expression for a_{ji} in terms of the second virial coefficient.

$$a_{ji} = -RT \int_{1/T=0}^{1/T} \frac{\partial B_{2ji}}{\partial\, 1/T}\, d(1/T)$$

$$= -RT \left[B_{2ji} - B_{2ji}(T = \infty) \right]$$

$$= -RT \left[B_{2ji} - b_{ji} \right] \qquad (19)$$

Thus, as is well known, a_{ji} arises from the attractive portion of the ji pair potential. Equations (16) and (17) show that, at low densities, the simple equations qualitatively describe the local compositions resulting from the Boltzmann factor.

We can now get some idea of the high-density radial distribution function assumed by the simple equations of state. The van der Waals equation assumes that there is no effect of density. The Redlich-Kwong and Peng-Robinson models assume that the low-density Boltzmann factor results in an overprediction of the high-density configurational energy, but this can be corrected by a simple function of density, F_v/ρ. We emphasize that this is a nonunique interpretation.

More important to the discussion at hand, the vdW-1 mixing rules assume that the local compositions are independent of density since the correction factor is the same for all ji pairs. This point has been noted by Sandler (55) who also states that this behavior is inconsistent with the investigations of Chandler and Weeks (61) and Weeks et al. (62) who have shown that the structure of a high-density fluid is largely determined by the repulsive forces.

However, these simple equations of state, together with the vdW-1 mixing rules, have resulted in reliable and frequently accurate predictions for a wide variety of nonpolar mixtures including those whose components have large differences in their intermolecular forces (63). It is important to pursue the nature and impact of these differences to guide improvements to the vdW-1 mixing rules for mixtures where they are obviously inadequate (e.g., polar-nonpolar binaries like methanol-benzene).

We therefore review the assumptions leading to Equation (19): a_{ji} is not strictly a low-density parameter. In fact, it is more correctly a high-density parameter since its value is determined by fitting vapor pressure or phase equilibrium data, for which the variation of the liquid-phase fugacities usually dominate. Further, we also note that the second virial coefficients predicted by the cubic equations are usually less negative than the experimental values, indicating that the effect of the attractive forces is underpredicted at low densities.

Perhaps a better interpretation of Equations (16) and (17) is that the a_{ji} arises from an effective pair potential which permits a reasonable description of the internal energy at high densities. However, these simple models do not reveal the form they assume for microscopic quantities, since we cannot deduce the form of a function from a single value of its integral.

This uncertainty in the ji distribution functions for the vdW-1 case makes an analysis of the density-dependent local-composition models very tenuous indeed. One argument, offered by Sandler and co-workers, suggests that the local compositions caused by the attractive forces are highest at low densities and decrease with density since high-density structure is determined by the repulsive forces. According to this argument the density-dependent local-composition models (29 - 31) are qualitatively wrong since they assume that the local-composition effect is zero at low densities and increases with density. Further comments are needed to put this statement into perspective.

It should first be noted that the density-dependent local-composition models do not predict random behavior at low densities since they describe second virial coefficient behavior correctly (Equation 19). Now the question is: Is it qualitatively correct to assume that the local-composition effect increases with density?

Sandler's study, consistent with other studies, shows us that the Wilson Boltzmann factors do overestimate the local composition effect at high densities. The computer simulations of Nakanishi and co-workers have shown that there is some local-composition effect at high densities due to the attractive forces and this clustering is approximately described by the NRTL model. Thus there is some justification for the model at high densities and the density-dependent local-composition models at least meets macroscopic boundary conditions at low and high densities.

We stress that "local-composition" is an ambiguous and intermediate quantity. It is perhaps more appropriate to evaluate models according to their description of more meaningful properties like the configurational energy.

The Future

vdW-1 mixing rules are reasonably based in theory and have enabled reliable predictions in many nonpolar systems. Further work using vdW-1 mixing rules as a basis for extension appears to be warranted.

Perturbation theory suggests that separate mixing rules are needed for different intermolecular effects. Approaches based on perturbation theory should contribute to the continued development of new mixing rules.

Most local composition mixing rule formulations now resemble Wilson's equation, which both overestimates the local-composition effect and becomes less accurate for mixtures with increasing size differences. New ideas are needed to improve these deficiencies. For example, Ely (64) has recently shown for square-well mixtures

$$n_{ji} \simeq e^{-\beta\varepsilon_{ji}} \left\{ \delta_{ji} - \beta\rho_j \left(\frac{\partial\mu_j}{\partial\rho_i} \right)_{\text{Hard Sphere}} + 4\pi R^3{}_{ji} \right\} \tag{20}$$

which can be substituted into the energy equation (Equation 3) to derive the mixture equation of state. Ely's development incorporates hard-sphere size effects into the local-composition formulation. Further work in this direction is strongly encouraged.

Conclusions

The capability of equations of state to describe asymmetric mixtures has improved substantially over the past five years. The old rule of thumb that one must use activity coefficient models for mixtures with polar components is beginning to fade away. This trend will benefit chemical process design and optimization, as more than one approach has been needed, in some cases, to simulate one entire process.

Equations of state with local-composition mixing rules offer short-term potential to describe highly polar-asymmetric systems in the same (empirical) manner as activity coefficient models. These models now largely resemble Wilson's equation and new ideas are needed to develop improved formulations.

Computer simulation studies are being used to provide information on local fluid structure. Some improvements to mixing rules have been suggested, however, have not yet been evaluated on real systems. Further work to improve mixing rules for application to real systems based on computer simulation is badly needed. While comparisons based on local compositions can provide some guidance in developing better models, analysis of macroscopic properties, such as activity coefficients and internal energy, is recommended to avoid potential ambiguities and pitfalls. A unified formulation of the energy, pressure and compressibility equations in terms of local particle numbers now exists and can straightforwardly be used to develop mixture equations of state for new formulations.

Acknowledgments

We wish to thank Professors S. Sandler, J. M. Prausnitz and L. L. Lee and Dr. J. F. Ely for sending us preprints of their unpublished papers, and Professor Prausnitz for his helpful comments.

List of Symbols

a	attractive constant in van der Waals-based equation of state
b	size constant in van der Waals-based equation of state
B	second virial coefficient
E	internal energy
g	radial distribution function
k	Boltzmann's constant
n	local particle number
N	number of molecules
R	limit of integration
T	absolute temperature
u	intermolecular pair potential

V	volume
X	fluid-phase mole fraction
ß	1/KT
δ	Kronecker delta
ε	energy parameter in potential function
μ	chemical potential
ρ	density
σ	hard-sphere diameter

Literature Cited

1. Sandler, S. I. Proceedings of the National Science Foundation Workshop, "Thermodynamic Needs for the Decade Ahead - Theory and Experiment," Washington, D. C., October 28-29, 1983.
2. Ely, J. F.; Baker J. K. "A Review of Supercritical Extraction;" NBS Technical Note 1070, 1983.
3. Chappelear, P. S.; Chen, R. J.; Elliot, D. E. Hydrocarbon Processing, 1977, 215-217.
4. George, B. A. Proceedings of the Sixty-first Annual Convention, Gas Processors Association, Dallas, March 15-17, 1982.
5. Miller, E. J.; Geist, J. M. Presented at the Joint Meeting of the Chemical Industry Engineering Society of China and the AIChE, Beijing, China, 1982.
6. Abrams, D. S.; Seneci, F.; Chueh, P. L.; and Prausnitz, J. M. Ind. Eng. Chem. Fundam., 1975, 14:52-54.31
7. van der Waals, J. D. Z. Phys. Chem., 1890, 5:133-173.
8. Soave, G. Chem. Eng. Sci., 1972, 27:1197.
9. Peng, D. Y.; Robinson, D. B. Ind. Eng. Chem. Fundam., 1976, 15:59-64.
10. Leland, T. W.; Chappelear, P. S.; Gamson, B. W. AIChEJ, 1962, 8:482-489.
11. Reid, R. C.; Leland, T. W. AIChEJ, 1965, 11:228-237.
12. Leland, T. W.; Chappelear, P. S. Ind. Eng. Chem., 1968, 60(7):15-43.
13. Henderson, D.; Leonard, P. J. Proc. Natl. Acad. Sciences, 1970, 67:1818-1823
14. Henderson, D.; Leonard, P. J. Proc. Natl. Acad. Sciences, 1971, 68:632-635
15. Vidal, J. Chem. Eng. Sci., 1978, 33:787-791.
16. Stotler, H. H.; Benedict, M. Chem. Eng. Prog. Symp. Ser., No. 6, 1953, 49:25-36.
17. Benedict, M.; Webb, G. B.; and Rubin, L. C. J. Chem. Phys., 1940, 8:334.
18. Benedict, M.; Webb, G. B.; and Rubin, L. C. Chem-Eng. Prog., 1951, 47:419.
19. Prausnitz, J. M.; Gunn, R. D. AIChEJ, 1958, 4:430-435.
20. Plocker, U.; Knapp, H.; Prausnitz, J. M. Ind. Eng. Chem. Proc. Des. Dev., 1978, 17:324.
21. Radosz, M.; Lin, H. M.; and Chao, K. C. Ind. Eng. Chem. Proc. Des. Dev., 1982, 21:653-658.
22. Lee, T. J.; Lee, L. L.; Starling, K. E. "Equations of State in Engineering and Research," 1979; Meeting of the American Chemical Society, September 11-14, 1978, 125-141.
23. Henderson, D. "Equations of State in Engineering and Reasearch," 1979, Meeting of the American Chemical Society, September 11-14, 1978, 1-30.

24. Donohue, M. D.; Prausnitz, J. M. AIChEJ, 1978, 24:849-860.
25. Wilson, G. M. Contribution No. 29, 1972, Center for Thermochemical Studies, Brigham Young University, March.
26. Huron, M. J.; Vidal, J. Fluid Phase Equilibria, 1979, 3:255-271.
27. Heyen, G. Proceedings of the 2nd World Congress of Chemical Engineering, October 4-9, 1981, 41-46.
28. Won, K. W. Presented at the Annual AIChE Meeting, New Orleans, November 8-12, 1981.
29. Whiting, W. B.; Prausnitz, J. M. Paper presented at Spring National Meeting, American Institute Chemical Engineers, Houston, Texas, April 5-9, 1981.
30. Whiting, W. B.; Prausnitz, H. M. Fluid Phase Equilibria, 1982, 9:119-147.
31. Mollerup, J. Fluid Phase Equilibria, 1981, 7:121-138.
32. Won, K. W. Fluid Phase Equilibria, 1983, 10:191-210.
33. Wilson, G. M. J. Am. Chem. Soc., 1964, 86:127-130.
34. Mathias, P. M.; Copeman, T. W. Presented at the Third International Conference on Fluid Properties and Phase Equilibria for Chemical Process Design, Callaway Gardens, Georgia, April 10-15, 1983. Also published in Fluid Phase Equilibria, 13:91-108.
35. Dimitrelis, D.; Prausnitz, J. M. Presented at Fall Annual Meeting, American Institute of Chemical Engineering, Los Angeles, CA, Nov. 14-19, 1982.
36. Mollerup, J. Fluid Phase Equilibria, 1983, 15:189-207
37. Ludecke, D.; Prausnitz, J. M. Fluid Phase Equilibria, 1985, 22:1-19.
38. Mansoori, G. A.; Carnahan, N. F.; Starling, K. E.; Leland, T. W. J. Chem. Phys., 1971, 54:1523-1525.
39. Larsen, E. R.; Prausnitz, J. M. AIChEJ, 1984, 30:732-738.
40. Li, M. H.; Chung, F. T.; Lee, L. L.; Starling, K. E. Presented at the Fall Annual Meeting, American Institute of Chemical Engineers, San Francisco, CA, November 25-30, 1984.
41. Chung, T. H.; Khan, M. M.; Lee, L. L.; Starling, K. E. Fluid Phase Equilibria, 1984, 17:351.
42. Renon, H.; Prausnitz, J. M. AIChEJ, 1968, 14:135-144.
43. McDermott, C.; Ashton, N. Fluid Phase Equilibria, 1977, 1:33-35.
44. Kemeny, S.; Rasmussen, P. Fluid Phase Equilibria, 1981, 7:197-203.
45. Abrams, D. S.; Prausnitz, J. M. AIChEJ 1975, 21:116-128.
46. Lee, L. L.; Chung, R. H.; and Starling, K. E. Fluid Phase Equilibria, 1983, 12:105-124.
47. Mansoori, G. A.; Leland, Jr., T. W. J. Chem. Soc., Faraday Trans. II, 1972, 68:320-344.
48. Seaton, N. A.; Glandt, E. D. Presented at the Fall Annual Meeting, American Institute of Chemical Engineers, San Francisco, CA, November 25-30, 1984.
49. Gierycz, P.; Nakanishi, K. Fluid Phase Equilibria, 1984, 16:255-273.
50. Nakanishi, K.; Tanaka, H. Fluid Phase Equilibria, 1983, 13:371-380.
51. Nakanishi, K.; Taukabo, K. J. Chem. Phys., 1979, 70:5848-5850.

52. Gierycz, P.; Tanaka, H.; Nakanishi, K. Fluid Phase Equilibria, 1984, 16:241-253.
53. Hu, Y.; Ludecke, D.; Prausnitz, J. M. Fluid Phase Equilibria, 1984, 17:217-241.
54. Shing, K. S.; Gubbins K. E. Mol. Phys., 1983, 49:1121-1138.
55. Sandler, S. I. Fluid Phase Equilibria, 1985, 19:233-257.
56. Lee, K. H.; Lombardo, M.; Sandler, S. I. "The Generalized van der Waal's Partition Function II. Application to Square-Well Fluid," Fluid Phase Equilibria, 1985, in press.
57. Lee, K. H.; Sandler, S. I.; Patel, N. C. "The Generalized van der Waal's Partition Function III. Local-Composition Models for a Mixture of Equal Size Square-Well Molecules," Fluid Phase Equilibria, 1985, in press.
58. Mansoori, G. A.; Ely, J. F. "Statistical Mechanical Theory of Local Compositions," submitted to Fluid Phase Equilibria, 1984,
59. Fischer, J. Fluid Phase Equilibria, 1983, 10:1-7.
60. Lebowitz, J. L.; Rowlinson, J. S. J. Chem. Phys., 1964, 41:133-138.
61. Chandler, D.; Weeks, J. D. Phys. Rev. Lettrs., 1970, 24:849
62. Weeks, J. D.; Chandler, D.; Anderson, H. C. J. Chem. Phys., 1971, 54:5237.
63. Oellrich, L.; Plocker, V.; Prausnitz, J. M.; Knapp, H. International Chemical Engineering, 1981, 21:1-16.
64. Ely, J. F., 1985, to be submitted for publication.

RECEIVED November 5, 1985

18

Calculation of Fluid-Fluid and Solid-Fluid Equilibria in Cryogenic Mixtures at High Pressures

Ulrich K. Deiters

Lehrstuhl für Physikalische Chemie II, Ruhr-Universität, D-4630 Bochum 1, Federal Republic of Germany

A non-cubic equation of state, which includes quantum corrections and special corrections for molecular size differences, is used to correlate fluid-fluid phase equilibria (vapour-liquid, liquid-liquid, and gas-gas) in binary cryogenic mixtures (containing H_2, CO, noble gases, etc.) up to 2000 bar. Volumetric properties of these mixtures are predicted successfully. The cubic Redlich-Kwong equation of state is shown to perform less satisfactorily. The thermodynamic functions of the mixtures are calculated from modified versions of either one-fluid theory or mean density approximation. An expression for the Gibbs energy of a pure solid is derived; from this it is possible to predict solid-fluid equilibria, including solid-liquid-gas three-phase lines, from experimental fluid-fluid equilibrium data, and vice versa.

The calculation of phase equilibria in binary mixtures at elevated pressure requires much more computational effort than calculations of low pressure equilibria. At high pressure it is no longer permissible to regard one or more of the phases in equilibrium as ideal, and theoretical approaches based on activity coefficients become inefficient, because appropriate standard states are lacking. On the other hand, equations of state can form a proper basis for phase equilibrium calculation methods for several reasons:

- They are applicable to liquid, vapour, and supercritical states alike, so that conceptual difficulties in the vicinity of critical points, which are so often encountered with activity coefficient methods, are completely avoided.
- Because of their built-in consistency it is possible to calculate vapour-liquid equilibria, liquid-liquid equilibria, and all transitions between these two kinds from the same input information.
- Calculations with equations of state not only yield the equilibrium phase compositions, but also densities and heats of vaporization.

It is the aim of this work to demonstrate that modern equations of state are not only capable of correlating fluid-fluid equilibria, but also of predicting PVT data and solid-fluid equilibria of mixtures (solid-supercritical fluid equilibria, melting) successfully.

0097-6156/86/0300-0371$06.00/0

Equations of State

Pure Substances. For the calculations described below we have used two van der Waals type equations of state, i.e. equations consisting of distinct repulsion and attraction terms, namely:

- the Redlich-Kwong (RK) equation (1) as a representative of the widely used class of cubic equations of state:

$$P = P_{rep} - P_{att} = \frac{RT}{V_m - b} - \frac{a}{\sqrt{T}V_m(V_m+b)} \qquad (1)$$

(Here a and b are the (substance-specific) parameters of attraction and repulsion.)

- and our own equation of state (2):

$$P = \frac{RT}{V_m}\left[1 + cc_\circ\frac{4\xi-2\xi^2}{(1-\xi)^3}\right] - \frac{Ra}{b}\rho^2\tilde{T}_{eff}\left[\exp(\tilde{T}_{eff}^{-1}) - 1\right]I_1(\rho,c) \qquad (2)$$

with

$$\xi = \frac{\pi\sqrt{2}}{6}\rho \qquad \rho = \frac{b}{V_m} \qquad \text{(reduced densities)}$$

$$\tilde{T}_{eff} = \frac{\tilde{T}-\lambda\rho}{Y(\rho,c)} \qquad \tilde{T} = \frac{Tc}{a} \qquad \text{(reduced temperatures)}$$

$$c_\circ = 0.6887 \qquad \lambda = -0.07c$$

The other expressions in Equation 2 are functions of ρ and c, which have been explained elsewhere (2). This equation of state is clearly non-cubic; its three parameters a, b, and c may be understood as characteristic temperature, covolume, and anisotropy.

At low temperatures, some light gases such as helium, neon, hydrogen etc. show deviations from classical mechanics, which are usually referred to as quantum effects. The larger part of these deviations can be explained by restrictions of the particle motions: at high densities each molecule is more or less confined to a cell by the repulsion of its neighbour molecules. Then the translational eigenstates are no longer closely spaced, so that it is not permissible to calculate the partition function by integration. It is possible, however, to represent the quantum corrected partition function as a product of the classical partition function and a correction function (3):

$$Q_{quant} = Q_{class}\,q_{corr}^{3N} \qquad (3)$$

For small deviations from classical behaviour the correction function can be approximated by

$$q_{corr} = 1 - \frac{y}{2} \qquad (4)$$

with $y = \Lambda/L$ (reduced wavelength)

$$\Lambda = \sqrt{\frac{h^2}{2\pi mkT}} \qquad \text{(thermal de Broglie wavelength)}$$

$$L = \sqrt[3]{\frac{1}{8N} V_f} \qquad \text{(cell size)}$$

The cell size depends on the free volume of the fluid. For large reduced wavelengths the correction function can be obtained from a series expansion; the coefficients are given in Table I:

$$\ln q_{corr} = \sum_{j=1}^{13} r_j y^j \tag{5}$$

Table I. Expansion Coefficients of the Quantum Correction Function

j	r_j	j	r_j
0	0	7	+8.663 291 244 9
1	-0.499 951 749 27	8	-8.727 837 827 2
2	-0.128 672 732 48	9	+5.742 346 562 0
3	+0.026 360 453 309	10	-2.444 507 627 4
4	-0.552 534 071 60	11	+0.648 968 116 33
5	+2.240 993 820 8	12	-0.097 758 376 489
6	-5.570 201 592 1	13	+0.006 387 718 839

The free volume depends on the equation of state. For van der Waals type equations, which consist of a repulsion and an attraction term, the free volume is defined by

$$\ln \frac{V_f}{V} = \int_{\infty}^{V} \left(\frac{P_{rep}}{nRT} - \frac{1}{V} \right) dV \tag{6}$$

The free volume associated with the equation of state 2 is given by

$$V_f = V \exp\left(-cc_0 \frac{4\xi - 3\xi^2}{(1-\xi)^2} \right) \tag{7}$$

From Equations 3, 5, and 6 it is evident that the quantum corrected version of any van der Waals type equation of state is given by

$$P = P_{rep}(1 - \sum_j j r_j y^j) - P_{att} \tag{8}$$

Since the dominating expansion coefficients are negative, the quantum correction amounts to an increase of the repulsion pressure. This is in agreement with the perturbation theory results of Singh and Sinha (4). The influence of the quantum correction on the representation of $\overline{P}VT$ data and critical data of pure substances has been discussed elsewhere (3, 5). It should be noted that the introduction of quantum corrections always leads to non-cubic equations of state.

Mixtures. One of the most successful methods of generalizing a pure substance equation of state to mixtures is the one-fluid theory. It amounts to applying the pure substance equation of state to a mixture, using concentration dependent parameters. The concentration dependence is usually contained in a set of interpolation formulas, referred to as mixing rules. For the Redlich-Kwong equation the following quadratic mixing rules have been used:

$$a = x_1^2 a_{11} + 2x_1 x_2 a_{12} + x_2^2 a_{22} \tag{9}$$

$$b = x_1^2 b_{11} + 2x_1 x_2 b_{12} + x_2^2 b_{22} \tag{10}$$

These mixing rules do not imply random mixing (6), nor are they compatible with the van der Waals mixing theory; the Redlich-Kwong attraction parameter is related to the molecular potential well depth and diameter by $a \sim \varepsilon^{3/2}\sigma^3$, whereas the attraction parameter of the van der Waals equation of state is proportional to $\varepsilon^1\sigma^3$. A detailed discussion of the physical meaning of attraction parameters in several equations of state and implications for improved mixing rules have been given by Mansoori in this symposium (7). Here we only want to make the point that other than van der Waals type mixing rules have successfully been used for a long time, and that it is not necessary for a successful mixing rule to be of the van der Waals type.

For Equation 2 the following mixing rules have been used (8):

$$c = x_1 c_{11} + x_2 c_{22} \tag{11}$$

$$b = x_1^2 b_{11} + 2x_1 x_2 b_{12} + x_2^2 b_{22} \tag{12}$$

$$a = x_1 a_{11} + x_2 a_{22} + \frac{2x_1 s_{11} q_2 \Delta a}{1 + \sqrt{1 + 4q_1 q_2 [\exp(\frac{-\Delta a}{T}) - 1]}} \tag{13}$$

with $\Delta a = \frac{2a_{12}}{s_{12}} - \frac{a_{11}}{s_{11}} - \frac{a_{22}}{s_{22}}$ (exchange energy)

and $q_i = \frac{x_i s_{ii}}{x_1 s_{11} + x_2 s_{22}}$ (surface fraction)

The calculation of the effective surface ratios s has been described elsewhere (9) (compare Equation 24). The square root term in Equation 13 accounts for non-randomness according to a quasichemical model (8). It contributes to the equation of state at low temperatures only. In the high temperature limit Equation 13 becomes approximately equivalent to averaging $\varepsilon\sigma^2$; the actual value of the σ-exponent depends somewhat on the size ratio.

The contributions of the quantum effects to the equation of state are averaged according to the following formula:

$$P = P_{rep}(1 - x_1 \sum_j j r_j y_1^j - x_2 \sum_j j r_j y_2^j) - P_{att} \tag{14}$$

with $y_i = \Lambda_i / L$

The reduced wavelengths are calculated separately for each component of the mixture. The effective cell size L is obtained from the mean reduced density using Equations 12, 2, and 7.

If the molecules of a mixture differ very much in size (diameter ratio above 1.6), the mixing rules given above are no longer sufficient. It is possible, however, to extend Equation 2 to mixtures with even larger size ratios by substituting the rigid sphere compressibility factor by an appropriately modified rigid sphere mixture expression (10). Here the expression of Mansoori, Carnahan, Starling, and Leland (11) has been used.

$$P_{rep} = \frac{RT}{V_m}\left[1 + cc_0\frac{E\xi+(3F-E+2)\xi^2-F\xi^3}{(1-\xi)^3}\right] \tag{15}$$

with $E = \dfrac{3(x_1+x_2R_{21})(x_1+x_2R_{21}^2)}{x_1+x_2R_{21}^3} + 1$

and $F = \dfrac{(x_1+x_2R_{21}^2)^3}{(x_1+x_2R_{21}^3)^2} - 1 \qquad R_{21} = \sqrt[3]{\dfrac{b_{22}}{b_{11}}}$ (diameter ratio)

The averaging procedures implied by Equation 14 and the use of Equation 15 constitute a departure from one-fluid theory. Conversely, if formal validity of one-fluid theory is assumed, these two equations would lead to density and temperature dependent parameters.

The parameters a_{12} and b_{12} are estimated from the usual geometric, resp. arithmethic, mean rules. Their exact values are always determined by fitting the equation of state to binary phase equilibrium data.

Phase Equilibria

Fluid-Fluid Equilibria. The thermodynamic conditions of phase equilibrium in a binary fluid mixture are (12):

$$\left.\begin{aligned} P' &= P'' \\ T' &= T'' \\ \mu_i' &= \mu_i'' \qquad i = 1, 2 \end{aligned}\right\} \tag{16}$$

The chemical potentials μ are obtained as derivatives of the Gibbs energy at constant pressure and temperature:

$$\mu_i = \left(\frac{\partial G}{\partial n_i}\right)_{P,T,n_{j\neq i}} \tag{17}$$

In the case of a fluid mixture, the Gibbs energy $G^f(P,T)$ is obtained from the following formula:

$$G^f(P,T) = n_1G_1^\circ(P^\circ,T) + n_2G_2^\circ(P^\circ,T) - \int_{V^\circ}^{V} P(V,T)\,dV + PV - nRT + nRT(x_1\ln x_1 + x_2\ln x_2) \tag{18}$$

The G° denote Gibbs energies of the pure components in the perfect gas state (very small pressure P°, very large volume V°). These terms cancel in the course of the phase equilibrium calculation, if chemical reactions are excluded. The integrand is the equation of state of the mixture. Equation 18 is applied to vapours, liquids, and supercritical states alike. Because of the continuity principle it is possible to integrate the equation of state "through" the vapour pressure line (see Figure 1). Together with Equation 18, 16 represents a system of non-linear equations, which can be solved numerically for the equilibrium phase compositions (13).

The conditions of phase equilibrium can also be formulated using fugacities; the resulting equations are equivalent to those given above, but can be made to appear less complicated, if the functions of the standard state are suppressed and residual properties are used. But in this work the use of chemical potentials and Gibbs energies is preferred for two reasons: Equation 18 is better adapted to our phase equilibrium algorithm (13), and the generalization of our computational method to mixtures with chemical reactions is more straightforward.

Solid-Fluid Equilibria. Since the continuity principle cannot be extended to solid phases, it is neither possible nor desirable to use the same equation of state for fluid and solid phases without creating conceptual difficulties. In some cases a successful correlation of solid-gas equilibria has been accomplished by fitting a fluid equation of state to solid state data, thus treating the solid-gas equilibrium formally as a liquid-vapour equilibrium. But as this approach renders the description of solid-liquid-gas three-phase lines impossible and is not capable of explaining the complete phase diagram of a mixture, it is not used in this work. Instead, a separate equation of state is employed for the solid state. Usually the following relation provides a sufficient description of solid PVT behaviour:

$$V^s(P,T) = V^s(P^{sf},T)\left[1 - \varkappa(P - P^{sf})\right] \tag{19}$$

P^{sf} in Equation 19 denotes the sublimation or melting pressure (whichever is more appropriate at the temperature T); $\varkappa$ is the isothermal solid compressibility. In most cases there will not be sufficient experimental data available to construct a more sophisticated solid equation of state. In order to obtain an expression for the Gibbs energy of a pure solid at arbitrary pressure, which can be used for phase equilibrium calculations, we have to link Equation 19 to 18 in a consistent manner (10):

$$G^s(P,T) = G^f(P^{sf},T) + \int_{P^{sf}}^{P} V^s(P,T)\,dP \tag{20}$$

This equation is obtained by (compare Figure 1)

- determining the Gibbs energy of the fluid phase at the sublimation/melting point,
- realizing that at this point solid and fluid phase have the same molar Gibbs energy,
- integrating the solid equation of state from P^{sf} up to the pressure desired.

Equation 20 in conjunction with 18 enables us to calculate phase equilibria between a fluid mixture and a pure crystalline phase.

Other authors have used a similar concept for treating solid-fluid phase equilibria (14, 15). The formalism proposed here is thermodynamically equivalent, but avoids the use of activity coefficients and metastable or hypothetical reference states. Furthermore our formalism accounts for the compressibility of the solid phase, the influence of which on phase equilibria is not always negligible.

Modified Mean Density Approximation

An alternative mixing theory, which in its original form is only slightly inferior to the one-fluid theory (16), is the mean density approximation. It has the advantage, however, that it can be modified without too much mathematical effort. For example, the introduction of density dependent attraction parameters into Equation 2 would yield an unwieldy formalism within one-fluid theory, but can be made easily within the mean density approximation. Our proposition is to replace the integral in Equation 18, which represents a Helmholtz energy of compression, by

$$A_{comp} = -\frac{\sum_i \sum_j x_i x_j g_{ij} \int P(V_m, T; a_{ij}, b, c_{ij}, \ldots)\, dV_m}{\sum_i \sum_j x_i x_j g_{ij}} \tag{21}$$

where the weight factors g are the contact values of the (i,j)-radial distribution function of a rigid sphere mixture. The average covolume parameter b is again obtained by Equation 12. According to scaled particle theory (17), the values of the g_{ij} are functions of composition and density according to

$$g_{ij} = \frac{1}{1-\xi}\left((C_{ij}+1)^3 - C_{ij}^3\right) \tag{22}$$

with

$$C_{ij} = \frac{\xi}{1-\xi}\,\frac{R_{i1}}{1+R_{ij}}\,\frac{x_1+x_2R_{21}^2}{x_1+x_2R_{21}^3} \tag{23}$$

For a more detailed explanation of the theoretical background of the modified mean density approximation see (18). Furthermore, two other contributions to this symposium are also concerned with modified mean density approximation (19, 20).

At high densities this theory yields the same statistical weight factors for the attractive interaction as Equation 13, but - in contrast to this equation - it also leads to a correct low pressure limit (18). Therefore this modified mean density approximation (Equation 21) is interesting for mixtures of molecules of very different sizes, in which also a large range of densities must be accounted for.

$$\begin{aligned} &\lim_{\xi\to 0} \frac{g_{22}}{g_{11}} = 1 &\qquad &\lim_{\xi\to 1} \frac{g_{22}}{g_{11}} = R_{21}^2 = s_{22} \\ &\lim_{\xi\to 0} \frac{g_{12}}{g_{11}} = 1 &\qquad &\lim_{\xi\to 1} \frac{g_{12}}{g_{11}} = \left(\frac{2}{1+R_{12}}\right)^2 = s_{12} \end{aligned} \tag{24}$$

In this case the surface ratios s have been calculated according to Eduljee and Sandler (9, 21, 22).

It might be argued that the averaging procedure Equation 21 should be written in terms of pressures rather than in terms of Helmholtz energies. It would be difficult, however, to perform a formal integration on the pressure equation to obtain A, G, or μ, whereas the differentiation of the Helmholtz energy equation to obtain the pressure is simple. Finally we want to point out that the modified mean density approximation cannot be applied to cubic equations of state without loosing the "cubic character" of the mixture equation of state.

Results

In all the following calculations two adjustable binary parameters, a_{12} and b_{12}, have been used. Because of the great sensitivity of phase equilibrium calculations for dense mixtures, one adjustable parameter is usually not sufficient. It must be noted, however, that the adjustable parameters are determined only once for a binary mixture; they are considered independent of temperature, density, and composition. Furthermore, the same parameters have always been used even for the calculation of different thermodynamic properties. If not stated otherwise, one-fluid theory has been used.

The hydrogen/methane system is an especially fascinating system for testing computational methods: The experimental data (23) cover a wide range of reduced temperatures, because of the high pressures both coexisting phases have high densities and therefore large deviations from ideality, and at low temperature quantum effects become important. Figure 2 shows three isotherms of the H_2/CH_4 system. The binary parameters of the equations of state have been determined from the middle (130 K) isotherm only. It is evident that Equation 2 leads to a rather good representation of the experimental material, whereas the RK equation is good at low pressures, but has difficulties at low temperatures and high pressures. The 100 K isotherm has also been calculated with the modified mean density approximation (using Equation 2). Its results agree well with the experimental data in the critical region, where the one-fluid theory shows some deviations; at low pressures, the results of the mean density approximation and of the one-fluid theory coincide.

We have used Equation 2 to calculate the critical line of the H_2/CH_4 system. The results are shown in Figure 3. Evidently this equation of state, together with the extensions and mixing rules described above, gives reliable extrapolations of phase envelopes to different temperatures.

As a severe test of the consistency of our approach we have used Equation 2 in connection with 19 to predict the solid-liquid-gas three-phase equilibrium, which has been observed in this system at low temperatures (23). It turns out that the slg three-phase line is indeed represented successfully; the deviations at high pressure are mostly due to the inaccuracy of the experimental volumetric data of solid methane (24). Even so the temperature minimum of the slg line, which has been observed experimentally, is reproduced by the calculation. Again we must point out that the parameters of the equation of state have not been fitted to the solid-fluid equilibrium, but are the same as for the VLE calculation mentioned above.

Furthermore we have tried to calculate volumetric data of homogeneous H_2/CH_4 mixtures from the equations of state 1 and 2, again using

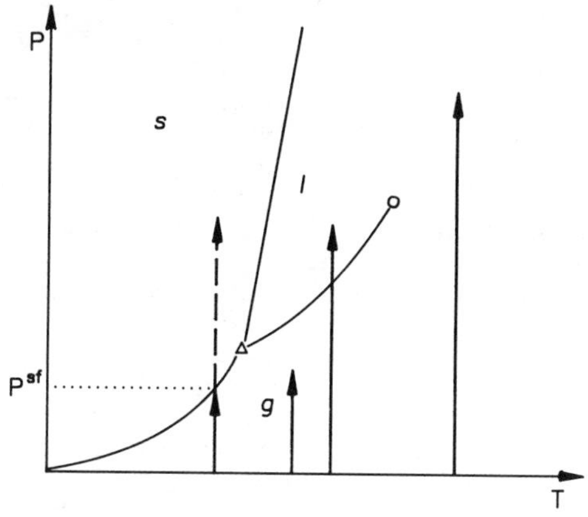

Figure 1. Calculation of the Gibbs energy of a pure substance. Key: o, critical point; Δ, triple point; ——, sublimation, melting, or vapour pressure curve; ⟶, integration of the fluid equation of state; ⟶, integration of the solid equation of state.

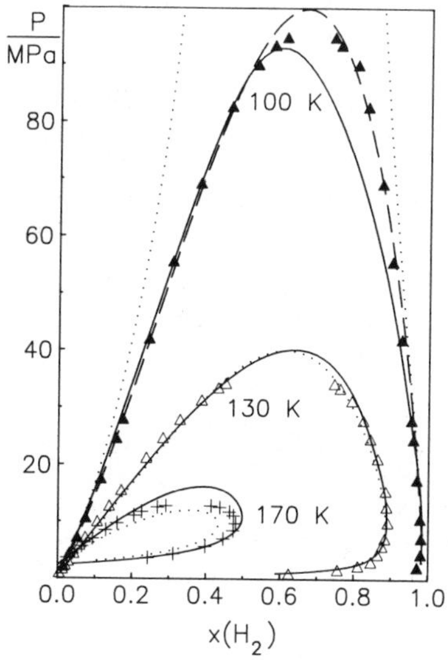

Figure 2. P-x diagram of the hydrogen/methane system. Key: ▲, Δ, +, experimental data (23); ..., calculated with the Redlich-Kwong equation of state; —— calculated with Equation 2 using one-fluid theory; --- using modified mean density approximation.

the same set of parameters as before. It turns out that both equations represent the experimental data (25) quite well for mixtures rich in hydrogen, but that the RK equation shows large deviations at other compositions. Apparently the cubic equation of state cannot fit phase equilibrium data and volumetric data simultaneously, whereas the other equation of state does not suffer from this limitation (Figure 4). Evidently the RK equation does not properly represent the molar volumes of liquid methane. Of course it is possible to find a parameter set which would lead to a better representation of the methane PVT data, but then the agreement of calculated and experimental phase equilibrium data would deteriorate. We consider this an important disadvantage of the RK equation, because the aim of this work is not the correlation of single thermodynamic properties, but the simultaneous description of several properties with the same computational method.

As another example of a phase equilibrium calculation based on the modified mean density approximation, we present the P-x diagram of the hydrogen/carbon monoxide system at 70K (Fig 5) In view of the high pressures involved, the one-fluid theory as well as the mean density approximation can be said to represent the experimental data successfully, but it is evident that in this case the mean density approximation is superior. At "low" pressure, however, both methods agree very well. Because of its higher CPU time requirements the modified mean density approximation should be applied to high density phase equilibria only.

In principal, there are no differences between liquids and gases, and our formulation should apply to sublimation and melting diagrams alike. This is demonstrated in Figure 6, which contains an isothermal melting diagram of the tetrafluoromethane/krypton system. The agreement between the experimental data (27) and the calculation (using Equation 2) is quite good. The strong curvature of the equilibrium curve indicates that this system is far from ideality. The systematic deviations of the calculated line at low pressures are probably due to the formation of a solid solution instead of a pure crystalline krypton phase (27). The equilibrium line originating at the melting point of CF_4 has been calculated from the same parameter set as the other equilibrium line. Unfortunately no volumetric data of solid CF_4 for the required temperature and pressure range are reported in literature, so that its solid molar volume had to be fitted to some binary data.

The binary system ethene/naphthalene shows an interesting interaction between solid-fluid and fluid-fluid equilibria. At moderate temperatures there is a solid-supercritical fluid equilibrium (Figure 7), to which we have fitted the binary parameters of Equation 2/15. The agreement between the experimental data and the calculated equilibrium states is quite good. Reversing the procedure described for the hydrogen/methane system, we have now used the parameters established from the solid-fluid equilibrium to predict the vapour-liquid equilibrium. The resulting phase diagram is shown in Figure 8: The critical line of the mixture, which would normally join the critical points of ethene and naphthalene, is interrupted by the slg three-phase line. The solid-supercritical fluid phase equilibria contained in Figure 7 belong to the P-T region marked "sg". The section of the critical line, which is outside the slg area, is predicted accurately by Equation 2. Furthermore, the slg line originating at the triple point of naphthalene is represented quite well; the deviations from the experimental data (28, 29), which are probably due to the lack of reliable compressibility data of naphthalene near its melting point, are never larger than 4 K.

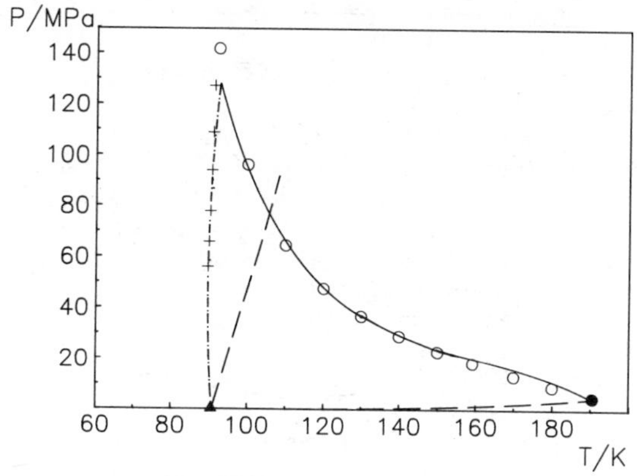

Figure 3. P-T phase diagram of the hydrogen/methane system. Key: ●, ▲, critical point and triple point of methane; ---, methane melting or vapour pressure line; o, experimental binary critical points (23); +, exp. slg three-phase states; ——, calculated critical line (with Equation 2/14); -·-, calculated three-phase line. Reproduced with permission from (32). Copyright 1985, Elsevier.

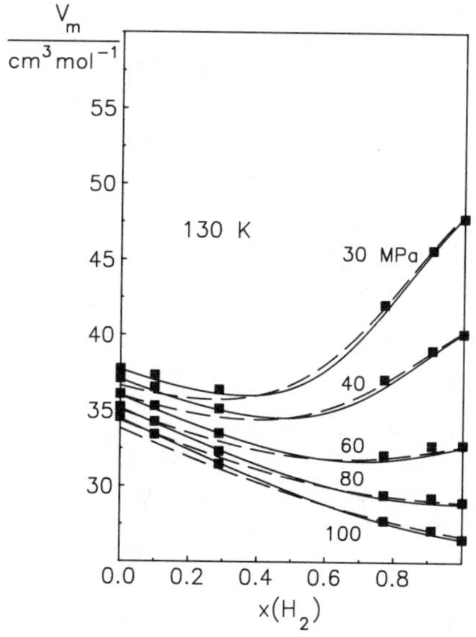

Figure 4. Volumetric properties of the hydrogen/methane system. Key: ■, experimental data (25); ——, calculated with Equation 2/14; ---, calculated with the Redlich-Kwong equation.

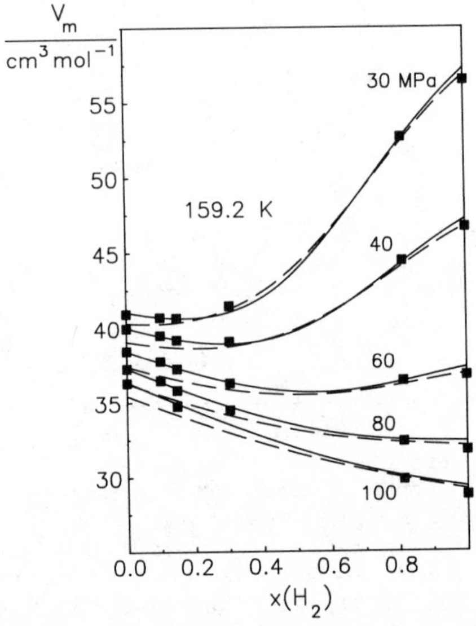

Figure 4. Volumetric properties of the hydrogen/methane system. Key: ■, experimental data (25); —, calculated with Equation 2/14; - - -, calculated with the Redlich-Kwong equation.

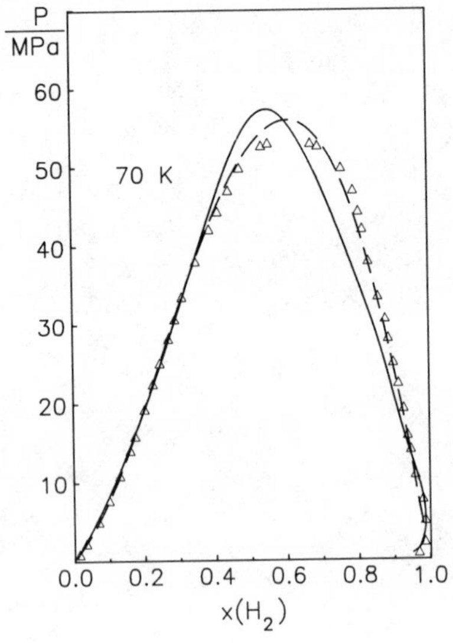

Figure 5. P-x diagram of the hydrogen/carbon monoxide system. Key: Δ, exp. data (26); ——, calculated with Equation 2/14 using one-fluid theory; ---, using modified mean density approximation.

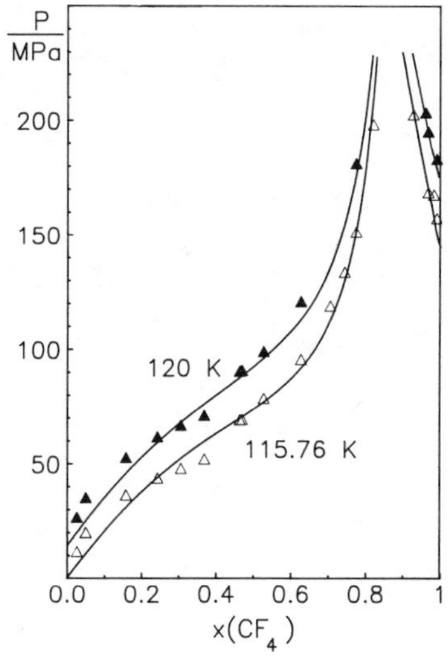

Figure 6. Isothermal melting diagram of the CF_4 /krypton system. Key: ▲, Δ, experimental data (27); ——, calculated with Equation 2.

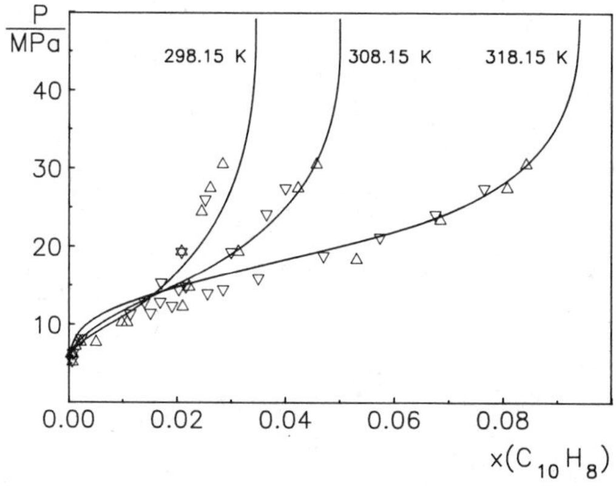

Figure 7. Isothermal solid-supercritical fluid phase equilibria of the ethene/naphthalene system. Key: Δ, ∇, exp. data (28, 29); ——, calculated with Equation 2/15.

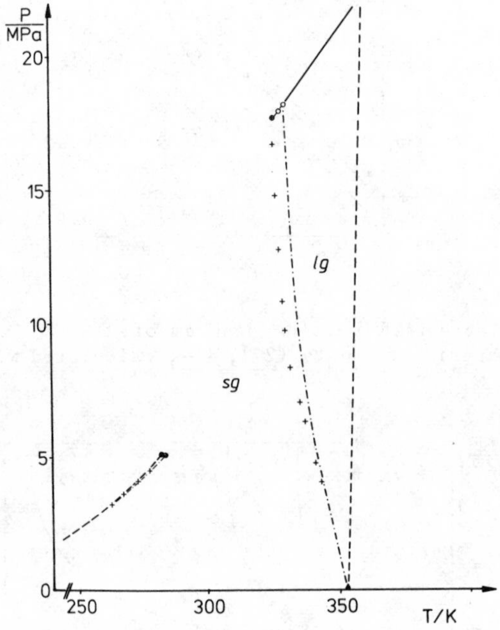

Figure 8. Phase diagram of the ethene/naphthalene system. Key: ▲, triple point of naphthalene; ●, critical point of ethene; ---, melting or vapour pressure lines; o, exp. binary critical point; +, exp. slg three-phase state; ——, calc. critical line (Equation 2/15); -·-, calc. slg three-phase line.

Finally we want to demonstrate that it is possible to predict vapour-liquid equilibria from liquid-liquid phase separations even in difficult cases. In many tetrafluoromethane/alkane mixtures a liquid-liquid phase separation occurs, where the critical line has a shallow temperature minimum. Figure 9 shows the critical lines (calculated with Equation 2) of the systems CF_4 /propane, CF_4 /butane, and CF_4 /pentane, together with some experimental data (30, 31). The agreement is quite good. The CF_4 /alkane systems are interesting from a phase-theoretical point of view: In the CF_4 /propane system the vapour-liquid and liquid-liquid critical curves are separate; in the CF_4 /pentane system a continuous transition between vapour-liquid and liquid-liquid equilibria occurs. The behaviour of the CF_4/butane system is similar to that of CF_4/propane, but here tem the critical endpoint occurs at a temperature above the critical temperature of CF_4. This is perhaps unusual, but in agreement with phase theory. An isothermal P-x diagram of the CF_4/butane system taken at a temperature slightly above the temperature minimum (Figure 10) contains three critical points: one for the vapour-liquid equilibrium, two for the liquid-liquid equilibria. Having fitted the binary parameters to the high pressure liquid-liquid equilibrium, we find that Equation 2 generates the lower liquid-liquid and the vapour-liquid equilibria in good agreement with the experimental data (30), whereas the results of the RK equation are only qualitatively correct.

We conclude that the non-cubic equation of state 2, eventually enhanced by Equations 14 and 15, does not only permit successful correlations of fluid-fluid equilibria, but also predictions of fluid-fluid equilibria of a different type, volumetric data, and solid-fluid equilibria in a wide temperature and pressure range. The Redlich-Kwong equation performs well at low densities, but fails at high pressures and at predictions of densities. In some cryogenic phase equilibria at high pressures the results of the calculation can be significantly improved, if the one-fluid theory is replaced by the modified mean density approximation.

Symbols

a	attraction parameter
A	Helmholtz energy
b	covolume parameter
c	anisotropy parameter
g	contact value of RDF
G	Gibbs energy
h	Planck constant
k	Boltzmann constant
L	effective cell size
m	molecular mass
n	amount of substance
N	number of molecules
P	pressure
q	surface fraction
Q	partition function
r	expansion coefficient
R	universal gas constant
R_{ij}	diameter ratio
s	effective surface
T	temperature
V	volume
x	molar fraction
y	reduced wavelength
$\varkappa$	isothermal compressibility
Λ	thermal de Broglie wavelength
μ	chemical potential
ρ, ξ	reduced densities

Subscripts

i	component index
m	molar property
rep	repulsion
att	attraction
f	free volume

Superscripts

°	perfect gas state
s	solid
f	fluid
′,″	phase indicator

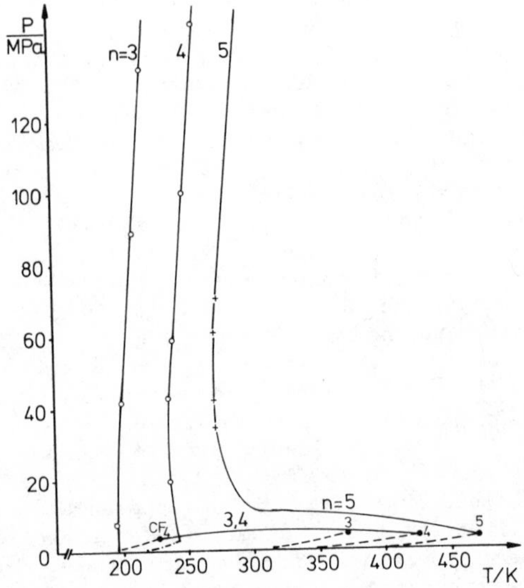

Figure 9. Phase diagram of some CF_4/alkane systems. The curves are marked with the carbon number n. Key: •, pure substance crit. point; ---, vapour pressure line; o, +, exp. binary crit. point, (30, 31, resp.); ——, calculated critical lines (Equation 2). For the CF_4/pentane system, the critical line originating at the CF_4 critical point has been omitted.

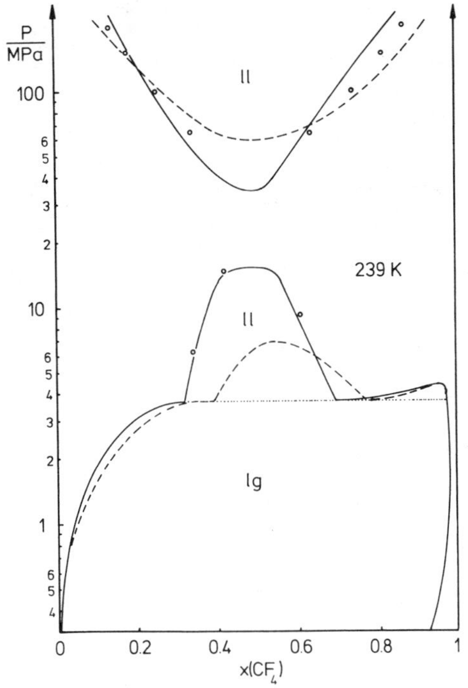

Figure 10. P-x diagram of the CF_4 /butane system (logarithmic scale). o, exp. data (30); ——, calculated with Equation 2; --- calc. with the Redlich-Kwong equation; ..., three-phase state. Reproduced with permission from (10). Copyright 1984, Verlag Chemie, Weinheim.

Acknowledgment

The author wishes to thank Prof. Dr. G.M. Schneider for friendly support of this work. Financial support by the Deutsche Forschungsgemeinschaft is gratefully acknowledged.

Literature Cited

1. Redlich, O.; Kwong, J. N. S. Chem. Reviews 1949, 44, 233.
2. Deiters, U. K. Chem. Eng. Sci. 1981, 36, 1139, 1147.
3. Deiters, U. K. Fluid Phase Equil. 1983, 10, 173.
4. Singh, Y.; Sinha, S. K. Phys. Rep. 1981, 79, 213.
5. Deiters, U. K. Fluid Phase Equil. 1983, 13, 109.
6. Rowlinson, J. S.; Swinton, F. L. "Liquids and Liquid Mixtures"; Butterworths: London, 1982; p. 279 ff.
7. Mansoori, G. A., this symposium.
8. Deiters, U. K. Chem. Eng. Sci. 1982, 37, 855.
9. Deiters, U. K. Fluid Phase Equil. 1983, 12, 193.
10. Deiters, U. K.; Swaid, I. Ber. Bunsenges. Phys. Chemie 1984, 88, 791.

11. Mansoori, G. A.; Carnahan, N. F.; Starling, K. E.; Leland, T. W. J. Chem. Phys. 1971, 54, 1523.
12. Haase, R. "Thermodynamik der Mischphasen"; Springer-Verlag: Berlin, 1956.
13. Deiters, U. K. Fluid Phase Equil. 1985, 19, 287.
14. Preston, G. T.; Prausnitz, J. M. Ind. Eng. Chem. Process Des. Develop. 1970, 9, 264.
15. Teller, M.; Knapp, H. Ber. Bunsenges. Phys. Chemie 1983, 87, 532.
16. Hoheisel, C.; Deiters, U. K.; Lucas, K. Mol. Phys. 1983, 49, 159.
17. Lebowitz, J. L.; Helfand, E.; Praestgaard, E. J. Chem. Phys. 1965, 43, 774.
18. Deiters, U. K. Habilitation Thesis, Ruhr-Universität, Bochum, FRG, 1985.
19. Ely, J. F., this symposium.
20. Li, M. H.; Chung, F. T. H.; So, C.-K.; Lee, L. L.; Starling, K. E., this symposium.
21. Eduljee, G. H. Fluid Phase Equil. 1983, 12, 190.
22. Sandler, S. I. Fluid Phase Equil. 1983, 12, 189.
23. Tsang, C. Y.; Clancy, P.; Calado, J. C. G.; Streett, W. B. Chem. Eng. Commun. 1980, 6, 365.
24. Breitling, S. M.; Jones, A. D.; Boyd, R.H. J. Chem. Phys. 1971, 54, 3959.
25. Machado, J. R. S.; Streett, W. B., unpublished data.
26. Tsang, C. Y.; Streett, W. B. Fluid Phase Equil. 1981, 6, 261.
27. Jeschke, P.; Schneider, G. M. J. Chem. Thermodynamics 1982, 14, 743.
28. Diepen, G. A. M.; Scheffer, F. E .C. J. Phys. Chem. 1953, 57, 575.
29. van Gunst, C. A.; Scheffer, F. E .C.; Diepen, G. A .M. J. Phys. Chem. 1953, 57, 578.
30. Jeschke, P.; Schneider, G. M. J. Chem. Thermodynamics 1982, 14, 547.
31. Mukherjee, A. Ph.D. Thesis, University of London, 1978.
32. Deiters, U. K. Fluid Phase Equil. 1985, 20, 275.

RECEIVED November 12, 1985

19

Optimal Temperature-Dependent Parameters for the Redlich-Kwong Equation of State

Randall W. Morris and Edward A. Turek

Amoco Production Company, Tulsa, OK 74102

Redlich-Kwong "a" and "b" parameters for several light hydrocarbons, carbon dioxide, nitrogen, and hydrogen sulfide have been optimized at many temperatures using accurate correlations of PVT data available in the literature. The regressed parameter values have been fit as functions of temperature for convenient use in calculations. The accuracy of volumetric calculations using the new parameters is compared to that obtained using the original Redlich-Kwong equation and the Soave and Yarborough modifications. The new parameters significantly improve the accuracy of volumetric calculations using the Redlich-Kwong equation.

One application of equations of state in the petroleum industry is in the numerical modeling of petroleum reservoirs. Since these mathematical models are very complex, a relatively simple equation of state is needed for computer solution of the model to be practical. The Redlich-Kwong equation (1) is a common form of equation of state used in such situations. The equation,

$$P = \frac{RT}{V-b} - \frac{a}{T^{0.5}V(V+b)} \qquad (1)$$

relates the pressure, molar volume, and temperature of a fluid using the two parameters, "a" and "b". For a mixture, the values of these parameters are determined by combining the parameters of the individual components. Thus, in general, mixture calculations will yield the best results when the individual component parameters provide the best fits to the properties of those components.

Accurate correlations of PVT data are available from IUPAC (2-4) for carbon dioxide, nitrogen, and methane, the National Bureau of Standards (5-8) for ethane, propane, isobutane, and normal butane, and the Texas A&M University Thermodynamics Research Center (9) for hydrogen sulfide. We have used data from these correlations to evaluate the accuracy of the Redlich-Kwong equation of state and to

0097-6156/86/0300-0389$06.00/0

develop new parameters. Since the qualitative results were the same for all substances studied, a detailed analysis of the results will be presented only for carbon dioxide.

Background

In the original development of the Redlich-Kwong equation of state, the parameters "a" and "b" were specified to be constant for a given fluid. For a pure substance, they were defined in terms of the critical temperature and critical pressure of the substance. These values of the parameters for carbon dioxide are represented by the horizontal lines in Figures 1 and 2, and will be referred to as the "standard" Redlich-Kwong parameters in the following discussion. Errors in volumetric calculations for CO_2 using these parameters are shown in Figures 3 and 4. Note the large errors in the calculation of liquid volumes.

In the Soave modification (10,11) of the Redlich-Kwong equation, the "a" parameter is replaced with a function of temperature which is adjusted to fit vapor pressures. The "b" parameter retains its standard value. This modification is intended to improve phase equilibrium calculations and results in little improvement in the volumetric calculations. Figure 1 shows the temperature dependence of the Soave "a" parameter consistent with Equation 1. Figures 5 and 6 show the errors in volumetric calculations for CO_2 using the Soave parameters.

Zudkevitch and Joffe (12) allowed both parameters to vary with temperature below the critical temperature, determining their values from the pressure, liquid density, and fugacity coefficient at saturation for a substance at a given temperature. They used the values of the parameters determined at the critical temperature for temperatures above the critical. Yarborough (13) applied this method using generalized correlations to calculate vapor pressure, saturated liquid density, and fugacity coefficient. The resulting correlation expresses the equation of state parameters in dimensionless form as functions only of reduced temperature and acentric factor. Values of these parameters for CO_2 are included in Figures 1 and 2 for comparison to the standard and Soave parameters. Figures 7 and 8 show the errors in volumetric calculations for CO_2 using these parameters. Note that the accuracy of the liquid volume calculations is improved considerably.

New Parameters

While the Yarborough parameters offer substantial improvement over the standard and Soave parameters at temperatures below the critical temperature, a further improvement is possible by allowing the parameters to vary with temperature above the critical also. We have done this for several substances, listed above, which are commonly encountered at supercritical temperatures. Values of the parameters were determined by optimizing the fit of the Redlich-Kwong equation to single-phase volumetric data. We also determined new values of the parameters at subcritical temperatures using both single-phase volumetric data and saturation data.

Optimal values of the parameters were determined at many temperatures. For each temperature, 197 pressure-volume data points were

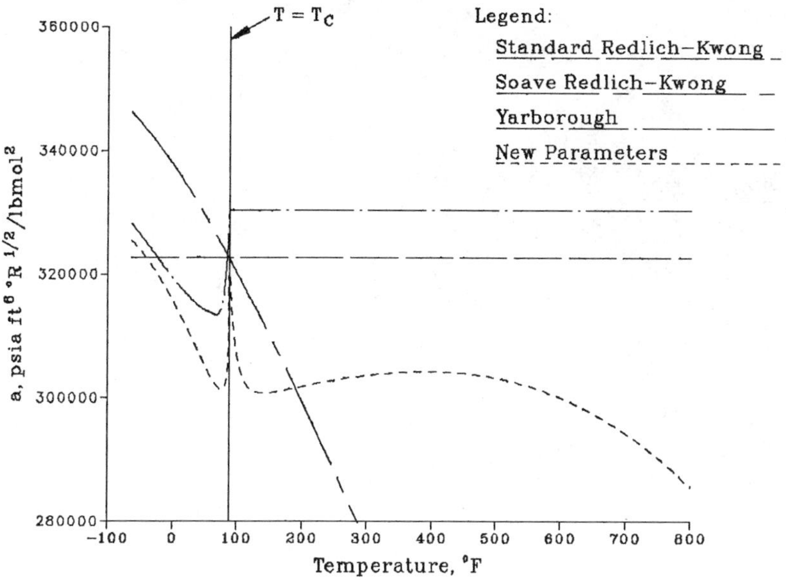

Figure 1. CO_2 "a" Parameter for the Redlich-Kwong Equation

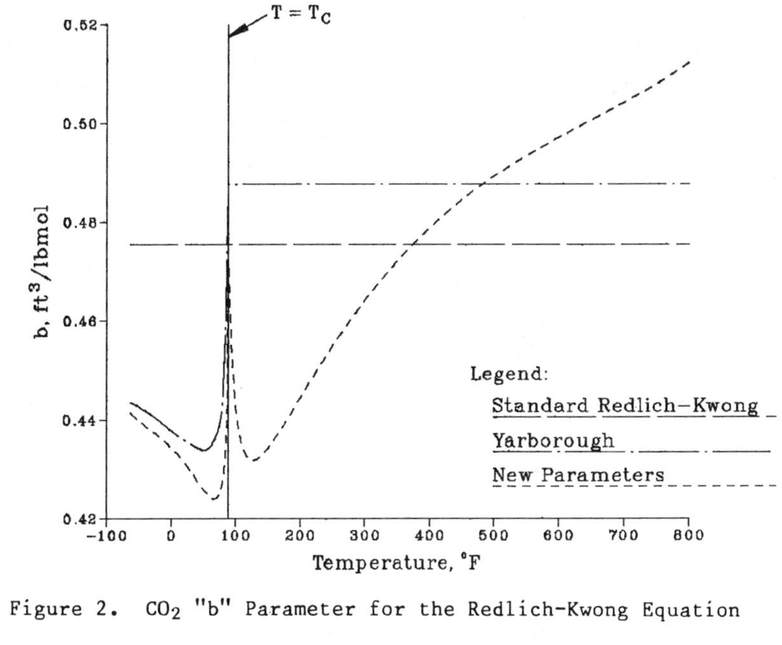

Figure 2. CO_2 "b" Parameter for the Redlich-Kwong Equation

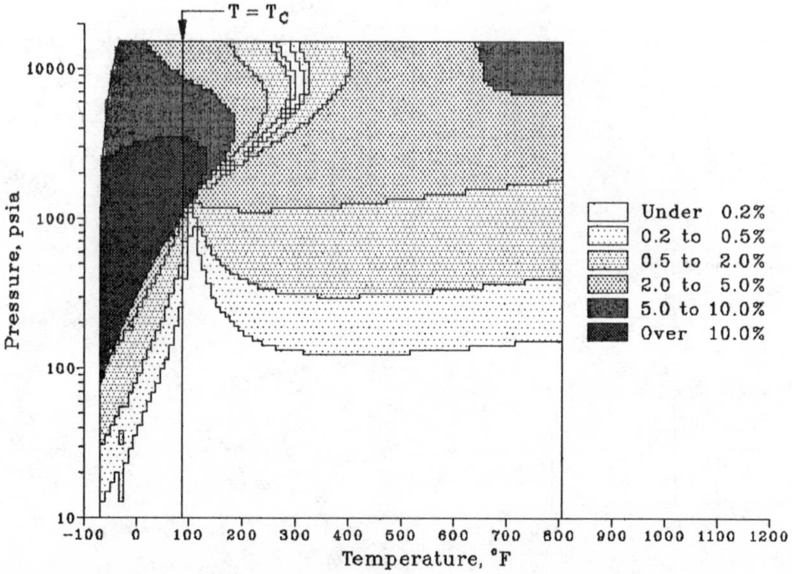

Figure 3. Error in Molar Volume Calculated for CO_2 using the Redlich-Kwong Equation with Standard Parameters

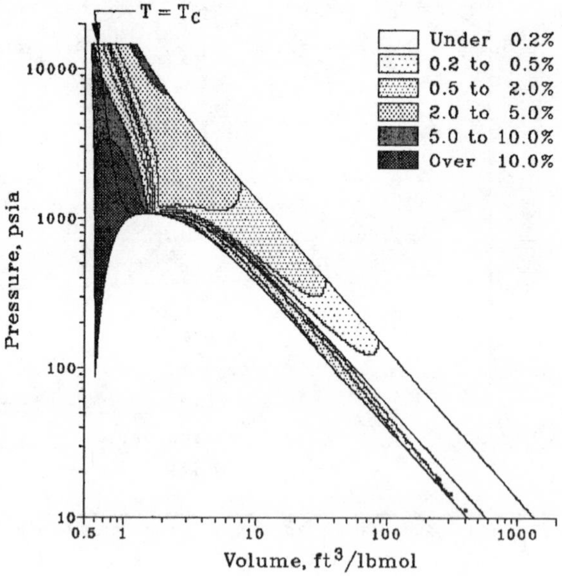

Figure 4. Error in Molar Volume Calculated for CO_2 using the Redlich-Kwong Equation with Standard Parameters

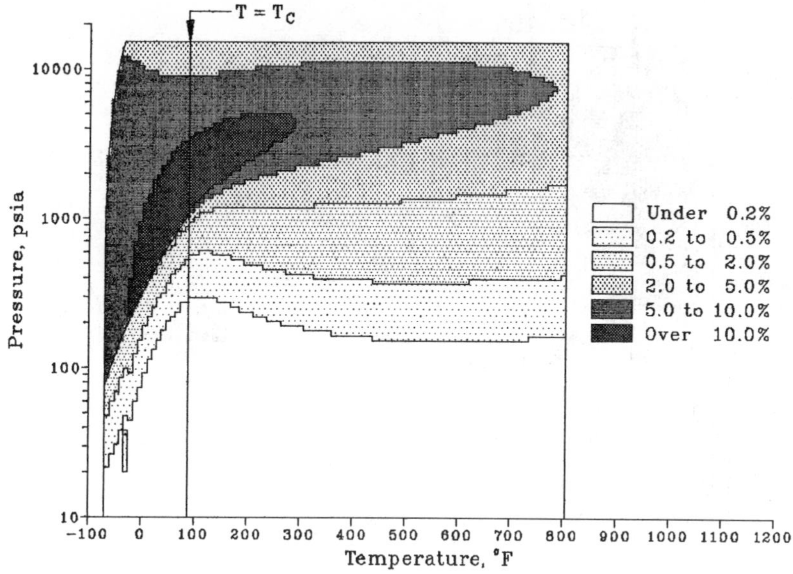

Figure 5. Error in Molar Volume Calculated for CO_2 using the Redlich-Kwong Equation with Soave Parameters

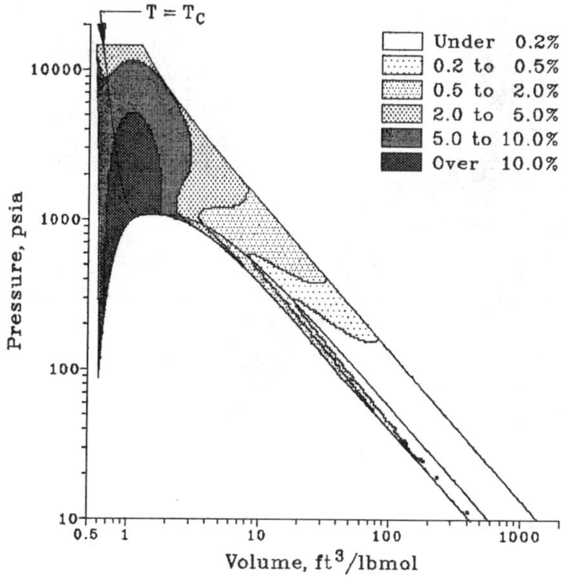

Figure 6. Error in Molar Volume Calculated for CO_2 using the Redlich-Kwong Equation with Soave Parameters

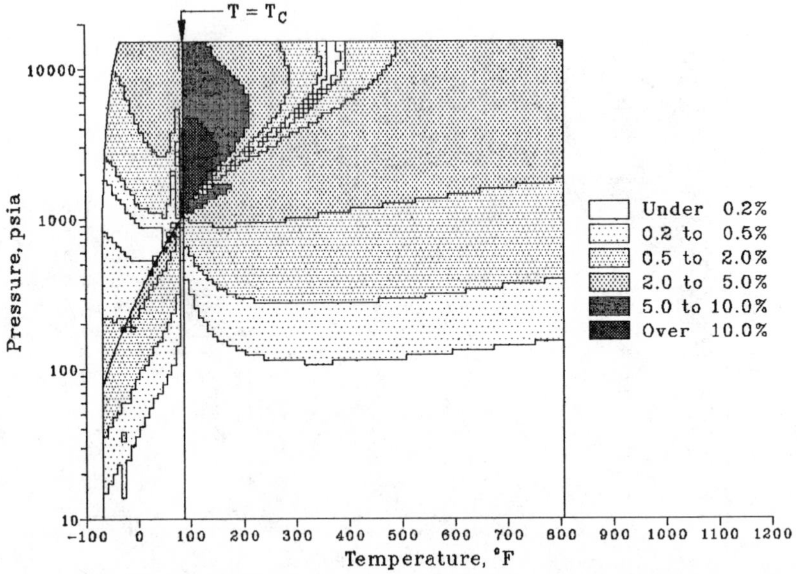

Figure 7. Error in Molar Volume Calculated for CO_2 using the Redlich-Kwong Equation with Yarborough Parameters

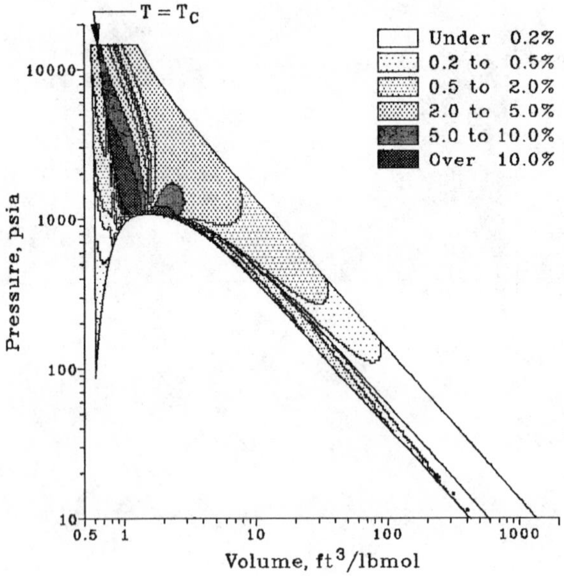

Figure 8. Error in Molar Volume Calculated for CO_2 using the Redlich-Kwong Equation with Yarborough Parameters

obtained from the correlations. The distribution of these points was linear with respect to log-log of volume in order to give the desired distribution of liquid and vapor volumes. Any of the 197 points that fell in the two-phase region for the given temperature were eliminated. For temperatures below the critical temperature, the saturation pressure and saturated liquid and vapor volumes were also used resulting in a maximum of 200 data points at a given temperature. Volumetric data were weighted according to the deviation of the Z factor from 1, since the fit to data exhibiting ideal gas behavior was found to be insensitive to the values of the equation of state parameters. Weightings of the saturation property deviations were adjusted by trial to give a reasonable balance with the volumetric data. The final objective function minimized at each temperature was

$$F = \Sigma[(Z-1)^2\Delta V]^2 + (16\Delta P_{sat})^2 + (25\Delta V_{sat,l})^2 + (2\Delta V_{sat,v})^2 \quad (2)$$

where the Δ quantities are fractional deviations, i.e.,

$$\Delta X \equiv (X_{calc} - X_{exp})/X_{exp}$$

for a given quantity X. Parameters were constrained during the regression to be consistent with the experimental critical temperature, i.e., parameter values yielding three roots in the equation of state above the experimental critical temperature or only one root below the experimental critical temperature were rejected.

The optimized parameter values were fit as functions of temperature for each substance. Since petroleum reservoir applications of equations of state are generally at a constant temperature, the parameters must be determined only once for a large number of calculations. For this reason, the functional forms used to fit the optimized parameter values could be made as complex as necessary for the desired accuracy without adversely affecting computing times. It should also be noted that the method used to determine these parameters results in a discontinuity in the temperature derivatives of the parameters at the critical temperature. This is not a problem for constant temperature applications but may need to be considered for calculations requiring the temperature derivatives of the equation of state parameters, e.g., enthalpy calculations. The fits to the optimized parameters are described in the Appendix.

The new parameters for CO_2 are compared to the standard values, and the Soave and Yarborough values in Figures 1 and 2. Errors in volumetric calculations using the new parameters are shown in Figures 9 and 10. Note that the large errors in calculated volumes are now confined to the near proximity of the critical point where they are unavoidable with the Redlich-Kwong equation of state. The inclusion of volumetric data in the parameter regressions at subcritical temperatures still allows a reasonable fit to the saturation pressures as shown in Figure 11. Saturation pressures for CO_2 calculated using the new parameters show an average error of about 0.3%.

Improvements in the accuracy of volumetric calculations for the other substances studied are similar to those obtained for CO_2. Errors in volumetric calculations at selected conditions for all substances studied are shown in Tables I through VIII along with average errors over the entire range of pressures and volumes considered.

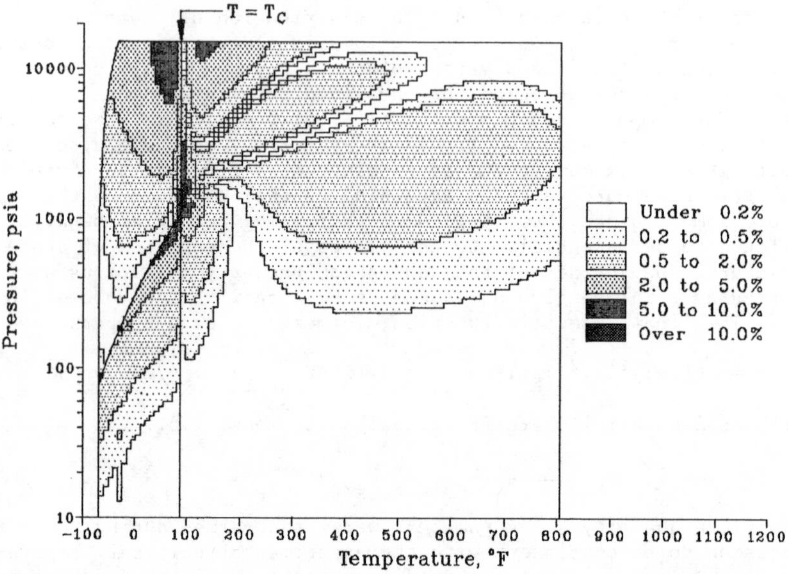

Figure 9. Error in Molar Volume Calculated for CO_2 using the Redlich-Kwong Equation with New Parameters

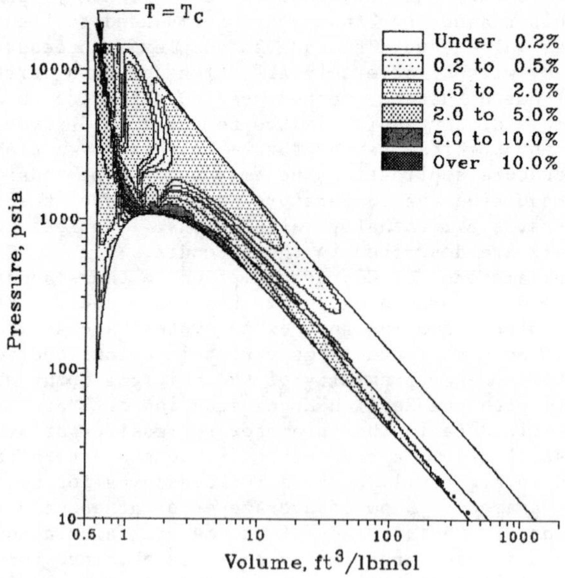

Figure 10. Error in Molar Volume Calculated for CO_2 using the Redlich-Kwong Equation with New Parameters

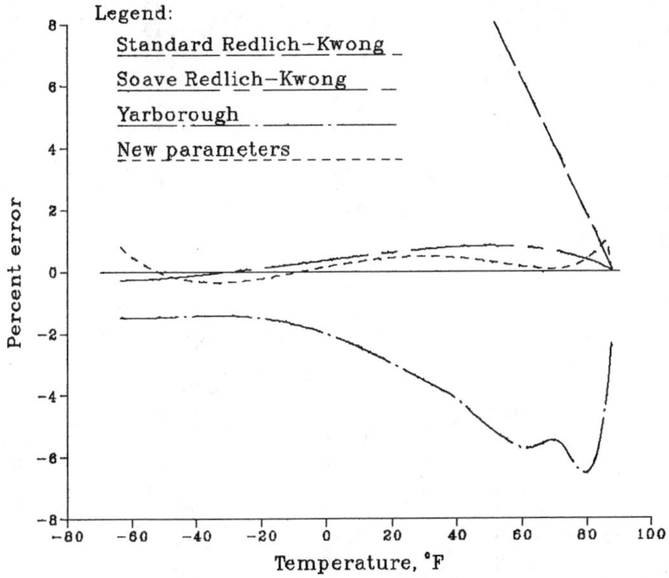

Figure 11. Error in Vapor Pressure Calculated for CO_2 using the Redlich-Kwong Equation

Table I. Percent Error in Calculated Volume for Carbon Dioxide

T_r	P_r	Standard Parameters	Soave Parameters	Yarborough Parameters	New Parameters
0.85	0.5	16.68	11.49	0.17	0.27
0.85	2.0	11.66	8.65	1.20	1.72
0.85	10.0	5.91	4.90	3.33	3.96
1.05	0.5	0.10	0.42	0.61	1.07
1.05	2.0	12.92	16.46	14.70	2.13
1.05	10.0	3.89	4.39	6.09	4.52
1.15	0.5	0.59	0.49	0.93	0.27
1.15	2.0	0.16	11.47	0.39	0.38
1.15	10.0	2.79	4.50	4.84	3.84
1.50	0.5	0.89	0.70	1.00	0.41
1.50	2.0	3.38	4.14	3.64	1.07
1.50	10.0	1.28	5.34	0.22	0.37
2.00	0.5	0.75	0.71	0.78	0.34
2.00	2.0	2.67	2.91	2.70	0.91
2.00	10.0	4.88	5.03	3.95	0.03
Average		3.15	4.10	2.19	1.02

Table II. Percent Error in Calculated Volume for Nitrogen

T_r	P_r	Standard Parameters	Soave Parameters	Yarborough Parameters	New Parameters
2.00	0.5	0.32	0.26	0.32	0.29
2.00	2.0	1.06	1.20	1.08	0.72
2.00	10.0	2.78	1.16	2.95	0.56
3.00	0.5	0.30	0.20	0.30	0.04
3.00	2.0	1.10	0.71	1.12	0.09
3.00	10.0	3.83	0.79	3.92	0.14
Average		1.85	0.88	1.91	0.20

Table III. Percent Error in Calculated Volume for Methane

T_r	P_r	Standard Parameters	Soave Parameters	Yarborough Parameters	New Parameters
1.15	0.5	0.35	0.27	0.29	1.01
1.15	2.0	7.56	8.55	7.08	0.63
1.15	10.0	0.38	0.50	0.37	0.22
1.50	0.5	0.23	0.08	0.21	0.65
1.50	2.0	0.31	1.88	0.30	1.62
1.50	10.0	0.63	1.87	0.04	0.81
2.00	0.5	0.21	0.22	0.20	0.28
2.00	2.0	0.36	1.35	0.38	0.56
2.00	10.0	0.62	2.39	1.01	0.21
3.00	0.5	0.21	0.20	0.21	0.02
3.00	2.0	0.70	0.80	0.73	0.01
Average		1.12	1.72	1.11	0.75

Table IV. Percent Error in Calculated Volume for Ethane

T_r	P_r	Standard Parameters	Soave Parameters	Yarborough Parameters	New Parameters
0.85	0.5	11.11	9.17	1.53	0.31
0.85	2.0	6.83	5.74	0.76	2.34
0.85	10.0	1.96	1.60	3.82	5.13
1.05	0.5	0.17	0.06	0.33	0.59
1.05	2.0	11.24	12.73	11.66	2.48
1.05	10.0	1.44	1.65	2.04	4.87
1.15	0.5	0.46	0.05	0.56	0.07
1.15	2.0	2.90	8.58	2.73	0.40
1.15	10.0	1.10	1.87	1.66	3.95
1.50	0.5	0.57	0.29	0.61	0.49
1.50	2.0	1.67	2.58	1.77	1.13
1.50	10.0	0.71	2.81	0.30	0.61
Average		2.45	3.14	1.53	1.03

Table V. Percent Error in Calculated Volume for Propane

T_r	P_r	Standard Parameters	Soave Parameters	Yarborough Parameters	New Parameters
0.85	0.5	13.72	10.37	1.33	0.41
0.85	2.0	8.88	6.97	0.78	2.31
0.85	10.0	3.32	2.69	3.77	5.08
1.05	0.5	0.29	0.06	0.59	0.68
1.05	2.0	12.01	14.37	13.09	2.93
1.05	10.0	1.98	2.31	3.28	4.98
1.15	0.5	0.54	0.22	0.74	0.17
1.15	2.0	1.12	9.40	0.98	0.60
1.15	10.0	1.24	2.40	2.46	2.62
1.50	0.5	0.49	0.70	0.55	0.31
1.50	2.0	2.85	2.83	3.01	1.72
1.50	10.0	1.56	3.27	0.66	1.03
Average		3.65	4.24	2.00	1.33

Table VI. Percent Error in Calculated Volume for Isobutane

T_r	P_r	Standard Parameters	Soave Parameters	Yarborough Parameters	New Parameters
0.85	0.5	14.88	10.68	0.77	0.37
0.85	2.0	9.65	7.25	1.29	2.32
0.85	10.0	3.55	2.76	4.34	5.26
1.05	0.5	0.60	0.17	0.99	0.24
1.05	2.0	11.81	14.69	13.21	4.11
1.05	10.0	1.85	2.26	3.53	4.10
1.15	0.5	0.91	0.00	1.17	0.76
1.15	2.0	0.11	9.67	0.27	0.67
1.15	10.0	0.91	2.31	2.48	0.55
Average		4.42	4.82	2.20	1.64

Table VII. Percent Error in Calculated Volume for Normal Butane

T_r	P_r	Standard Parameters	Soave Parameters	Yarborough Parameters	New Parameters
0.85	0.5	16.28	11.68	1.31	0.37
0.85	2.0	10.92	8.28	0.68	2.19
0.85	10.0	4.65	3.77	3.69	5.00
1.05	0.5	0.34	0.12	0.78	0.66
1.05	2.0	13.06	16.21	14.60	3.75
1.05	10.0	2.84	3.28	4.69	4.35
1.15	0.5	0.68	0.29	0.96	0.37
1.15	2.0	0.50	11.04	0.32	0.69
1.15	10.0	1.85	3.36	3.58	1.22
Average		5.07	5.58	2.44	1.59

Table VIII. Percent Error in Calculated Volume for Hydrogen Sulfide

T_r	P_r	Standard Parameters	Soave Parameters	Yarborough Parameters	New Parameters
0.85	0.5	9.75	8.05	0.68	0.39
0.85	2.0	6.28	5.32	0.98	2.06
0.85	10.0	2.34	2.02	3.29	4.37
1.05	0.5	0.28	0.49	0.14	0.63
1.05	2.0	11.22	12.58	11.53	3.23
1.05	10.0	2.20	2.40	2.69	2.79
1.15	0.5	0.08	0.40	0.16	0.07
1.15	2.0	6.42	11.85	6.23	0.68
1.15	10.0	2.14	2.86	2.59	3.64
1.50	0.5	0.24	0.58	0.27	0.11
1.50	2.0	0.19	3.89	0.27	0.20
1.50	10.0	1.46	4.85	1.80	2.84
Average		2.87	3.79	1.79	0.93

Conclusion

The accuracy of volumetric calculations using the Redlich-Kwong equation of state was improved significantly by allowing the "a" and "b" parameters to vary with temperature both above and below the critical temperature. By utilizing saturation data as well as volumetric data in the development of the new parameters, these improvements were made without sacrificing the fit to vapor pressures. The optimized parameters developed in this work were fit as functions of temperature for convenient use in calculations.

Nomenclature

a = parameter for equation 1, psia ft^6 $^oR^{0.5}/lbmol^2$

b = parameter for equation 1, $ft^3/lbmol$

b^* = limiting value for parameter b, $ft^3/lbmol$

F = objective function for parameter optimization

ΔP = fractional deviation in pressure calculation

ΔT = $(T/T_c) - 1$

ΔV = fractional deviation in molar volume calculation

P = pressure, psia

P_c = critical pressure, psia

P_r = reduced pressure, P/P_c

R = gas constant, psia $ft^3/lbmol\ ^oR$

T = temperature, ^oR

T_c = critical temperature, ^oR

T_r = reduced temperature, T/T_c

V = molar volume, $ft^3/lbmol$

Z = compressibility factor, PV/RT

Subscripts

c = critical point
calc = calculated
exp = experimental
l = liquid
r = reduced
sat = saturation
v = vapor

Appendix

The optimized Redlich-Kwong "a" and "b" parameters determined in this work were fit as functions of temperature. The functions used in these fits are empirical and no physical significance should be attached to their forms. Since the fits were developed separately over a period of time, some variation in functional form may be seen. For convenience, all the functions are expressed in terms of the variable ΔT, defined as

$$\Delta T \equiv (T/T_c) - 1$$

i.e., the dimensionless displacement from the critical temperature. Critical temperatures and ranges of fits of the parameters are shown in Table IX.

Table IX. Critical Temperatures and Ranges of Parameter Fits

Substance	Critical temperature, °R	Range of fit, °F
Carbon Dioxide	547.578	-63 to 800
Nitrogen	227.16	-75 to 650
Methane	343.00	-75 to 650
Ethane	549.594	-200 to 600
Propane	665.730	-200 to 600
Isobutane	734.130	-200 to 600
Normal Butane	765.288	-200 to 600
Hydrogen Sulfide	672.354	-90 to 600

The equation of state parameters were fit as dimensional quantities. Parameter "a" is in psia $ft^6\ {}^oR^{0.5}/lbmol^2$. Parameter "b" is in $ft^3/lbmol$. Separate fits were required for subcritical and supercritical parameters in order to fit the "spike" in the parameter values at the critical temperature.

In order to insure the proper behavior at temperatures slightly above the critical, a check must be applied to the calculated values of the parameters. The quantity "b*", the minimum allowable value of the supercritical "b" parameter is calculated from the "a" parameter at the same temperature with the following equation:

$$b^* = b_1 a/[T_c(1+b_2\Delta T)]^{1.5} \qquad (3)$$

where: $b_1 = 1.8886163E-02$

$b_2 = 0.99974$

If ΔT is less than 0.004 or if the calculated value of the "b" parameter is less than "b*", "b*" must be used as the value of the "b" parameter. This check may be omitted for nitrogen and methane since their fits do not extend down to their critical temperatures. The parameter fits for all substances studied are given below.

Carbon Dioxide Parameters, $\Delta T \leqq 0$

$$a = a_0\Delta T^3 + a_1\Delta T^2 + a_2\Delta T + a_3 + a_4(a_5-\Delta T)^{a_6} + a_7/(a_8-\Delta T)$$

$$b = b_0\Delta T^3 + b_1\Delta T^2 + b_2\Delta T + b_3 + b_4(b_5-\Delta T)^{b_6} + b_7/(b_8-\Delta T)$$

a_0 = 1.1323818E+05	b_0 = -2.3973118E+00
a_1 = -1.9341561E+05	b_1 = -1.8705039E+00
a_2 = -2.0363902E+05	b_2 = -7.0399570E-01
a_3 = -5.2494786E+07	b_3 = 4.6945137E-01
a_4 = 5.2775697E+07	b_4 = -1.9580259E-01
a_5 = 1.6289565E-04	b_5 = 3.3775759E-06
a_6 = -7.9739155E-05	b_6 = 3.1643771E-01
a_7 = 4.5206914E-05	b_7 = 1.3103149E-11
a_8 = 8.4787834E-09	b_8 = 1.2868684E-09

Carbon Dioxide Parameters, $\Delta T \geqq 0$

$$a = a_0\Delta T^3 + a_1\Delta T^2 + a_2\Delta T + a_3 + a_4(\Delta T - a_5)^{a_6}$$

$$b = b_0\Delta T^3 + b_1\Delta T^2 + b_2\Delta T + b_3 + b_4(\Delta T - b_5)^{b_6}$$

a_0 =	-1.2327970E+04	b_0 =	4.9137761E-02
a_1 =	-3.9510605E+03	b_1 =	-1.5082045E-01
a_2 =	1.5753912E+04	b_2 =	1.9277928E-01
a_3 =	2.9883762E+05	b_3 =	4.0844576E-01
a_4 =	1.2638522E-02	b_4 =	1.9737045E-04
a_5 =	-1.2669980E-01	b_5 =	-3.8637696E-02
a_6 =	-6.9947616E+00	b_6 =	-1.7936190E+00

Nitrogen and Methane Parameters

$$a = a_0\Delta T^3 + a_1\Delta T^2 + a_2\Delta T + a_3 + a_4/(\Delta T - a_5)$$

$$b = b_0\Delta T^3 + b_1\Delta T^2 + b_2\Delta T + b_3 + b_4/(\Delta T - b_5)$$

	Nitrogen	Methane
a_0 =	-1.6347508E+01	5.7694890E+03
a_1 =	1.2161843E+03	-3.2063240E+04
a_2 =	-2.2892243E+04	3.0940267E+04
a_3 =	1.2959794E+05	1.6229084E+05
a_4 =	-4.8072688E+04	-4.6636347E+00
a_5 =	-7.5945730E-01	1.1820010E-01
b_0 =	-2.8577341E-04	3.0568302E-03
b_1 =	3.8658153E-03	-1.5876295E-02
b_2 =	-1.7061211E-02	2.2593112E-02
b_3 =	5.2704092E-01	4.8294387E-01
b_4 =	-1.2361065E-01	-4.0107835E-04
b_5 =	-1.2508721E+00	8.0512599E-02

Ethane, Propane, Isobutane, and Normal Butane Parameters, $\Delta T \leqq 0$

$$a = a_1\Delta T^5 + a_2\Delta T^4 + a_3\Delta T^3 + a_4\Delta T^2 + a_5\Delta T + a_6 + a_7(a_8 - \Delta T)^{a_9}$$

$$b = b_1\Delta T^5 + b_2\Delta T^4 + b_3\Delta T^3 + b_4\Delta T^2 + b_5\Delta T + b_6 + b_7(b_8 - \Delta T)^{b_9}$$

	Ethane	Propane	Isobutane	Normal Butane
a_1 =	0.0	1.3807587E+05	3.6060539E+05	4.0124834E+05
a_2 =	0.0	0.0	0.0	0.0
a_3 =	6.0408944E+04	0.0	0.0	0.0
a_4 =	-3.3524012E+05	-6.8075273E+05	-8.9042232E+05	-9.9038190E+05
a_5 =	-2.4440624E+05	-5.1847288E+05	-7.4744128E+05	-8.2677046E+05
a_6 =	-5.3770245E+06	-4.1533054E+08	1.1845554E+06	1.2222223E+06
a_7 =	5.8143088E+06	4.1613493E+08	3.6937924E+04	6.7757452E+04
a_8 =	4.5467943E-05	3.2664789E-05	1.7804777E-04	1.1836024E-04
a_9 =	-9.6079783E-04	-2.5461817E-05	-1.8159933E-01	-1.3348188E-01

b_1 =	0.0	0.0	0.0	0.0
b_2 =	0.0	2.8826091E-01	0.0	0.0
b_3 =	-3.8152843E-01	0.0	-5.0788946E-01	-5.0388863E-01
b_4 =	-7.6440305E-01	-7.2744096E-01	-1.2234509E+00	-1.2016256E+00
b_5 =	-4.9300521E-01	-5.4966388E-01	-7.8285224E-01	-7.5953251E-01
b_6 =	8.3401715E-01	-7.0736939E+02	-4.8981625E+02	-4.9144994E+02
b_7 =	-2.5418456E-01	7.0819594E+02	4.9086550E+02	4.9249286E+02
b_8 =	2.8828651E-05	3.7042340E-05	6.0889312E-05	5.0910144E-05
b_9 =	7.9453607E-02	-2.4660292E-05	-5.1127477E-05	-5.1232547E-05

Ethane, Propane, Isobutane, and Normal Butane Parameters, $\Delta T \geqq 0$

$$a = a_1\Delta T^5 + a_2\Delta T^4 + a_3\Delta T^3 + a_4\Delta T^2 + a_5\Delta T + a_6 + a_7(a_8+\Delta T)^{a_9}$$

$$b = b_1\Delta T^5 + b_2\Delta T^4 + b_3\Delta T^3 + b_4\Delta T^2 + b_5\Delta T + b_6 + b_7(b_8+\Delta T)^{b_9}$$

	Ethane	Propane	Isobutane	Normal Butane
a_1 =	0.0	0.0	0.0	0.0
a_2 =	0.0	1.2332546E+06	0.0	5.2547626E+06
a_3 =	0.0	0.0	3.7513525E+06	0.0
a_4 =	-2.9168739E+04	-1.7413558E+06	-4.1274194E+06	-4.3472555E+06
a_5 =	6.6705082E+04	1.2077341E+06	1.4272780E+06	2.4222201E+06
a_6 =	4.6578248E+05	-5.4364187E+08	1.2027989E+06	-3.2439274E+07
a_7 =	1.2357826E+00	5.4425460E+08	2.5675094E+02	3.3361438E+07
a_8 =	4.3979460E-02	1.5085713E-02	4.7010838E-02	1.3513673E-02
a_9 =	-3.2057354E+00	-1.3187677E-04	-2.1024705E+00	-3.6386297E-03
b_1 =	2.3099616E-01	0.0	0.0	0.0
b_2 =	-2.9478506E-01	1.3021642E+00	0.0	1.1231303E+01
b_3 =	0.0	0.0	3.3538827E+00	-7.6112989E+00
b_4 =	0.0	-1.8915504E+00	-3.6241228E+00	0.0
b_5 =	1.9013417E-01	1.4616927E+00	1.4281118E+00	1.0382612E+00
b_6 =	6.4692946E-01	-1.1210554E+01	1.1243492E+00	1.1002003E+00
b_7 =	1.1226280E-03	1.1849368E+01	1.3058917E-05	5.5989194E-04
b_8 =	1.5821016E-02	9.8746681E-03	6.0297320E-02	3.5280578E-02
b_9 =	-1.0175663E+00	-6.5869382E-03	-3.3703921E+00	-1.7457633E+00

Hydrogen Sulfide Parameters, $\Delta T \leqq 0$

$$a = a_1\Delta T^3 + a_2\Delta T^2 + a_3\Delta T + a_4 + a_5(a_6-\Delta T)^{a_7}$$

$$b = b_1\Delta T^3 + b_2\Delta T^2 + b_3\Delta T + b_4 + b_5(b_6-\Delta T)^{b_7}$$

a_1 =	2.7553548E+05	b_1 =	7.5061764E-02
a_2 =	-1.5271335E+05	b_2 =	-2.6513067E-01
a_3 =	-2.2153932E+05	b_3 =	-2.7365804E-01
a_4 =	-7.8818355E+09	b_4 =	-7.1403337E+03
a_5 =	7.8822161E+09	b_5 =	7.1407259E+03
a_6 =	3.7436399E-04	b_6 =	2.4591169E-04
a_7 =	-9.8433542E-07	b_7 =	-1.4554375E-06

Hydrogen Sulfide Parameters, $\Delta T \geqq 0$

$$a = a_1\Delta T^3 + a_2\Delta T^2 + a_3\Delta T + a_4 + a_5(a_6+\Delta T)^{a_7}$$

$$b = b_1\Delta T^3 + b_2\Delta T^2 + b_3\Delta T + b_4 + b_5(b_6+\Delta T)^{b_7}$$

a_1 = 8.9992497E+04	b_1 = 1.6468196E-01
a_2 = -2.1018269E+05	b_2 = -2.7842604E-01
a_3 = 9.3022318E+04	b_3 = 1.7715814E-01
a_4 = 4.1713206E+05	b_4 = -1.8220125E+02
a_5 = 5.4872735E+02	b_5 = 1.8259959E+02
a_6 = 3.4032689E-02	b_6 = 8.3692118E-03
a_7 = -1.1257065E+00	b_7 = -9.1864978E-05

Literature Cited

1. Redlich, O.; Kwong, J. N. S. Chemical Reviews 1949, 44, 233-244
2. Angus, S.; Armstrong, B.; de Reuck, K. M. "International Thermodynamic Tables of the Fluid State-3, Carbon Dioxide"; Pergamon: Oxford, 1976
3. Angus, S.; de Reuck, K. M.; Armstrong, B. "International Thermodynamic Tables of the Fluid State-6, Nitrogen"; Pergamon: Oxford, 1979
4. Angus, S.; Armstrong, B.; de Reuck, K. M. "International Thermodynamic Tables of the Fluid State-5, Methane"; Pergamon: Oxford, 1978
5. Goodwin, R. D.; Roder, H. M.; Straty, G. C. "Thermophysical Properties of Ethane from 90 to 600 K at Pressures to 700 Bar"; National Bureau of Standards Technical Note 684, National Bureau of Standards: Washington, D.C., 1976
6. Goodwin, R. D; Haynes, W. M. "Thermophysical Properties of Propane from 85 to 700 K at Pressures to 70 MPa"; National Bureau of Standards Monograph 170, National Bureau of Standards: Washington, D.C., 1982
7. Goodwin, R. D; Haynes, W. M. "Thermophysical Properties of Isobutane from 114 to 700 K at Pressures to 70 MPa"; National Bureau of Standards Technical Note 1051, National Bureau of Standards: Washington, D.C., 1982
8. Haynes, W. M.; Goodwin, R. D. "Thermophysical Properties of Normal Butane from 135 to 700 K at Pressures to 70 MPa"; National Bureau of Standards Monograph 169, National Bureau of Standards: Washington, D.C., 1982
9. Chen, S. S.; Kreglewski, A. "Selected Values of Properties of Chemical Compounds"; Thermodynamics Research Center Data Project: Texas A&M University, 1976; Tables 14-2-(1.01)-j, 14-2-i, 14-2-k, and 14-2-kb
10. Soave, G. Chem. Eng. Sci. 1972, 27, 1197-1203
11. Graboski, M. S; Daubert, T. E. Ind. Eng. Chem. Process Des. Dev. 1978, 17, 448-454
12. Zudkevitch, D.; Joffe, J. AIChE J. 1970, 16, 112-119
13. Yarborough, L. In "Equations of State in Engineering and Research"; Chao, K. C.; Robinson, R. L., Eds.; ADVANCES IN CHEMISTRY SERIES No. 182, American Chemical Society: Washington, D.C., 1979; pp. 385-439

RECEIVED November 8, 1985

20

Phase Behavior of Mixtures of San Andres Formation Oils with Acid Gases

Application of a Modified Redlich–Kwong Equation of State

Joseph J. Chaback and Edward A. Turek

Amoco Production Company, Tulsa, OK 74102

A modified Redlich-Kwong equation of state, with appropriate parameter adjustments, was successful in reproducing the volumetric properties and complex phase equilibria exhibited by San Andres Formation oil-acid gas systems. These parameter adjustments are carried out by trial and error to match a portion of experimental PVT and phase equilibria data available for these oils and their mixtures with acid gases. The remaining data are used to test the selected parameters. The similarity of the C_7+ fraction throughout the geologic formation allowed application of a common C_7+ compositional distribution for all the San Andres oils. Thus, for any oil from the subject formation, only the overall fluid composition and reservoir temperature are required to calculate volumetric properties and phase behavior.

Development and evaluation programs for miscible enhanced oil recovery (EOR) processes require fluid properties for both the reservoir oil and mixtures of the displacing gas and the reservoir oil. Western Texas contains numerous candidate fields for carbon dioxide flooding. In most cases these fields have been pressure-depleted and waterflooded. Yet these fields still represent a substantial resource base. The net reserves potentially recoverable with carbon dioxide approaches 700 million barrels. Six of these fields have been the subject of extensive experimental PVT and phase equilibrium studies. These fields produce from the San Andres Formation.

These experimental investigations include first-contact and multiple-contact studies. First-contact studies are those in which only mixtures of reservoir oil and carbon dioxide are studied. In multiple-contact studies, two phase equilibrium is achieved for an oil-carbon dioxide mixture. Then either the oil-rich or CO_2-rich phase is recontacted with fresh CO_2 or fresh reservoir oil, respec-

0097–6156/86/0300–0406$08.00/0

tively. Only first-contact data are discussed here; a subsequent paper will discuss the extension to multiple contact results.

The overall objective of the present investigation was to provide a Redlich-Kwong equation of state description which would be representative of all the oils produced from an entire formation. Thus, if successful, only the oil composition and reservoir temperature would ultimately be required to predict the important CO_2-oil phase equilibria. Such an approach was taken in order to minimize the need for additional costly experimental studies on fluids from wells not already treated in the experimental program. This is a severe test of the equation of state. The successes and shortcomings of fluid descriptions so developed are discussed below.

Development of the Fluid Descriptions: Overview

The strategy employed in the investigation consists of first developing an equation of state description for the reservoir oil prior to contact with a displacing gas. This uncontacted or base oil description includes a C_7+ heavy hydrocarbon distribution and characterizing parameters and customized methane-C_7+ interaction parameters for that distribution. The description was then extended to mixtures of CO_2 with the base reservoir oil.

In addition, the effect of compressing the C_7+ description was examined. For the base oil, the 34 components resulting from chromatographic analysis were regrouped to create descriptions with 14, 7, 3 and 2 C_7+ pseudocomponents. For CO_2-rich systems C_7+ descriptions with 14 and 3 pseudocomponents were developed. These cases allowed a comparison to be made between an extended C_7+ description and a more compressed description suited to compositional reservoir model calculations where the total number of components must be restricted because of computer time and memory constraints.

Descriptions were developed using as much of the available phase equilibrium data as possible. For these San Andres fields a large quantity of base oil PVT information was accumulated prior to initiation of EOR studies. Typically, this information includes reservoir fluid compositional analysis, a differential vaporization analysis (DVA), and pertinent operating parameters from the well. Even though much of these data were obtained thirty to forty years ago, they still were found to form a consistent data set.

In addition, more recent work associated with enhanced gas drive studies was incorporated into the base oil data collection. These experimental data were evaluated for consistency of reported composition and fluid properties. Those data found consistent were used to develop a representative C_7+ heavy hydrocarbon distribution (simulated true boiling point analysis) and equation of state parameters including methane-C_7+ interaction parameters. Next, the base oil description was refined to match properties of CO_2-base oil mixtures. Obtaining satisfactory agreement of calculated and experimental results necessitated adjustment of carbon dioxide-C_7+ and carbon dioxide-methane interaction parameters. Moreover, where hydrogen sulfide was present in significant amounts in the displacing gas, specialized H_2S-C_7+ and H_2S-CO_2 interaction parameters were also developed.

In all cases, the parameter adjustments were made by trial and error using an heuristic adjustment method; that is, changes were made interactively at a computer console rather than by some formalized mathematical optimization approach. As a consequence of the cumbersome nature of the adjustment procedure, only base oil and some first-contact CO_2-oil data were used in the data adjustment process. The remainder of the first-contact data (including results for three-phase (L-L-V) fluid equilibria) were used only to test the parameters.

The Modified Redlich-Kwong Equation of State

Throughout the discussion reference is made to various physical properties and equation of state parameters, for example, critical pressure and acentric factor. These properties are incorporated into the Redlich-Kwong equation (Equation 1) as shown in Equations 2 and 3.

$$P = \frac{RT}{V - b} - \frac{a}{\sqrt{T}\, V(V + b)} \tag{1}$$

where a and b are equation parameters defined by:

$$a \equiv \sum_{i=1}^{NC} \sum_{j=1}^{NC} x_i x_j (1 - C_{ij}) \left[\frac{\Omega_{ai} R^2 T_{ci}^{2.5}}{P_{ci}} \cdot \frac{\Omega_{aj} R^2 T_{cj}^{2.5}}{P_{cj}} \right]^{1/2} \tag{2}$$

$$b \equiv \sum_{i=1}^{NC} \sum_{j=1}^{NC} x_i x_j \frac{(1 + D_{ij})}{2} \left[\frac{\Omega_{bi} R T_{ci}}{P_{ci}} + \frac{\Omega_{bj} R T_{cj}}{P_{cj}} \right] \tag{3}$$

The omega parameters, Ω_{ai}, Ω_{bi}, are correlated with reduced temperature and acentric factor:

$$\Omega_a = \begin{cases} f(T_r,\omega) & T_r \leqq 1 \\ f(1,\omega) & T_r > 1 \end{cases} \tag{4}$$

$$\Omega_b = \begin{cases} g(T_r,\omega) & T_r \leqq 1 \\ g(1,\omega) & T_r > 1 \end{cases} \tag{5}$$

The f and g functions shown in Equations 4 and 5 are different correlations based on reduced temperature and acentric factor as described by Yarborough (1). Symbols used in these expressions are defined in the Glossary at the end of the text. Table I shows such properties and identifies pseudocomponents as they appear in computer output for a typical case: Oil D mixed with a H_2S/CO_2 displacing gas. Note that oils are given letter designations consistent with those of Turek, et al. (2), whenever appropriate. As described by Morris and Turek (3), the Ω_a and Ω_b for n-butane and lighter components are functions of temperature for values of reduced temperature above and below the critical temperature. Those for pentane and heavier components are a function of temperature only below the critical temperature as shown in Equations 4 and 5 above.

TABLE I. COMPONENT PROPERTIES FOR THE GENERALIZED REDLICH-KWONG EQUATION OF STATE FOR OIL D, SAN ANDRES FORMATION

COMPONENT[1]	COMP. ID NO.	Ω_A [2]	Ω_B [2]	T_c, DEG R	P_c, PSIA	MOLE WT	DENSITY AT 60°F	CRIT VOL, FT3/LB MOL	ACENTRIC FACTOR
Methane	2	0.45268	0.08890	343.00	666.45	16.043	0.0	1.58450	0.0105
Carbon Dioxide	3	0.40340	0.07975	547.58	1070.74	44.009	0.0	1.51300	0.2310
Ethane	4	0.41017	0.08141	549.59	706.54	30.070	0.35630	2.35600	0.0992
Hydrogen Sulfide	5	0.41158	0.08112	672.35	1306.65	34.080	0.78920	1.56500	0.0911
Propane	6	0.41509	0.07987	665.73	616.04	44.097	0.50690	3.20400	0.1523
I-Butane	7	0.42619	0.08094	734.13	527.94	58.124	0.56250	4.15000	0.1852
N-Butane	8	0.42826	0.08060	765.29	550.56	58.124	0.58380	4.08600	0.1996
I-Pentane	9	0.43724	0.08127	828.69	483.00	72.146	0.62420	4.90000	0.2223
N-Pentane	10	0.44083	0.08049	845.19	489.50	72.146	0.63060	4.87000	0.2539
N-Hexane	11	0.44836	0.07907	913.79	440.00	86.172	0.66330	5.93000	0.3007
C7S	27	0.44672	0.07982	985.43	528.30	92.920	0.76245	5.68100	0.2689
C8S	28	0.45181	0.07828	1039.48	479.87	105.640	0.78373	6.44100	0.3127
C9S	29	0.45612	0.07670	1088.86	436.69	118.370	0.80201	7.20900	0.3542
C10S	30	0.46028	0.07498	1131.15	392.42	132.390	0.81030	8.13200	0.3989
C11S	31	0.46395	0.07319	1168.41	352.55	147.070	0.81460	9.13900	0.4449
C12S	32	0.46709	0.07142	1202.51	319.55	161.750	0.81869	10.15600	0.4892
C13S	33	0.46719	0.07095	1248.74	335.51	171.200	0.87164	10.31500	0.4952
C14S	34	0.46984	0.06931	1277.12	306.55	185.900	0.87144	11.34200	0.5360
C15S	35	0.47200	0.06773	1303.39	281.87	200.540	0.87254	12.36300	0.5751
C16-17S	36	0.47417	0.06560	1339.72	252.93	221.540	0.87753	13.80000	0.6272
C18-20S	37	0.47582	0.06237	1392.45	215.18	256.300	0.88733	16.17500	0.7081
C21-24S	38	0.47370	0.05688	1456.08	175.89	305.210	0.90421	19.48000	0.8500
C25-29S	39	0.46662	0.05210	1522.75	139.98	366.510	0.92769	23.56000	0.9750
C30-40S	40	0.43883	0.04238	1613.14	97.35	487.900	0.96365	31.76400	1.2500

[1] "S" denotes C_7+ pseudocomponent.

[2] Ω_A and Ω_B are evaluated at 105°F. These parameters are temperature dependent.

In general the CO_2 interaction parameters are functions of temperature. Only those CO_2 interaction parameters developed during the data matching exercise are temperature independent. Table II contains the unlike molecular pair interaction parameters for this example San Andres system. The parameter sets in Tables I and II apply directly only to the Oil D system. However, the pseudocomponent properties and reservoir temperatures are similar for the various San Andres oils; hence, the parameters in Table I and II could probably be applied to San Andres oils of different composition without incurring serious error. The appropriate C_7+ pseudocomponent composition would have to be used for such cases.

Preparation of Compositional Data

The carbon number distribution within the C_7+ fraction is obtained via temperature programmed gas chromatography (simulated true boiling point analysis). Distributions for several fields and an average used to represent the San Andres Formation are shown in Figure 1. The chromatographic distribution is subsequently adjusted to match the cryoscopically determined molecular weight of the C_7+ fraction. First, the C_{40}+ carbon number is set to 50 and the C_{40}+ mol% to 5.0. These adjustments are made to reflect the inevitable loss of some heavier C_7+ components during the chromatographic elution. A final adjustment is then applied to match the cryoscopic mole weight. This adjustment applies an exponential correction to the distribution which is proportional to carbon number.

This adjusted C_7+ distribution is mated with the appropriate overall composition, nitrogen through hexanes, with the C7+ fraction lumped as a single component. The lumped C_7+ fraction is distributed among the carbon number fractions shown in Figure 1. The development continues with the selection of the appropriate set or array of PNA (Paraffins, Aromatics, and Naphthenes) proportions for the extended pseudocomponent analysis. This C_7+ characterization method is an extension of ideas described in the AGA monograph by Bergman et al. (4). The subsequent adjustment of some pseudocomponent properties to match experimental data is, in effect, a further refinement of these PNA ratios.

Consolidation of the extended C_7+ description to 14, 7, 3, or 2 pseudocomponents is carried out next. For descriptions with 14 or 7 pseudocomponents this consolidation procedure is designed to give approximately equal compositions for all pseudocomponents. Thus, a wider range of less abundant heavier carbon number fractions is combined to form the heavier pseudocomponents. An example showing these 14 pseudocomponents is shown in Table I. For three pseudocomponents consolidation of the distribution at carbon number 12 and carbon number 20 is selected. For two components division at carbon number 20 and 12 was tested for base oil calculations only. As discussed below, division at carbon number 20 gave better results.

Development of Base Oil Parameters

In addition to the base oil data provided by Turek et al. (2) and

TABLE II. INTERACTION PARAMETERS (C_{ij}, D_{ij}) FOR THE GENERALIZED REDLICH-KWONG EQUATION OF STATE FOR OIL D, SAN ANDRES FORMATION, EVALUATED AT 105°F

ID NUMBER[1]	C_{ij} [2]	D_{ij} [2]
2 3	*-0.090000[3]	-0.038260
2 4	0.005000	0.0
2 5	0.084700	0.0
2 6	0.010000	0.0
2 7	0.010000	0.0
2 8	0.010000	0.0
2 9	0.010000	0.0
2 10	0.010000	0.0
2 11	0.010000	0.0
2 (27-40)	*0.010000	0.0
3 4	0.156698	-0.033048
3 6	0.152101	-0.029926
3 7	0.149255	-0.027993
3 8	0.148010	-0.027147
3 9	0.146046	-0.025813
3 10	0.143313	-0.023956
3 11	0.139264	-0.021206
3 27	*0.090000	-0.023074
3 28	*0.090000	-0.020501
3 29	*0.090000	-0.018062
3 30	*0.090000	-0.015436
3 31	*0.090000	-0.012733
3 32	*0.090000	-0.010130
3 33	*0.090000	-0.009777
3 34	*0.090000	-0.007380
3 35	*0.090000	-0.005082
3 36	*0.090000	-0.002021
3 37	*0.090000	0.002733
3 38	*0.090000	0.011071
3 39	*0.090000	0.018416
3 40	*0.090000	0.034575
3 5	*0.160000	0.0
5 4	0.090500	0.0
5 6	0.084600	0.0
5 7	0.082000	0.0
5 8	0.080000	0.0
5 9	0.078000	0.0
5 10	0.075000	0.0
5 11	0.070539	0.0
5 (27-40)	*0.150000	0.0

[1]Component ID Nos. are identified with specific compounds in Table I.
[2]Default C_{ij} and D_{ij} for carbon dioxide are functions of temperature and acentric factor. See References (1) and (2).

[3]An * denotes generalized San Andres C_{ij} Value. D_{ij}'s are all default values.

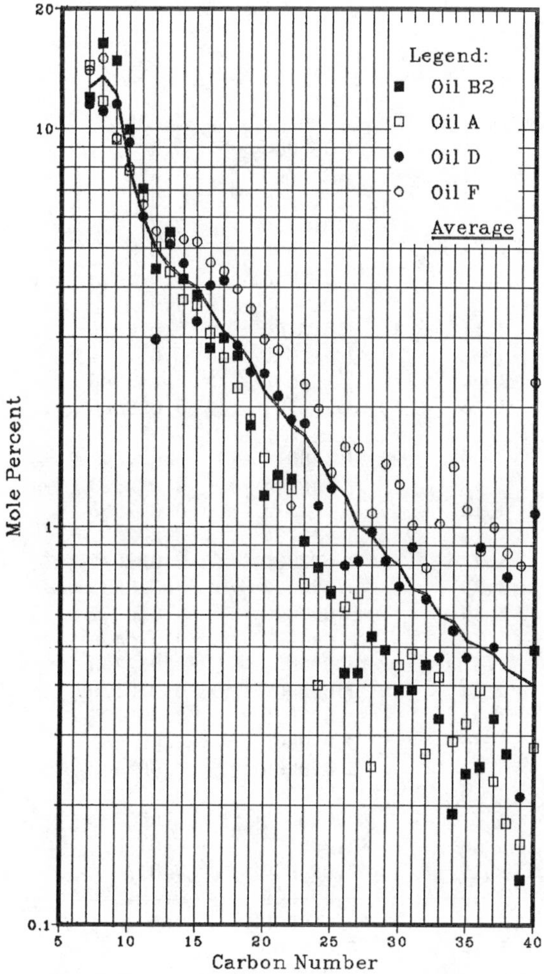

Figure 1. Comparison of Heavy Hydrocarbon Distributions. Results for Fluids from the San Andres Formation: Analyses for Several Fields and Smoothed Average.

Yarborough (5), extensive results for San Andres fluids have been accumulated from other enhanced gas drive studies as well as from historical measurements during the pressure depletion phase of reservoir operation. Table III compares the fluid compositions used

TABLE III. COMPARISON OF CANDIDATE DATABASE COMPOSITIONS

All Compositions are in Mol%

	Database Case[1]								
Comp.	B3[2]	DSX	B	F1	F	A	B1	D	B2
N_2	0.55	1.08	1.35	0.47	0.54	2.15	0.52	0.00	0.41
C_1	15.61	9.63	18.19	19.03	16.12	17.78	15.43	15.24	13.06
CO_2	0.61	1.72	0.29	2.63	4.68	4.74	0.56	0.00	0.49
C_2	7.43	4.60	5.25	7.86	13.50	8.57	7.26	3.69	7.04
H_2S	0.00	0.41	0.00	0.00	0.00	1.66	0.00	0.00	0.04
C_3	6.32	5.41	4.41	7.60	10.03	7.71	6.14	4.66	7.18
iC_4	0.48	1.29	1.10	1.29	1.92	0.94	0.49	0.00	1.59
nC_4	4.62	4.11	3.57	3.91	6.52	4.71	4.10	3.89	4.57
iC_5	1.18	1.96	1.77	1.65	1.47	1.58	1.36	1.49	2.26
nC_5	2.93	2.66	3.34	1.71	1.95	3.21	2.69	1.80	2.72
C_6	4.69	6.48	3.39	3.44	3.16	3.17	3.61	2.37	3.21
C_7+	55.58	60.65	57.34	50.41	40.11	43.79	57.84	66.86	57.48
C_7+ Properties									
Mol Wt	250	232	242	215	243	253	236	232	240
Sp Gr	0.8822	0.8880	0.8861	0.8805	0.8839	0.889	0.8870	0.8874	0.8870

[1]Database cases are given letter designations consistent with those used by Turek, et al. (2)
[2]B3 refers to an alternate analysis of Oil B1 obtained by an independent laboratory.

in these studies. These data were screened, as described below, for inclusion in a San Andres Formation base oil database by making fluid property predictions using default equation of state parameters developed from previous experience with hydrocarbon mixtures. Phase equilibria were calculated with the Redlich-Kwong equation of state; the Standing-Katz correlation (6) was used to calculate the liquid densities. As such, this method has been found to generally provide accurate results for oil bubble points and densities but cannot be used for CO_2-oil mixtures. Its value here lies in establishing consistency between measured fluid composition and fluid properties.

In general, a case was rejected for inclusion in the database if the reported liquid density was in serious disagreement with the Standing-Katz value (that is, greater than 3%) or if the bubble point pressure could not be matched with reasonable adjustment of the C_7+ acentric factors (that is, less than 20%). Only reservoir oil density data for Oil B2 were rejected on these grounds, and, as a consequence, Oil B2 was not included in the database.

Refined C_7+ descriptions for each of the selected database cases were then prepared. These descriptions were based upon C_7+ distribution, molecular weight, and PNA distribution. Further adjustments by trial and error were made to optimally match the measured base oil properties. These data included single phase densities, usually obtained at 1500-2000 psia and reservoir temperature and the bubble point pressure also at reservoir temperature. To bring about agreement of calculated and experimental results, the pseudocritical pressures and methane interaction parameters with selected C_7+ hydrocarbon fractions were adjusted.

Changes in the critical pressure affect primarily the density while changes in the methane-heavy hydrocarbon interaction parameter affect primarily the bubble point pressure. These two parameters are nearly independent in their effect; thus, agreement of calculated and experimental values for these data is easily achieved. As noted above, adjusting the critical pressure of a C_7+ pseudocomponent is tantamount to changing its PNA distribution to reflect the specific characteristics of the fluid. The resulting adjusted parameter sets for each database entry were arithmetically averaged to provide a general San Andres Formation description for the base oil.

Table IV shows these averaged parameters for 7, 3 and 2 C_7+ pseudocomponents. Those for 14 pseudocomponents are given in Table I and include acentric factor and pseudocritical pressure values for components Nos. 38, 39 and 40. To create a C_7+ fraction with

TABLE IV. ADJUSTED EQUATION OF STATE PARAMETERS FOR SAN ANDRES FLUIDS USING 7, 3 and 2 C_7+ PSEUDOCOMPONENTS

	C_{ij} x 1000			P_c's, psia					
				7		3		2	
	7	3	2	C22-27	C28+	C13-20	C21+	C7-20	C21+
AVE	10	12	16	156.3	103.3	257.99	125.55	334.0	127.7

C_{ij} denotes interaction parameters for methane with each of the 14 pseudocomponents.

P_c's are critical pressures for the pseudocomponent carbon number grouping.

two pseudocomponents, the C_7+ distribution was split at carbon number 20 or 12. Results for division at carbon number 20 are shown in Table IV. One effect of reducing the number of pseudocomponents is to increase the size of the interaction parameter with methane. This increase is a consequence of the increased solubility of methane in the more severely compressed and, therefore, lighter pseudocomponents. The change in pseudocomponent character is also reflected by the increase in critical pressure.

Table V compares results for the various C_7+ compressions. Note the insensitivity of the AAD (average absolute deviation) to changes in C_7+ compression and that for only two pseudocomponents division at carbon number 20 gives better results. In addition to interpolation within the database cases themselves, the averaged parameter sets were also tested by prediction of historical measurements of reservoir oil properties. Table VI lists the

TABLE V. COMPARISON OF EXPERIMENT WITH CALCULATIONS USING AVERAGE PARAMETERS FOR THE DATABASE CASES USING 14, 7, 3 AND 2 C_7+ PSEUDOCOMPONENTS

	Bubble Point Pressure, psia						Single Phase Densities, gm/cc					
OIL	Exp	14	7	3	2A[1]	2B[1]	Exp	14	7	3	2A	2B
B1	800	801	800	804	819	810	0.811	0.817	0.817	0.820	0.829	0.818
DSX	628	582	580	579	572	575	0.832	0.825	0.823	0.821	0.817	0.819
B	1064	1008	1005	1007	1013	1019	0.815	0.815	0.814	0.816	0.819	0.817
F1	1033	1039	1036	1025	999	1003	0.788	0.785	0.784	0.779	0.762	0.768
F	1000	990	988	989	995	989	0.776	0.777	0.777	0.778	0.781	0.779
A	1190	1247	1244	1251	1267	1262	0.792	0.796	0.790	0.799	0.810	0.805
B1	749	788	787	787	788	784	0.815	0.815	0.815	0.814	0.813	0.813
D	620	669	668	667	671	663	0.827	0.830	0.830	0.828	0.827	0.825
AAD %		4.0	4.1	4.2	5.0	4.6		0.44	0.40	0.65	1.37	0.94

[1]2A denotes division of the C_7+ distribution into pseudocomponents at carbon number 12; 2B refers to division at carbon number 20.

screened historical data and compares the reported compositions. Cases No. 558, 238, 241 and 444 were accepted as reported. Calculated results for these cases are shown in Table VII.

Note the excellent agreement between calculated and experimental values. Recall these data were obtained 30 to 40 years ago; agreement is probably within experimental uncertainty. Of special interest are results for oil formation volume factor (FVF). It is well represented at both the bubble point and 2000 psia. This agreement suggests that the average description fits not only fluid density at a point but fluid compressibility over a range of pressure as well. Note further that the wells cited in these cases span a range of temperatures from 94° to 114°F while the parameters were developed from data at 105° and 106°F. Thus, the averaged parameters may be extrapolated without loss of precision at least 10°F above or below the temperature at which they were developed.

Development of Fluid Descriptions for Carbon Dioxide-Oil Mixtures

The preceding development of base oil descriptions for the forma-

TABLE VI. COMPARISON OF HISTORICAL DATABASE COMPOSITIONS

All Compositions are in Mol%

Case	213	214	215	216	217	238	241	244	444	553	558	561	642
N_2	0.39	---	---	---	---	2.33	2.87	0.13	0.80	2.16	1.58	0.75	---
C_1	6.64	4.00	3.21	2.81	3.22	18.30	22.94	18.07	22.61	16.70	18.24	2.74	19.41
CO_2	0.50	---	---	---	---	0.66	0.15	3.86	4.04	0.09	0.23	0.52	---
C_2	7.78	3.94	4.06	3.62	4.22	8.20	8.20	6.75	7.60	7.77	8.61	5.15	7.72
H_2S	1.05	---	---	---	---	0.08	0.00	1.33	0.92	0.03	0.00	0.52	---
C_3	9.67	10.54	8.40	11.07	9.13	7.37	6.88	5.88	6.98	7.25	7.31	6.11	7.61
iC_4	1.34	1.99	2.83	2.06	3.45	1.15	1.34	1.46	1.22	1.35	1.04	1.37	2.18
nC_4	5.59	6.41	6.89	5.92	5.32	4.19	4.25	3.78	4.14	4.38	4.38	3.91	4.72
iC_5	2.26	0.15	0.23	0.12	0.73	1.94	1.67	1.48	1.65	2.03	1.47	2.15	1.22
nC_5	2.57	6.23	6.28	5.19	1.87	2.44	2.07	2.81	2.22	1.71	1.89	2.24	2.60
C_6	2.70	4.58	4.74	4.46	5.18	3.48	3.39	3.31	3.46	1.23	1.95	2.50	3.14
C_7+	59.51	62.16	63.36	64.75	63.88	49.86	46.24	51.64	44.36	55.30	53.30	72.04	51.40
C_7+ Properties													
Mol Wt	238	225	223	216	219	233	242	236	271	225	223	223	230
Sp Gr	0.8794	0.8778	0.8775	0.8852	0.8822	0.8888	0.8816	0.8805	0.8990	0.8899	0.8905	0.8849	0.8888
Reservoir T, (°F)	95	94	94	94	93	108	114	107	109	108	110	102	108
Bubble Pt. (psia)	419	320	264	436	284	1122	1478	1072	1288	1380	1198	175	1493
GOR at Bubble Pt. (SCF/bbl)	204	162	150	175	149	320	384	328	394	326	365	103	420

TABLE VII. SAN ANDRES RESULTS FOR HISTORICAL CASES: COMPARISON OF CALCULATED AND EXPERIMENTAL RESULTS

C_7+ Compression	AAD%				
	BPP	GOR	FVF(BPP)	FVF(2000)	DVA(resid)
14	5.44	6.56	1.59	1.62	1.13
7	5.31	6.56	1.59	1.63	1.17
3	5.79	5.99	1.47	1.51	1.11
2	6.61	6.62	1.53	1.58	1.36

where BPP denotes bubble point pressure, psia,
GOR denotes gas/oil ratio, SCF/bbl of residual oil,
FVF(BPP) denotes formation volume factor at the bubble point, bbl reservoir fluid/bbl residual oil
FVF(2000) denotes formation volume factor at 2000 psia,
DVA denotes DVA residual oil specific gravity.

tions of interest suggests that hydrocarbon phase behavior is well described by the modified Redlich-Kwong equation of state. Thus, extension of those reservoir oil descriptions to carbon dioxide-oil mixtures ought to be brought about by changing only carbon dioxide-related interaction parameters. Parameters related to hydrocarbon pseudocomponents ought to remain unchanged.

Early in the investigation it was recognized that the physical situation was too complex to be represented by such an approach. Attempts to fit the CO_2-oil mixture data obtained by Turek, et al. (2) with only CO_2 parameter adjustments proved unsuccessful. In particular, the liquid-liquid branch of the saturation pressure-composition (P-X) locus was poorly represented. Often single phase behavior was predicted at compositions where two-phase behavior was observed.

However, prior to the present investigation, calculations with Oil B2-CO_2 mixtures led to the realization that large acentric factors for the heaviest pseudocomponents (characteristic of the Edmister correlation (7)) used in conjunction with the characterization procedure described by Yarborough (1) could bring the liquid-liquid branch of the P-X locus into better agreement with experiment. See Figure 2. For this investigation the decision was made to adjust the C_7+ acentric factors assigned through the characterization procedure discussed earlier and give up some precision in the calculation of base oil density in order to enhance the prediction of phase behavior. Note that Edmister-like acentric factors are typically larger by some 5-10% than those provided by our C_7+ characterization procedure utilizing PNA distributions.

The phase equilibrium data for Oils B1, B2, D, F and A form the first-contact data set for matching. The compositions of the base oils used in these carbon dioxide mixture studies are reported in Table VI. The matching procedure was started with Oil B2-CO_2 mixtures because of earlier success in matching calculated and

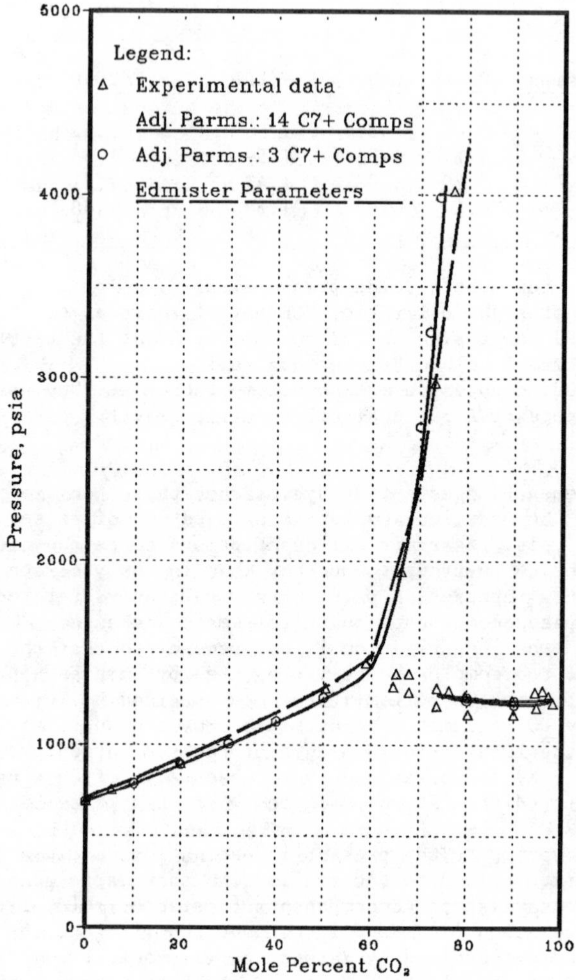

Figure 2. CO_2-Oil B2 System: Comparison of Saturation Loci at 106°F. Experimental Values Versus Calculations for 14 and 3 C7+ Pseudocomponents.

experimental results for that system. An overall compromise fit to the first-contact data for Oil B2 was achieved through trial and error adjustments of the acentric factors for the C_{20}+ pseudocomponents as well as through adjustments to the carbon dioxide-C_7+ and carbon dioxide-methane interaction parameters.

Results for Oil B2 are collected in Table VIII for density, phase distribution and coexisting phase composition. The 2.5% error in base oil density incurred by the new parameters is still within acceptable limits. Moreover, they now allow the carbon dioxide-rich phase equilibria to be calculated with satisfactory precision as well. Further results with these new parameters are given in Figure 2 which shows a comparison of experimental saturation loci with that calculated in this investigation and using the Edmister correlation. Note the excellent agreement for all properties except three-phase consolute pressure loci.

Carbon dioxide compositions are well represented. Only for the mixture with 95 mol% carbon dioxide is there serious disagreement between calculated and experimental K-values. However, comparison with other oil-rich fluid compositions from San Andres oil-CO_2 systems shows that the reported CO_2 composition of 81.4 mol% in the oil-rich phase is probably in error. See Table VIII. Note also the similarity of results for calculations with only 3 pseudocomponents to those with 14 which once again underscores the insensitivity of these results to C_7+ breakdown.

Tables IX and X extend the application of these new acentric factors to all of the remaining first-contact data for San Andres fluids obtained by Turek, et al. (2). Because these are now predictions rather than interpolations, some loss of accuracy is expected. Typically, Table IX shows the base oil density to be in error by approximately 3%. On the other hand, the calculated oil-rich phase densities remain in excellent agreement with experiment. Note the trend in calculated results for oil-rich and CO_2-rich liquid density for the Oil D-CO_2 system. There is a substantial improvement in density prediction as the carbon dioxide composition is increased to values consistent with the flooding process. CO_2-rich liquid densities, on the other hand, are still in error by 5-10%. Again, the table shows the similarity of results for 3 and 14 C_7+ components.

Except for the Oil F case, Table X shows that the CO_2 K-values and individual carbon dioxide phase compositions are in reasonable agreement with experimental values. For most cases they agree within 5 or 6%. For the Oil F case the reported composition of 60 mol% is suspect; it is possible that the individual phase compositions are also in error.

While Figures 3 and 4 show that the P-X locus is well represented along two-phase boundaries, three-phase equilibria are poorly represented. The three-phase region fails to extend across the entire experimentally observed composition range. Furthermore, the pressure range for the three-phase region is too narrow, that is, the upper consolute pressure is too near the lower consolute pressure. As shown in Figure 4, for mixtures containing Oil A this failure is especially pronounced.

Similarly, phase distribution results are not in good agreement with experiment. Usually too much oil-rich liquid is pre-

TABLE VIII. FIRST CONTACT CO_2-OIL DATA USED FOR PARAMETER ADJUSTMENT
RESULTS FOR 14 AND 3 C_7+ PSEUDOCOMPONENTS
BASIS: OIL B2 DATA AT 106°F AND 2200 PSIA

Composition,	Density, gm/cc					
	Oil-Rich Liquid			CO_2-Rich Liquid		
Mol% CO_2	Exp't	Calc(14)	Calc(3)	Exp't	Calc(14)	Calc(3)
Base Oil	0.820[1]	0.841	0.842	---	---	---
75	0.865	0.865	0.867	0.766	0.756	0.752
85	0.892	0.886	0.888	0.785	0.769	0.766
95	0.919	0.918	0.917	0.786	0.771	0.769

	Liquid-Liquid Phase Equilibria											
	Phase Distribution (Vol% Oil-Rich Liquid)			Phase Composition (Mol% CO_2)						CO_2 K-Value		
				Oil-Rich Phase			CO_2-Rich Phase					
	Exp't	Calc(14)	Calc(3)	Exp't	Calc(14)	Calc(3)	Exp't	Calc(14)	Calc(3)	Exp't	Calc(14)	Calc(3)
75	75.3	75.5	74.2	66.7	69.3	68.6	86.0	86.6	87.0	1.29	1.25	1.27
85	38.1	39.9	40.1	67.7	69.5	68.9	92.0	91.0	91.2	1.36	1.31	1.32
95	9.6	9.2	9.6	81.4[2]	66.2	66.1	98.1	96.3	96.3	1.21	1.46	1.46

[1]Estimated from Standing-Katz correlation.
[2]Value believed to be in error.

TABLE IX. COMPARISON OF CALCULATED AND EXPERIMENTAL FIRST CONTACT DATA FOR SAN ANDRES FLUIDS

	Density, gm/cc						Phase Distribution (Vol% Oil-Rich Phase)		
	Oil-Rich Phase			CO_2-Rich					
Mol% CO2	Exp't	Calc(14)	Calc(3)	Exp't	Calc(14)	Calc(3)	Exp't	Calc(14)	Calc(3)
Oil B1 at 2015 psia, 106°F									
Base Oil	0.815	0.837	0.838	---	---	---	---	---	---
50	0.827	0.848	0.848	---	---	---	---	---	---
78	0.873	0.870	0.871	0.766	0.738	0.734	46.2	62.2	61.1
95	0.920	0.913	0.914	0.767	0.751	0.748	8.5	10.0	9.4
Oil D at 2000 psia, 105°F									
Base Oil	0.827	0.854	0.853	---	---	---	---	---	---
25	0.835	0.857	0.857	---	---	---	---	---	---
50	0.840	0.861	0.861	---	---	---	---	---	---
71.2	0.861	0.867	0.869[1]	0.743	0.733	0.731[1]	68.5	88	84[1]
80	0.878	0.880	0.880	0.773	0.752	0.748	40.3	61	59.9
85	0.892	0.888	0.888	0.782	0.758	0.754	31.1	44	43.4
95	0.924	0.915	0.913	---	0.757	0.754	8.1	11	11.0
Oil F[2] at 2015 psia, 105°F									
Base Oil	0.776	0.799	0.801	---	---	---	---	---	---
30	0.789	0.809	0.811	---	---	---	---	---	---
50	0.790	0.815	0.817	---	---	---	---	---	---
60[3]	0.811	0.825	0.828	0.737	0.686	0.683	73.6	91	88
Oil A[2] at 2015 psia, 105°F									
Base Oil	0.792	0.819	0.823	---	---	---	---	---	---
50	0.813	0.831	0.833	---	---	---	---	---	---
75	0.864	0.866	0.870	0.749	0.705	0.703	40.8	56.5	56.1
95	0.928	0.918	0.920	0.767	0.742	0.740	7.3	8.5	7.62

[1]Calculated result for 71.98 mol% CO_2.
[2]Compositions are in mol% added CO_2.
[3]Calculated results for 63.5 mol% added CO_2.

TABLE X. COMPARISON OF CALCULATED AND EXPERIMENTAL FIRST CONTACT DATA FOR SAN ANDRES FLUIDS

	Phase Composition, mol% CO_2								
	Oil-Rich Phase			CO_2-Rich Phase			CO_2 K-Value		
Mol% CO_2	Exp't	Calc(14)	Calc(3)	Exp't	Calc(14)	Calc(3)	Exp't	Calc(14)	Calc(3)
Oil B1 at 2015 psia, 106°F									
Base Oil	---	---	---	---	---	---	---	---	---
50	---	---	---	---	---	---	---	---	---
78	65.1	68.4	67.8	85.1	88.0	88.3	1.31	1.29	1.30
95	68.3	66.2	65.7	95.3	96.0	96.4	1.40	1.45	1.47
Oil D at 2000 psia, 105°F									
Base Oil	---	---	---	---	---	---	---	---	---
25	---	---	---	---	---	---	---	---	---
50	---	---	---	---	---	---	---	---	---
71.2	65.5	67.8	67.4[1]	86.5	86.8	87.3[1]	1.32	1.28	1.30[1]
80	---	69.2	68.6	91.0	90.0	90.3	---	1.30	1.32
85	65.9	69.3	68.7	93.9	91.9	92.1	1.42	1.33	1.34
95	66.2	66.4	66.2	---	96.5	96.6	---	1.45	1.46
Oil F[2] at 2015 psia, 105°F									
Base Oil	---	---	---	---	---	---	---	---	---
30	---	---	---	---	---	---	---	---	---
50	---	---	---	---	---	---	---	---	---
60[3]	59.2	63.5	62.9	68.0	77.8	78.0	1.15	1.23	1.24
Oil A[2] at 2015 psia, 105°F									
Base Oil	---	---	---	---	---	---	---	---	---
70	---	---	---	---	---	---	---	---	---
75	65.1	64.8	64.3	82.3	85.3	85.5	1.26	1.32	1.33
95	64.4	64.1	63.8	92.6	95.9	96.3	1.44	1.50	1.51

[1]Calculated result for 71.98 mol% CO_2.
[2]Compositions are in mol% added CO_2.
[3]Calculated results for 63.5 mol% added CO_2.

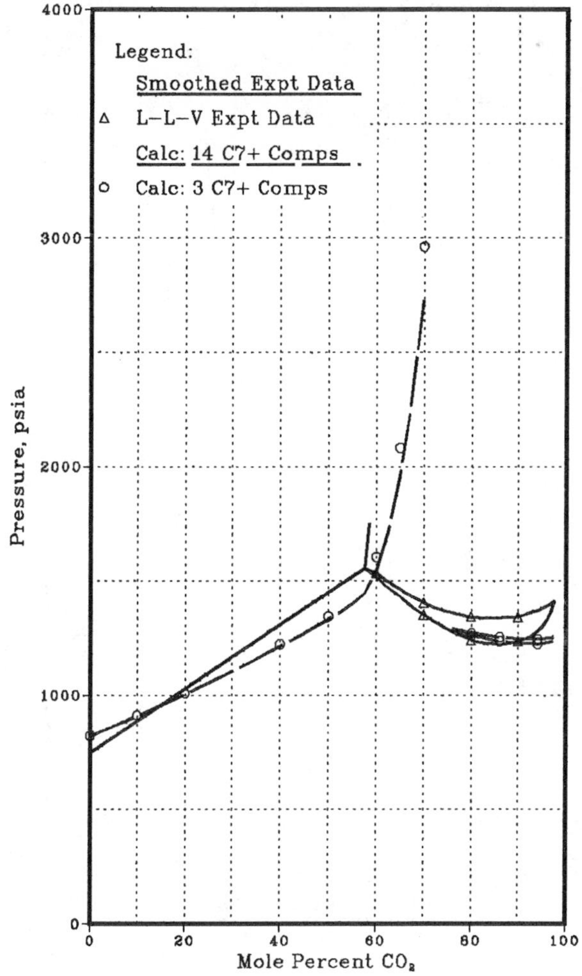

Figure 3. CO_2-Oil B1 System: Comparison of Saturation Loci at 106°F. Experimental Values Versus Calculations for 14 and 3 C7+ Pseudocomponents.

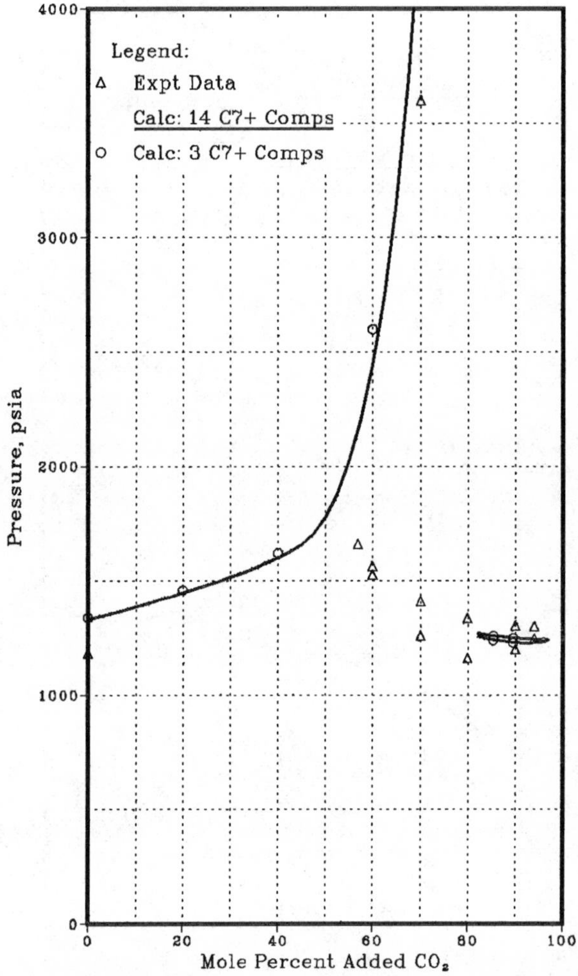

Figure 4. Carbon dioxide-Oil A system: Comparison of Saturated Loci at 105F. Experimental values versus calculations for 14 and 3 C7+ pseudocomponents.

dicted. Errors are large especially at the lower carbon dioxide compositions where the premature termination of the three phase region occurs. Poor phase volume predictions may be attributed, at least in part, to this failure to calculate three-phase behavior across the entire experimentally observed composition range. Note that the predictions improve somewhat at higher carbon dioxide compositions where three-phase behavior is predicted. See Table IX.

Additional insight into this discrepancy is obtained through comparison of the calculated and measured P-X loci. In the liquid-liquid region the quality lines are very steeply sloped and closely spaced as they emanate from the upper consolute locus. See Figure 5. Because the quality lines are so steeply sloped in the liquid-liquid region, small compositional errors can lead to large errors in the calculated P-X locus and in the locations of quality lines. Such errors can lead to poor results for calculated phase distributions as noted in Table IX. Moreover, the likelihood of errors is increased for cases where the distortion in quality lines is made worse because of a poor fit to the experimental saturation pressure locus. The results for Oil F, shown in Figure 5, are an example of this behavior.

A further example of this behavior is shown in Figure 3. The calculated saturation locus is again more gently sloped than the experimental locus and, so again, cuts into a region that experimentally contains two coexisting phases. Thus, too much oil-rich liquid is expected. Less understandable is the same prediction for Oil A. Figure 4 shows that the calculated P-X locus lies to the right of the experimental one. Too little oil-rich liquid is expected. Hence, the error in the slope of the quality lines exists even when the shape of the saturation locus is correct.

Extension of the San Andres Description to Acid Gas-Oil Mixtures

Because of interest in hydrogen sulfide mixed with carbon dioxide as a displacing medium, this section explores calculation of experimental data obtained from mixtures of carbon dioxide and hydrogen sulfide in proportions of 80-20 or 70-30 mol%. Table XI reviews the available base oil compositions. Yarborough (5) and Core Laboratories (in a private study for Amoco) have obtained data on 80-20 CO_2/H_2S mixtures. The phase behavior observed by Core Laboratories is shifted slightly toward higher acid gas compositions due in part to a slight difference in the preparation of the acid gas; their acid gas was found to contain 21 mol% H_2S while Yarborough had 20 mol%. In an earlier Core Laboratories study data were obtained for a 70-30 proportion of CO_2 and H_2S. Because this Core Laboratory study reported no composition, an oil of composition similar to Oil DS is assumed.

As in the carbon dioxide-oil studies, the philosophy of the present approach was that parameter adjustments for oil-H_2S interactions were all that would be needed to bring about agreement between experimental and calculated values. Tables I and II show the recommended parameters for 14 pseudocomponents; a three pseudocomponent description was not prepared. Through trial and error adjustments, an overall fit to the P-X locus was obtained for an 80-20 ratio. Experimental and calculated results are compared in

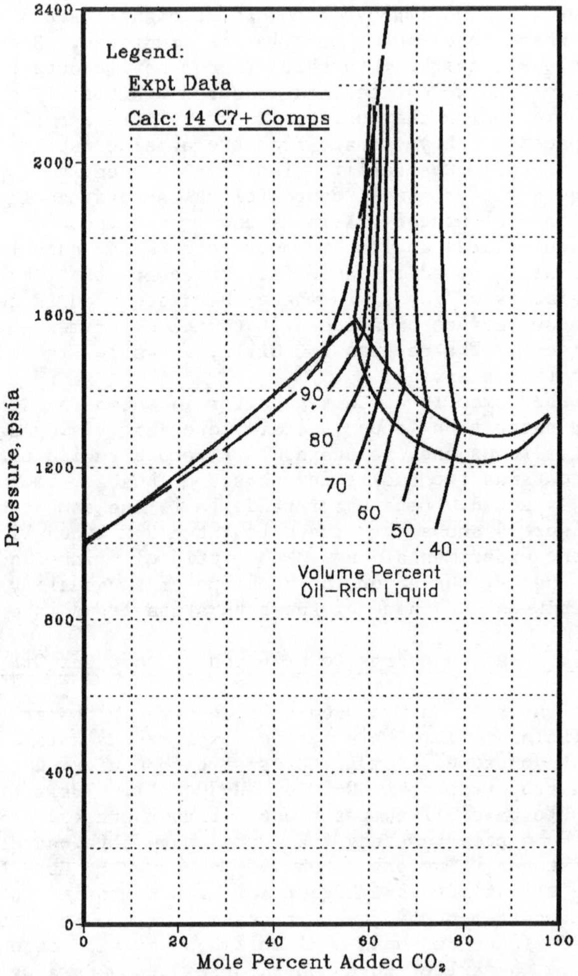

Figure 5. CO_2-Oil F System: Intersection of Calculated Saturation Locus with Experimental Quality Lines at 105°F.

TABLE XI. COMPARISON OF SAN ANDRES OILS USED IN SOUR GAS STUDIES

Component	Oil DS	Oil DSX
N_2	---	1.08
C_1	10.87	9.63
CO_2	1.10	1.72
C_2	3.66	4.60
H_2S	0.10	0.41
C_3	4.35	5.41
iC_4	1.16	1.29
nC_4	4.36	4.11
iC_5	1.82	1.96
nC_5	2.73	2.66
C_6	2.77	6.48
C_7+	67.08	60.65
C_7+ Properties		
Mol Wt	223	232
Sp Gr	0.8863	0.8880

Figure 6. Calculations for Figure 6 were made using the composition of Oil DS shown in Table XI. These parameters were then used to predict the P-X locus for the 70-30 mixture. Comparison of calculated and experimental results are shown in Figure 7, using Oil DS composition.

Note that for the 80-20 acid gas the three-phase region is found to extend across the entire composition range from the saturation locus to virtually pure acid gas. This behavior contrasts with the very narrow region of three-phase equilibria observed for carbon dioxide systems. For the 70-30 CO_2/H_2S mixture three-phase behavior is also nearly as well represented with the three-phase region extending almost to the liquid-liquid branch.

Table XII compares two-phase equilibrium results for a composition near the droplet point of about 71 mol% added 80/20 acid gas. Note the excellent agreement for phase split and equilibrium phase compositions. The oil-rich liquid density is in error by about 2%, characteristic of San Andres Formation oil-CO_2 mixture calculations generally.

Figures 8 and 9 compare constant composition volumetric expansion (CCVE) data for the 80-20 and 70-30 CO_2/H_2S mixtures. For the 80-20 case, Figure 8 displays CCVE behavior for an acid gas concentration below the observed mixture critical point. The calculations for 25 and 50 mol% added acid gas (using the Oil DSX composition) are in good agreement with experimentally observed values. However, for compositions richer than the critical point (about 72 mol% added acid gas) CCVE behavior is not so well represented.

As shown in Figure 6, the calculated critical point is too high by 1800 psia. A similar error is expected for the 70-30 case. This large error distorts the predicted quality lines away from

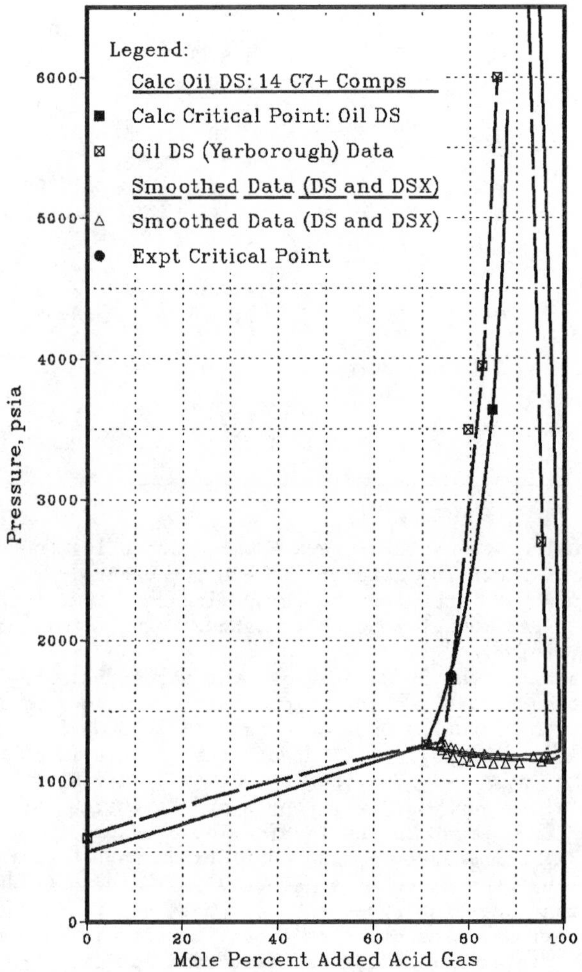

Figure 6. Acid Gas-Oil DS and DSX Systems: Comparison of Calculated Versus Experimental Saturation Pressure Loci at 105°F for 20% H_2S/80% CO_2 in the Acid Gas.

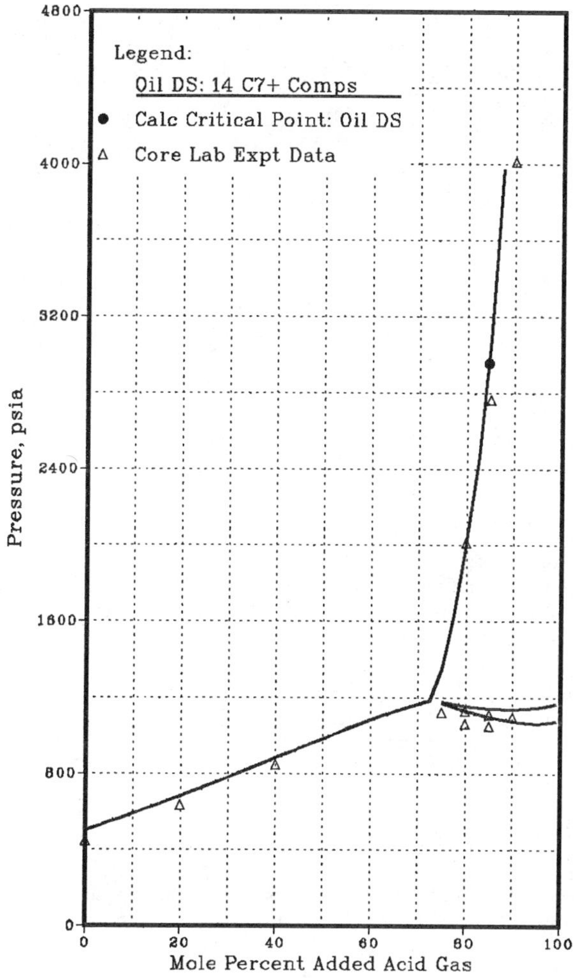

Figure 7. Acid Gas-Oil DS System: Comparison of Calculated Versus Experimental Saturation Pressure Loci at 105°F for 30% H_2S/70% CO_2 in the Acid Gas.

TABLE XII. LIQUID-VAPOR EQUILIBRIUM COMPOSITIONS FOR OIL DS ACID GAS MIXTURE: COMPARISON OF EXPERIMENT[1] WITH CALCULATIONS AT 105°F AND 1220 PSIA. BASIS: 71% ACID GAS: 80/20 CO_2/H_2S RATIO

Component	Liquid Exp't	Liquid Calc	Vapor Exp't	Vapor Calc
C_1	3.26	2.94	9.54	7.14
CO_2	52.20	56.04	76.90	76.05
C_2	1.59	1.06	1.11	1.00
H_2S	15.13	14.28	9.79	14.08
C_3	1.79	1.30	0.83	0.65
iC_4	0.41	0.35	0.17	0.12
nC_4	1.44	1.31	0.50	0.37
C_5	0.44	0.55	0.18	0.11
nC_5	0.54	0.83	0.20	0.13
C_6	1.72	0.84	0.24	0.08
C_7+	21.48	20.50	0.57	0.27
C_7+ Properties				
Mole Wt.	223	223	---	---
Sp. Gr.	0.8841	0.8863	---	---
Liquid Phase Density (gm/cc) at				
1800 psia	0.826	0.844	---	---
1500 psia	0.824	0.839	---	---
Volume Percent Liquid	93	92.3	7	7.7

[1]Experimental data are taken from Yarborough (5).

those occurring experimentally because the quality lines converge at the critical point.

For example, in Figure 9 results for the 70-30 mixture are shown at 80 and 90 mol% added acid gas. Although no critical point was determined experimentally for the 70-30 case, based on the critical point location for the 80-20 mixture these compositions are likely to lie on the dew point side of the critical composition. Because the calculated critical point occurs at a composition richer in acid gas than expected experimentally, it is not surprising that at 80 mol% added acid gas the calculated results shown in Figure 9 trend toward a bubble point, that is, 100 volume percent liquid, while the experimental data show a trend toward a dew point, that is, zero volume percent liquid. At 90 mol% added acid gas both the experimental and calculated critical compositions are exceeded, and a dew point is now calculated. But the calculated dew point pressure remains too high.

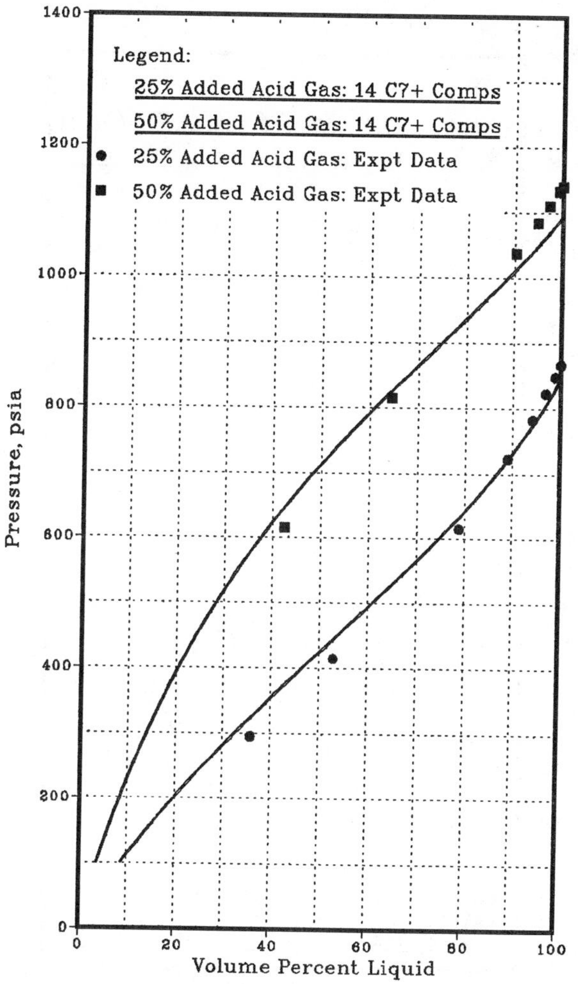

Figure 8. Acid Gas-Oil DSX System: Comparison of Calculated Versus Experimental Phase Distributions at 105°F for 25% and 50% Added Acid Gas with 20% H_2S/80% CO_2 in the Acid Gas.

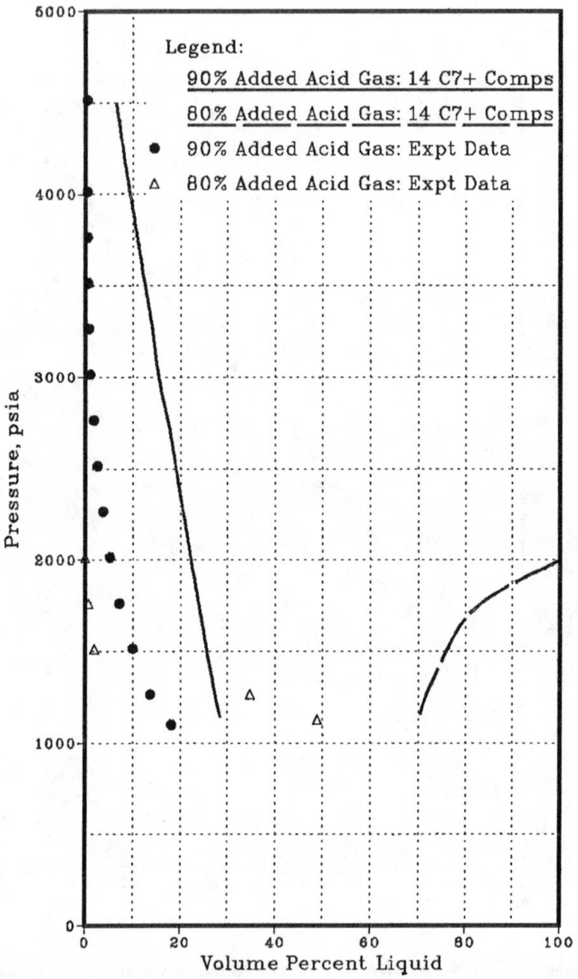

Figure 9. Acid Gas-Oil DS System: Comparison of Calculated Versus Experimental Phase Distributions at 105°F for 80% and 90% Added Acid Gas with 30% H_2S/70% CO_2 in the Acid Gas.

Conclusions

With appropriate parameter adjustments of the type discussed here, the modified Redlich-Kwong equation of state is capable of reproducing the volumetric properties and complex phase equilibria exhibited by San Andres Formation oil-acid gas systems. The adjusted parameters lead to prediction of liquid-vapor criticality, although the calculated critical pressure is too high by a significant amount. Two-phase saturation loci (L-L of L-V) are more accurately reproduced than those for three phases (L-L-V). Finally, changing the representation of the C_7+ fraction from 14 to 3 pseudocomponents does not significantly affect the precision of the calculated results.

Glossary of Symbols

C_{ij}, D_{ij} = interaction parameters; $1 \leqq i \leqq NC$, $1 \leqq j \leqq NC$

$C_{ij} = C_{ji}$ and $D_{ij} = D_{ji}$

NC = number of components

R = gas constant

P = pressure of fluid

P_{ci} = critical pressure of component i

T = temperature of fluid

T_{ci} = critical temperature of component i

V = specific volume of fluid

x_i = mole fraction of component i

Literature Cited

1. Yarborough, L. In "Equations of State in Engineering and Research"; Chao, K. C.; Robinson, R. L., Eds.; ADVANCES IN CHEMISTRY SERIES No. 182, American Chemical Society: Washington D. C., 1979, pp. 385-439.

2. Turek, E. A.; Metcalfe, R. S.; Fishback, R. E. AIME Preprints 1984, Society of Petroleum Engineers Paper No. 13117.

3. Morris, R. W.; Turek, E. A. In "Equations of State Theories and Applications"; Chao, K. C.; Robinson, R. L., Eds.; ACS SYMPOSIUM SERIES, To be published, American Chemical Society: Washington, D. C., 1985.

4. Bergman, D. F.; Tek, M. R.; Katz, D. L. "Retrograde Condensation in Natural Gas Pipelines"; AGA Project PR-26-69, 1975; pp. 143-148.

5. Yarborough, L. Internal Amoco Production Co. report.

6. Standing, M. B.; Katz, D. L. AIME Pet. Div. Tech. 1942, 146, 159-165.

7. Edmister, W. C. Pet. Refiner 1958, 37, 173-178.

RECEIVED November 8, 1985

21

Application of Cubic Equations of State to Polar Fluids and Fluid Mixtures

Gus K. Georgeton, Richard Lee Smith, Jr., and Amyn S. Teja

School of Chemical Engineering, Georgia Institute of Technology, Atlanta, GA 30332-0100

The limitations of cubic equations of state for phase equilibrium predictions involving polar fluids have been widely attributed to the inability of these equations to correlate the vapor pressures and densities of these fluids. This work examines the behavior of two and three constant cubic equations (in particular, those of Peng and Robinson, and Patel and Teja) in calculations of vapor-liquid equilibria, liquid-liquid equilibria and critical states of mixtures containing polar components.

In spite of their limitations, cubic equations of state are widely used in phase equilibrium calculations since they represent a satisfactory compromise between accuracy and speed of computation. The theoretical and practical limitations of these equations have been discussed by a number of workers including Abbott (1) and Vidal (2). In particular, the inability of two-constant cubic equations to accurately predict liquid densities has been documented in some detail. Vidal (2) has reviewed a number of three-constant cubic equations and shown that the addition of a third constant in general leads to improved density predictions.

In this work, we have compared representatives of two and three constant equations for their ability to correlate densities, vapor pressures, vapor-liquid equilibria, liquid-liquid equilibria and critical states of mixtures containing polar components.

Pure Fluid Calculations

As shown by Abbott, a general cubic equation may be written as follows:

$$P = \frac{RT}{V - \beta} - \frac{\alpha}{V^2 + \delta V + \varepsilon} \tag{1}$$

0097-6156/86/0300-0434$06.00/0

where particular choices for α, β, δ and ε for two of the more popular two constant cubic equations (the Redlich-Kwong-Soave and the Peng-Robinson equations) are given in Table I. Also shown in Table I are these constants for some recent three-constant cubic equations, namely those proposed by Schmidt and Wenzel (3), Harmens and Knapp (4), Heyen (5) and Patel and Teja (6). These three constant equations afford a means for generalizing the Soave and Peng-Robinson equations and of interpolating among them. Thus, the Schmidt-Wenzel equation may be viewed as an interpolation between the Soave equation (which gives good liquid density predictions for argon and methane, when $\omega \simeq 0$) and the Peng-Robinson equation (which gives good liquid density predictions for n-heptane, when $\omega \simeq 1/3$). Similarly, the Patel-Teja equation reduces to the Soave equation when $c = 0$, to the Peng-Robinson equation when $c = b$ and to the Schmidt-Wenzel equation when $c = 3\omega b$. Good density predictions can be obtained because of this added flexibility and, as shown below, these predictions also extend to polar components.

As was demonstrated by Soave (7), cubic equations may be used successfully in phase equilibrium calculations if the constant 'a' is made temperature dependent and if this temperature dependence is obtained from the vapor pressure of the pure component. Two of the common temperature functions proposed for 'a' are discussed below. The first is a quadratic, originally proposed by Soave and later used successfully in the Peng-Robinson and other equations of state. This is given by:

$$\alpha\,[T_R] = a_c\{1 + F(1 - T_R^{1/2})\}^2 \quad (2)$$

Both Soave and Peng and Robinson, however, correlated F with the acentric factor ω and their equations therefore worked best for nonpolar substances. Patel and Teja (6) later showed that if F is determined from the actual vapor pressures of the pure components, then the equation may also be used for polar substances. The functional form of the temperature dependence is, however, incorrect at high reduced temperatures -- since $\alpha[T_R]$ must decrease monotonously with T_R. Heyen suggested an exponential form for $\alpha[T_R]$ which is given by:

$$\alpha\,[T_R] = a_c \exp\{K\,(1 - T_R^{\,n})\} \quad (3)$$

This form is useful for supercritical gases such as H_2 and He and, indeed, because of the additional constant available for fitting, it correlates vapor pressures of many polar compounds extremely well (Table II). However, as shown in Figure 1 for ammonia, the two functional forms (including their first, second and third derivatives with temperature) show similar behavior over most practical ranges of temperature. Moreover, as will be shown below, any small improvement in pure component vapor pressure representation is not always apparent when mixture phase behavior is calculated. For most practical purposes, therefore, the two forms are equivalent in their ability to predict phase equilibria.

In order to compare the correlative and predictive capabilities of the two and three constant cubics we show below

TABLE I. Choice of Constants for the General Cubic Equation

	α	β	δ	ε
2-constant				
SRK	a[T]	b	b	0
PR	a[T]	b	2b	$-b^2$
3-constant				
SW	a[T]	b	$b + 3\omega b$	$-3\omega b^2$
HK	a[T]	b	bc	$-(c-1)b^2$
H	a[T]	b[T]	b[T] + c	$-$ b[T] c
PT	a[T]	b	b + c	$-$ bc

TABLE II. The Effect of Different Forms of α on Vapor Pressures

Compound	PR- ΔP(AAD%)	PT-1 Constant* ΔP(AAD%)	PT-2 Constant** ΔP(AAD%)
1-Butanol	10.81	7.13	1.33
1-Pentanol	14.70	10.17	0.47
1-Octanol	17.96	9.64	2.98
Acetic Acid	6.69	2.36	1.32
Butanoic Acid	7.69	8.56	4.54
Octanoic Acid	12.71	5.87	1.33

* Patel-Teja equation with Soave-type function for $\alpha[T_R]$ containing 1 constant.

** Patel-Teja equation with Heyen-type function for $\alpha[T_R]$ containing 2 constants.

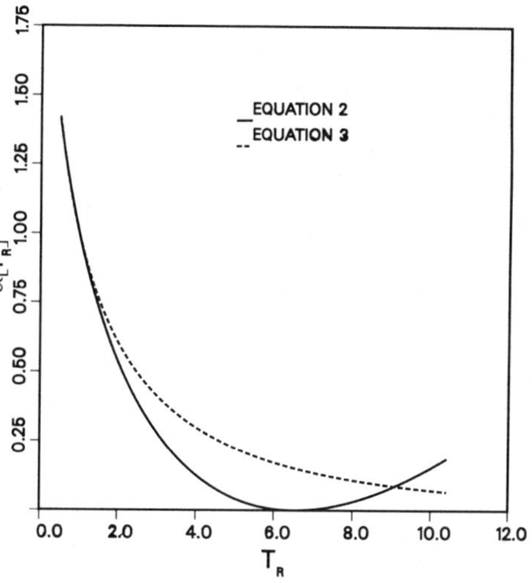

Figure 1. Comparison of the Two Forms of α for Ammonia.

the results of our calculations of liquid densities, vapor pressures, vapor-liquid and liquid-liquid equilibria, and critical points using the Peng-Robinson (PR) and Patel-Teja (PT) equations.

The Patel-Teja equation may be written as:

$$P = \frac{RT}{V - b} - \frac{a[T]}{V(V + b) + c(V - b)} \tag{4}$$

The equation of state constants may be obtained from the critical point by setting:

$$\left(\frac{\partial P}{\partial V}\right)_{T_c} = 0 \; ; \; \left(\frac{\partial^2 P}{\partial V^2}\right)_{T_c} = 0 \; ; \; \frac{P_c V_c}{RT_c} = \zeta_c \tag{5}$$

The third constant in the equation of state allows ζ_c to be chosen freely, unlike the case of the two constant equations. Acceptable prediction of both low and high pressure densities requires that, in general, ζ_c be greater than the experimental critical compressibility (Abbott (1)) and we have confirmed this for a large number of substances.

Application of equation (5) leads to:

$$a_c = \Omega_a (R^2 T_c^2 / P_c) \tag{6}$$

$$b = \Omega_b (RT_c/P_c) \tag{7}$$

$$c = \Omega_c (RT_c/P_c) \tag{8}$$

where

$$\Omega_c = 1 - 3\zeta_c \tag{9}$$

$$\Omega_a = 3\zeta_c^2 + 3(1 - 2\zeta_c)\Omega_b + \Omega_b^2 + 1 - 3\zeta_c \tag{10}$$

and Ω_b is the smallest positive root of:

$$\Omega_b^3 + (2 - 3\zeta_c)\Omega_b^2 + 3\zeta_c^2 \Omega_b - \zeta_c^3 = 0 \tag{11}$$

It should be noted here that if $\zeta_c = 0.3074$, these equations reduce to the Peng-Robinson equation and if $\zeta_c = 0.3333$, they reduce to the Soave equation. The Patel-Teja equation therefore retains many of the useful features of the Soave and Peng-Robinson equations.

If, for convenience, we choose the Soave-type function for $\alpha[T_R]$ (equation 2) then the Patel-Teja equation contains four substance-specific parameters T_c, P_c, ζ_c and F (for nonpolar substances, ζ_c and F can be correlated with ω, so that the equation contains only three substance-specific parameters, as shown elsewhere (6)). In general, ζ_c and F are obtained from a fit of vapor pressure and saturated liquid density data. Details of the objective functions and fitting procedures are given elsewhere (6). The constants are given in Table III for 27 substances, including polar substances such as alcohols and acids, along with absolute average deviations in vapor pressures and saturated liquid densities for both the PT and PR equations of state. Table IV contains values of F and ζ_c for compounds determined previously (6,8). It is interesting to note that ζ_c is found to be close to the SRK value of 0.3333 for small spherical molecules such as argon, methane, oxygen and nitrogen; and that it is close to the PR value of 0.3074 for moderately large hydrocarbons such as n-heptane. What is perhaps surprising is that a value close to 0.3074 is obtained for many moderately sized molecules, including those that are polar. The PR and PT equations therefore give very similar predictions of liquid densities (and of density-derived properties such as partial molar volumes) for moderately sized molecules. A significant improvement in liquid density prediction is seen when the PT equation is used to calculate densities of substances whose ζ_c is very different from 0.3074. This is shown in Figure 2 for water and in Figure 3 for methanol-water mixtures.

Table III shows comparisons between the vapor pressures and liquid densities of polar molecules predicted by the PR and PT equations. Not unexpectedly, allowing F to be determined from the actual vapor pressure, rather than the vapor pressure at T_R=0.7 (i.e. from the acentric factor) and an n-alkane correlation, leads to significant improvements in predictions, especially for polar fluids. Vapor pressure predictions can be further improved if the exponential form (with two substance-specific constants) is used, as shown in Table II. However, for many substances over practical ranges of reduced temperature, the two forms of $\alpha[T_R]$ give equivalent results. All calculations shown below therefore utilize the Soave-type temperature function with one substance-specific constant F.

One disadvantage of the PT equation is that F (and ζ_c) must be known for all pure fluids of interest. For nonpolar fluids, these constants can be related to the acentric factor as follows:

$$F = 0.452413 + 1.30982\omega - 0.295937\omega^2 \tag{12}$$

$$\zeta_c = 0.329032 - 0.0767992\omega + 0.0211947\omega^2 \tag{13}$$

For these fluids, the PT equation therefore uses the same input information as the PR equation (but is better able to predict densities of large molecules for which ζ_c < 0.3074). These correlations do not, however, hold for polar fluids. We have therefore attempted to develop such correlations for classes of

TABLE III. Substance Dependent Constants ζ_c and F for Several Compounds

Substance	ζ_c	F	P-T ΔP(AAD%)	P-T Δρ(AAD%)	P-R ΔP(AAD%)	P-R Δρ(AAD%)
Ammonia	0.283	0.642740	1.80	3.13	0.74	15.12
Benzene	0.311	0.698911	1.70	3.30	1.97	3.45
Toluene	0.306	0.753893	0.72	3.37	0.93	3.05
O-Xylene	0.305	0.812845	0.70	3.40	0.76	3.66
m-Xylene	0.301	0.816962	0.87	3.44	0.83	5.49
p-Xylene	0.300	0.807023	0.82	3.44	0.92	5.73
Methanol	0.274	0.965347	1.96	3.87	5.39	21.73
Ethanol	0.292	1.171714	1.13	3.99	1.12	10.89
1-Propanol	0.302	1.211304	4.95	3.79	5.23	5.46
1-Butanol	0.305	1.221182	7.13	3.41	10.81	3.98
1-Pentanol	0.308	1.240459	10.17	3.08	14.70	2.88
1-Hexanol	0.330	1.433586	8.42	8.29	14.64	13.02
1-Heptanol	0.301	1.215380	13.62	3.31	22.16	5.95
1-Octanol	0.308	1.270267	9.64	3.44	17.96	2.70
Acetic Acid	0.258	0.762043	2.36	3.32	6.69	32.62
Propanoic Acid	0.295	1.146553	5.64	4.80	5.25	42.81
Butanoic Acid	0.329	1.395151	8.56	5.30	7.69	8.88
Pentanoic Acid	0.292	1.174746	5.58	3.04	7.54	10.48
Hexanoic Acid	0.291	1.272986	5.56	3.44	36.73	8.47
Octanoic Acid	0.292	1.393678	5.87	2.93	12.71	10.96
Decanoic Acid	0.290	1.496554	3.83	4.65	18.33	11.74
Acetone	0.283	0.701112	1.34	3.23	2.66	15.73
Diethyl Ether	0.308	0.787322	0.60	3.67	0.65	3.19
Carbon Tetrachloride	0.314	0.694866	1.16	3.92	1.32	4.33
Ethyl Acetate	0.296	0.842965	2.12	3.45	1.60	8.49
nPropyl Acetate	0.295	0.882502	2.70	3.17	2.35	8.39
Diethyl Ketone	0.294	0.826046	1.39	2.95	1.27	8.79

TABLE IV. Compilation of Previously Determined Values of ζ_c and F

Compound	ζ_c	F
Argon	0.328	0.450751
Nitrogen	0.329	0.516798
Oxygen	0.327	0.487035
Methane	0.324	0.455336
Ethane	0.317	0.561567
Ethylene	0.313	0.554369
Propane	0.317	0.648049
Propylene	0.324	0.661305
Acetylene	0.310	0.664179
n-Butane	0.309	0.678389
i-Butane	0.315	0.683133
1-Butene	0.315	0.696423
n-Pentane	0.308	0.746470
i-Pentane	0.314	0.741095
n-Hexane	0.305	0.801605
n-Heptane	0.305	0.868856
n-Octane	0.301	0.918544
n-Nonane	0.301	0.982750
n-Decane	0.297	1.021919
n-Undecane	0.297	1.080416
n-Dodecane	0.294	1.115585
n-Tridecane	0.295	1.179982
n-Tetradecane	0.291	1.188785
n-Heptadecane	0.283	1.297054
n-Octadecane	0.276	1.276058
n-Eicosane	0.277	1.409671
Carbon Dioxide	0.309	0.707727
Carbon Monoxide	0.328	0.535060
Sulfur Dioxide	0.310	0.797391
Hydrogen Sulfide	0.320	0.583165
Water	0.269	0.689803
Cyclohexane	0.303	0.665434
Quinoline	0.310	0.859036
m-Cresol	0.300	1.000087
Diphenyl Methane	0.305	1.082667
1-Methyl Naphthalene	0.297	0.827417

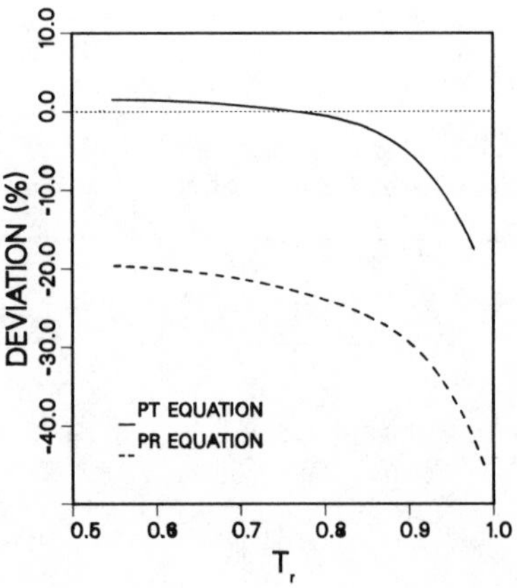

Figure 2. Comparison of the Liquid Densities for Saturated Water.

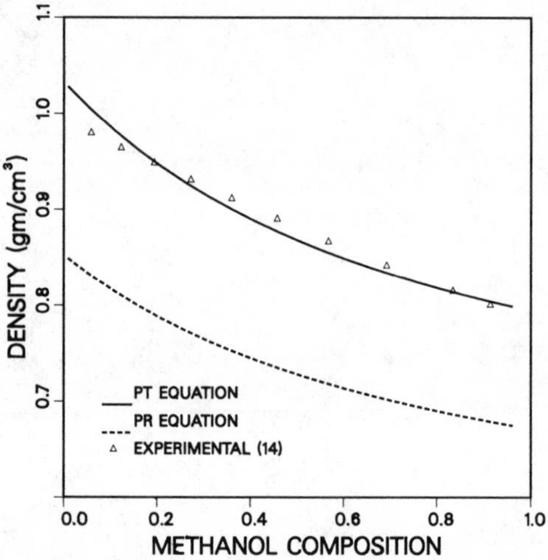

Figure 3. Comparison of the Saturated Liquid Densities for Methanol-Water Mixtures at 298.15K.

polar fluids by examining the differences between F and ζ_c of these fluids and those of n-alkanes containing the same number of carbon atoms. (This approach is similar in principle to the homomorph concept of Bondi and Simkin (9)).

The normal alkanes show a great regularity in their behavior with the number of carbon atoms. Properties such as T_c, P_c, F and ζ_c can therefore be correlated accurately with the number of carbon atoms in the molecule. This has been demonstrated for T_c and P_c elsewhere (10) and is shown for F and ζ_c in Figures 4 and 5. Properties of related fluids -- such as isoparaffins, olefins, napthenic compounds -- can then be determined from the n-alkane correlations if an effective carbon number (ECN) is used. The ECN is the number of carbon atoms in an n-alkane having the same normal boiling point as the fluid of interest. The ECN concept has been used successfully by Ambrose (11) and more recently by Chase (12) and Willman and Teja (10) to predict vapor pressures. It should be added that the ECN can be, and usually is, non-integral for compounds other than the n-alkanes.

Since polar fluids do not follow the n-alkane correlations, we have plotted F and ζ_c for classes of these fluids (alcohols, carboxylic acids) against carbon number and these correlations are also shown in Figures 4 and 5. Not suprisingly, the n-alkane and polar series correlations exhibit the largest differences at low carbon numbers. As the carbon number increases (beyond n-hexanol in the case of the alcohol series and n-butyric acid in the case of the carboxylic acid series), the curves approach the n-alkane correlation.

In general, the difference between F and ζ_c for polar substances and those for the n-alkanes containing the same number of carbon atoms decreases as the hydrocarbon part of a polar molecule becomes more dominant. These differences can also be correlated with the ECN. Figures 6 and 7 show plots of ψ and θ where these are defined by:

$$\psi = (F_{polar}/F_{HC}) - 1 \qquad (14)$$

and

$$\theta = \zeta_{c,HC}\,\psi - \zeta_{c,polar} \qquad (15)$$

As can be readily seen from the diagrams, ψ and θ correlate well with the ECN.

Therefore, for any pure component, the four parameters of the PT equation are obtained with only a knowledge of the ECN, which is obtained from the boiling point. T_c and P_c are obtained directly for each compound, as in Reference 10, with ζ_c and F obtained directly from correlations only for the alkanes. For alcohols and acids, the base values (alkanes) are determined along with the corrections due to the polarity effects. These values are calculated from correlations with ECN, and used to generate the ζ_c and F for the polar compounds using equations 14 and 15. It is expected that the constants for secondary and tertiary alcohols, for example, may be obtained from

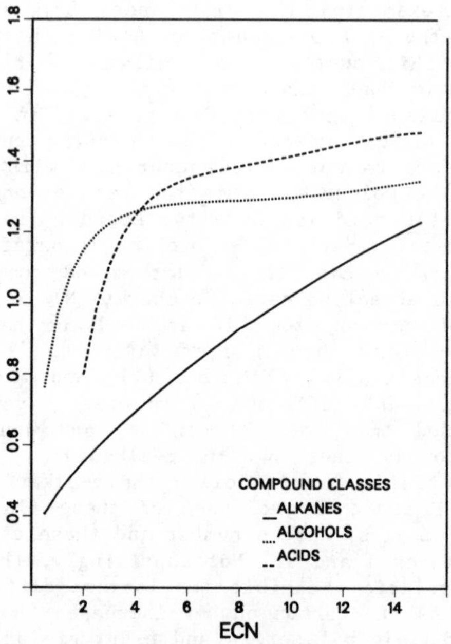

Figure 4. Correlation of F for Several Classes of Compounds.

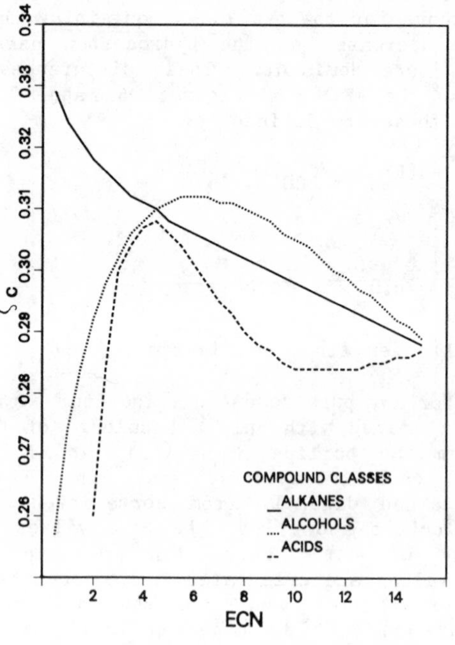

Figure 5. Correlation of ζ_c for Several Classes of Compounds.

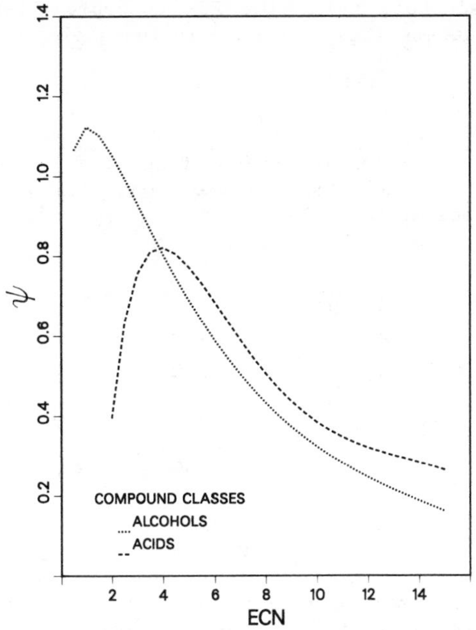

Figure 6. Correlation of the Correction Parameter ψ for Alcohols and Acids.

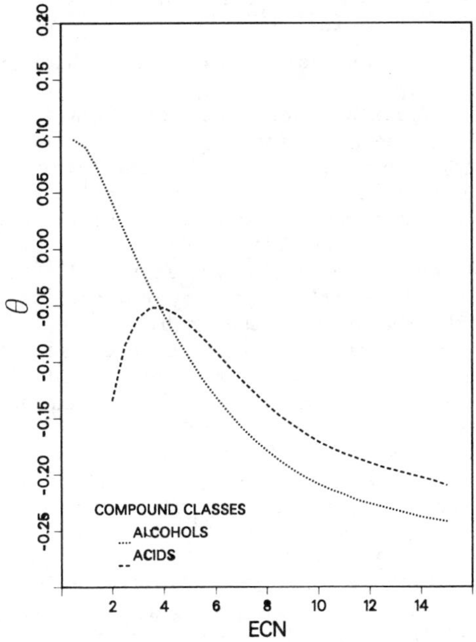

Figure 7. Correlation of the Correction Parameter θ for Alcohols and Acids.

those of the n-alcohols using the ECN concept. This approach therefore allows generalized constants to be developed for classes of polar fluids.

Mixture Calculations

Equation 4 can be used for the calculation of mixture properties and phase equilibria if the mixture constants a_m, b_m, c_m, are calculated using the mixing rules:

$$a_m = \Sigma\Sigma\, x_i x_j a_{ij} \tag{16}$$

$$b_m = \Sigma\, x_i b_i \tag{17}$$

$$c_m = \Sigma\, x_i c_i \tag{18}$$

The choice of this model is, of course, completely arbitrary -- particularly in the case of c_m. We have chosen a linear mixing rule for c_m because it is simple and also because, setting $c_m = b_m$ allows the equations to reduce to the PR case. This would not be possible if a different mixing rule is used for c_m.

The cross-interaction term a_{ij} in equation (16) is evaluated using the combining rule:

$$a_{ij} = \xi_{ij}\, (a_{ii}\, a_{jj})^{1/2} \tag{19}$$

where ξ_{ij} is a binary interaction parameter which is evaluated from experimental data.

The PT and PR equations were used to calculate vapor-liquid equilibria for a number of binary mixtures containing polar components. A single binary interaction coefficient was used in each case. Overall, the errors in the correlation of bubble point pressures and vapor phase compositions could be reduced to similar values for these two equations with the use of binary interaction coefficients. However, as shown in Figure 8, the VLE curve calculated using the PR equation for a system such as toluene + 1-pentanol is distorted because of the inability of the PR equation to predict the vapor pressure of 1-pentanol accurately. The distortion in the P-x-y diagram is reduced considerably when the PT equation is used and obviously depends on how well the function $\alpha[T_R]$ represents the vapor pressure of the pure components. Thus the greatest distortion will occur when the PR equation is used to calculate VLE for systems containing 1-hexanol, 1-heptanol, 1-octanol, hexanoic acid, decanoic acid, octanoic acid, etc. as can readily be determined from Table 3. This is also shown in Figure 9 for the cyclohexane + 1-hexanol system. Not surprisingly, improved density representation has little effect on the VLE calculations. However, at high pressures, density effects should become significant although very few high pressure data are available for mixtures containing polar components with $\zeta \ll 0.3074$.

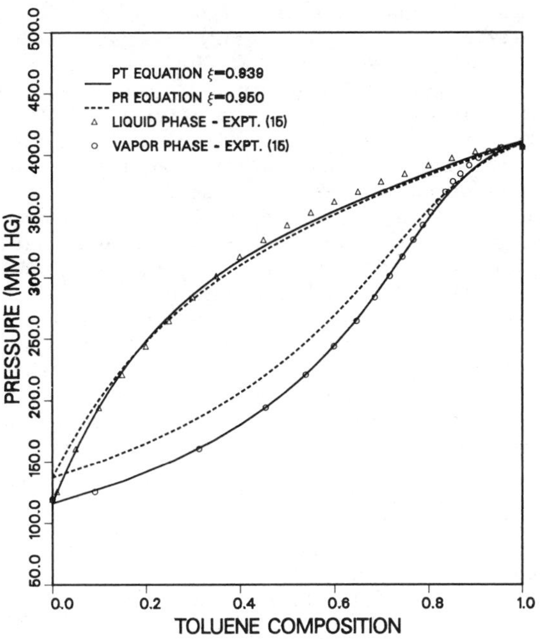

Figure 8. VLE in the Toluene - Pentanol System at 90°C.

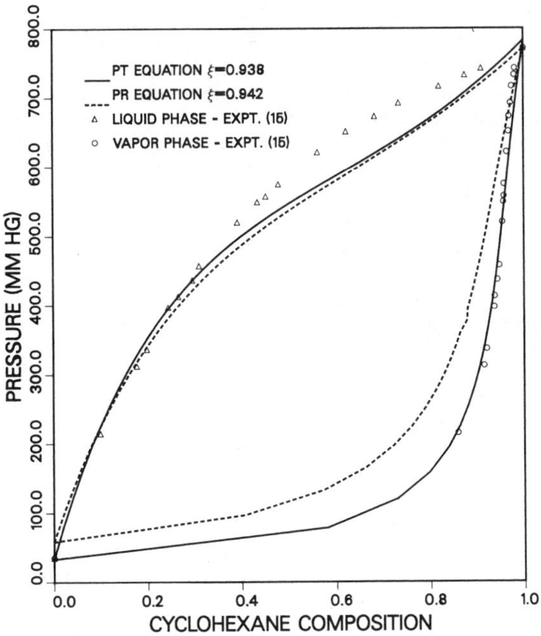

Figure 9. VLE in the Cyclohexane-Hexanol System at 81.2°C.

The PT and PR equations were also used to calculate liquid-liquid equilibria for several binary systems containing polar components. Again, the effect of improved densities was not apparent in the calculated LLE curves. However, both equations were very poor in their ability to predict LLE behavior with the van der Waals mixing rules, as can be seen in Figure 10 for the methanol-heptane system and it is entirely possible that significant differences would be observed with, for example, density-dependent mixing rules. Note that both equations of state predict one liquid phase which is essentially pure methanol.

The effects of improved densities can be readily observed when critical loci are calculated. This has been discussed in detail elsewhere (13). We show here our calculations for the argon + water system (Figures 11-12). Both the PR and the PT equation correlate the T_c vs x_1 and the P_c vs T_c behavior in this highly non-ideal system fairly well up to pressures of 50 MPa. The V_c vs x_1 behavior is, however, considerably better predicted by the PT equation, as evidenced in Figure 11. One consequence of this is obvious if the P_c-V_c-T_c curve is examined (Figure 12). As can be seen from the diagram, the P_c-V_c-T_c curves (and hence the PVTx surfaces) calculated using the PR and PT equations are distorted near the pure component ends of the curves due to their poor representation of pure-substance critical volumes. This distortion is greater for the PR equation than for the PT equation and cannot be overcome simply by setting $\zeta_c = Z_c^{expt}$ (13) since this distorts the rest of the curve.

Thus, high pressure phase equilibria -- particularly in the critical region -- must reflect this distortion in the PVTx surface. Unfortunately, few data exist in this region for polar mixtures. Nevertheless, we can say with some justification that a three-constant cubic equation is better able to predict the properties of mixtures at high pressures when density effects are significant. It may also be advantageous to use three constant cubics with density-dependent mixing rules since the overall representation of the PVTx behavior is more realistic than in the case of two-constant cubics.

Conclusions

In this work, the Peng-Robinson (PR) and Patel-Teja (PT) equations were used as representatives of the class of cubic equations of state. Pure component vapor pressures and densities, and VLE and LLE for mixtures of polar compounds were examined along with the applicability of the equations in the critical region. In general, cubic equations of state can be extended readily to polar components if actual vapor pressure data are used to obtain $\alpha[T_R]$. However, the choice of functional form of α is unimportant at moderate conditions at which most data exist. A significant improvement in densities is obtained if the cubic equation uses a third constant (as in the PT equation), which is related to a calculated critical compressibility. Separate correlations for determining substance dependent parameters are necessary for polar compounds, but these parameters can be

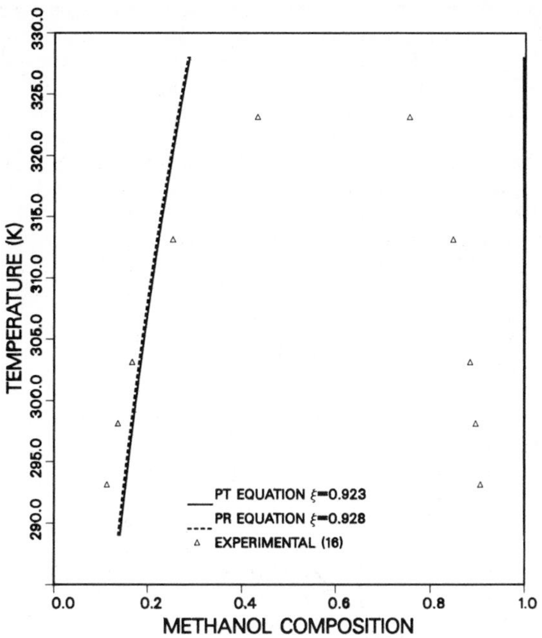

Figure 10. LLE in the Methanol-Heptane System.

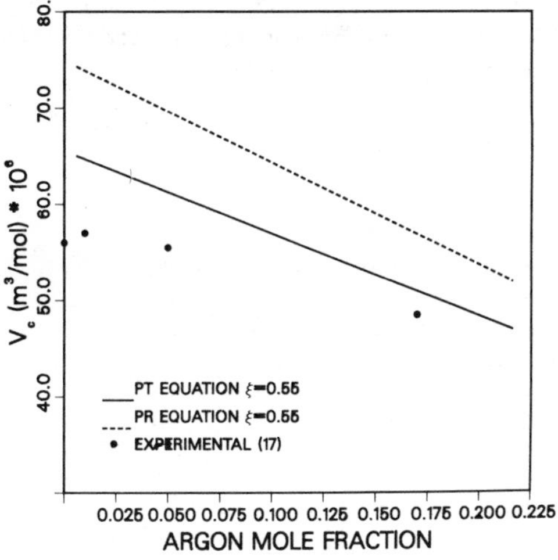

Figure 11. V_c-x Projection of the Argon-Water System in the Critical Region.

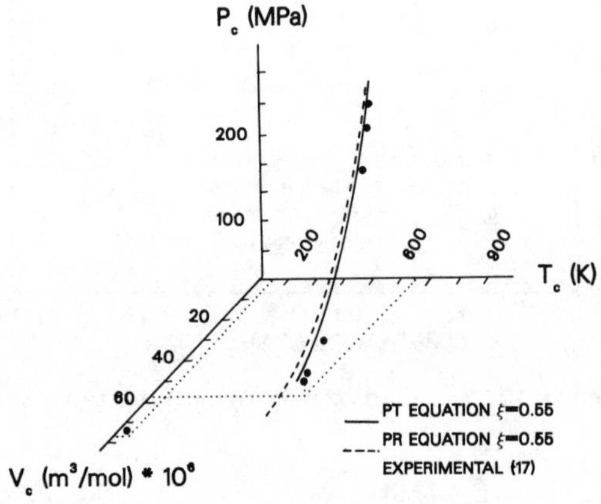

Figure 12. PVT Diagram of the Argon-Water System in the Critical Region.

obtained with only a knowledge of the boiling point. The effect of improved densities is not reflected in low pressure VLE and LLE predictions in terms of absolute deviations. However, for VLE, the curves predicted with the PT equation are less skewed than those predicted with the PR equation. An improvement due to better density representation might be noticed if density dependent mixing rules are used. The critical locus, nevertheless shows the dramatic effect of improved densities, particularly if viewed in three dimensions.

Literature Cited

1. Abbott, M. M. ADVANCES IN CHEMISTRY SERIES No.182, American Chemical Society; 1979; p. 47.

2. Vidal, J. Fluid Phase Equilibria, 1983, 13, 15.

3. Schmidt, G.; Wenzel, H. Chem. Eng. Sci., 1980, 35, 1503.

4. Harmens, A.; Knapp, H. Ind. Eng. Chem., Fund. 1980, 19, 291.

5. Heyen, G. "Phase Equilibria and Fluid Properties in the Chemical Industry", 2nd International Conference Berlin (West), Proceedings, Part 1 1980.

6. Patel, N. C.; Teja, A. S. Chem. Eng. Sci. 1982, 37, 463.

7. Soave, G. Chem. Eng. Sci. 1972 27, 1197.

8. Han, C. H. M.S. Thesis, Georgia Institute of Technology, Atlanta, 1984.

9. Bondi, A.; Simkin, D. J. AIChE J. 1957, 3, 473.

10. Willman, B. T.; Teja, A. S. Ind. Eng. Chem., Proc. Des. Dev. In Press.

11. Ambrose, D. NPL Report Chem. 57, Teddington, England, 1976.

12. Chase, J. D. Chem. Eng. Prog. 1984, 80 (4), 63.

13. Teja, A. S.; Smith, R. L.; Sandler, S. J. Fluid Phase Equilibria 1983, 14, 265.

14. Carr, C.; Riddick, J. A. Ind. Eng. Chem. 1951, 43, 692.

15. Gmehling, J.; Onken, U.; Arlt, W. "Vapor-Liquid Equilibrium Data Collection", Vol I, Part 2b, DECHEMA, Frankfurt, Germany, 1978.

16. Sorensen, J. M.; Arlt, W. "Liquid-Liquid Equilibrium Data Collection", Vol V, Part 1, DECHEMA, Frankfurt, Germany, 1979.

17. Hicks, C. P.; Young, C. L. Chem. Rev. 1975, 75, 119.

RECEIVED November 4, 1985

22

Parameters from Group Contributions Equation and Phase Equilibria in Light Hydrocarbon Systems

Ali I. Majeed[1] and Jan Wagner[2]

[1]**Nawazish International, Wetumka, OK 74883**
[2]**School of Chemical Engineering, Oklahoma State University, Stillwater, OK 74078**

The Parameters From Group Contribution equation of state is applied to pure fluids and mixtures with emphasis on the representation of phase equilibria. Group parameters and group-interaction parameters were derived from published pure component data using a nonlinear, multiproperty fitting program. Comparisons of calculated and experimental vapor-liquid equilibrium phase compositions, volumetric properties, and enthalpy departures demonstrated the applicability of the PFGC equation to systems of hydrocarbons and hydrocarbons with water. Phase behavior of mixtures of light hydrocarbons with acid gases is also described over a fairly wide range of temperature and pressure.

The use of equations of state to describe the phase behavior and thermodynamic properties of light hydrocarbon systems is well established. Among the more widely used correlations are the Soave-Redlich-Kwong (1), Peng-Robinson (2), and Starling-Benedict-Webb-Rubin (3) equations. These equations were developed to describe the behavior of nonpolar or weakly polar substances. When restricted to these types of systems all three equations yield phase behavior predictions suitable for many applications in process design. At least two of these equations, the Soave-Redlich-Kwong and the Peng-Robinson equations, have been extended to hydrocarbon-water systems in an effort to describe vapor-liquid-liquid phase behavior.

Many problems encountered in the gas processing industry involve nonideal liquid solutions. For example, dehydration and hydrate inhibition processes may involve mixtures of light hydrocarbons with aqueous methanol or glycol solutions. The Parameters From Group Contributions, or PFGC, equation is an equation of state analogy to an activity coefficient equation which is capable of describing vapor-liquid-liquid equilibrium in systems exhibiting nonideal liquid behavior. As the name implies, the parameters in this equation are derived from group contribution techniques rather than correlations with critical properties. This approach also offers potential advantages in

0097–6156/86/0300–0452$06.50/0

applications to systems involving undefined mixtures of petroleum and synthetic liquids. The functional groups in these types of mixtures can be identified by modern analytical techniques such as NMR spectroscopy. Reduced reliance on critical properties correlations in terms of specific gravity, average boiling point, characterization parameters, etc., should lead to improved predictions of thermodynamic properties of synthetic and natural hydrocarbon systems.

In this paper we would like to share some of our experiences and results in developing and evaluating the PFGC equation for light hydrocarbon systems. Our primary emphasis has been on the prediction of phase behavior via equations of state for applications to process design.

The PFGC Equation of State

The Parameters From Group Contributions (PFGC) equation of state was introduced by Cunningham and Wilson (4) in 1974. The details of the model formulation and derivations of the equation are given by Cunningham (5). A brief summary is included here for convenience.

The basis for the PFGC equation of state lies in the assumption that the form of empirical relationships which have successfully described the excess Gibbs free energy of mixing can also be used as the basis for modeling the Helmholtz free energy. In addition, the void spaces between molecules in a mixture are assumed to be identifiable as an additional component designated as "holes". The volume is evaluated as the molecular volume occupied by the component divided by the total volume. Including an arbitrary parameter for one mole of holes, volume fractions can be converted to mole fractions. Using these relationships for mole fractions permits the Helmholtz free energy of mixing to be expressed as a function of composition and volume.

The Helmholtz free energy of mixing is composed of two contributions:

1. A modified Flory-Huggins equation to account for entropy effects due to differences in molecular size, and
2. A modified Wilson equation which represents the individual groups in a mixture.

Molecular activity coefficients corrected for differences in molecular size are calculated as the sum of group activity coefficients which, in turn, are determined by group composition rather than by molecular composition.

The Helmholtz free energy is written as

$$\frac{A}{RT} = \sum_I N_I \ln \frac{n_I b_I}{V} + (V - nb)\frac{s}{b} \ln\left(\frac{V - nb}{V}\right)$$
$$-nb\left(\frac{c}{b_I}\right) \sum_i \psi_i \ln\left(\frac{V - nb + nb\sum_j \psi_i \lambda_{ij}}{V\lambda_{ii}}\right) \qquad (1)$$

where upper case subscripts refer to molecular properties, lower case subscripts refer to group properties, and

n = total number of moles

b = total molecular volume = $\sum_I x_I b_I$

b_I = volume of one mole of molecules of type I = $\sum_i m_{Ii} b_i$

b_i = volume of one mole of groups of type i

m_{Ii} = number of groups of type i in molecule I

b_H = volume of one mole of holes

V = total volume

C = a universal constant in the modified Wilson equation

$s = \sum_I x_I s_I$ = external degrees of freedom parameter for the mixture

$s_I = \sum_i m_{Ii} s_i$ = external degrees of freedom parameter for molecules of type I

s_i = external degrees of freedom parameter for groups of type i

E_{ij} = interaction energy between groups i and j

$\lambda_{ij} = \mathrm{EXP}(-E_{ij}/kT)$ = interaction parameter between groups i and j

$\psi_i = \dfrac{\sum_I x_I m_{Ii} b_i}{b}$ = group fraction of type i

The PFGC equation of state follows directly from the expression for the Helmholtz free energy of mixing by using the following thermodynamic relationship:

$$\frac{P}{RT} = -\frac{\partial}{\partial V}\left(\frac{A}{RT}\right)_{T,n} \tag{2}$$

Thus, in terms of compressibility factor, the PFGC equation of state can be written as

$$Z = \frac{P\nu}{RT} = 1 - \frac{s\nu}{b}\ln\left(1 - \frac{\nu}{b}\right) - s$$

$$+ b \left(\frac{c}{b_H}\right) \sum_i \psi_i \left(\frac{b - b \sum_j \psi_j \lambda_{ij}}{\nu - b + b \sum_j \psi_j \lambda_{ij}}\right) \tag{3}$$

where ν is the molar volume.

Taking c/b_H as a "universal" constant, there are three basic parameters in the PFGC equation:

(1) E_{ij}, the interaction energy between groups
(2) b_i, the volume of one mole of groups of type i
(3) s_i, a parameter proportional to the external degrees of freedom per group i

Note that three parameters must be known for groups instead of molecules. The interaction energy is evaluated as

$$E_{ij} = a_{ij} \frac{E_{ii} + E_{jj}}{2} \tag{4}$$

where a_{ij} is an interaction coefficient. When a_{ij} is unity, the mixture properties are nearly ideal. If a_{ij} is less than one, deviations from ideality are in the positive direction; for a_{ij} greater than one, deviations from ideality are in the negative direction. The group interaction energy is slightly temperature dependent. For convenience, the following form is used (6)

$$E_{ii} = E_{ii}^{(0)} + E_{ii}^{(1)} \left(\frac{283.2}{T,K} - 1\right) + E_{ii}^{(2)} \left(\frac{283.2}{T,K} - 1\right)^2 \tag{5}$$

Thus, the PFGC equation of state in its final form has five adjustable parameters: s_i, b_i, $E_{ii}^{(0)}$, $E_{ii}^{(1)}$, and $E_{ii}^{(2)}$. In addition, there is one binary interaction coefficient for each pair of groups. Application of the equation of state to the prediction of thermodynamic properties is straight-forward once appropriate group parameters are available.

The PFGC equation of state written in terms of compressibility factor exhibits cubic-type behavior. Several iterative schemes can be applied to find the liquid-like and/or vapor-like roots of Equation 3. Moshfeghian, et al. (6) used a direct substitution method. However, a third-order iteration method using a Richmond convergence scheme has been found to improve both the speed and reliability of the calculations. The basic algorithm is given by Lapidus (7) as

$$x_{i+1} = x_i - \frac{2f(x_i)\, f'(x_i)}{2[f'(x_i)]^2 - f(x_i)\, f''(x_i)} \tag{6}$$

Setting $x = b/\nu$, Equation 3 can be written as

$$f(x) = (1 - s)x - \frac{Pb}{RT} - s \ln (1-x) + b \left(\frac{c}{b_H}\right) \sum_i \psi_i \left(1 - \sum_j \lambda_{ij}\right) \frac{x^2}{1 - x + x \sum_j \psi_j \lambda_{ij}} \tag{7}$$

The first and second derivatives are

$$f'(x) = 1 + s \frac{x}{1 - x} + b \left(\frac{c}{b_H}\right) \sum_i \psi_i \left[\frac{1}{(1 - x + x \sum_j \psi_j \lambda_{ij})^2} - 1\right] \tag{8}$$

and

$$f''(x) = \frac{s}{(1 - x)^2} + 2b \left(\frac{c}{b_H}\right) \sum_i \psi_i \frac{1 - \Sigma \psi_j \lambda_{ij}}{(1 - x + x \sum_j \psi_j \lambda_{ij})^3} \tag{9}$$

Other thermodynamic functions and properties can also be derived from the equation of state. For phase equilibrium calculations, the fugacity coefficient for component I in a multicomponent mixture is given by

$$\ln \phi_I \equiv \frac{1}{RT} \int_v^\infty \left[\left(\frac{\partial P}{\partial n_I}\right)_{T,v,n_J} - \frac{RT}{V}\right] dV - \ln (z) \tag{10}$$

Using this definition with Equation 3 leads to

$$\ln \phi_I = \left[\frac{s_I}{b} \nu - \frac{s}{b^2} b_I \nu - S_I\right] \ln \left(1 - \frac{b}{\nu}\right)$$

$$+ \left(\frac{c}{b_H}\right) \left\{ \sum_i m_{Ii} b_i \ln \left(\frac{\nu - b + b \sum \psi_j \lambda_{ij}}{\nu}\right)\right.$$

$$\left. - b \sum_i \psi_i \frac{b_I - \Sigma m_{Ij} b_j \lambda_{ij}}{\nu - b + b \sum_j \lambda_{ij}} \right\} \tag{11}$$

Expressions have also been derived to the chemical potential and the isothermal effect of pressure on enthalpy (6,8,9).

To obtain values for the group parameters in the PFGC equation of state, several steps have been followed. First, groups were selected to represent the components of interest. Data for those components which were identified by a single group, such as methane and carbon dioxide, were used to obtain pure group parameters for these groups/components. Parameters for other types of groups are more difficult to obtain and require a considerable amount of judgement in the fitting procedures. For example, preliminary estimates of the methyl and methylene group parameters were obtained from thermodynamic data for ethane and ethylene, respectively. The final methyl and methylene group parameters, as well as the group binary interaction coefficient, were derived from a simultaneous regression using data for butane, hexane, octane, and heptadecane. Parameters were fit to minimize the average absolute error in pure component vapor pressure. However, care was exercised to maintain reasonable quality for volumetric and enthalpy departure predictions.

Mixtures of different components were used to obtain group binary interaction coefficients. These interaction coefficients were selected to minimize average absolute percentage errors in equilibrium K-values for each component in the mixture. Group parameters are listed in Table I.

Applications

Vapor Pressure. A basic requirement of any equation of state which will be used for phase equilibrium calculations is that pure component vapor pressures must be predicted accurately. Table II provides a comparison of predicted and experimental vapor pressures for several paraffin hydro-carbons. Errors in vapor pressure predictions using the Soave-Redlich-Kwong (SRK) equation of state are included for comparison. Percentage errors in calculated vapor pressure of isopentane as a function of reduced temperature are shown in Figure 1. This type of error distribution is typical of the PFGC equation of state and is very similar to that of the cubic SRK equation. However, average absolute percentage errors in pure component vapor pressure are on the order of two to ten times greater using the PFGC equation.

One limitation of the PFGC equation with the current set of groups is the inability of this group contribution method to distinguish the effect of molecular structure on physical properties. This is evident in the average absolute errors in predicted vapor pressures for 2-methylpentane and 3-methylpentane. In terms of group composition, these two components are identical.

The PFGC equation does predict pure component vapor pressures fairly well, especially considering the number of parameters available. For example, only eleven parameters are used to describe the pure component properties of all the normal alkanes -- two sets of five group parameters plus one binary interaction coefficient. In contrast, the SRK equation of state includes three parameters for each pure component.

Molar Volumes. A basic problem with cubic equations of state is the ability to predict both phase behavior and volumetric properties well (17). The PFGC equation exhibits this same behavior. Absolute average percentage errors in calculated liquid and vapor volumes are summarized in Table III for typical paraffins, olefins, aromatics, and nonhydrocarbons. Errors in calculated volumes using the SRK equation of state are included for comparison.

TABLE I. TYPICAL GROUP PARAMETERS FOR THE PFGC EQUATION

No	Group	b m^3/kg-mole	s	$E^{(0)}$ K	$E^{(1)}$ K	$E^{(2)}$ K
2.	CH_4	.03727	1.8982	-71.3501	-32.5201	4.2153
3.	$-CH_3$	.02078	1.9780	-176.5500	-32.5722	3.3501
4.	$-CH_2-$	.01665	.4956	-150.0167	-43.2389	0.2599
5.	>CH-	.01505	.9260	-55.6278	-19.1556	0.2100
6.	>C<	.01614	-2.6435	-81.5054	-23.0062	0.0343
7.	$=CH_2$	.02093	1.3235	-133.2944	-28.1383	0.8857
8.	$>CH_2$	.02336	.6333	-110.3583	-42.0811	0.1697
9.	=CH-	.01610	.3471	-142.8333	-107.6667	17.2222
10.	=C<	.01634	.2272	-174.0336	-62.3567	32.4755
11.	-C=	.01940	-.2419	92.0556	-61.1222	-.0046
12.	N_2	.02777	2.3659	-65.7222	-18.4444	1.6667
13.	CO_2	.02079	3.0920	-320.1728	-94.5814	0.07422
14.	CO	.02530	2.5992	-75.7056	-31.7222	3.6022
15.	H_2S	.02527	3.4335	-338.6667	-96.0556	8.8889
16.	H_2O	.01248	2.2000	-1472.9444	-1544.0556	476.9444
17.	>C=	.01689	.1938	-153.7611	-69.8889	26.7175
23.	=C=	.03972	-.6500	-0.4167	-56.3144	.0006
24.	>C-	.03801	-.4428	-22.2261	-9.7502	.0092

TABLE I. (Continued)

λ_{ij}

	CH_4	$-CH_3$	$-CH_2-$	>CH-	>C<	$=CH_2$	=CH-	=CH-	=C<	-C=	N_2	CO_2	CO	H_2S	H_2O
CH_4	1.0														
$-CH_3$	.945	1.0													
$-CH_2-$	.830	1.010	1.0												
>CH-	.700	1.0	1.0	1.0											
>C<	1.0	.013	.405	1.0	1.0										
$=CH_2$	1.050	.951	.935	.850	1.0	1.0									
$>CH_2$	.850	1.0	1.115	1.0	1.0	1.0	1.0								
=CH-	.800	1.0	1.0	1.0	1.0	1.0	1.0	1.0							
=C<	.100 ·	1.0	1.328	.300	1.0	1.0	1.0	1.0	1.0						
-C=	1.0	.947	.962	.850	1.0	1.005	1.0	1.0	1.0	1.0					
N_2	.950	.800	8.60	.400	1.0	.900	.320	.680	1.0	.300	1.0				
CO_2	.765	.850	.850	.800	1.0	.900	.920	.920	.800	.900	.780	1.0			
CO	1.0	1.0	1.0	1.0	1.0	1.0	1.0	1.0	1.0	1.0	1.0	1.0	1.0		
H_2S	.722	.850	1.0	1.0	1.0	.840	1.0	1.0	.880	1.0	.500	.905	1.0	1.0	
H_2O	.263	.335	.330	.250	1.0	.390	1.0	.250	1.0	.250	.318	.550	.080	.530	1.0
>C=	1.0	1.0	1.101	1.0	1.0	1.0	1.0	1.0	1.0	1.193	1.0	1.0	1.0		1.0
=C=	1.0	1.0	1.0	1.0	1.0	1.0	1.0	1.0	1.0	1.0	1.0	1.0	1.0		1.0
>C-	.050	1.0	.648	1.0	1.0	1.0	1.0	1.0	1.0	1.0	1.200	.050	1.0	.200	1.0

TABLE II. COMPARISON OF VAPOR PRESSURE PREDICTIONS

Component	Average Absolute Percent Error PFGC	SRK	No. of Points	Ref.
Methane	2.38	1.50	21	10
Ethane	3.60	1.59	38	10
Propane	5.32	1.06	45	10
i-Butane	1.89	.86	25	11
n-Butane	5.66	1.62	52	10
i-Pentane	1.16	1.13	38	12
n-Pentane	6.46	1.55	55	10
n-Hexane	1.95	1.33	44	11
n-Heptane	4.47	1.05	61	10
n-Octane	3.45	1.22	62	10
n-Nonane	3.72	1.16	27	13
n-Decane	2.15	1.15	27	13
n-Tetradecane	5.78	2.31	27	13
n-Pentadecane	7.86	2.94	27	13
n-Hexadecane	9.90	2.29	27	13
n-Heptadecane	13.16	4.07	27	
2-Methylpentane	14.80	1.07	58	10
3-Methylpentane	6.43	1.40	58	10
2,3-Dimethylbutane	13.10	1.13	59	13
Nitrogen	1.86	0.78	60	14
Carbon Dioxide	6.94	.47	48	15
Hydrogen Sulfide	0.94	.91	30	16

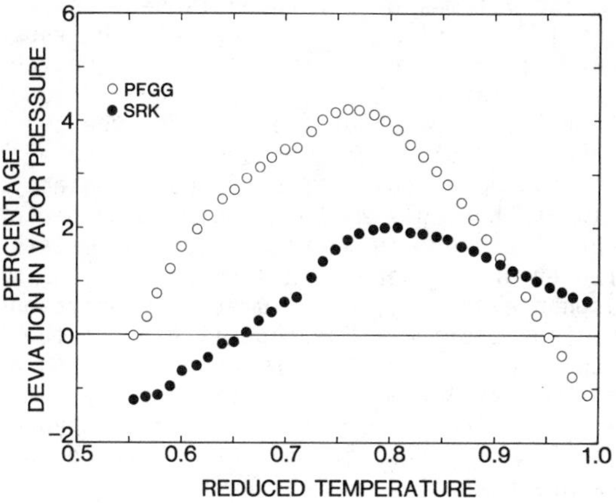

Figure 1 Deviations in vapor pressure predictions for iso-pentane (16).

TABLE III. AVERAGE ABSOLUTE PERCENT ERRORS IN MOLAR VOLUMES

Component	Liquid			Vapor			
	PFGC	SRK	No. Pts.	PFGC	SRK	No. Pts.	Ref.
Methane	7.11	4.51	62	3.79	2.19	95	18
Ethane	3.25	6.59	58	4.94	3.01	77	19
iso-Pentane	14.14	12.42	38	6.19	7.48	83	12
n-Hexane	2.86	19.44	44	8.40	3.02	69	11
Ethylene	3.81	8.11	63	9.50	3.22	108	20
Propylene	1.76	8.60	63	4.72	2.56	64	21
1-Butene	2.48	12.96	27	27.50	3.19	27	11
Benzene	6.55	14.16	48	12.33	1.85	48	11
Toluene	16.05	16.98	61	5.04	1.74	61	10
Nitrogen	1.77	3.91	81	9.32	1.30	116	14
Carbon Monoxide	17.40	4.67	25	6.31	2.36	67	11
Carbon Dioxide	12.20	15.08	60	13.30	4.05	71	15
Hydrogen Sulfide	4.86	6.85	27	4.95	5.24	66	16
Sulfur Dioxide	7.83	19.54	34	14.2	7.61	64	11

Figure 2 illustrates the performance of the two equations for predicting the molar volumes of saturated isopentane. The SRK equation predicts liquid volumes that are too high, especially near the critical point. In contrast, the PFGC equation predicts molar liquid volumes that are too low. In general, absolute errors in liquid volumes are about the same for the two equations.

Errors in predicted vapor volumes are of the same order of magnitude for the two equations. Predictions of molar volumes can be improved somewhat, but only at the expense of vapor pressure or phase behavior predictions. In fitting group parameters we have tried to minimize errors in vapor pressure predictions while maintaining reasonable accuracy of volumetric predictions.

Vapor-Liquid Equilibrium. The capabilities of the PFGC equation of state in predicting the phase behavior and phase compositions has been evaluated for a variety of systems, including those containing both hydrocarbon and nonhydrocarbon components. Examples of errors in predicted K-values and liquid volume fractions are presented in Tables IV and V. Inspection of these tables shows a fairly good agreement between experimental and calculated phase behavior for both hydrocarbon-hydrocarbon and hydrocarbon-non-hydrocarbon binary systems.

One of the difficulties in using the group contribution technique involves the use of the same group binary interaction coefficients for both pure components and mixtures. For example, the binary interaction coefficient between the $-CH_3$ and $=CH_2$ groups in toluene was optimized using vapor pressure data for toluene. However, the same interaction coefficient is a key variable in obtaining good predictions of toluene K-values in mixtures of paraffins. This problem is characteristic of a group contribution technique where the number of groups are much smaller than the number of components. The group parameters and the binary interaction coefficients represent a delicate balance between the ability to predict pure component thermodynamic properties and the representation of multicomponent vapor-liquid equilibria.

The performance of the PFGC equation in predicting the phase behavior and volumetric of multicomponent natural gas and retrograde condensate types of systems has been discussed by Wagner et al. (59). For heavier mixtures, the phase behavior of a synthetic oil with high carbon dioxide contents represents an extreme test of the PFGC equation. Using group interaction coefficients derived from carbon dioxide - hydrocarbon binary systems, the PFGC equation predicted reasonably well the saturation pressures of the mixture with carbon dioxide contents up to 97 mole percent (69) as shown in Figure 3.

Aqueous Light Hydrocarbon Systems

For applications of the PFGC equation to aqueous systems, two group interaction coefficients were defined for the various phases present. One binary group interaction coefficient is used for both the vapor and hydrocarbon-rich liquid phases; a second group interaction coefficient is used for the aqueous liquid phase.

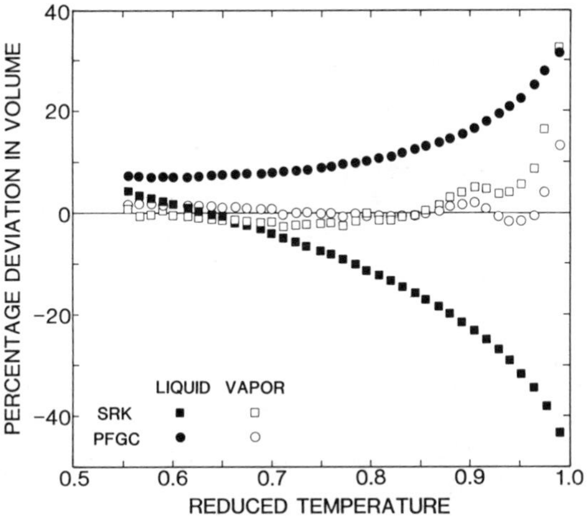

Figure 2 Deviations in saturated molar volume predictions for iso-pentane (16)

TABLE IV. DEVIATIONS IN K-VALUE PREDICTIONS FOR HYDROCARBON-HYDROCARBON BINARIES

System	Temperature Range (°F)	Pressure Range (psia)	No. of Points	Percent Abs. Avg. Error			Reference No.
				K_1	K_2	L/F	
$CH_4(1) - C_2H_6(2)$	-225.0 → -99.8	28.0 → 748.0	118	4.21	9.98	9.16	22
$CH_4(1) - C_3H_8(2)$	-176.0 → 32.0	50.0 → 1450.0	81	7.32	17.80	8.44	23
$CH_4(1) - nC_4H_{10}(2)$	-200.0 → 40.0	20.1 → 1822.0	105	8.33	22.42	18.14	24
$CH_4(1) - nC_5H_{12}(2)$	-140.0 → 50.0	50.0 → 2200.0	64	6.42	17.63	11.02	25
$CH_4(1) - nC_6H_{14}(2)$	-131.24 → 32.0	19.9 → 2675.0	105	7.44	32.05	11.50	26
$CH_4(1) - nC_8H_{18}(2)$	79.0 → 302.0	146.9 → 3865.0	35	11.60	12.51	8.76	27
$CH_4(1) - nC_{10}H_{22}(2)$	100.0 → 460.0	40.0 → 5000.0	157	9.13	25.42	10.89	28
$C_2H_6(1) - C_3H_8(2)$	0.0 → 200.0	100.0 → 752.0	78	2.11	2.78	9.24	29
$C_2H_6(1) - nC_4H_{10}(2)$	150.0 → 250.0	509.0 → 805.0	19	3.30	4.07	16.93	30
$C_2H_6(1) - nC_5H_{12}(2)$	40.0 → 340.0	50.0 → 955.0	68	5.56	7.92	21.99	31
$C_2H_6(1) - nC_6H_{14}(2)$	150.0 → 350.0	25.0 → 1146.0	39	7.94	13.16	5.22	32
$C_2H_6(1) - nC_{10}H_{22}(2)$	50.0 → 460.0	100.0 → 1715.0	112	14.53	22.66	6.18	33

C_3H_8(1) - nC_5H_{12}(2)	160.0 → 370.0	100.0 → 650.0	69	6.19	7.15	16.60	34
C_3H_8(1) - $nC_{10}H_{22}$(2)	40.0 → 460.0	25.0 → 1028.0	58	4.23	19.89	11.25	35
nC_5H_{10}(1) - $nC_{10}H_{22}$(2)	340.0 → 460.0	50.0 → 714.0	36	5.33	12.88	5.53	36
nC_5H_{12}(1) - Cyclohexane (2)	102.2 → 175.8	14.7 → 14.7	28	22.78	19.20	79.01	37
nC_5H_{12}(1) - Methylcyclopentane (2)	97.16 → 158.72	14.7 → 14.7	44	9.82	5.85	15.32	37
nC_5H_{12}(1) - Methylcyclohexane (2)	99.77 → 209.12	14.7 → 14.7	47	10.18	7.81	14.30	37
nC_8H_{18}(1) - 2-methylpentane (2)	50.0 → 104.0	0.23 → 6.60	48	4.23	7.71	10.42	38
nC_8H_{18}(1) - 3-methylpentane (2)	50.0 → 104.0	0.31 → 6.01	48	3.10	4.68	12.12	38
Benzene (1) - CH_4(2)	298.22 → 422.4	288.19 → 3519.1	18	28.39	32.27	6.11	39
Benzene (1) - nC_3H_8(2)	100.0 → 400.0	20.0 → 850.0	73	7.75	9.92	25.70	40
Toluene (1) - CH_4(2)	-100.0 → 0.0	100.0 → 3500.0	77	22.24	18.37	5.62	40

TABLE V. SYSTEM DEVIATIONS IN K-VALUE PREDICTIONS FOR NONHYDROCARBON-HYDROCARBON BINARIES

System	Temperature Range (°F)	Pressure Range (psia)	No. of Points	Percent Abs. Avg. Error			Reference No.
				K_1	K_2	L/F	
CO_2(1) - CH_4(2)	-100.0 → 29.0	161.0 → 1146.0	45	8.51	4.89	9.61	41
CO_2(1) - C_2H_6(2)	-58.0 → 68.0	90.0 → 914.0	54	3.01	19.80	20.41	42
CO_2(1) - C_3H_8(2)	40.0→ 160.0	100.0 → 950.0	67	4.49	5.95	9.90	43
CO_2(1) - nC_4H_{10}(2)	100.0 → 280.0	60.0 → 1150.0	54	7.45	7.07	9.11	44
CO_2(1) - nC_5H_{12}(2)	40.1 → 220.0	33.0 → 1397.0	47	8.98	12.20	6.31	45
CO_2(1) - nC_6H_{14}(2)	104.0 → 248.0	113.0 → 1682.0	40	9.48	15.90	5.94	46
CO_2(1) - $nC_{10}H_{22}$(2)	40.0 → 460.0	50.0 → 2732.0	88	11.32	20.80	7.78	47
CO_2(1) - $nC_{16}H_{34}$(2)	372.9 → 735.08	284.5 → 749.5	31	11.21	12.21	8.16	48
CO_2(1) - Benzene (2)	77.0 → 104.0	129.6 → 1124.1	17	2.07	24.82	7.77	49
CO_2(1) - Toluene (2)	100.6 → 399.0	48.4 → 2218.0	34	10.63	19.20	11.76	50
CO_2(1) - H_2S(2)	-2.34 → 194.0	293.9 → 1175.7	85	3.53	8.30	21.42	51
CO_2(1) - N_2(2)	-67.0 → 32.0	255.0 → 1907.0	30	4.34	17.87	12.43	52

N_2(1) - CH_4(2)	-240.0 → 130.0	40.5 → 710.0	101	4.74	6.19	11.03	53
N_2(1) - C_2H_6(2)	-99.67 → 62.33	152.7 → 1913.7	31	13.70	13.74	10.19	54
N_2(1) - nC_4H_{10}(2)	100.0 → 280.0	236.0 → 3402.0	28	14.23	13.61	17.25	55
N_2(1) - nC_5H_{10}(2)	-200.0 → 0.0	350.8 → 4506.5	21	22.05	26.14	13.24	56
N_2(1) - $nC_{10}H_{22}$(2)	100.0 → 280.0	80.0 → 5000.0	92	27.62	18.58	12.96	57
N_2(1) - Benzene (2)	167.0 → 257.0	900.86 → 4454.4	17	10.60	10.10	6.07	58
N_2(1) - CO_2(2)	-67.0 → 32.0	255.0 → 1907.0	30	17.87	4.34	12.43	52
H_2S(1) - CH_4(2)	-120.0 → 200.0	200.0 → 1600.0	59	5.08	12.46	7.49	53
H_2S(1) - C_2H_6(2)	-99.80 → 50.0	9.45 → 442.0	45	2.08	4.68	13.34	54
H_2S(1) - nC_4H_{10}(2)	100.0 → 250.0	69.4 → 1150.0	77	3.21	6.62	16.02	55
H_2S(1) - nC_5H_{12}(2)	40.0 → 340.0	20.0 → 1302.0	60	4.66	6.04	5.29	56
H_2S(1) - $nC_{10}H_{22}$(2)	40.0 → 340.0	20.0 → 1935.0	50	8.96	28.82	21.66	57

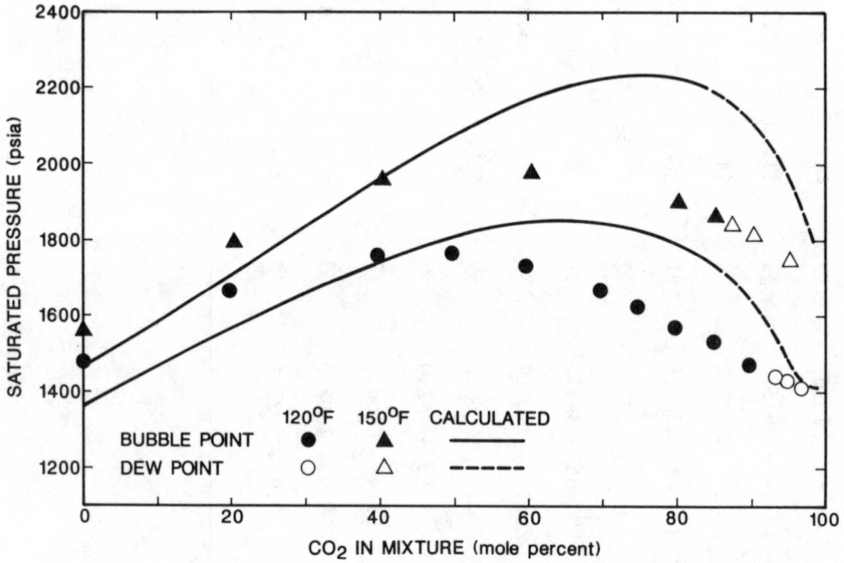

Figure 3 Experimental and predicted bubble point pressures for a synthetic oil with varying carbon dioxide content (140)

Data for mixtures of water with light hydrocarbons, carbon dioxide, hydrogen sulfide, nitrogen, and carbon monoxide were used to derive binary group interaction coefficients for the aqueous phase. These interaction coefficients were fit to minimize the average absolute errors in the composition of each phase. Some of the nonhydrocarbon-water group interaction coefficients were found to be linearly dependent on absolute temperature.

Table VI summarizes the errors in predicted phase compositions for several binary water-hydrocarbon and water-nonhydrocarbon systems. For paraffins and olefins, the PFGC equation gives excellent predictions of the vapor, hydrocarbon liquid, and aqueous liquid phases up to approximately 9000 psia. Vapor-liquid-liquid predictions for the binary systems of water with carbon dioxide and hydrogen sulfide are good up to 3000 psia. The ability of the PFGC equation to handle these nonideal aqueous mixtures is a result of using an activity coefficient model as a theoretical basis for the equation of state.

Other Applications

We have recently linked the PFGC equation of state with Parrish and Prausnitz (68) hydrate model as modified by Menton, Parrish and Sloan (69). The preliminary results of this effort are very promising (59), and we are continuing our evaluation of the prediction of hydrate formation conditions below the ice point. Procedures are also being developed to extend the PFGC equation to mixtures of undefined components in natural gas condensates and crude oils.

Summary

The PFGC equation of state has the capability to predict vapor-liquid equilibrium and volumetric properties for a variety of light hydrocarbon and hydrocarbon-water systems. In general, the quality of the predictions is quite good considering the small number of group parameters available to describe a large number of components. In more practical terms, the PFGC equation is suitable for process design/simulation calculations for many light hydrocarbon systems. Although the Soave-Redlich-Kwong and Peng-Robinson equations of state are more reliable for "normal" hydrocarbon systems, the PFGC equation has an advantage in aqueous systems with both hydrocarbons and nonhydrocarbons.

Acknowledgment

The authors appreciate the financial support of the Mobil Foundation and the School of Chemical Engineering, Oklahoma State University, which made this work possible.

TABLE VI. H_2O BINARY SYSTEMS DEVIATIONS IN PHASE CONCENTRATION PREDICTIONS

System	Temperature Range (°F)		Pressure Range (psia)		No. of Points	Percent Abs. Avg. Error in Smaller Component Conc.			Reference No.
						Vapor Phase	Hydrocarbon Liquid	Water	
$H_2O(1)$ - $CH_4(2)$	302.0	680.0	711.3	14226.0	45	7.33			61
$H_2O(21)$ - $C_2H_6(2)$	100.0	460.0	200.0	10000.0	130	4.71			62
$H_2O(1)$ - $C_3H_8(2)$	42.3	310.0	72.0	3000.0	240	7.16	12.90	10.51	63
$H_2O(1)$ - $nC_4H_{10}(2)$	99.9	280.0	52.2	491.6	7	6.98	6.60	11.21	64
$H_2O(1)$ - $nC_4H_{10}(2)$	100.0	280.0	52.2	934.7	115		21.80		65
$H_2O(1)$ - $nC_5H_{12}(2)$	100.0	600.0	120.0	3000.0	32	9.81	31.66	18.50	66
$H_2O(1)$ - $nC_6H_{14}(2)$	392.0	437.0	536.6	789.0	3	24.2			65
$H_2O(1)$ - $nC_7H_{16}(2)$	392.0	473.0	413.4	984.8	5	16.6			65
$H_2O(1)$ - $nC_9H_{20}(2)$	392.0	536.0	301.7	1305.4		18.67			65
$H_2O(1)$ - $nC_{10}H_{22}(2)$	392.0	564.8	264.0	1446.0	7		21.43		65
$H_2O(1)$ - $CO_2(2)$	122.0	662.0	2900.0	50759.0	68	12.46		24.52	67
$H_2O(1)$ - $H_2S(2)$	100.0	600.0	50.0	2000.0	18	12.01		3.55	65
$H_2O(1)$ - $N_2(2)$	100.0	600.0	50.0	2000.0	16	7.84		3.81	65
$H_2O(1)$ - $CO(2)$	100.0	600.0	50.0	200.0	18	9.68		14.90	65

Literature Cited

1. Soave, G., Chem. Eng. Sci., 1972, Vol. 27, 1197
2. Peng, D. Y. and Robinson, D. B. Canadian Jour. Chem. Engr., 1976, Vol. 54, December, 595.
3. Starling, K. E. and Han, M. S. Hydrocarbon Processing, 1972, Vol. 51, No. 5, 129.
4. Cunningham, J. R. and Wilson, G. M. Paper presented at GPA Meeting, Tech. Section F, Denver, Colorado, March 25, 1974.
5. Cunningham, J. R., M.S. Thesis, Brigham Young University, Provo, Utah, 1974.
6. Moshfeghian, M., Shariat, A., and Erbar, J. H., Paper presented at the AIChE National Meeting, Houston, Texas, April 1-5, 1979.
7. Lapidus, L., "Digital Computation for Chemical Engineers", McGraw-Hill, New York, 1962.
8. Moshfeghian, M., Shariat, A., and Erbar, J. H. Paper presented at NBS/NSF Symp. on Thermo. of Aqueous Sys., Airlie House, Virginia, 1979.
9. Moshfeghian, M., Shariat, A., and Erbar, J. H. ACS Symposium Series 113.
10. Engineering Sciences Data Unit, Chemical Engineering Series, Physical Data, Volume 5, 251-259 Regent Street, London WIR 7AD, August, 1975.
11. Canjar, L. and Manning, F. "Thermodynamic Properties and Reduced Correlations for Gases," Gulf Publishing Corporation, Houston, Texas, 1966.
12. Arnold, E. W., Liou, D. W., and Eldridge, J. W., J. Chem. Eng. Data, 1965, Vol. 10, No. 2, 88.
13. Selected Values of Properties of Hydrocarbons and Related Compounds, API Research Project 44, Texas A&M University, Updated, 1979.
14. Strobridge, T. R., "The Thermodynamic Properties of Nitrogen from 64 to 300 K between 0.1 and 200 Atmospheres", Nat. Bur. Stand., Tech. Note 129, 1962.
15. IUPAC "International Thermodynamic Tables of the Fluid State", Chem. Data Series No. 7, Carbon Dioxide, Pergamon Press, London, 1980.
16. West, J. R., Chem. Engr. Prog., 1948, 44(4), 287.
17. Videl, J., Proceedings of the Third International Conference on Fluid Properties and Phase Equilibria for Process Design, Callaway Gardens, Georgia, April 10-15, 1983.
18. Goodwin, R. D., "The Thermophysical Properties of Methane from 90 to 500 K at Pressures 700 Bar", Nat. Bur. Stand. (U.S.), Tech. Note 653, 1974.
19. Goodwin, R. D., "The Thermophysical Properties of Ethane from 90 to 600 K at Pressures 700 Bar", Nat. Bur. Stand. (U.S.), Tech. Note 684, 1976.
20. ASHRAE Handbook, Fundamentals 1981, American Society of Heating, Refrigerating and Air Conditioning Engineers, Third Printing, 1982, 17.
21. IUPAC "International Thermodynamic Tables of the Fluid State", Chem. Data Series No. 25, Propylene, Pergamon Press, London, 1980.

22. Wichterle, I. and Kobayashi, R. J. Chem. Eng. Data, 1972, 17(1), 9.
23. Akers, W. W., Burns, J. F, and Fairchild, W. R. Ind. Eng. Chem., 46(12), 1954, 2531.
24. Douglas, G. E., Chen, R. J. J., Chappelear P. S., and Kobayashi, R. J. Chem. Eng. Data, 1974. 19(1), 71.
25. Kahre, L. C., J. Chem. Eng. Data, 1975, 20(4), 363.
26. Lin, Y. N., Chen, R. R. J., Chappelear, P. S., and R. Kobayashi, J. Chem. Eng. Data, 1977, 22(4), 402.
27. Kohn, J. P. and Bradish, W. F., J. Chem. Eng. Data, 1964, 9(1), 5.
28. Reamer, H. H., Olds, R. H., Sage, B. H., and Lacey, W. N. Ind. Eng. Chem., 1942, 34(12), 1526.
29. Matschke, D. E. and Thodos, G. J. Chem. Eng. Data, 1962, 7(2), 232.
30. Thodos, G. and Mehra, V. S. J. Chem. Eng. Data, 1965, 10(4), 307.
31. Reamer, H. H., Sage, B. H., and Lacey, W. N. Ind. Eng. Chem., 1940, 5(1), 44.
32. Zais, E. J. and Silberberg, I. L. J. Chem. Eng. Data, 1970, 15(2), 253.
33. Reamer, H. H. and Sage, B. H. J. Chem. Eng. Data, 1962, 7(2), 161.
34. Sage, B. H. and Lacey, W. N. Ind. Eng. Chem., 1940, 32(7), 992.
35. Reamer, H. H. and Sage, B. H., J. Chem. Eng. Data, 1966, 11(1), 17.
36. Reamer, H. H. and Sage, B. H., J. Chem. Eng. Data, 1964 9(1), 24.
37. Myers, H. S., Petroleum Refinery, 1957, 36(3), 175.
38. Liu, E. K. and Davison, R. R., J. Chem. Eng. Data, 1981, 26(1), 85.
39. Lin, H., Sebastion, H. M., Simnick, J. J., and Chao, K. C. J. Chem. Eng. Data, 1979, 24(2), 146.
40. Glanville, J. W., Sage, B. H., and Lacey, W. N. Ind. Eng. Chem., 1950, 42(3), 508.
41. Chang, H. L. and Kobayashi, R. J. Chem. Eng. Data, 1967, 12(4), 517.
42. Donnelly, H. G. and Katz, D.L. Ind. Eng. Chem., 1954, 46(3), 511.
43. Fredenslund, A. and Mollerup, J. J. Chem. Soc., Faraday Transactions I, 1974, 70, 1653.
44. Reamer, H. H. and Sage, B. H. Ind. Eng. Chem., 1951, 43(11), 2515.
45. Besserer, G. J. and Robinson, D. B. J. Chem. Eng. Data, 1973, 18(3), 298.
46. Besserer, G. J. and Robinson, D. B. J. Chem. Eng. Data, 1975, 20(1), 93.
47. Robinson, D. B., Kalra, H., Ng, H. J., and Kubota, H. J. Chem. Eng. Data, 1978, 23(4), 317.
48. Sage, B. H. and Reamer, H. H. J. Chem. Eng. Data, 1963, 8(4), 508.
49. Yorizane, M., Yoshimura S., and Masuoka, H. Kagaku Kogaku, 1966, 30(12), 1093.
50. Katayama, T. and Ohgaki, K. J. Chem. Eng. Data, 1976, 21(1), 53.

51. Robinson, D. B. and Ng, H. J. J. Chem. Eng. Data, 1976, 23(4), 325.
52. Katz, D. L. and Carson, D. B. Trans. AIME, 1942, 146, 150.
53. Zenner, H. G. and Dana, L. I. Chem. Engr. Progr. Symp. Ser. No. 44, 1963, 59, 36.
54. Kobayashi, R., Chappealear, P. S., and Stryjek, R. J. Chem. Eng. Data, 1974, 19(4), 334.
55. Schlinder, D. L., Swift, G. W., and Kurata, F. Hydrocarbon Processing, 1966, 45(11), 205.
56. Robinson, D. B., Kalra, H., Ng, H. J., and Miranda, R. D. J. Chem. Eng. Data, 1978, 23(4), 321.
57. Akers, W. W., Kehn, D. M., and Kilgore, C. H. Ind. Eng. Chem., 1954, 46(12), 2536.
58. Yorizane, M., Sadamota, S., and Yoshimura, S. Kagaku Kogaku, 1968, 32(3), 257.
59. Wagner, J., Erbar, R. C., and Majeed, A. I. Proceedings of the 64th Annual GPA Convention, Houston, Texas, March 18-20, 1985, 129.
60. Turek, E. A.; Metcalfe, R.S.; Yarborough, L.; and Robinson, R. L., Jr. Soc. of Pet. Engr. Jour. 1984, 308.
61. Robinson, D. B., and Besserer, G. J. Gas Processors Associations, Research Report RR-7, 1972.
62. Robinson, D. B., Kalra, H., Ng, H. J., and Kubota, H. J. Chem. Eng. Data, 1980, 25(1), 51.
63. Reamer, H. H., Selleck, F. T., Sage, B. H., and Lacy, W. N. Ind. Eng. Chem. 1953, 45, 1810.
64. Yarborough, L. and Vogel, J. L. Chem. Engr. Prog., Symp. Series, No. 81, 1964, Vol. 63, 1.
65. Robinson, D. B., Kalra, H., and Rempis, H. GPA Research Report, RR-31, May, 1981.
66. Reamer, H. H., Sage, B. H., and Lacey, W. N. Ind. Eng. Chem., 1953, 45(8), 1805.
67. Kobayashi, R. and Katz, D. L. Ind. Eng. Chem., 1953, 45(2), 440.

RECEIVED November 8, 1985

COMPUTATION AND ALGORITHM

23

Convergence Behavior of Single-Stage Flash Calculations

Marinus P. W. Rijkers[1] and Robert A. Heidemann

Department of Chemical and Petroleum Engineering, University of Calgary, Calgary, Alberta T2C 1R4, Canada

Flash calculation procedures have been investigated to determine when a flash routine (1) converges to the trivial solution or (2) converges very slowly or not at all. The two mixtures used as the basis for the study are a CO_2 containing natural gas and a CO_2-H_2S-CH_4 mixture that shows liquid-liquid separations. The Peng-Robinson equation was used to describe all the equilibrium phases. The computation method is the successive substitution procedure with four different approaches for updating the K values.

When a single equation of state is used to model all the equilibrium phases in a flash calculation, a trivial solution can be reached in which the calculated phases are identical and have all the properties of the feed stream. The necessary conditions for equilibrium, i.e.,

$$g_i = \ln f_i^V - \ln f_i^L = 0 \quad ; \quad i=1,\ldots,N \qquad (1)$$

are obviously satisfied by such a solution.

Most of the practical equations of state have at most three volume roots at any pressure and temperature. Whenever three volume roots are obtained for the feed mixture it is possible to avoid the trivial solution since a "liquid" feed and "vapor" feed can be assigned different volumes at the flash conditions and the fugacities in the liquid-like and vapor-like phases will surely be different. The problems with trivial solutions can occur only at T and P conditions where the equation of state for the feed mixture has only one positive real volume root.

There have been several suggestions made for obtaining from an equation of state a "liquid-like" or "vapor-like" volume, as needed, in regions where there is only one volume root (1-4). These ideas are reported to be helpful in avoiding trivial solutions in

[1]Current address: Technische Hogeschool Delft, The Netherlands.

0097-6156/86/0300-0476$06.00/0

simulations where many equilibrium calculations are performed in sequence.

An alternative approach in dealing with successive equilibrium computations has been to use converged solutions at one or more points to initiate a new computation (5-7). It is assumed that initiating near the solution has the advantage of reducing the number of iterations required to reach convergence and of increasing the probability of convergence to the correct, non-trivial, solution.

It has also been reported that, in general, flash routines can converge very slowly in some parts of the P-T co-existence region. Michelsen (8,9) has presented an analysis of the conventional successive substitution method and has shown very clearly that rapid convergence could not be expected at conditions close to the critical point of the mixture being flashed.

In this paper, a study has been performed to determine the effect of the initiation procedure on whether or not the trivial solution is reached. Also, several alternative flash calculation algorithms have been examined to see what convergence behavior can be expected in all parts of the phase diagram.

Initiation

The calculations performed here are all based on use of K values. For substance i, present in the liquid and vapor phases with mole fractions x_i and y_i, respectively, the phase distribution coefficients are defined by

$$K_i = y_i/x_i \tag{2}$$

The K_i are eventually to be found from the Peng-Robinson equation of state (10).

The trivial solution is characterized by K_i=1 for all i. Initiating computations with K_i values near unity would appear to risk convergence to the trivial solution.

A widely used procedure is to initiate K_i from Raoult's law, $K_i=P_i^s/P$ where P_i^s is the vapor pressure of substance i at the equilibrium temperature. A proposal ascribed to G.M. Wilson permits estimation of these Raoult's Law K values from the pure component critical properties and acentric factors, i.e.

$$\ln K_{i_W} = \ln(P_{c_i}/P) + 5.37\ (1 + \omega_i)(1-T_{c_i}/T) \tag{3}$$

It would be impossible to examine all possible ways for initiating the K_i. We have chosen to vary the initial K values from unity through the Wilson values according to

$$\overline{\ln K} = \alpha\ \overline{\ln K_W} \tag{4}$$

where α is a scalar. The overlines are meant to show that ℓn K and

$\ell n\ K_W$ are vector quantities. The K values obtained from (4) are unity when $\alpha = 0$ and are the Wilson Raoult's law values when $\alpha = 1$. We have found it useful to vary α in the interval $0 < \alpha \leqq 1.5$ and to observe the converged flash calculations to see for what values of α the trivial solution is obtained.

Flash Calculations

Given the K values it is possible to compute how a given feed mixture separates into two phases. The mass balance equations are organized as described, for instance, by Null (11), i.e., for one mole of feed with mole fractions z_i,

$$(1-V)x_i + VK_ix_i = z_i, \tag{5}$$

hence, $$x_i = z_i/[1 + (K_i-1)V] \tag{6}$$

and $$y_i = K_iz_i/[1 + (K_i-1)V] \tag{7}$$

The liquid and vapor mole fractions must sum to unity, or

$$h(V) = \sum (y_i-x_i) = \sum (K_i-1)z_i/[1+(K_i-1)V] = 0 \tag{8}$$

Finally, equation (8) is an equation in one unknown, V, which can be solved by the Newton-Raphson procedure. The resulting value of V, when substituted into equations (6) and (7), gives the equilibrium phase compositions.

The behavior of equation (8) has been analyzed by Nghiem et al. (12). They point out that $h(V) = 0$ has one and only one meaningful root, $0 \leqq V \leqq 1$, if and only if $h(0) \geqq 0$ and $h(1) \leqq 0$. If either of these inequalities is violated, the mixture will be in only one phase.

In particular, if $h(0) < 0$, then $V=0$ and the mixture is all in the liquid phase with mole fractions $x_i = z_i$. The calculated mole fractions in the vapor phase, $y_i = K_ix_i$, will not sum to unity. If further calculations are required, it is necessary to normalize; i.e. to set

$$y_i = K_ix_i / \sum_j K_jx_j \tag{9}$$

Similarly, if $h(1) > 0$, then $V=1$, $y_i = z_i$ and meaningful liquid mole fractions are found from

$$x_i = (y_i/K_i)/\sum_j (y_j/K_j) \tag{10}$$

In all the flash calculations performed in this study the above procedures were followed each time the K_i were specified. After obtaining the phase mole fractions from equations (6) and (7) (or (9) or (10) if necessary) it remained to update the K_i. In equation of state computations the K_i are certainly composition dependent.

Updating the K Values

Having obtained the phase mole fractions from the K values and mass balance equations, the fugacities are calculated from the Peng-Robinson equation, and the equilibrium conditions of equation (1) are checked. If they are not satisfied with sufficient precision the K_i must be corrected.

In the conventional successive substitution procedure, as described, for instance by Anderson and Prausnitz (13), the new K values are obtained from

$$K_i = \phi_i^L / \phi_i^V \tag{11}$$

This equation is derived from the definition of the fugacity coefficients
i.e.,

$$\phi_i^V = f_i^V / (y_i P) \tag{12}$$

and

$$\phi_i^L = f_i^L / (x_i P) \tag{13}$$

and by equating f_i^V and f_i^L.

At intermediate stages in the calculations, however, the fugacities are not equal. The values obtained for f_i^V and f_i^L follow from phase mole fractions that, in turn, are calculated from the K values used in the preceding iteration. Mehra, et al. (14) observed that equation (11) can be rewritten in forms that emphasize its iterative character, particularly

$$K_i^{(n+1)} = [(y_i/x_i)(f_i^L/f_i^V)]^{(n)} \tag{14}$$

Or, since $y_i/x_i = K_i$ and, since from equation (1) $g_i = \ln(f_i^V/f_i^L)$, then

$$\overline{\ln K}^{(n+1)} = \overline{\ln K}^{(n)} - \bar{g}^{(n)} \tag{15}$$

where superscripts (n) and (n+1) indicate the iteration level.

The convergence properties of four K value updating algorithms were investigated.

1. Conventional Successive Substitution (COSS)

Equation (15) is the most widely used in equation of state calculations to up-date the K values. This procedure is referred to here as the "Conventional Successive Substitution Algorithm" (COSS).

2. Accelerated Successive Substitution 1. (ACSS1)

Mehra et al (14) and Nghiem and Heidemann (15) have proposed that equation (15) be modified by introducing a scalar multiplier at each iteration, $\lambda^{(n)}$, as follows;

$$\overline{\ell n\ K}^{(n+1)} = \overline{\ell n\ K}^{(n)} - \lambda^{(n)}\ \bar{g}^{(n)} \qquad (16)$$

Mehra viewed λ as a variable step length in a steep descent method for minimizing the Gibbs free energy. Nghiem showed that the formulae proposed by Mehra for evaluating λ were derivable from a class of Quasi-Newton algorithms described by Zeleznik (16) as applied to solving the simultaneous equations $\bar{g} = 0$ with $\overline{\ell n\ K}$ as the vector of independent variables.

The simplest procedure proposed by Mehra yields

$$\lambda^{(n+1)} = \lambda^{(n)}\ \left|\bar{g}^{(n)T}\bar{g}^{(n)}/\bar{g}^{(n)T}(\bar{g}^{(n+1)} - \bar{g}^{(n)})\right| \qquad (17)$$

which is to be initiated with

$$\lambda^{(0)} = \lambda^{(1)} = 1.0 \qquad (18)$$

This is the procedure referred to here as Accelerated Successive Substitution 1 (ACSS1).

3. Accelerated Successive Substitution 2. (ACSS2)

Rijkers (17) has proposed an alternative to equation (17) for accelerating the successive substitution procedure. His analysis follows Mehra's (14) approach but treats the minimization problem in a somewhat different manner. Rijkers' algorithm is

$$\lambda^{(n+1)} = \lambda^{(n)}\ \left|\bar{h}^{(n)T}\bar{h}^{(n)}/\bar{h}^{(n)T}(\bar{h}^{(n+1)} - \bar{h}^{(n)})\right| \qquad (19)$$

with

$$\lambda^{(0)} = \lambda^{(1)} = 1.0 \qquad (20)$$

The vector $\bar{h}$ in Rijkers' algorithm is defined by

$$\bar{h} = \underline{U}^{-1}\ \bar{g} \qquad (21)$$

where the matrix $\underline{U}^{-1}$ has elements

$$u_{ij}^{-1} = (\partial n_{jV}/\partial \ell n K_i)_{T,P,K_k}$$

$$= V(1 - V)(x_i y_i/z_i)[\delta_{ij} + x_j y_j/(z_j S)] \qquad (22)$$

where δ_{ij} is the kroneker delta and where S is defined by

$$S = 1 - \sum x_k y_k/z_k \qquad (23)$$

This algorithm is referred to here as "Accelerated Successive Substitution 2" (ACSS2).

4. Newton-Raphson Updating. (NR)

A fourth alternative for up-dating the K values is the Newton-Raphson procedure. The two accelerated successive substitution procedures are Quasi-Newton algorithms for solving

$$\bar{g}(\overline{\ell n\ K}) = \bar{0} \tag{24}$$

The Newton-Raphson procedure is

$$\underline{J}\ \overline{\Delta \ell n\ K} = -\ \bar{g} \tag{25}$$

where $\underline{J}$ is the Jacobian matrix. It is most convenient to express $\underline{J}$ as the product

$$\underline{J} = \underline{W}\ \underline{U}^{-1} \tag{26}$$

where $\underline{W}$ has elements

$$w_{ij} = (\partial \ell n f_i^L/\partial n_j^L) + (\partial \ell n f_i^V/\partial n_j^V) \tag{27}$$

and $\underline{U}^{-1}$ is as defined in equation (22).

The Trivial Solution and Stability

Reference was made earlier to Michelsen's (8,9) analysis of the successive substitution procedure. One conclusion to be drawn from his analysis is that successive substitution cannot converge to the trivial solution at any point where the homogeneous phase is thermodynamically unstable.

There are many equivalent ways to state the criteria for stability in a homogeneous phase. Mechanical stability requires that

$$(\partial P/\partial v)_T < 0 \tag{28}$$

Diffusional stability requires that matrix $\underline{B}$ with elements

$$b_{ij} = (\partial \ell n f_i/\partial n_j)_{T,V,n} \tag{29}$$

should be positive definite.

The limit of stability is located where the determinant of $\underline{B}$ is zero and whenever det($\underline{B}$) < 0 the mixture is unstable. However, for $\underline{B}$ to be positive definite requires that the determinant of all the principal minors of $\underline{B}$ should also be positive (Amundson (18)).

Figures 1 and 2 show the phase envelopes for two mixtures. The first mixture is a CO_2-containing natural gas. The second is a CH_4-H_2S-CO_2 ternary mixture that shows liquid-liquid separations. Also drawn in Figures 1 and 2 are the curves corresponding to $(\partial P/\partial v)_T = 0$

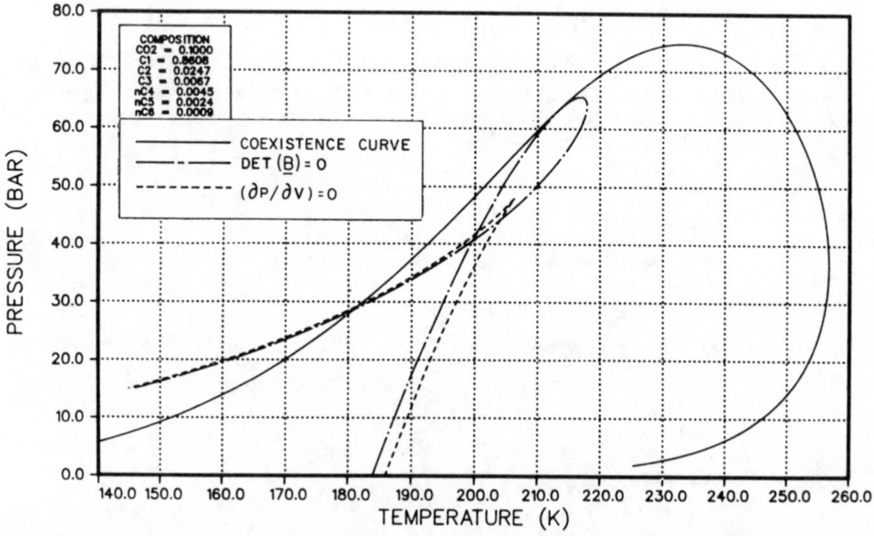

Figure 1. Phase Boundary and Stability Limits for Mixture I

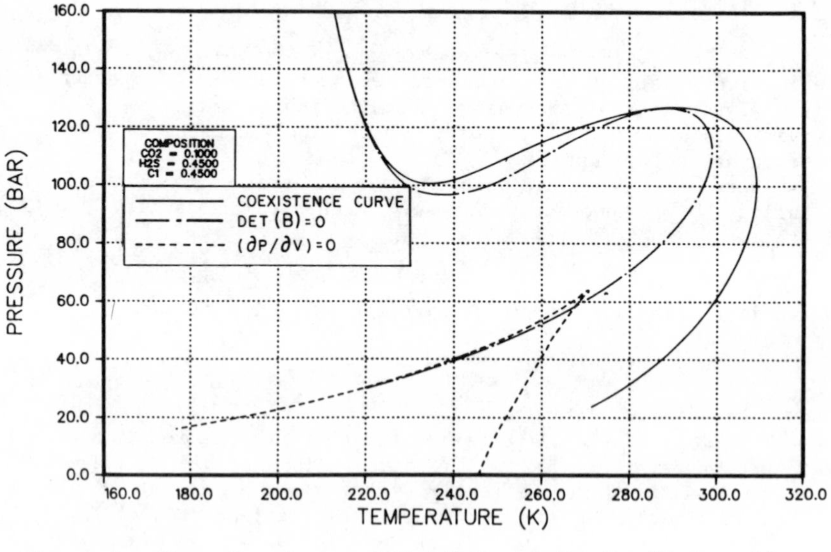

Figure 2. Phase Boundary and Stability Limits for Mixture II

and det($\underline{B}$) = 0. Inside these curves the mixtures are unstable and, according to Michelsen's analysis, it would be impossible for successive substitution flash algorithms to converge to trivial solutions.

The region where $(\partial P/\partial v)_T > 0$ corresponds to the region where the equation of state has three volume roots. The boundary curve of the region has been constructed by fixing the volume, calculating the temperature that makes $(\partial P/\partial v)_T = 0$ and then calculating the pressure from the equations of state evaluated at the volume and temperature. The cusp is the "mechanical critical point" for the mixture and is well inside the two phase region.

The diffusional stability limit, det($\underline{B}$) = 0, was evaluated in the same manner as the limit of mechanical stability and as described by Heidemann and Khalil (19). The critical point of the mixture lies on this curve.

For the natural gas mixture, the two boundaries for the unstable region enclose only a small part of the two-phase region. The unstable region for the CH_4 - H_2S - CO_2 mixture includes most of the phase diagram. Inside these unstable regions the trivial solution is impossible with successive substitution.

Initiation and the Trivial Solution

For the two mixtures of Figures 1 and 2 a large number of flash calculations have been performed. For the mixture of Figure 1, the two-phase region was intersected at pressures of 20, 65 and 70 bar and flash calculations were performed with varying initial K values over the whole temperature range within the two-phase region. For the mixture of Figure 2, equivalent calculations were performed at 40 and 100 bar.

Figures 3 through 6 and Figures 7 and 8 contain the results for the two mixtures at the various pressures. The initial K values were varied, as described earlier, by letting the "Phase Distribution Coefficient Initiation Parameter", α, take on values between zero and 1.5. The value of α is shown on the ordinate of the Figures and the temperature is shown on the abscissa. The areas of the T - α plane are labelled to show the solution reached.

For the first mixture at 20 bar, the two-phase region extends from 170 to 254 K. Between 170 and 194 K the mixture is either mechanically or diffusionally unstable or both.

As expected, the conventional successive substitution algorithm did not converge to the trivial solution within this temperature interval for any α value. This is indicated by the corresponding cross-hatched region of Figure 3.

Also shown in Figure 3 is the region of low α values, outside the unstable temperature interval, where COSS converged to the trivial solution. To avoid the trivial solution it is only necessary to have α large enough. Near the phase boundary, 254 K, larger α values are required. However, no trivial solutions were found at any temperature so long as α was greater than about 0.45. In particular, if the Wilson K values are used (α = 1), the trivial solution would never be reached at 20 bar and any temperature.

Figure 4 shows convergence behavior with COSS applied to the first mixture at 65 bar. The two-phase region is contained between 215 and 248 K. (The pressure is above the "mechanical critical pressure", hence the mixture is not mechanically unstable at any temperature.) Figure 4 indicates that trivial solutions were reached at temperatures between the phase boundary and the stability limit when the K values were initiated too near unity (α below about 0.6). At temperatures greater than 218 K, the trivial solution was avoided for $\alpha > 0.35$. Certainly, no trivial solutions were encountered when COSS was initiated at the Wilson K values ($\alpha = 1$).

Although convergence to the trivial solution is impossible in the unstable region, the number of iterations required could become large. In Figure 4, open boxes are used to indicate points where convergence was not achieved. Convergence failure is discussed in more detail later.

The calculations for mixture I at 65 bar were repeated using the first accelerated successive substitution procedure (ACSS1). The convergence behavior is shown in Figure 5. It will be seen that ACSS1 and COSS behave similarly as regards the tendency to converge to the trivial solution. Since both procedures can be regarded as steep descent algorithms, neither can reach the trivial solution in the unstable temperature interval. Aside from some isolated points, ACSS1 and COSS converge to the correct solution at the same values of α.

Figure 6 shows the results of ACSS1 flash calculations for mixture I at 70 bar. The isobar at this pressure does not intersect the unstable region. Potentially, therefore, the trivial solution could be reached at any temperature within the two-phase region between about 221 and 244 K. The figure shows, however, that if α is large enough, (i.e., $\alpha > 0.6$) the trivial solution is never reached. The results are of the same type as seen in Figures 3-5 and demonstrate that initiating with the Wilson K values is sufficient to avoid trivial solutions even along a difficult retrograde isobar.

Convergence results for the mixture of Figure 2 at 40 and 100 bar, using COSS, are shown in Figures 7 and 8. The isobar at 40 bar extends from a low-temperature liquid-liquid region, through a vapor-liquid region, to the phase boundary at about 287 K. The mixture is thermodynamically unstable at temperatures below 259 K, therefore it is impossible for COSS to converge to the trivial solution at any temperature below 259 K, regardless of the initial K values used. The plot in Figure 7 contains only a very small region (which is not cross-hatched) bounded by $259 < T < 287$ K and $\alpha < 0.3$ in which trivial solutions were found.

As shown in Figure 7, two different two-phase solutions were possible in the temperature interval $188 < T < 202$ K. Depending on the initial K values, it was possible to reach either a vapor-liquid solution or a liquid-liquid solution. There is obviously a three-phase region in this vicinity and its existence complicates flash calculations when the number of phases specified is less than the number of phases which should actually exist at equilibrium (according to the model). No attempt was made to locate the three-phase region. It should be noted, however, that convergence

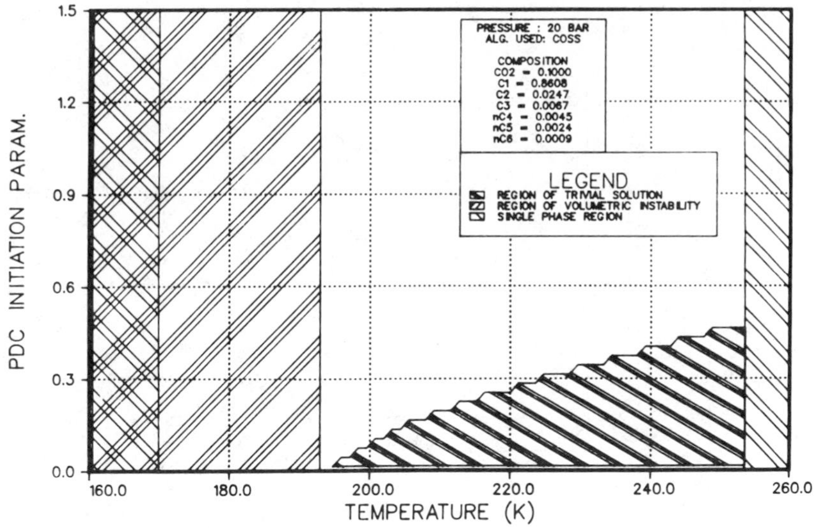

Figure 3. Region of Solution for Mixture I at 20 Bar

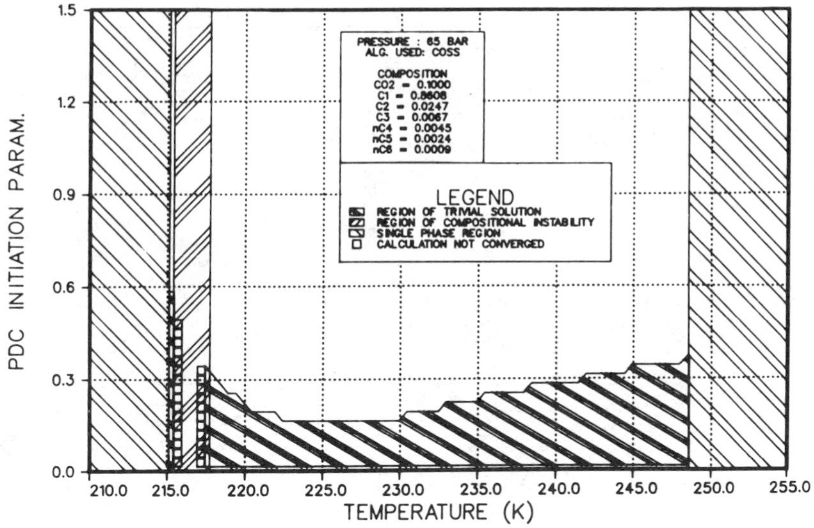

Figure 4. Region of Solution for Mixture I at 65 Bar Using COSS

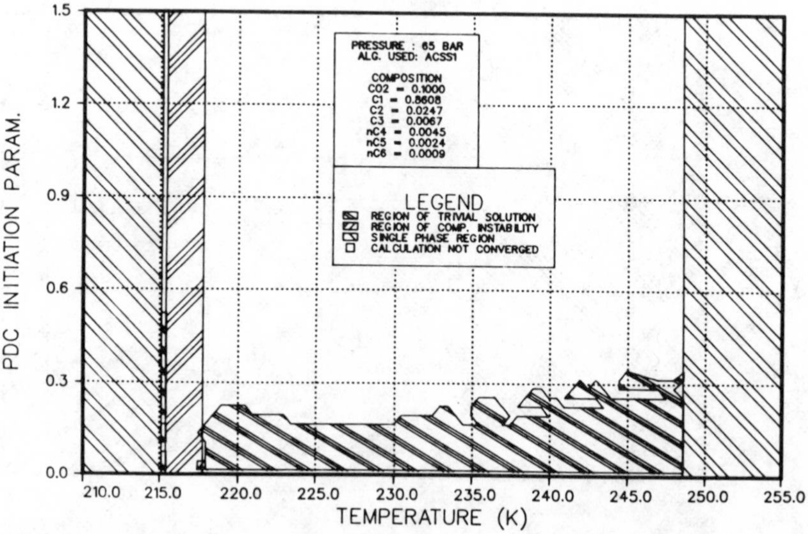

Figure 5. Region of Solution for Mixture I at 65 Bar Using ACSS1

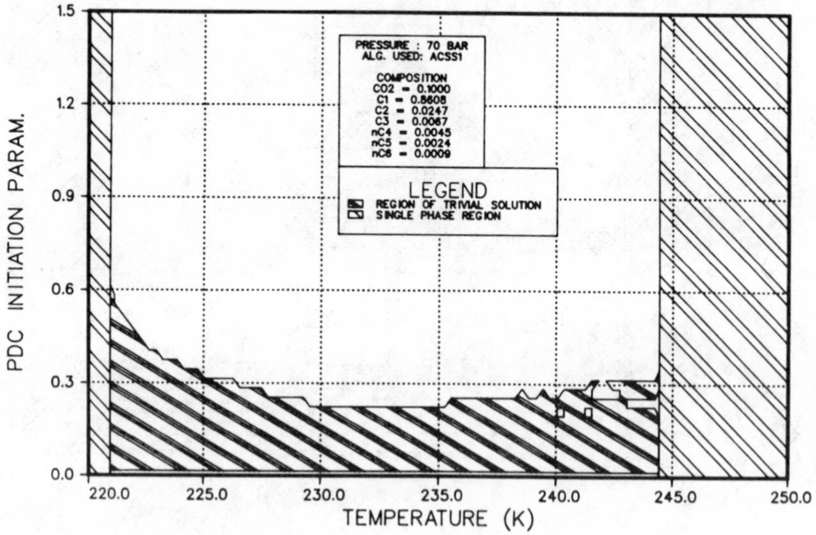

Figure 6. Region of Solution for Mixture I at 70 Bar

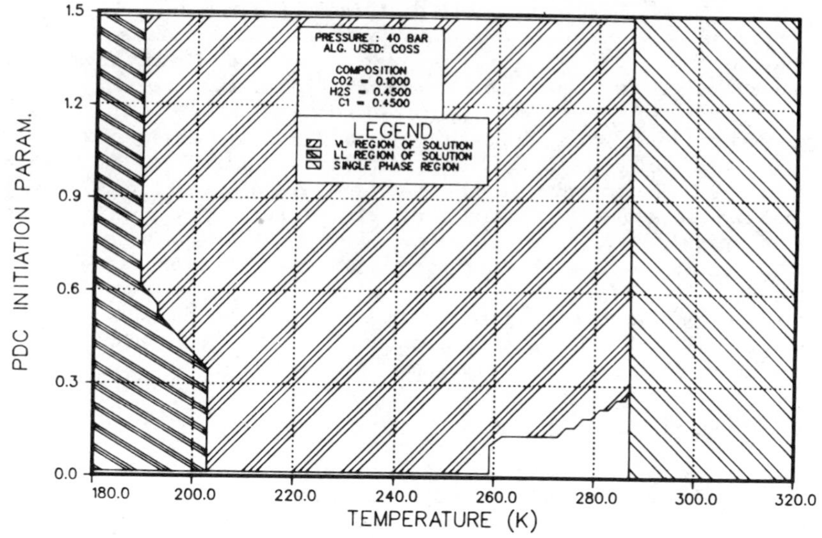

Figure 7. Region of Solution for Mixture II at 40 Bar

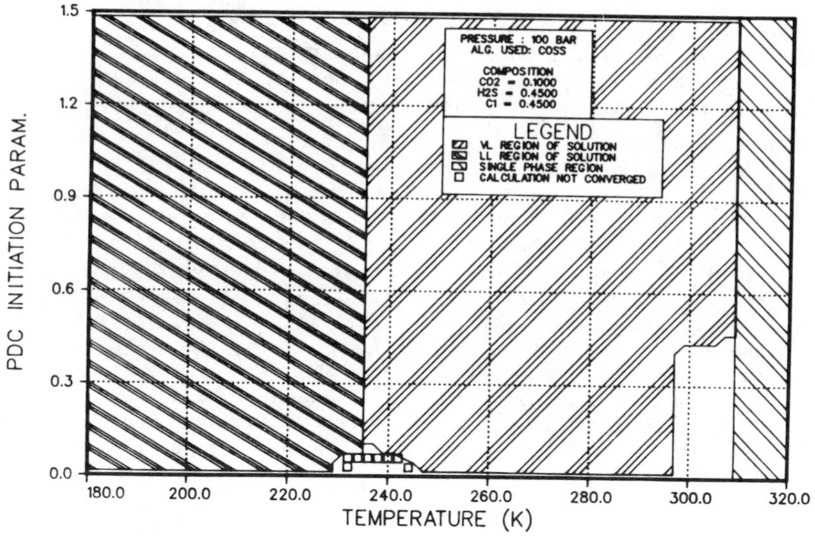

Figure 8. Region of Solution for Mixture II at 100 Bar

to a non-trivial solution was obtained at all temperatures and at all α values in the vicinity of the three phase region.

Figure 8 shows convergence properties of the mixture of Figure 2 at 100 bar. The pressure is just below the dip in the phase boundary curve at about 235 K so that the mixture remains in two phases from very low temperatures to about 309 K. There is a temperature interval from about 228 to 249 K where the mixture is outside the limit of diffusional stability. Likewise, from about 297 to 309 K the mixture is outside the limit of stability. Only in these two separate intervals are trivial solutions possible when flash calculations are performed using COSS. These trivial solutions are avoided if α is greater than about 0.45, as shown in Figure 8.

Also indicated in Figure 8 are a few points where convergence was not achieved. These points are all in the intermediate stable temperature interval with very low α values.

Slow Convergence Problems

In some regions of the phase diagrams the successive substitution algorithms converged very slowly. In order to reduce computation time a dual algorithm was used. The computations were begun with the successive substitution procedure, but if convergence was not achieved within a reasonable number of iterations the Newton-Raphson algorithm for updating the K values was begun.

This procedure did not change the conclusions as to whether or not the trivial solution was ultimately to be reached so long as the flash calculation was well on its way to convergence and so long as there was to be a split into two phases of finite amount.

The second condition is necessary because the Jacobian matrix of equation (26) is undefined when V = 0 or V = 1. The problem arises with matrix $\underline{U}^{-1}$ given by equation (22).

The dual algorithm improved convergence behavior considerably. In preparing Figure 4, 3283 independent flash calculations were required. With COSS alone, 238 of these calculations did not converge within 100 iterations. With the dual algorithm only 27 poorly initiated calculations were left unconverged. Figure 7 required 3179 flash calculations, 382 of which did not converge with COSS alone. All calculations converged with the dual algorithm.

The reason for convergence failure even of the Newton-Raphson procedure requires some discussion. What was achieved in these cases was an oscillatory behavior between a solution like the trivial solution and one like the correct solution. These failures were generally associated with initial guesses for the K values near unity (α near zero).

Speed of Convergence

The speeds of convergence of four different algorithms for updating the K values are compared in Figures 10 through 13. The mixture of Figure 1 is the basis for the comparison. Flash calculations were performed at 930 points throughout the two-phase region for this mixture and at each point the calculation was initiated with the Wilson K values (α = 1).

Figure 9 shows the COSS results. At each of the 930 points the number of iterations to convergence has been inserted in the diagram. Through most of the two-phase region, convergence was obtained in 25 iterations or fewer and the average iteration count to convergence was just over 15. At only three points of the 930 was convergence unattainable in 100 iterations and these points are very close to the critical point of the mixture.

Figures 10 and 11 show iteration counts for the ACSS1 and ACSS2 algorithms, respectively. These two algorithms perform quite comparably, with average iterations to convergence of 9.84 and 9.85. ACSS1 is apparently superior to ACSS2 on the grounds that convergence was obtained within 100 iterations at all 930 points with ACSS1 and ACSS2 failed to converge at one point.

The total execution time at 930 points was 181.8, 141.6, and 166.3 seconds for COSS, ACSS1 and ACSS2, respectively. On this account ACSS1 appears also to be superior to ACSS2 and to successive substitution without acceleration (COSS).

Figure 12 shows the results obtained when the Newton-Raphson algorithm was employed to update K values. As before, these calculations were all initiated with the Wilson K values. A striking result is that the average number of iterations per correct solution is only 4.71, less than half the average of the other three procedures. However, the iteration count gives only a partial picture of the efficiency of the algorithm and two negative factors must be considered.

The first negative feature is that the N-R updating procedure for the K values failed to reach the correct solution at 40 of 930 points. This poor convergence behavior was a result of an undefined Jacobian at the phase boundaries. If equations (6) and (7) did not yield a two phase solution, small quantities of L or V were introduced. Consequently the x_i and y_i were evaluated from equations (6) or (7) and were normalized by applying equations (9) or (10). This approach clearly did not solve the problem: At 25 points the trivial solution was reached and at the remaining 15 points the procedure did not converge.

The second point to consider is that each N-R iteration requires considerable matrix manipulation and is much more consuming of computer time than a successive substitution iteration. The total time required for the 930 points of Figure 12 is 424.3 seconds; more than twice the time required for the COSS procedure.

The failure to converge is no doubt a more serious concern than the computer time consumed in the N-R procedure. Computing time certainly depends on the skill of the programmer and no doubt a N-R routine can be written which will be more efficient than ours.

Discussion

All examples presented here support the conclusion that the successive substitution algorithm cannot converge to the trivial solution, however initiated, at points where the homogeneous phase is unstable. This property is due to the nature of successive substitution as a steep descent procedure for minimizing the Gibbs free energy.

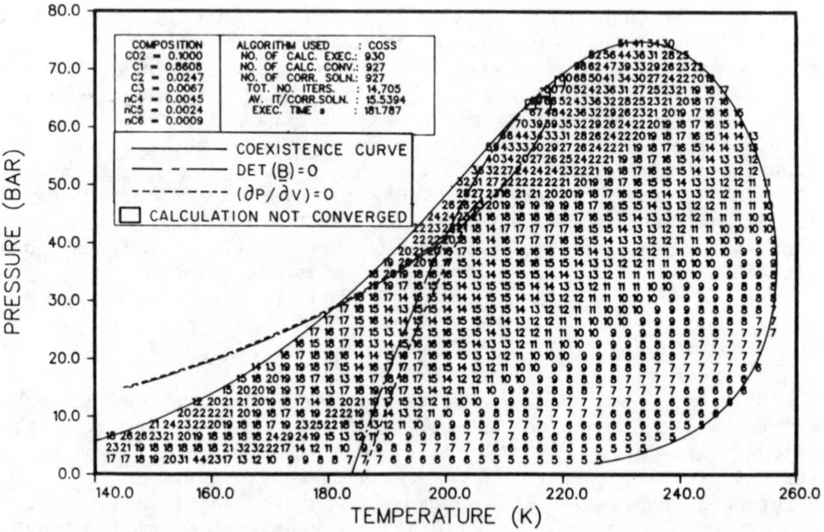

Figure 9. Iteration Counts for Mixture I. COSS Initiated with Wilson K Values

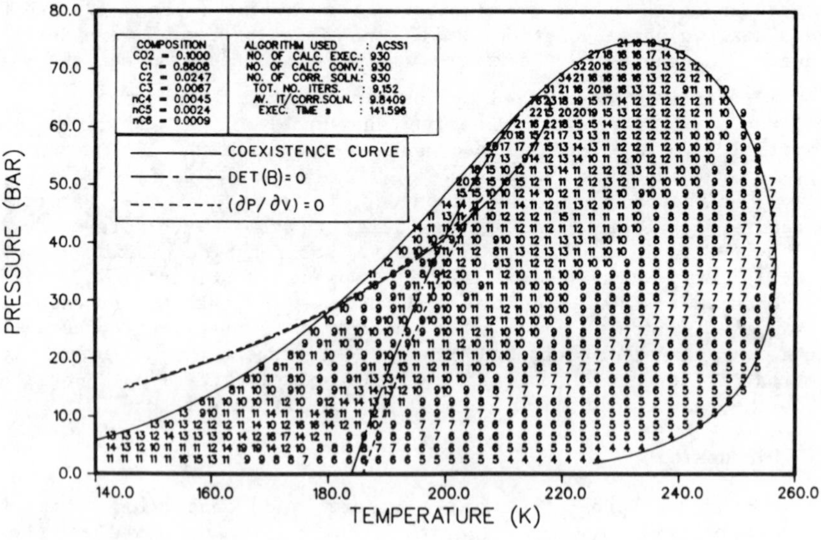

Figure 10. Iteration Counts for Mixture I. ACSS1 Initiated with Wilson K Values

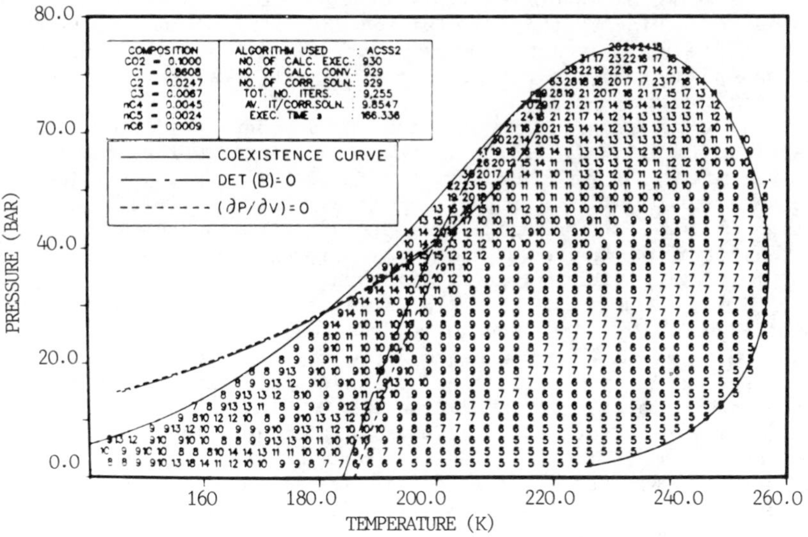

Figure 11. Iteration Counts for Mixture I. ACSS2 Initiated with Wilson K Values

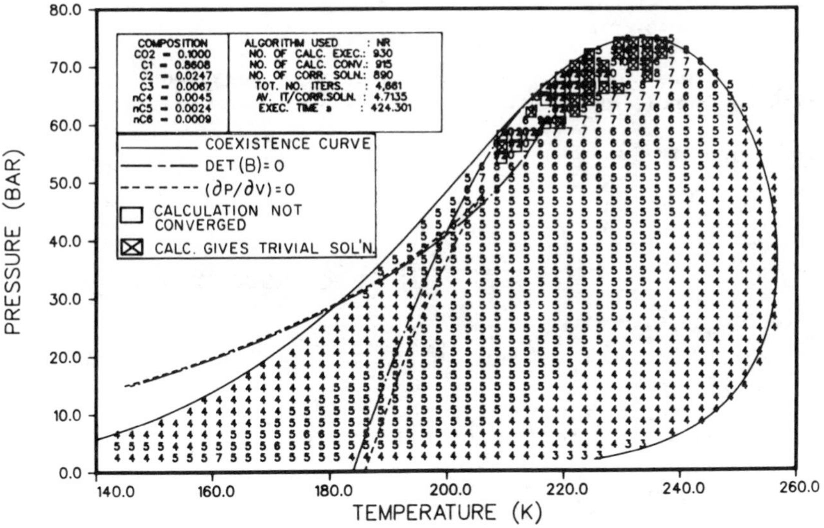

Figure 12. Iteration Counts for Mixture I. Newton-Raphson Initiated with Wilson K Values

Resorting to Newton-Raphson updating of K values, however, amounts to giving up the minimization character of the algorithm in favor of using an equation solving technique. The N-R scheme can (it has been demonstrated) converge to the trivial solution even at unstable points.

In the examples it appears that convergence to trivial solutions is unlikely if the K values are initiated appropriately. For the mixture of Figure 1, initiating with the Wilson K values was adequate to avoid convergence to the trivial solution anywhere inside the two-phase region. This is true even though the region of instability is a very small part of the two-phase region.

Initiating with K values near unity can result in convergence to trivial solutions. This implies that there may be some risks involved in using K values at a converged solution to initiate a flash calculation at a nearby point, particularly near the critical.

The results regarding speed of convergence indicate a definite advantage in accelerating the successive substitution algorithm. The first procedure, ACSS1, appears sufficient. The Newton-Raphson procedure used here certainly reduces the iteration count dramatically but has associated with it an increased risk that the trivial solution or non-convergence will be reached. Also, it is not certain that a reduced number of iterations implies a reduction in computer time.

Other types of flash calculations than the constant T and P flash examined in this manuscript are important. The occurrence of trivial solutions in isenthalpic flash routines, for instance, requires further study.

Acknowledgments

This work was supported by the National Science and Engineering Research Council of Canada.

Literature Cited

1. Poling, B.E.; Grens, E.A.; Prausnitz, J.M. Ind. Engg. Chem. Proc. Des. Dev. 1981, 20, 127.
2. Gundersen, T. Comp. Chem. Engg 1982, 6, 245.
3. Coward, I.; Gayle, S.E.; Webb, D.R. Trans. I. Chem. E. 1978, 56, 19.
4. Veeranna, D.; Rihani, D.N. Fluid Phase Equil 1984, 16, 41.
5. Asselineau, L.; Bogdanic, G.; Vidal, J. Fluid Phase Equil 1979, 3, 273.
6. Michelsen, M.L. Fluid Phase Equil 1980, 4, 1.
7. Mehra, R.K.; Heidemann, R.A.; Aziz, K. Soc. Pet. Engrs J 1982, 22, 61.
8. Michelsen, M.L. Fluid Phase Equil 1982, 9, 1.
9. Michelsen, M.L. Fluid Phase Equil 1982, 9, 21.
10. Peng, D-Y.; Robinson, D.B. Ind. Engg. Chem. Fundamentals 1976, 15, 59.
11. Null, H.R. "Phase Equilibrium in Process Design"; Wiley Interscience Inc.: New York, 1970.

12. Nghiem, L.X.; Aziz, K.; Li, Y.K. Soc. Pet. Engrs Jl 1983, 23, 521.
13. Anderson, T.F.; Prausnitz, J.M. Ind. Engg. Chem. Proc. Des. Dev. 1980, 19, 9.
14. Mehra, R.K; Heidemann, R.A.; Aziz, K. Can. J. Chem. Eng. 1983, 61, 590.
15. Nghiem, L.X.; Heidemann, R.A. In 2nd European Symposium on Enhanced Oil Recovery 1982, p.303 Ed. Techniqs., Paris, 1982.
16. Zeleznik, F.J. J. Assoc. Comp. Mach. 1968, 15, 265.
17. Rijkers, M.P.W. M.Sc. Thesis, The University of Calgary, Calgary, Alberta, Canada, 1985.
18. Amundson, N.R. "Mathematical Methods in Chemical Engineering"; Prentice-Hall, Inc.: Englewood Cliffs, N.J., 1966.
19. Heidemann, R.A.; Khalil, A.M. AIChE J. 1980, 26, 769.

RECEIVED November 8, 1985

24

Four-Phase Flash Equilibrium Calculations for Multicomponent Systems Containing Water

R. M. Enick, G. D. Holder, J. A. Grenko, and A. J. Brainard

Chemical and Petroleum Engineering Department, University of Pittsburgh, Pittsburgh, PA 15261

A technique for predicting one- to four-phase flash equilibrium is presented for multicomponent systems containing water, such as CO_2/crude oil/water mixtures which characterize the carbon dioxide miscible flooding of petroleum reservoirs. The Peng-Robinson equation of state is used to describe the aqueous and hydrocarbon phases. An accelerated and stabilized successive substitution method is used to obtain convergence, even in the near critical region. Additional hydrocarbon phases are introduced by using a fugacity based testing scheme. Water-free flash calculations are first performed, yielding one, two or three phase equilibrium. A comprehensive search strategy is then used to consider eleven general classifications of systems which may result, including water-rich liquid/hydrocarbon-rich liquid/CO_2-rich liquid/vapor equilibrium. Improved methods for obtaining initial estimates of additional phases are presented and a reliable scheme of searching for additional hydrocarbon-rich phases is introduced which considers all three possible phase identities.

General Objectives. Multiple phase behavior is often encountered in the gas miscible flooding of petroleum reservoirs. In water-free carbon dioxide/crude oil systems, for example, three phases often exist in equilibrium at low temperatures.(1) However, water is almost universally present, either interstitially or because it is injected for mobility control;(2) hence its presence should also be considered in phase behavior studies. Because of the relatively low miscibility of water with both carbon dioxide and hydrocarbons at reservoir conditions, an aqueous phase will almost always exist, resulting in the possible presence

0097-6156/86/0300-0494$07.50/0

of as many as four phases. In asphaltic crude systems, a fifth phase, a solid precipitate, may also form,(3) but predicting its presence or composition is beyond the scope of this study.

Generally, water-free flash calculations are made when these systems are modeled,(4) and an equation of state is used to calculate component fugacities. Temperature, pressure and overall composition are typically specified in the flash calculation technique in order to determine the amounts and compositions of each possible phase.

The objective of this study is to develop an efficient algorithm for CO_2/hydrocarbon/water systems in which the number of phases, one to four, is determined and in which the composition and amount of each phase, including the aqueous phase, is accurately described. Emphasis is placed on four-phase equilibrium and the effects of the presence of water on the phase behavior of a CO_2/hydrocarbon mixture.

Review of the Literature: Multiphase Flash Calculations. The model for the two phase flash problem was presented in 1952 by Rachford and Rice(5) and many improvements were introduced in subsequent studies,(6-8) but multiphase flash calculations were not addressed until 1969 when Deam and Maddox(9) presented the three-phase flash equilibrium problem.

Since that time many improvements in the algorithms used have led to more rapid convergence rates and to the correct identification of the number and composition of the stable equilibrium phases. These improvements have addressed the two distinct problems which characterize flash calculations. The first is defining a thermodynamic model, such as an equation of state, which gives results in agreement with experimental measurements. The second problem is finding a numerical solution to the flash calculation with the given thermodynamic model. The major contributions to these improvements include efficient numerical techniques such as the over-relaxation method, accelerated successive substitution, and the multi-variant Newton-Raphson technique, the development of equations of state, algorithms for obtaining initial estimates for the composition of each phase, and the application of Gibbs free energy minimization principles for evaluating the stability of any equilibrium phase.(10-25)

The solution scheme presented by Risnes and Dalen(25) serves as a foundation for this work. Improved algorithms have been developed for determination of the number of phases, the initial estimation of phase compositions, and the search strategy for determining if an additional phase should exist or if an existing phase should be eliminated. Emphasis is placed on quantitatively describing how the addition of water to a hydrocarbon based mixture affects the number and compositions of equilibrium phases.

Two Phase And Multiphase Flash Calculations

Governing Equations and Computational Algorithms. The equations employed to describe the phases, define thermodynamic equilibrium, facilitate the solution of two phase and multiphase solutions, initiate estimates of any phase's composition, and efficiently adjust these initial estimates are presented in this section. In addition to the system temperature, pressure and overall composition, the input data required for the solution of these flash equilibrium problems consist of the critical temperature, critical pressure and acentric factor of each component, as well as a binary interaction parameter for each pair of components.

All phases are described by an equation of state. The Peng-Robinson equation of state is presented in Table Ia along with its corresponding expressions for compressibility factor, fugacity coefficient and mixing rules, in Table Ib, Equations 1-13.

Table Ia. Peng-Robinson Equation of State(15)

$$P = \frac{RT}{v-b} - \frac{a}{v(v+b) + b(v-b)} \quad (1)$$

$$b = 0.07780\ RT_c/P_c \quad (2)$$

$$a(T) = 0.45724\ R^2T_c^2\alpha/P_c \quad (3)$$

$$\alpha^{0.5} = 1 + m(1-Tr^{0.5}) \quad (4)$$

$$m = 0.37464 + 1.54226\omega - 0.26992\omega^2 \quad (5)$$

Table Ib. Compressibility Factor and Fugacity Coefficient

$$Z^3 - (1-B)Z^2 + (A-3B^2-2B)Z - (AB-B^2-B^3) = 0 \quad (6)$$

where
$$Z = Pv/RT \quad (7)$$
$$A = aP/R^2T^2 \quad (8)$$
$$B = bP/RT \quad (9)$$

$$\ln \Psi_k = \frac{b_k}{b}(Z-1) - \ln(Z-B) -$$

Table Ib. Continued

$$\frac{A}{2\sqrt{2}B}\left(\frac{2\sum_i x_i a_{ik}}{a} - \frac{b_k}{b}\right) \ln\left(\frac{Z + (\sqrt{2}+1)B}{Z - (\sqrt{2}-1)B}\right) \quad (10)$$

$$b = \sum_i x_i b_i \quad (11)$$

$$a = \sum_i \sum_j x_i x_j a_{ij} \quad (12)$$

$$a_{ij} = (1-\sigma_{ij}) a_i^{0.5} a_j^{0.5} \quad (13)$$

The governing equations for two phase water-free equilibrium include overall and component material balances, mole-fraction constraints, and the thermodynamic equilibrium criteria, (Equations 14-17). These 2n+2 independent expressions define the two phase equilibrium problem of 2n+2 unknowns, L, V, x_i and y_i. The overall material balance,

$$L + V = N = 1 \quad (14)$$

component material balances,

$$Lx_i + Vy_i = Nz_i \quad (15)$$

mole fraction contraints,

$$\Sigma x_i = \Sigma y_i = 1 \quad (16)$$

and thermodynamic equilibrium criteria

(a) $fi_L = fi_V$ (17)

(b) system of predicted phases must minimize Gibbs energy of system

define the flash calculation problem.

Two phase flash calculations may be performed following the procedure. The equilibrium constants are defined as

$$K_i = y_i/x_i \quad (18)$$

Eliminating L from (15) and summing over all components gives

$$\Sigma x_i + V\,\Sigma(y_i - x_i) = \Sigma z_i = 1 \ (N=1) \quad (19)$$

The g(V) function, used to solve (14), (15), (16), and (18), is defined as

$$g(V) = \Sigma(y_i - x_i) = \Sigma \frac{(K_i - 1)z_i}{1 + (K_i - 1)V} \quad (20)$$

where V is the root of g(V) = 0.
In the single phase state,

a. if $V < 0$ then $V = 0$, $x_i = z_i$, $y_i = K_i x_i / \Sigma K_i x_i$ (21)

b. if $V > 1$ then $V = 1$, $y_i = z_i$, $x_i = (y_i/K_i)/\Sigma(y_i/K_i)$ (22)

The bring-back procedure is defined by the following expressions,

a. if $V = 0$ then $\Sigma(\gamma K_i - 1)z_i = 0$ (23)

$$K_i^{new} = \gamma K_i = K_i/\Sigma K_i z_i \quad (24)$$

$$\gamma = 1/\Sigma K_i z_i \quad (25)$$

b. if $V = 1$ then $\Sigma(1-1/\gamma K_i)z_i = 0$ (26)

$$K_i^{new} = \gamma K_i = K_i/\Sigma z_i/K_i \quad (27)$$

$$\gamma = \Sigma z_i/K_i \quad (28)$$

Values are assumed for the equilibrium constants defined by Equation 18. The n fugacity equations are then replaced by the equilibrium constant equations, enabling the straightforward solution for the unknowns. The moles of vapor, V, are first determined by using Newton's method to find the root of $g(V) = 0$. This function is a mathematical combination of the 2n+2 defining equations and at $g(V) = 0$ the amount of flashed vapor and liquid are known ($L=1-V$ for system of one total mole). Compositions are then calculated by simultaneously solving Equations 14, 15 and 16. If the root falls outside of the two-phase interval, $V < 0$ or $V > 1$, the system is in a single phase state. The composition of the non-existing phase is then calculated as if the system is at the saturation pressure. In order to hasten the re-entry of a phase lost in the correction process into the solution algorithm where it may eventually appear in the final answer, the "bring-back" procedure may be used (Equations 23-28). Since a common multiple, γ, applied to each of the K_i's does not effect the composition of the the corresponding phases (it does affect the amounts of each phase), this factor may be introduced to force the g(V) function to zero. This adjustment keeps the non-existing phase at the edge of the two-phase boundary, where its existence will subsequently be tested.

Two procedures for initializing K-value estimates may be used. Ideally, these estimates should assure that calculations start in the two phase region, are as close as possible to the actual values, and display sufficient contrast to avoid a trivial solution where all K_i's equal unity. The Wilson formula(14), is an empirical expression which provides K-value estimates which meet these criteria under most conditions.

$$K_i = Pri^{-1} \exp[5.3727(1+\omega)(1-Tri^{-1})] \quad (29)$$

In the near-critical region, however, the evaporation technique consistently yields better estimates. This may be attributed to the technique being based upon the same equation of state used to define two-phase equilibrium. The evaporation technique consists of the following sequence of calculations:

1. Assume L = 1.0
2. Calculate f_i's
3. Evaporate .01 mole; composition proportional to f_i's
4. Continue until $L \cong 0.50$
5. Perform material balance, $K_i = y_i/x_i$

Improved K-values may be obtained by the methods based on fugacity expressions

$$f_{iL} = x_i \Psi_{iL} P \tag{30}$$

$$f_{iv} = y_i \Psi_{iV} P \tag{31}$$

an iterative correction of K-values, by successive substitution is defined as

$$K_i^{t+1} = \Psi_{iL}/\Psi_{iV} = K_i^t R_i^t \quad \text{where } t = \text{iteration number} \tag{32}$$

$$R_i = f_{iL}/f_{iV} \tag{33}$$

In order to reduce the number of iterations in the near-critical region, where convergence may require hundreds of iterations, the basic successive-substitution method must be accelerated(25). This accelerated successive substitution is defined as

$$K_i^{t+1} = K_i^t R_i^{t\,(1-k_i)^{-1}} \tag{34}$$

$$\text{where } k_i = (R_i^t - 1)/(R_i^{t-1} - 1) \tag{35}$$

This acceleration technique may be implemented only after the basic successive-substitution method is monotonically approaching a solution, as indicated by an error norm, ρ, of approximately 10^{-4}. Furthermore, the acceleration must be performed alternately with the conventional technique, and any accelerated K-value which cause the error norm to increase must be rejected. The criterion for solution acceptance is based on the thermodynamic equilibrium criterion, Equation 17, which requires that the liquid and vapor phase fugacities of each component be equal. The solution acceptance is expressed as

$$\rho = \Sigma(R_i - 1)^2 < \varepsilon \tag{36}$$

A fugacity residual error norm, Equation 36, of 10^{-16} is attainable for many systems, including the example problems presented in a subsequent section.

The equations which govern multiphase equilibrium are different from the two-phase equations only in that two additional phases are introduced. For a four-phase system, these 4n+4 independent equations define the equilibrium problem of 4n+4 unknowns. These equations include the overall material balance,

$$L_1 + L_2 + L_3 + V = N = 1 \tag{37}$$

component material balances,

$$L_1 x_{1i} + L_2 x_{2i} + L_3 x_{3i} + V y_i = N z_i \tag{38}$$

mole fraction constraints,

$$\Sigma x_{1i} = \Sigma x_{2i} = \Sigma x_{3i} = \Sigma y_i = 1 \tag{39}$$

and thermodynamic equilibrium criteria

(a) $f_{iL_1} = f_{iL_2} = f_{iL_3} = f_{iV}$ (40)

(b) system of predicted phases must minimize Gibbs energy

The multiphase flash calculation procedure is also quite similar to that of the two phase system.
The equilibrium constants are defined in reference to the vapor phase as

$$K_{1i} = y_i/x_{1i} \quad K_{2i} = y_i/x_{2i} \quad K_{3i} = y_i/x_{3i} \tag{41}$$

The gas phase composition, for N = 1, is given by

$$y_i = z_i/[1 + L_1\left(\frac{1}{K_{1i}} - 1\right) + L_2\left(\frac{1}{K_{2i}} - 1\right) + L_3\left(\frac{1}{K_{3i}} - 1\right)] \tag{42}$$

Eliminating V from Equation 38 and summing over all components, gives

$$\Sigma y_i + L_1 \Sigma(x_{1i} - y_i) + L_2 \Sigma(x_{2i} - y_i) + L_3 \Sigma(x_{3i} - y_i) = \Sigma z_i = 1 \tag{43}$$

Defining the g_j functions, used to determine the phase distribution,

$$g_j\,(L_1, L_2, L_3) = \Sigma(x_{ji} - y_i) \tag{44}$$

Combining Equations 41, 42, 43 and 44

$$g_j = \Sigma \frac{(\frac{1}{K_{ji}} - 1)z_i}{1 + L_1 (\frac{1}{K_{1i}} - 1) + L_2(\frac{1}{K_{2i}} - 1) + L_3(\frac{1}{K_{3i}} - 1)} \quad (45)$$

The g_j functions are used to solve Equations 37, 38, 39, 41, where L_1, L_2 and L_3 are the roots of $g_j(L_1, L_2, L_3) = 0$. This is valid only if the gas phase is present and the y_i's are normalized. If the gas phase is not present, assume the oil-rich L_1 phase exists. The oil rich phase composition is then given by

$$x_{1i} = z_i/[1 + L_2(\frac{K_{1i}}{K_{2i}} - 1) + L_3(\frac{K_{1i}}{K_{3i}} - 1) + V(K_{1i}-1)] \quad (46)$$

Eliminating L_1 from Equation 38 and summing over all components, yields

$$\Sigma x_{1i}+L_2\Sigma(x_{2i}-x_{1i}) + L_3\Sigma(x_{3i}-x_{1i}) + V\Sigma(y_i-x_{1i}) = \Sigma z_i=1 \quad (47)$$

One may now define h_j functions, analagous to the g_j functions

$$h_j (L_2, L_3, V) = \Sigma(x_{ji}-x_{1i}) \quad (48)$$

Combining Equations 41, 46, 47 and 48

$$h_j = \Sigma \frac{(\frac{K_{1i}}{K_{ji}} - 1)z_i}{1 + L_2(\frac{K_{1i}}{K_{2i}} - 1) + L_3(\frac{K_{1i}}{K_{3i}} - 1) + V(K_{1i}-1)} \quad (49)$$

The h_j functions are used to solve Equations 37, 38, 39, 41, where L_2, L_3 and V are the roots of $h_j(L_2, L_3, V) = 0$. This is valid only if the hydrocarbon rich liquid phase is present and the x_{1i}'s are normalized.

The empirical equations used to provide initial estimates for an additional hydrocarbon phase are listed below.

$$x_{2i} = [(x_{1i}Tr_i^{\,2}/Pr_i)/\sum_{i=1}^{n-1}(x_{1i}Tr_i^{\,2}/Pr_i)] \quad L_2\text{phase} \quad (50)$$

$$y_i = [(x_{2i}Tr_i^{\,2}/Pr_i)/\sum_{i=1}^{n-1}(x_{2i}Tr_i^{\,2}/Pr_i)] \quad V\text{ phase} \quad (51)$$

$$x_{1i} = [(x_{2i}Pr_i/Tr_i^{\,2})/\sum_{i=1}^{n-1}(x_{2i}Pr_i/Tr_i^{\,2})] \quad L_1\text{phase} \quad (52)$$

These estimates normally meet the three previously described criteria for initial K-value estimates. These expressions indicate that the concentration of a component in a given phase is proportional to the amount of that component present in the mixture, as measured by its concentration in another phase, and by its tendency to be in a less or more dense phase, as estimated by its reduced properties. For example, if a two phase flash yields L_1-L_2 equilibria, components with a high reduced temperature and a low reduced pressure are likely to appear in a vapor phase whose existence is being evaluated. These initial estimates may be adjusted by using the following procedure. If, for example, the L_2 phase is being tested for, neglect variation of fugacity coefficient with composition. Then correct the composition of the L_2 phase

$$x_{2i} = x_{2i}^{t} \, (f_{iL_1}/f_{iL_2})^{t} \tag{53}$$

and normalize concentrations.

$$x_{2i}^{t+1} = x_{2i}/\Sigma x_{2i} \tag{54}$$

Because of the low miscibility of hydrocarbons and water, the initial estimate of K_{jn} values and K_{L_3i} values (n refers to water) may be obtained by implementing a two step procedure. Trace amounts of (.001 mole) water are added to L_1, L_2, and V phases, and the f_n values adjusted until

$$f_{nL_3} = f_{nL_1} = f_{nL_2} = f_{nV} \tag{55}$$

Then, a small amount of CO_2 and each hydrocarbon

$$x_{iL_3} = x_{1i}/[(Tr_i/Tr_n)^2(Pr_i/Pr_n)^5] \tag{56}$$

is introduced into the aqueous phase, L_3, until

$$f_{iL_3} = f_{ij} \tag{57}$$

Use x_{iL2} in equation 56 if L_1 not present; and use yi if both L_1, and L_2 not present. Initial K-values are then given by $K_{ji} = y_i/x_{ji}$

Since many hydrocarbons have an extremely small solubility in water, computational difficulties may occur as values of x_{L3i} approach zero and as values of K_{L3i} as approach infinity. To prevent any divergence problems, an arbitrary but reasonable maximum value of 10^{12} is set for K-values of hydrocarbons in water. Aqueous-phase concentrations are automatically set equal to zero if the K_{L3}i-value exceeds this limit.

The accelerated successive substitution technique for mulitphase problems may be derived in a similar maneuver as the two phase problem, Equations 30-36.

The iterative correction of K-values by successive substitution is

$$K_{ji}^{t+1} = K_{ji}^{t} f_{iL_j}/f_{iv} = K_{ji}^{t} R_{ij}^{t} \text{; } t = \text{iteration number} \quad (58)$$

$$R_{ij} = f_{iL_j}/f_{iv} \quad (59)$$

Bring-back factors for g_j functions are

$$\gamma_g = \Sigma y_i/K_{ji} \quad (60)$$

and the bring-back for h_j functions are

$$\gamma_{h_1} = \Sigma(K_{1i}/K_{ji})x_{1i} \quad \text{for disappearing liquid phase} \quad (61)$$

$$\gamma_{h_2} = 1/\Sigma K_{1i}x_{1i} \quad (62)$$

for disappearing gas phase, which must be multiplied to the other sets of K_{ji} constants

The relation between the γ factor and fugacity is given by

$$\gamma_g = \Sigma x_{ji}^{t} \frac{f_{iv}}{f_{iL_j}} = \overline{\frac{f_{iV}}{f_{iL_j}}} \quad (63)$$

$$\gamma_{h_1} = \gamma_{h_2} = \overline{\frac{f_{iv}}{f_{iL_j}}} \quad (64)$$

By combining the correction and bring back procedures, one obtains

$$K_{ji}^{t+1} = K_{ji}^{t} \left(\frac{f_{iL_j}}{f_{iV}}\right)/\left(\overline{\frac{f_{iL_j}}{f_{iV}}}\right) \quad (65)$$

Therefore, accelerated successive substitution may be expressed as

$$K_{ji} = K_{ji}^{t} R_{ij}^{t\,(1-k_{ji})^{-1}} \quad (66)$$

$$\text{where } k_{ji} = \frac{R_{ij}^{t-1}}{R_{ij}^{t-1} - 1} \quad (67)$$

The solution acceptance criterion for the multiphase system is

$$\rho = \Sigma\Sigma(R_{ji}-1)^2 < \varepsilon \quad (68)$$

Relationships for iterative correction using the conventional and accelerated successive-substitution techniques, Equations 53 and 64, are similar to the two phase system expressions, Equations 32 and 34. The bring-back correction factors are presented for both the g_j and h_j functions, Equations 58-62, serve the same purpose as the bring back factor for two-phase calculations, facilitating the re-entry of a phase that is lost in the correction process even though it appears in the final solution.

Search Strategy for Multiphase Solutions. The basic successive-substitution algorithm for solving multiphase problems involves a sequence of computations. The existence of any additional hydrocarbon phase is evaluated after performing a water-free two-phase flash calculation. Subsequently, water is introduced into the system and the resultant effect on the phase distribution is determined. This stepwise addition of phases and search for additional or disappearing phases is best described by Figure 1. This flow chart represents a search strategy for multiphase solutions, which systematically evaluates the number of phases present at the specified conditions.

Water-Free Flash Equilibrium. The initial step in the multiphase flash equilibrium calculation is the prediction of the two-phase water-free equilibrium. The results of this calculation fell into three categories: divergence, one-phase equilibrium and two-phase equilibrium.

If divergence occurs, the other method of obtaining initial estimates of K-values should be employed. In general, both the Wilson formula and the evaporation technique yield convergent solutions throughout the two phase region. The use of the Wilson formula, however, proves advantageous in many systems near the saturtion curves, whereas the evaporation technique consistently yields better initial estimates in the near critical region. Since the Wilson formula is much simpler in form, it is used first in all calculations. If divergence occurs or if false solutions appear, the evaporation technique may also be employed.

The flash calculation may also yield a single phase hydrocarbon-rich liquid, L_1, or vapor, V. This single phase solution may also be the carbon dioxide-rich upper liquid phase, L_2. In any case, a search for an additional hydrocarbon phase is required in order to establish whether the Gibbs energy of the system is at it lowest value. In this study, the alternative method of initializing K-values is employed if a single phase solution arises, and the two-phase result is consistently obtained if it does exist. If no additional phase is predicted or if a less stable two phase solution is obtained, the system may safely be categorized as single phase.

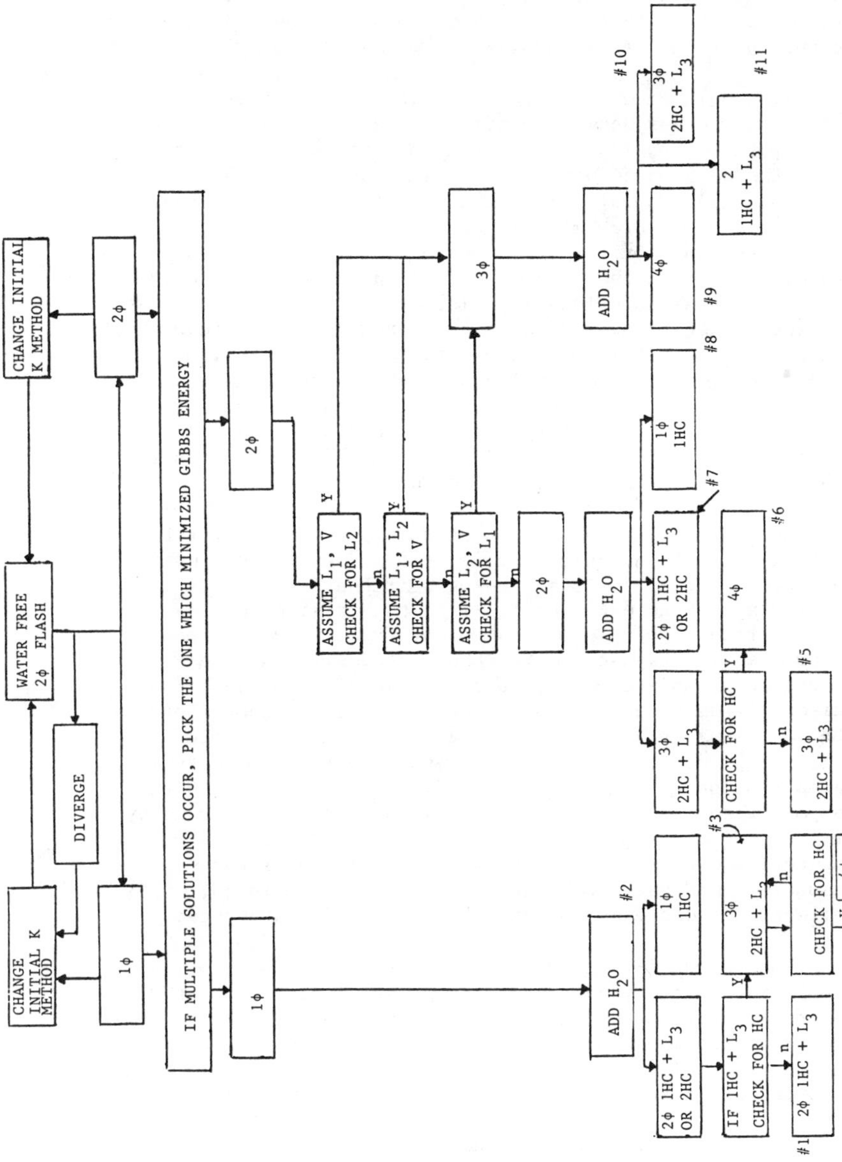

Figure 1. Multiphase solution search strategy.

Two phase equilibrium may also be predicted by the initial flash calculation. Although referred to as liquid and vapor, these systems may be L_1-L_2, L_2-V or L_1-V. Although two phase equilibrium may be established at this point, the system cannot be classified as such until it has been ascertained that a third hydrocarbon phase cannot exist at the specified conditions.

It should be emphasized that although the presence of additional phases usually minimizes the Gibbs energy of the system, there are instances where the system is more stable in a single phase solution as opposed to the two phase solution.(19) Specifically, this may occur with a single phase L_2 system which may satisfy all of the two phase L_1-V water-free equilibrium conditions except for the Gibbs energy minimization. Furthermore, near the three phase region, false two phase solutions may also appear. For example, at pressures just below the three phase region in multicomponent systems, L_1-V systems may also satisfy all equilibrium conditions (with the exception of Gibbs energy minimization) for L_2-V or L_1-L_2 systems. Two different solutions may often be obtained near the multiphase phase region using the two K-value initation techniques. The system which minimizes the Gibbs energy will be the stable solution.

An additional hydrocarbon phase may be tested for by forming a small amount (.01 mol) of fluid with a composition characteristic of the phase being searched for, yet displaying sufficient contrast with the existing phases to prevent trival solutions. Component fugacities may then be calculated and adjusted using the previously described procedures. This composition correction for the additional phases yields three possible results: (a) the composition of the additional phase being identical to one of the two original phases (b) a distinct phase with an average fugacity ratio (component fugacity in an existing phase divided by component fugacity in the additional phase) less than unity, or (c) a distinct phase whose average fugacity ratio is greater than unity. Only in the third case will the new phase grow and multiphase equilibrium be established, reflecting a decrease in the Gibbs energy of the system.

The search for an additional hydrocarbon phase may be simplified if the identity of the phase, i.e. L_1, L_2 or V, can be ascertained. This may be accomplished if the identities of the two initial phases can be established. Although calculations of the mixture critical point or phase classification schemes based on the mixture critical volume and cricondentherm are useful for this purpose, they are somewhat lengthy in nature. Therefore, in this study all three possible identities, L_2, V, and L_1 are considered sequentially by assuming the two phase system to be L_1-V, L_1-L_2, and L_2-V, respectively. If the additional phase being tested for does not exist at the given conditions, the initial estimate of its composition will be quickly adjusted to the composition of one of the original phases. Should all three assumptions yield no additional hydrocarbon phase, two-phase equilibrium is ensured. If an additional hydrocarbon phase is found to exist in any of the three tests, three phase equilibria is established and the search ended.

At this point, the water-free multiphase flash calculations are complete and the system may be in one- , two- , or three-phase equilibrium. If water is present, it is now added to the system.

<u>Introduction of Water to System.</u> The introduction of water into the hydrocarbon system will not simply result in the formation of an aqueous phase accompanied by insignificant changes in the hydrocarbon phase behavior. Water-free equilibrium may be shifted to the extent that hydrocarbon phases may disappear or appear, and in some cases the aqueous phase itself may not form. These effects are detailed in this section and outlined by the search strategy illustrated in Figure 1 and the generalized flow diagram of computations, Figure 2.

The effects of water will be discussed as they relate to the three general classifications of water-free equilibria: one- , two- and three-phase systems. In each case, the moles of water introduced into the system may be used as the initial estimate of L_3. The initialization of L_1, L_2, and V simply requires the normalization of the water-free phase distribution such that the sum of the mole fractions is unity.

The addition of water to a single phase system usually results in the formation of an aqueous phase, L_3, in equilibrium with the hydrocarbon phase. However, several other less obvious possibilities also exist. When a small amount of water is introduced into a high-temperature hydrocarbon phase, the aqueous phase may not appear due to the relatively high solubility of water in the hydrocarbon phase under such conditions. Furthermore, a second hydrocarbon phase may appear if the single phase mixture is near the two phase region, yielding two hydrocarbon phases in equilibrium. Another possible transition in phase distribution arises when water is added to a single phase mixture near its saturation point. For example, the introduction of water to a single phase L_1 mixture may shift the water-free equilibrium to the extent that the L_1 phase becomes saturated and a V phase appears. Therefore, when L_1-L_3 equilibrium exists, a bubble of the V phase must be formed and its existence tested. There is even a possibility of two additional hydrocarbon phases appearing, in which case four phases would coexist.

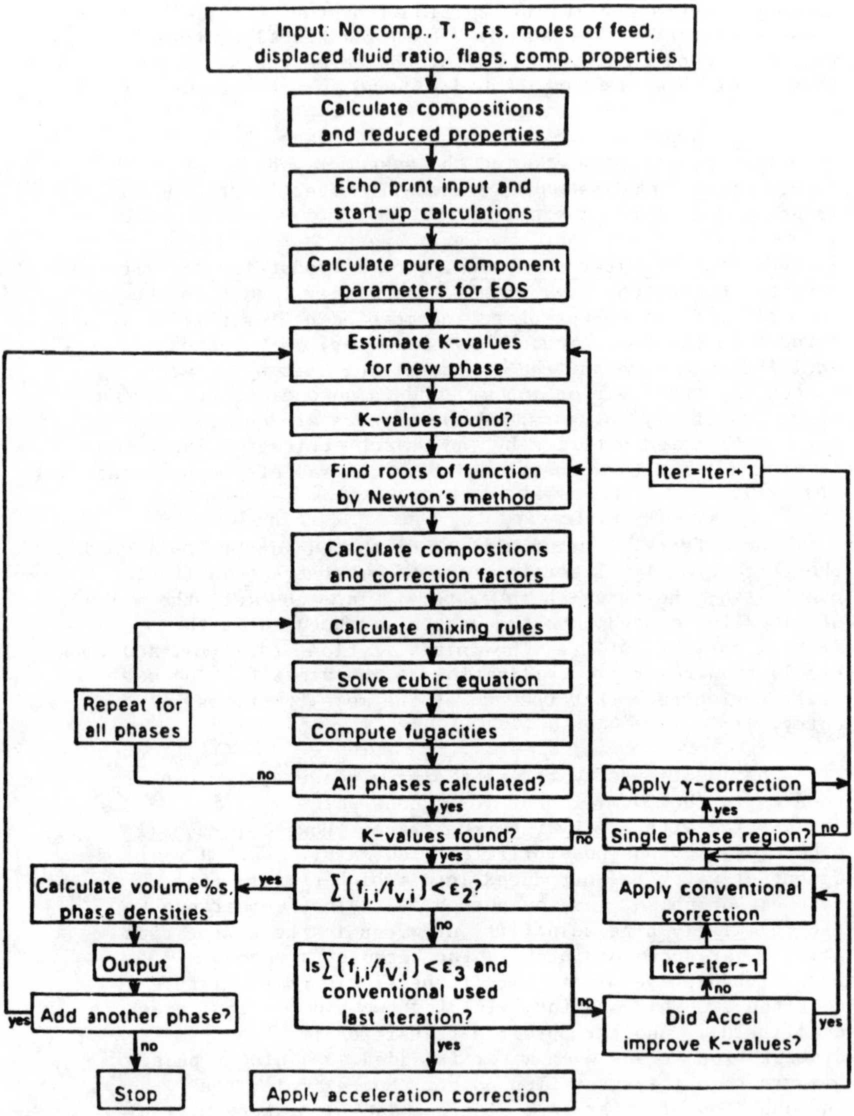

Figure 2. Generalized flow diagram of computational algorithm.

When water is added to a two phase system, the most obvious result is that an aqueous phase will be in equilibrium with two hydrocarbon phases. However, if the two-phase system is near the three phase boundary, the introduction of a water-rich liquid phase may cause an additional hydrocarbon phase to appear, resulting in four phase equilibrium. Therefore, the existence of an additional hydrocarbon phase must be evaluated whenever two hydrocarbon phases are in equilibrium with an aqueous phase. At elevated temperatures the aqueous phase may not form since the water may be completely distributed between the two hydrocarbon phases. Two phase equilibrium may also result when water is added to a two phase mixture near the dew point, causing a hydrocarbon phase to disappear. Lastly, in systems near saturation conditions the two phase equilibrium may be shifted such that one of the hydrocarbon phases disappears, leaving a single hydrocarbon phase.

Three classifications of results occur when water is introduced to a three phase system, the first being the addition of an aqueous phase, yielding four phase equilibrium. The second is characterized by a disappearing hydrocarbon phase, leaving two hydrocarbon phases in equilibrium with the aqueous phase. Thirdly, near the edge of the multiphase region, two hydrocarbon phases may disappear, leaving one hydrocarbon and one aqueous phase in equilibrium.

Results of Example Calculations A series of calculations for two systems will illustrate the nine most common cases which may result upon the introduction of water to one- , two- or three-phase systems, Figure 1. Table II describes the conditions for which each of these cases occurs.

Table II. Classification of Example Computations

Figure	Range(K)	Case
3	400.0-404.5	1
3	404.5-416.2	3
3	416.2-440.0	5
3	440.0-450.0	7
3	450.0-469.0	8
3	469.0-480.0	2
4	298.0-299.2	5
4	299.2-299.5	10
4	299.5-300.9	9
4	300.9-301.3	6
4	301.3-303.0	5

The first system is a multicomponent hydrocarbon/water system described by Peng and Robinson(14) and Heidemann(13). The second is a $CO_2/CH_4/nC_{16}H_{34}/H_2O$ system which serves as a model for multicomponent crude oil systems. The Peng-Robinson equation of state was used to describe all phases. Any carbon dioxide/water chemical reaction was ignored. The fugacity ratio error norm was set at 10^{-16} for all computations. Approximately 20 interations (Figure 2) were required for low to moderate pressure three-phase (2 hydrocarbon and 1 aqueous) systems. Four phase computations requited 25-35 iterations. High pressure calculations, especially near the critical point and saturation curves, converged in 35-50 iterations.

It has been established that when the Peng-Robinson equation of state is used in conjunction with the mixing role given in Equation 13 of Table I, it is not possible to predict both the solubility of hydrocarbons or CO_2 in the aqueous phase and the solubility of water in the hydrocarbon- or carbon dioxide-rich phase. Therefore, "optimal" interaction parameters available in literature for such systems are usually obtained by matching aqueous phase compositions, since such data are more reliable than vapor phase measurements. No attempt was made, therefore, to optimize these values, which listed in Table III.

Table III. Non-Zero Interaction Parameters

Figure 3	
Water/Hydrocarbons	0.48

Figures 4,5

	CO_2	CH_4	$C_{16}H_{34}$	H_2O
CO_2	-	0.100	0.125	0.100
CH_4		-	0.040	0.300
$C_{16}H_{34}$			-	0.500
H_2O				-

Recent progress has been made, however, in the development of new, simple mixing rule for asymetric systems (27). When it is used in conjunction with the modified "a" and "b" equation of state parameters, accurate predictions of mutual solubilities and aqueous phase densities may be attained. Such improvements may easily be incorporated into the flash calculation technique presented in this study, greatly enhancing its capability to accurately describe the phase densities and compositions of multicomponent, multiphase asymetric systems.(28)

Phase distribution results are presented in Figures 3, 4 and 5. Water-free results are indicated by the dashed lines and equilibrium established after the introduction of water is represented by the solid lines. Plots of density and composition, although not presented, are also continuous over phase boundaries, verifying the consistency of this multicomponent, multiphase, flash equilibrium technique.

Multicomponent Hydrocarbon/Water System. Although this system is not characteristic of reservoir conditions, it clearly illustrates several of the effects of water on phase equilibrium and also demonstrates that the computation scheme is general in nature; it is not limited to carbon dioxide/oil/water systems.

The results, Figure 3, not only duplicate those presented by Peng and Robinson(14) between 404.5 K and 450 K, but also extend to lower and higher temperatures. At 400 K the water-free mixture is a single phase, hydrocarbon rich mixture (L_1). The system remains at single phase L_1 as temperature is increased up to 416.2 K, at which point a vapor phase, V, forms. As temperature is elevated above this bubble point temperature, the amount of the vapor phase increases steadily as the L_1 phase diminishes until, at 469 K, the dew point is reached. Above this temperature, only a single V phase remains.

When water is introduced into the system, significant changes in phase behavior occur. At 400 K, a hydrocarbon rich liquid, L_1, is in equilibrium with an aqueous phase, L_3. These two phases remain in equilibrium with only slight changes in their distribution as temperature is increased. At 404.5 K, however a vapor phase appears. As the temperature is elevated above this bubble point temperature, the vapor phase steadily enlarges as both the L_1 and L_3 phase become smaller. At 440 K, the aqueous phase, L_3, disappears, leaving the two hydrocarbon phases in equilibrium. The dew point is reached at 450 K and only a single phase vapor remains with further increases in temperature.

In this example, the addition of water to the system causes the formation of an aqueous phase which disappears at 440 K, a 12 K reduction of the bubble point temperature and a 19 K reduction of the dew point temperature.

$CO_2/CH_4/nC_{16}H_{34}/H_2O$ System. The results of the multiphase flash calculations for this system are presented in Figure 4. At 298 K, the water-free system consists of a hydrocarbon-rich liquid phase, L_1, and a carbon dioxide-rich liquid phase, L_2. Only minor changes in the L_1-L_2 phase distribution are evidenced as temperature increases. At 299.2 K, however, a vapor phase forms and three phase L_1-L_2-V equilibrium is established. Further increases in temperature cause the vapor phase to grow and the L_2 phase to diminish by

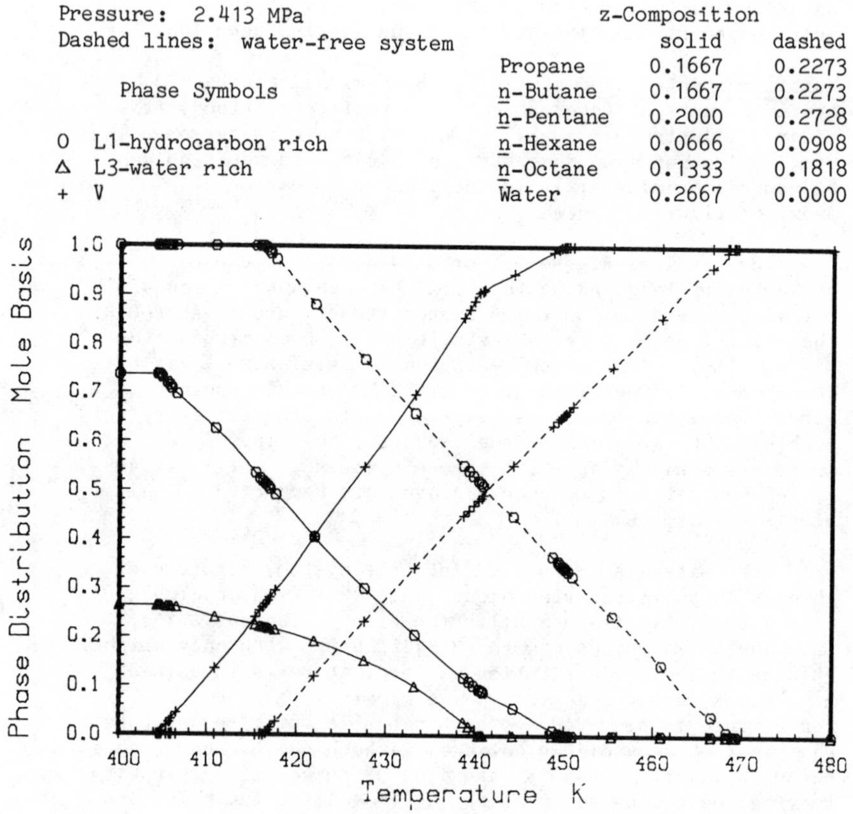

Figure 3. Effect of water on phase distribution; Peng-Robinson prediction.

Pressure: 7.239 MPa
Dashed lines: water-free system

Phase Symbols

O L1-hydrocarbon rich
X L2-carbon dioxide rich
Δ L3-water rich
\+ V

z-Composition	solid	dashed
Carbon dioxide	0.750	0.93750
Methane	0.025	0.03125
n-Hexadecane	0.025	0.03125
Water	0.200	0.00000

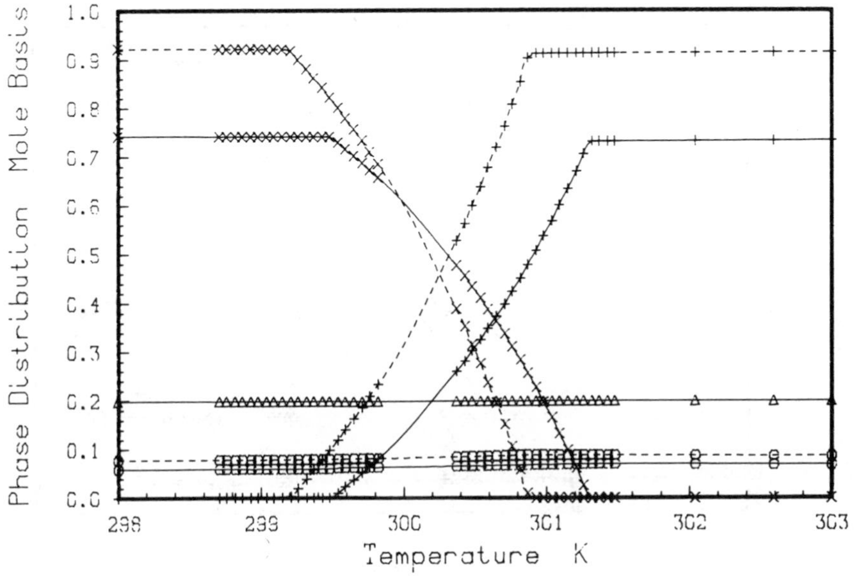

Figure 4. Effect of water on phase distribution; Peng-Robinson prediction; isobaric system, multiple phase region.

Temperature: 299.7 K
Dashed lines: water-free system

Phase Symbols

O L1-hydrocarbon rich
X L2-carbon dioxide rich
Δ L3-water rich
\+ V

	z-Composition	
	solid	dashed
Carbon dioxide	0.750	0.93750
Methane	0.025	0.03125
n-Hexadecane	0.025	0.03125
Water	0.200	0.00000

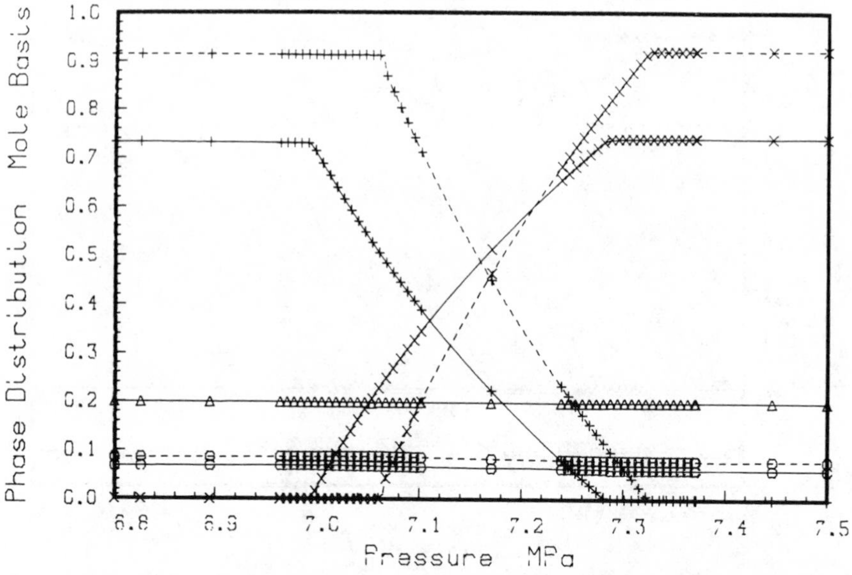

Figure 5. Effect of water on phase distribution; Peng-Robinson prediction; isothermal system, multiple phase region.

approximately the same amount since only insignificant changes occur in the L_1 phase. The three phase region ends at 300.9 K as the L_2 phase disappears and only the L_1 and V phases remain. Further increases in temperature cause only insignificant changes in the L_1-V phase distribution.

The addition of water to this model system causes distinct changes in the phase distribution. From 298 K to 299.5 K, the L_1 and L_2 phases are in equilibriuim with an aqueous phase, L_3. Only minor changes in the phase distribution occur over this range. At 299.5 K, a vapor phase forms and four-phase L_1-L_2-L_3-V equilibrium is established. As temperature increases, the V phase becomes larger and the L_2 phase reduces at about the same rate since only very small changes occur in the L_1 and L_3 phase distribution. The four-phase region ends as the L_2 phase eventually disappears at 301.3 K. Three-phase L_1-L_3-V equilibrium occurs at higher temperatures with only slight changes apparent in the phase distribution.

The introduction of water to this particular system causes the formation of an aqueous phase, a .3 K increase in the temperature at which the V phase appears and a .4 K increase in the temperature at which the L_2 phase disappears. In effect, the multiple hydrocarbon phase region becomes slightly wider and is shifted toward elevated temperatures in the presence of an aqueous phase.

In Figure 5 the same system is evaluated at a constant temperature over a range of pressure. In this figure, the addition of water not only results in the formation of an aqueous phase, but also increases the width of the multiple-hydrocarbon-phase region and shifts it toward lower pressures.

Conclusions

A multiphase equation of state flash equilibrium calculation technique has been developed in which:

1. The number of phases (one, two, three, or four) is determined, not assumed.
2. An improved method of searching for a third hydrocarbon phase is introduced which considers all three possible identities of the additional phase.
3. An efficient means of estimating the composition of the additional phase is presented which considers the relative amount of the component present in the mixture as well as its reduced properties.
4. A comprehensive search strategy which checks for additional or disappearing hydrocarbon and aqueous phases is used to consider eleven general classifications of systems which may result from the introduction of water into one, two and three phase water-free systems.

5. Example calculations not only illustrate the numerical efficiency of the solution technique, but also display continuity over phase region boundaries.

Legend of Symbols

a	equation of state coefficient
A	equation of state coefficient
b	equation of state coefficient
B	equation of state coefficient
f	fugacity
g	function of liquid fractions
G	Gibbs free energy
h	function of vapor fraction and extra liquid fractions
HC	hydrocarbon phase, L_1, L_2 or V
k	acceleration parameter
K	equilibrium constant
L	liquid moles
m	equation of state coefficient
n	number of components
N	total number of moles
P	pressure
R	gas constant or (when subscripted) fugacity ratio
T	temperature
v	molar volume
V	vapor moles
y	mole fraction in vapor phase
z	mole fraction in feed
Z	compressibility factor
α	equation-of-state coefficient
γ	bring-back factor
ε	tolerance
ρ	fugacity residual norm
ϕ	phase
Ψ	fugacity coefficient
ω	accentric factor

Subscripts

c	critical state
g	g-function
h	h-function
i	component number
j	component number or phase number
k	component number
L	liquid phase
n	component n, water (if present)
r	reduced
V	vapor phase
1	liquid phase number, hydrocarbon-rich
2	liquid phase number, CO_2-rich
3	liquid phase number, aqueous

Superscripts

t iteration number
´ initial estimate

Literature Cited.

1. Shelton, J.; Yarborough, L. Journal of Petroleum Technology September 1977, 1171-8.
2. Stalkup, F. Miscible Displacement; SPE of AIME: New York, 1983, pp. 6-30.
3. Stalkup, F." Journal of Petroleum Technology August 1978, pp. 1102-1112.
4. Stalkup, F. Op. Cit., pp. 71-96.
5. Rachford, H.; Rice, J." Journal of Petroleum Technology October 1952, p. 19.
6. Organik, E.; Meyer, H. Journal of Petroleum Technology May 1955, pp. 9-13.
7. Bennett, C.; Brasket, C.; Tierney, J. American Institute of Chemical Engineer´s Journal 1960, Vol. 6, No. 1, pp. 67-70.
8. Shelton, R.; Wood, R. The Chemical Engineer April 1965, pp. CE68-75.
9. Deam, J.; Maddox, R. Hydrocarbon Processing July 1969, pp. 163-4.
10. Erbar, J.H. Proceedings of the Vapor-Liquid Symposium, 1972, (Poland: Polish Academy of Science).
11. Dluziewski, J.; Alder, S.; Ozardesh, H. Chemical Engineering Progress, November 1973, Vol. 69, No. 11, pp. 79-80.
12. Lu, B.C.-Y.; Yu, P., Sugie, A.H. Chemical Engineering Science, 1974, Vol. 29, pp. 321-326.
13. Heidemann, R. American Institute of Chemical Engineers Journal, 1974, Vol. 20, pp. 847-55.
14. Peng, D.; Robinson, D. Canadian Journal of Chemical Engineering, December 1976, Vol. 54, pp. 595-98.
15. Peng, D.; Robinson, D. Industrial and Engineering Chemistry Fundamentals, 1976, Vol. 15, No. 1, pp. 59-64.
16. Wilson, G. Proceedings of the 65th National American Institute of Chemical Engineers Meeting, 1969, Paper No. 15C.
17. Fussell, L. Society of Petroleum Engineers Journal August 1979, pp. 203-10.
18. Fussell, D.; Yanosik, J. Society of Petroleum Engineers Journal June 1978, pp. 173-82.
19. Baker, L.; Pierce, A.; Luks, K. Society of Petroleum Engineers Journal October 1982, pp. 731-42.
20. Mehra, R.; Heidemann, R.; Aziz, K. Society of Petroleum Engineers Journal February 1982, pp. 61-8.
21. Fayers, F.J., ed., Enhanced Oil Recovery, "Phase Equilibrium Calculations in the Near Critical Region, by R. Risnes and V. Dalen," (New York: Elsevier Scientific Publishing Company, 1981), pp. 329-349.

22. Carnahan, B.; Luther, H.; Wilkes, J. Applied Numerical Methods; John Wiley & Sons: New York; p. 61.
23. Soave, G. Chemical Engineering Series, 1972, Vol. 27, pp. 1197-1203.
24. Rabinowitz, P., Numerical Methods for Non-Linear Algebraic Equations, "A Fortran Subroutine for Non-Linear Algebraic Systems, by M.J. Powell" (Longon: Gordon and Breach Science Publishers, 1970), pp. 115-161.
25. Risnes, R.; Dalen, V. Society of Petroleum Engineers Journal February 1984, pp. 87-95.
26. Abbott, M. American Institute of Chemical Engineers Journal, 1973, Vol. 19, No. 3, pp. 596-601.
27. Panagiotopolous, A.; Reid, R., "A New Mixing Rule for Cubic Equations of State for Highly Polar, Asymetric Systems," presented at the ACS Symposium on Equations of State-Theories and Applications (April 20-May 3, 1985) Miami, Florida.
28. Enick, R.M.; Holder, G.D.; Mohamed, R.S. 60th Annual SPE Technical Conference and Exhibition, 1985, paper SPE 14148.

RECEIVED November 5, 1985

INTERPRETATION AND EXTENSION

25

Interpretations of Trouton's Law in Relation to Equation of State Properties

Grant M. Wilson

Wiltec Research Co., Inc., 488 South 500 West, Provo, UT 84601

When corrections are made in Trouton's law for differences in molar volume of both liquid and vapor phases, then the resulting entropy deviation correlates with restricted motion in the liquid phase due to molecular size, shape, flexibility, and polarity. These entropy deviations also correlate with hard sphere equation of state properties which only depend on density without significant effect of temperature. Based on these results a new hard sphere equation of state is proposed which better fits the high density region of hard spheres based on molecular dynamics calculations than does the Carnahan - Starling equation.

Entropy deviations obtained by correcting the entropy of vaporization for differences in molar volume in both the liquid and vapor phases plot almost as a single line versus boiling point for a wide variety of non-polar and polar compounds. This provides a new means for estimating heats of vaporization and vapor pressures for a wide range of compounds based on a measured boiling point and liquid density.

An examination of the entropy deviation term shows that it correlates with restricted motion in the liquid phase resulting from molecular size, shape, flexibility, and polarity. For paraffin hydrocarbons, the main effect appears to be due to restricted motion as a result of the length and flexibility of chain molecules; while for hydrogen bonded compounds, restricted motion occurs as a result of hydrogen bonding. Compounds with a dipole moment also exhibit restricted motion as a result of interactions between the dipoles.

When deviation entropies are calculated at other temperatures besides the boiling point, they appear to correlate with hard sphere equation of state properties which only depend on density without any significant effect of temperature. As a result a modified hard sphere equation of state is proposed which better fits the high density region of hard spheres based on molecular dynamics

0097-6156/86/0300-0520$06.00/0

calculations than does the Carnahan-Starling equation. By allowing one of the parameters to vary, the same equation can be used to account for restricted motion in the liquid phase resulting from molecular size, flexibility, polarity, etc.

The hard sphere equation of state has not been widely used even though it appears to be more accurate than the Van der Waals repulsive term. The reason for this is probably a compensating effect for error in the attraction term when the Van der Waals term is used. An examination of data on a simple fluid shows that the attractive terms of the Redlich-Kwong and Peng-Robinson equations fit the low density region quite well, but they significantly diverge at high densities. This explanation of a compensating effect by the use of the Van der Waals repulsion term seems reasonable in view of these differences at high densities.

Discussion

Trouton's law states that for many non-polar fluids the ratio of $\Delta H/T_b$ is about 21 cal/mole-°K. Hildebrand (4) has shown that this ratio is more nearly constant at temperatures where the molar volumes in the gas phase are the same rather than when the pressures are the same. Kistiakowsky (5) proposed an equation which essentially corrects for differences in vapor molar volumes at the boiling point to give the following equation:

$$\frac{\Delta H}{T_b} = 8.75 + R\ln T_b \tag{1}$$

This equation does well for low molecular weight non-polar compounds, and thus is rather restricted in its use.

In contrast to these early observations which frequently serve as the basis for estimating heats of vaporization, it proves interesting to calculate an entropy deviation at the boiling point based on the difference between the actual entropy of vaporization and an ideal entropy for the change in molar volume in going from the liquid phase to the vapor phase as follows:

$$\frac{\Delta S^{DEV}}{R} = \frac{\bar{S}^{\circ} - \bar{S}^{L}}{R} - \left(\frac{\bar{S}^{\circ} - \bar{S}^{L}}{R}\right)^{I}$$

or (2)

$$\frac{\Delta S^{DEV}}{R} = \frac{\bar{S}^{V} - \bar{S}^{L}}{R} - \left(\frac{\bar{S}^{V} - \bar{S}^{L}}{R}\right)^{I} + \frac{\bar{S}^{\circ} - \bar{S}^{V}}{R} - \left(\frac{\bar{S}^{\circ} - \bar{S}^{V}}{R}\right)^{I}$$

At the boiling point, the vapor phase is nearly ideal; so $(\bar{S}^{\circ} - \bar{S}^{V})/R - (\bar{S}^{\circ} - \bar{S}^{V})^{I}/R$ is virtually zero. Thus, Equation 2 reduces to the following:

$$\frac{\Delta S^{DEV}}{R} = \frac{\bar{S}^V - \bar{S}^L}{R} - \left(\frac{\bar{S}^V - \bar{S}^L}{R}\right)^I \tag{3}$$

The difference, $(S^V - S^L)/R$, is given by $\Delta H/RT_b$; and $(S^V - S^L)^I/R$ is given by the following equation:

$$\left(\frac{S^V - S^L}{R}\right)^I = \int_{\bar{V}^L}^{\bar{V}^V} \left(\frac{\partial S}{\partial V}\right)_T dV = \int_{\bar{V}^L}^{\bar{V}^V} \frac{d\bar{V}}{\bar{V}} = \ln \frac{\bar{V}^V}{\bar{V}^L} \tag{4}$$

At one atmosphere $\bar{V}^V = RT/P$ where P equals one atmosphere. Equation 3 can therefore be written as follows:

$$\frac{\Delta S^{DEV}}{R} = \frac{\Delta H}{RT_b} - \ln \frac{RT}{P\bar{V}^L} \tag{5}$$

A plot of $\Delta S^{DEV}/R$ for hydrocabons and rare gases is given in Figure 1. Data in this and subsequent plots are based on Antoine constants for heats of vaporization; and the Rackett equation for liquid volumes using data and constants given in Reid et al (7). Figure 1 shows that $\Delta S^{DEV}/R$ varies linearly with T_b for paraffin hydrocarbons. Branched paraffins such as neo-pentane appear to plot on the same line, while cyclic paraffins plot about one-half to one unit below the line. A clue to the reason for variation of $\Delta S^{DEV}/R$ versus T_b of the paraffin hydrocarbons is given in Figure 1 by the fact that $\Delta S^{DEV}/R$ of the rare gases is a constant independent of T_b. The rare gases are spherical atoms while the paraffin hydrocarbons are long flexible molecules. Also, the deviation entropies of the aromatic hydrocarbons are closer to the rare gases compared to the paraffin hydrocarbons at the same T_b. This is consistent with less flexibility of the aromatic hydrocarbons. This suggests that the main reason for the variation of $\Delta S^{DEV}/R$ versus T_b is the increase in molecular flexibility as the chain length increases. Figure 2 shows $\Delta S^{DEV}/R$ of other compounds including water and ammonia. Surprisingly, water and ammonia plot on almost the same line as the paraffin hydrocarbons. In this case, the increase in entropy compared to the rare gases presumably is not due to molecular flexibility, but instead is due to restricted motion from orientation or other similar effects in the liquid phase due to hydrogen bonding. Curiously these effects appear to be about equal in magnitude to flexibility effects for non-polar compounds with equal boiling points. Thus whether $\Delta S^{DEV}/R$ is a result of flexibility or polar effects, the net result is about the same for compounds with the same boiling point. Figure 2 also shows a dashed line for $\Delta S^{DEV}/R$ of the diatomic halogen compounds. Again, this plot deviates significantly from the n-paraffin plot because the flexibility does not increase as the boiling point increases. However, there does appear to be some restriction of motion as the molecular weights and sizes of these diatomic molecules increase because $\Delta S^{DEV}/R$ does increase slightly versus T_b. Figure 3 shows

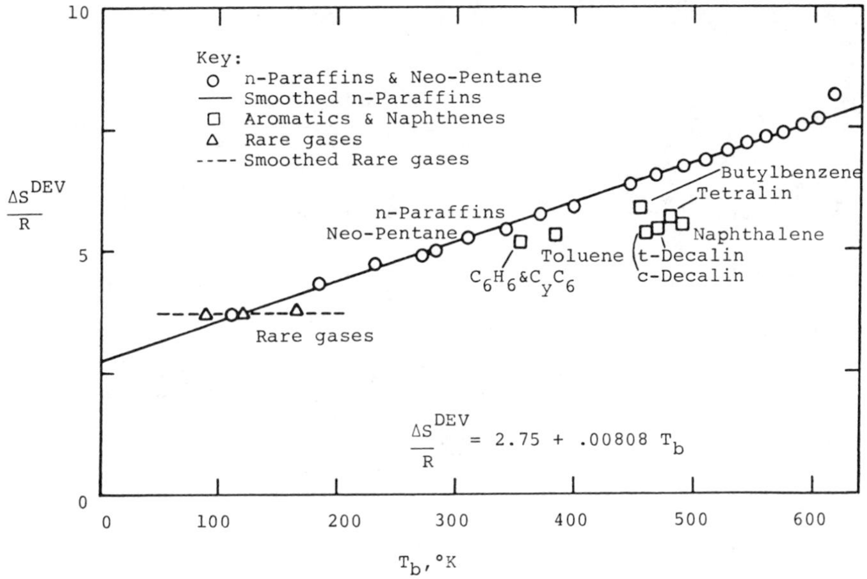

Figure 1. Entropy Deviation from Ideal Gas of Hydrocarbons and Rare Gases at their Boiling Points.

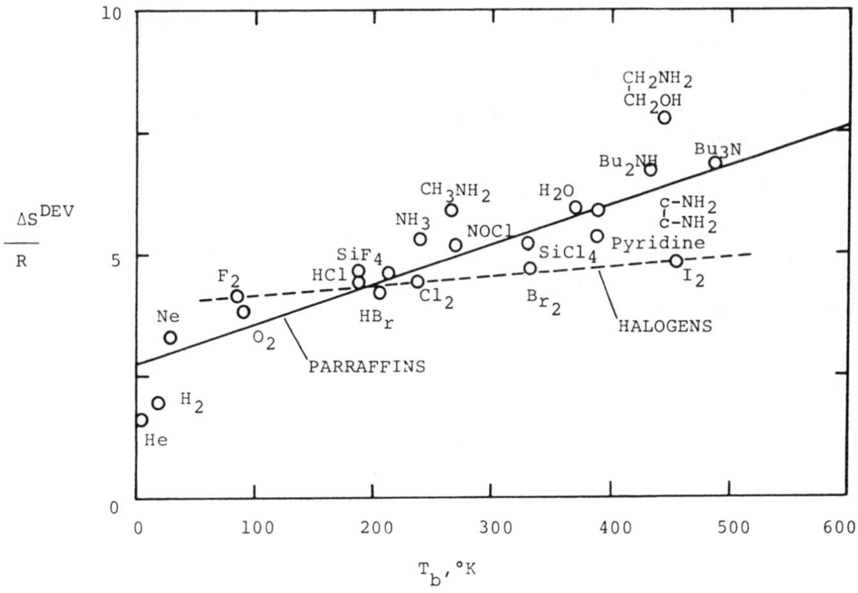

Figure 2. Entropy Deviation from Ideal Gas of Light Gases, Halogen Compounds, Amines, H_2S, and Water.

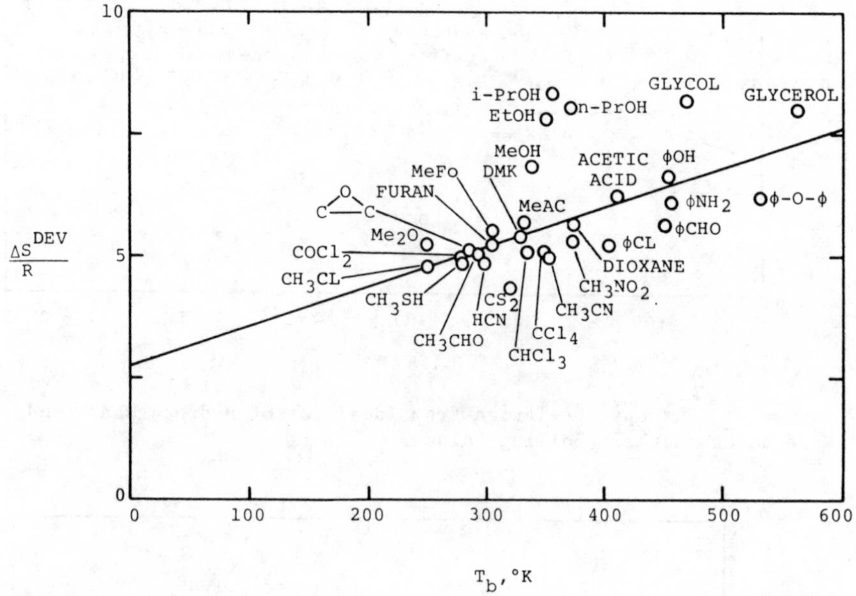

Figure 3. Entropy Deviation from Ideal Gas of Compounds with Various Functional Groups.

additional $\Delta S^{DEV}/R$ data. This plot shows that the alcohols, glycol, and glycerol exhibit $\Delta S^{DEV}/R$ values which are higher than the paraffin hydrocarbons by one to two units; thus suggesting that in these cases the orienting effects of hydrogen bonding of these compounds are greater than the flexibility effects which restrict the motion of non-polar compounds. However, even with these differences, the error in predicting the heat of vaporization would be 20% or less because $\Delta S/R$ is about 10 or more.

The results in Figures 1, 2, and 3 provide interesting features pertaining to entropy effects in the liquid phase. In addition, these figures can be used directly for estimating heats of vaporization. The estimated heats of vaporization combined with the Antoine equation with a generalized value for "C" provides an excellent means for estimating vapor pressures based only on a measured boiling point and liquid density. For this purpose the following generalized correlation for "C" is recommended.

$$C\,°K = -.002896T_b^{1.672} \tag{6}$$

When this value for "C" is used, the other parameters are as follows:

$$B = \frac{(T_b + C)^2}{T_b} \; \frac{\Delta H}{RT_b} \tag{7}$$

$$A = 6.633 + \frac{B}{T_b + C} \tag{8}$$

$$\ln(P,\text{mmHg}) = A - \frac{B}{T + C} \tag{9}$$

Also from T_b, $\Delta H/RT_b$, and V_b the critical constants of a compound can be estimated. This is almost analogous to the estimation of hydrocarbon critical constants based on density and boiling point, but with hydrocarbons the molecular weight can also be estimated. Unfortunately, this is not possible in general without knowing the class of compound involved.

Besides the immediate utility of Figures 1, 2, and 3 they provide insight into entropy effects in the liquid phase that prove worth pursuing. If the deviation entropy of a compound is determined along the saturation curve with corrections made for the vapor phase effects in Equation 2 at high reduced temperatures, then curves are obtained such as the curve for argon shown in Figure 4. Non-random entropy effects in non-polar liquids are very small as is evidenced by the fact that the product $\bar{U}\bar{V}^L$ is a constant; or $(\partial U/\partial V)_T \cong -U/V$ (2). Thus the main effect on $\Delta S^{DEV}/R$ will be the variation of liquid volume with temperature along the saturation curve. In fact, the curve in Figure 4 for argon bears considerable resemblance to the curve predicted for hard spheres based on the equation of Carnahan and Starling (3). An alternate curve is plotted in Figure 4 based on the following equation:

$$\frac{\Delta S^{DEV}}{R} = -\nu \ln\left(1-\frac{b}{\overline{V}}\right) \qquad (10)$$

Which involves two parameters in contrast to one used in the Carnahan-Starling equation. Equation 10 gives a hard sphere compressibility factor as follows:

$$Z = 1 + \nu \frac{b}{\overline{V} - b} \qquad (11)$$

Compressibility factors from this equation with $\nu = 3.5$ are plotted in Figure 5 where comparison is made with the Carnahan-Starling equation:

$$Z = \frac{1 + y + y^2 - y^3}{(1 - y)^3} \qquad (12)$$

$$y = \frac{b}{4\overline{V}} \ , \ b = \frac{2}{3} N \pi \sigma^3$$

Comparison is also made with results of molecular dynamics calculations with hard spheres by Alder and Wainwright (1). Figure 5 shows that Equation 11 follows the results of Alder and Wainwright at high densities much more closely than does the Carnahan-Starling equation. One could introduce a second parameter to the Carnahan-Starling equation in order to improve agreement, but the incentive seems rather low in view of the simplicity of Equation 11. Without a correction, the Carnahan-Starling equation corresponds to a much softer fluid at high densities than is actually predicted from molecular dynamics. The reason for this is not clear, and it appears to be in error because the minimum volume that the hard spheres can occupy is the close packed crystal volume shown at v/v_o= 1.0 in Figure 5. The formation of the crystal requires a transition from a fluid state to the crystalline state, and the minimum volume in the fluid state is about 14% larger than the volume of the close packed crystal. Therefore, Equation 11 which represents a modified hard sphere equation is proposed for the calculation of hard sphere properties at high densities.

Equation 11 was fitted to the data plotted in Figure 4 by fitting two points along the curve. This requires the assumption that temperature effects are negligible. Alternatively, both parameters can be fitted at each temperature by assuming that $(\partial U/\partial V)_T$ is equal to $-\Delta U/V$. When this is done, two simultaneous equations involving $\Delta S^{DEV}/R$ and $T(\partial S/\partial V)_T$ can be solved for b and V as follows:

$$\frac{\Delta S^{DEV}}{R} = -\nu \ln\left(1 - \frac{b}{\overline{V}}\right) \qquad (10\text{-a})$$

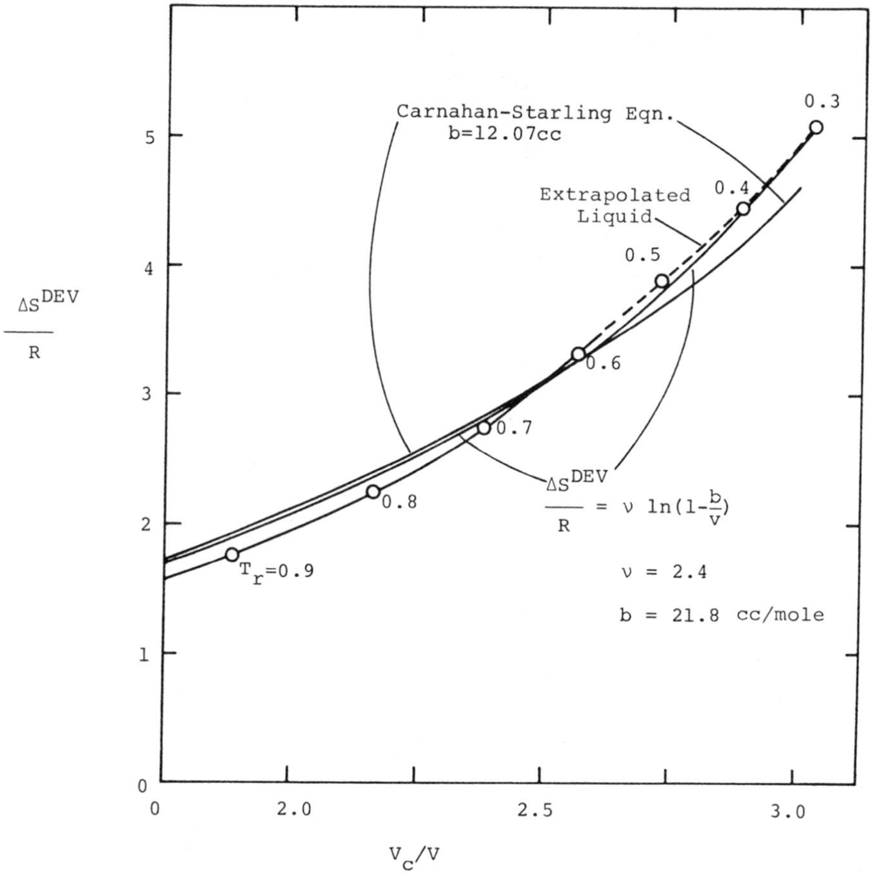

Figure 4. Deviation Entropy of Argon Along the Liquid Saturation Curve.

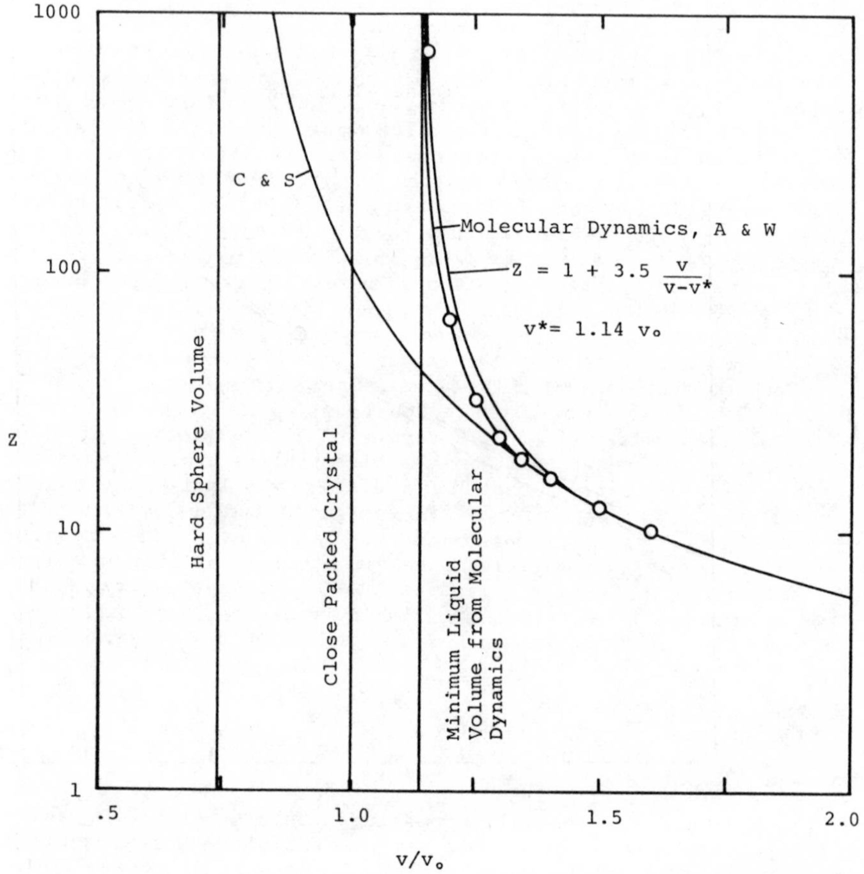

Figure 5. Compressibility Factor of Hard Spheres from Carnahan-Starling Equation Versus Molecular Dynamics.

$$\frac{\bar{V}}{R}\left(\frac{\partial S}{\partial V}\right) = -\frac{\Delta U}{RT} + \frac{P\bar{V}^L}{RT} = 1 + \nu\left(\frac{b/\bar{V}}{1-b/\bar{V}}\right) \qquad (13)$$

In Equation 13, P is the saturation pressure; so except at high reduced temperature, PV^L/RT is very small. Values of internal energy were determined from ΔH where a correction was made for ΔZ both in calculating ΔH and in converting ΔH to ΔU and for the energy content of the vapor. When this was done, values of b were obtained as are plotted in Figure 6. This figure shows that derived values of b in the modified hard sphere equation correspond quite closely to the minimum liquid volume predicted from the Rackett equation at zero reduced temperature. Also the volume of the molecules can be estimated from the Rackett equation by taking the ratio of 0.74/1.14 from Figure 5 times the minimum liquid volume from the Rackett equation. This ratio provides the lower dashed line in Figure 6 which is actually higher than derived values of b in the Carnahan-Starling equation. These results strongly suggest that the derived values of b for argon in the two equations are consistent with the interpretation given in Figure 5.

Figure 6 suggests that the extrapolated liquid volume based on the Rackett equation at zero reduced temperature corresponds closely to the value of b to be used in Equations 10 and 11. If this is assumed, then values of $\Delta S^{DEV}/R$ versus b/V can be plotted for other materials as are shown in Figure 7. These curves appear to conform rather closely except that they differ by nearly a constant multiplying factor depending on the compound. Curiously, even polar compounds such as ammonia seem to fit the pattern. These results suggest that the only difference between these compounds and argon is the value of ν used in Equation 10. For non-polar compounds, this seems reasonable because the thermal pressure is probably more closely related to the free volume per molecular segment than it is to the free volume per molecule. Equation 11 gives the following:

$$Z = 1 + \frac{1}{\frac{1}{\nu}\left(1-\frac{b}{\bar{V}}\right)} - \nu \qquad (14)$$

At high densities, the only significant term is $(1-b/V)/\nu$ which corresponds to the free volume per segment where ν represents a parameter proportional to the number of segments in the molecule. Thus it appears that Equation 11 can not only be used for hard spheres but also for long or bulky molecules such as those plotted in Figure 15. For polar compounds ν would not have the significance of number of segments; but if ν is also interpreted as a parameter proportional to restricted degrees of freedom in the liquid phase; then it still has meaning for polar compounds.

Figure 8 shows the result of simultaneously fitting b and ν for various compounds at $T_r = 0.6$. Results at other temperatures differ only slightly. This plot shows that ν is a linear function of b for the paraffin hydrocarbons. Also, the aromatic hydrocarbons plot on virtually the same line; thus showing that any reduction in degrees of freedom of the aromatic hydrocarbons is also reflected in a smaller b. The rare gases deviate slightly from the line but the

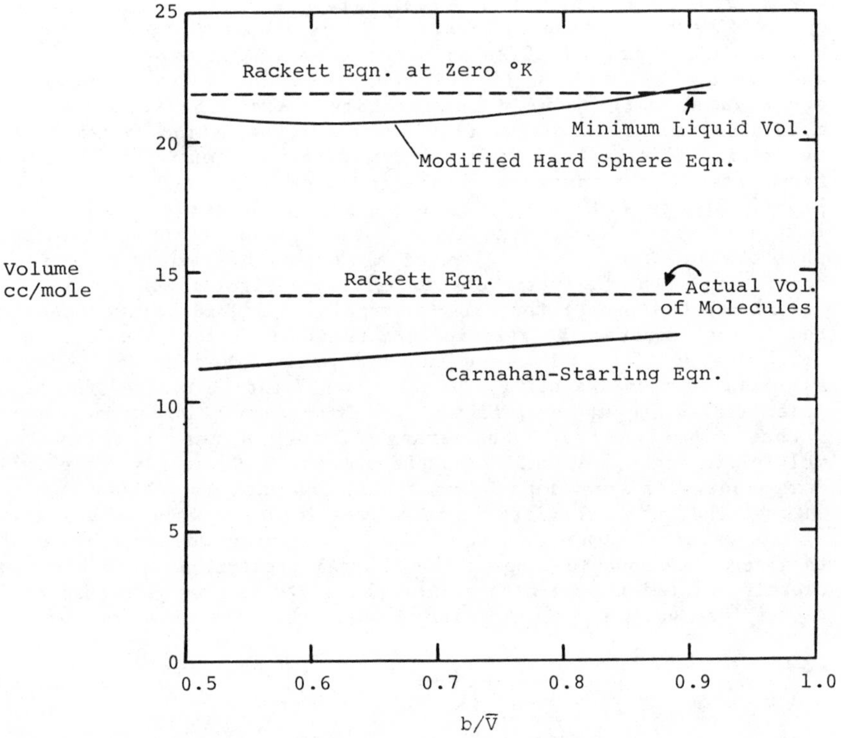

Figure 6. Molecular Volume of Argon Compared with Carnahan-Starling and Rackett Equation.

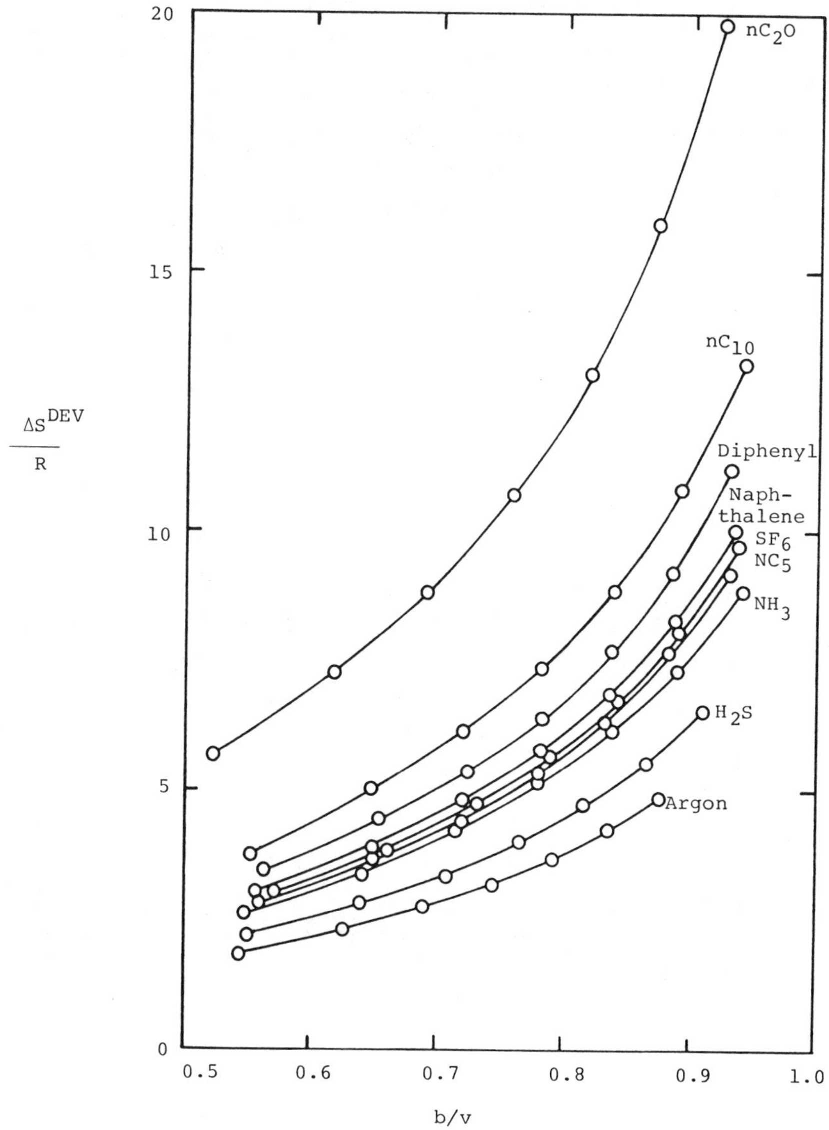

Figure 7. Entropy Deviation from Ideal Gas Versus b/V of Various Compounds.

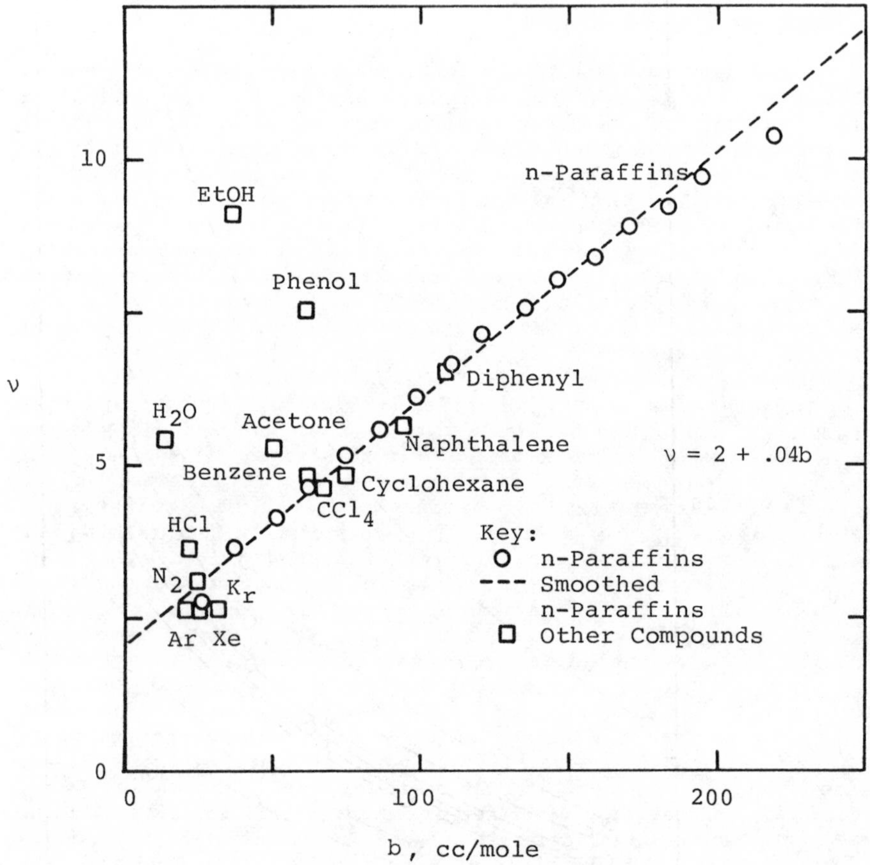

Figure 8. Degree of Freedom Parameter Versus b of n-Paraffins and Other Compounds, Modified Hard Sphere Equation.

difference is rather small. Greater differences are shown by polar molecules such as water, ethanol, phenol, acetone, and HCl which deviate significantly; thus showing a rather large restriction in molecular motion of the compounds in the liquid phase compared to the sizes of the molecules.

Why Van der Waals RT/(V-b)?

Many equations of state have been proposed over the years. Some of these contain the RT/(V-b) term from the Van der Waals equation while others replace it with another term including the hard sphere equation of Carnahan and Starling. It seems significant that the two equations commonly in use today; the Soave modification of the Redlich-Kwong equation and the Peng-Robinson equation of state use the original term of Van der Waals. Why should this be so when a repulsion term based on hard spheres seems much more reasonable? Also, from the results above it appears that real fluids conform fairly closely to a modified hard sphere equation of state. Molecular dynamicists also chide us for persisting in the use of what is considered to be an obsolete form. The answer is probably not because of resistance to new ideas. Instead it appears that both equations of state give better results for liquid properties when the original Van der Waals term is used compared with the use of the Carnahan-Starling equation. If this is so, it suggests that there is a compensation of errors between the attraction and repulsion terms in these equations of state. An examination of the internal energy of a simple fluid according to Lee-Kesler (7) in Figure 9 shows that there are significant differences at high densities between all modifications of the Redlich-Kwong equation and the actual curve. At lower densities, the Redlich-Kwong equation fits the curve quite accurately; but it doesn't bend back as the data do at high densities. By contrast, a Lennard-Jones fluid (6) appears to conform with the simple fluid at high densities even though the acentric factor of a Lennard-Jones fluid is about -0.058; which is not a simple fluid. A third curve corresponding to the Peng Robinson equation of state could be drawn in Figure 9. At low densities it conforms as well as does the Redlich-Kwong equation, and at high densities it deviates less severely than the Redlich-Kwong equation of state. These deviations at high densities indicate that both equations deviate significantly from the true energy curve at high densities. This problem is also evident from an examination of data in Figure 10 where $(\partial U/\partial V)_T$ is plotted versus $\bar{U}/\bar{V}$ for various compounds. This figure shows that non-polar compounds appear to conform closely to a slope of unity while polar compounds deviate significantly from this. A slope of unity is consistent with the upper end of the curve for a simple fluid at high densities shown in Figure 9. The Redlich-Kwong equation gives a slope of about 0.65 for the range of densities in the liquid region while the Peng-Robinson equation gives a slope of about 0.75; thus showing that both equations deviate significantly at high densities.

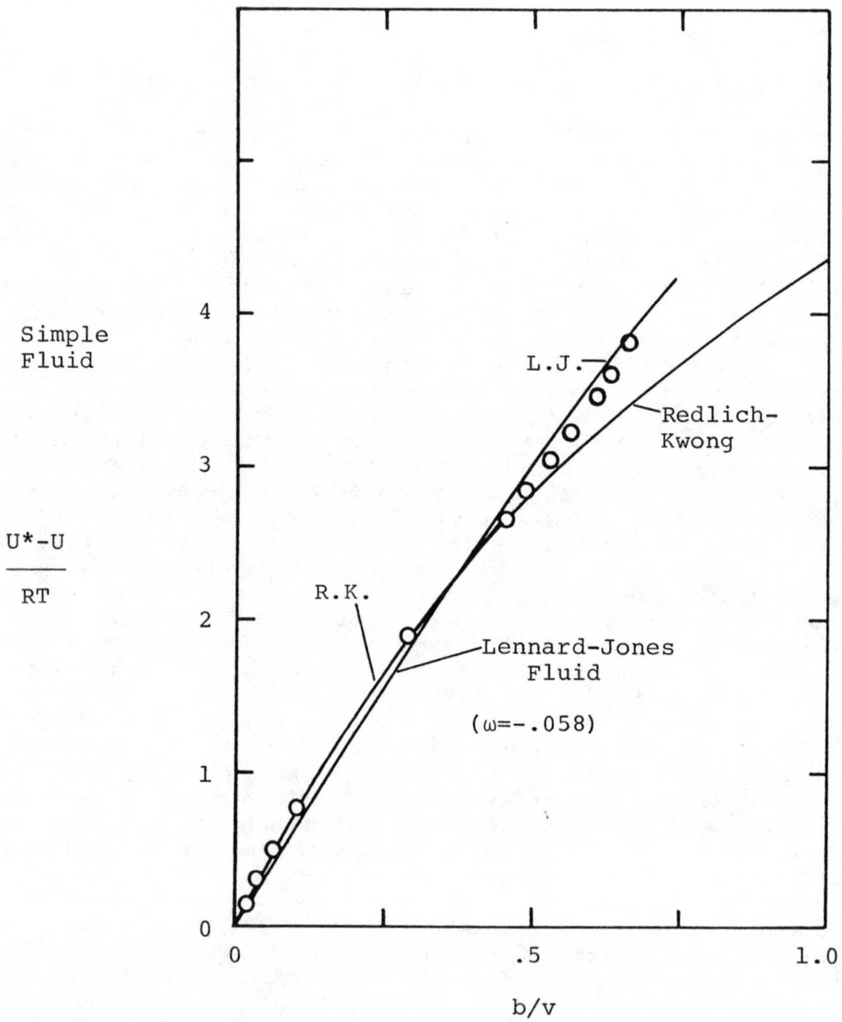

Figure 9. Internal Energy of a Simple Fluid (ω=0) at T_r= 1.0 (Lee-Kesler).

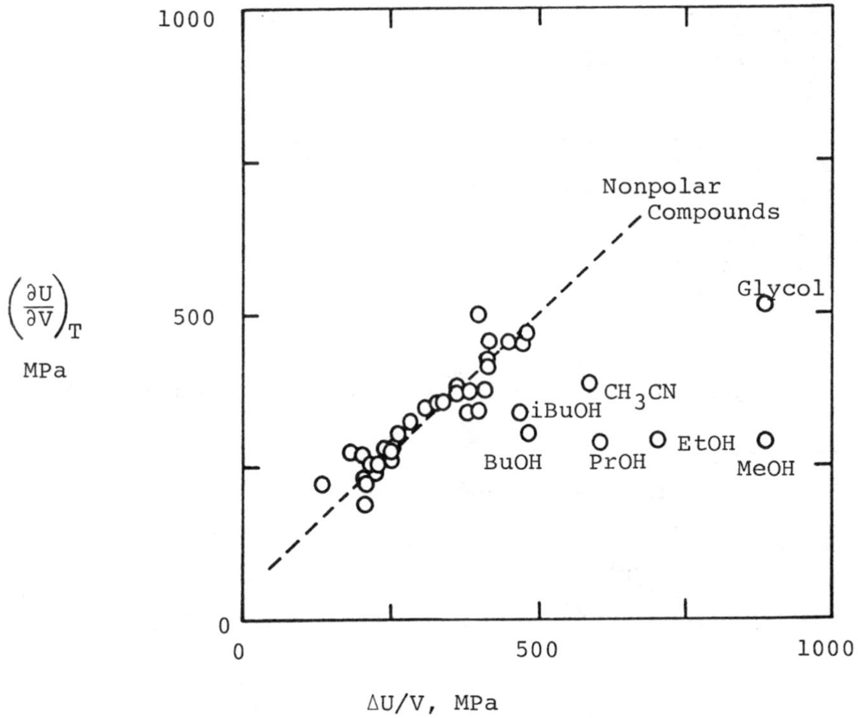

Figure 10. Variation of $(\partial U/\partial V)_T$ Versus $\Delta U/V$ at 20°C (from A. M. F. Barton, CRC Handbook of Soly. Parameters, 1983).

Notation

Upper Case Letters:

A, B, C = Antoine Constants
H = enthalpy
N = Avogadro's number
P = pressure
R = gas constant
S = entropy
T = absolute temperature
U = internal energy
V = volume
Z = Compressibility factor

Lower Case Letters:

b = volume analogous Van der Waals b
y = b/4V in Carnahan-Starling hard sphere equation

Superscripts:

DEV = deviation from ideal
I = ideal
L = liquid phase
V = vapor phase
$-$ = molar property or average property
^{o} = property at a low enough pressure to behave as an ideal gas

Subscripts:

b = at atmospheric boiling point
c = value at critical point
r = reduced property

Greek Letters:

Δ = incremental change
σ = hard sphere diameter
ν = degree of freedom parameter in proposed modified hard sphere equation of state

References

1. Alder, B. J., and Wainwright, T. E., J. Chem. Phys. 1960, 33, 1439.
2. Barton, A. F. M., CRC Handbook of Solubility Parameters and other Cohesion Parameters, CRC Press, Inc.; Boca Raton, Florida, (1983).
3. Carnahan, N. F., and Starling, K. E., J. Chem. Phys. 1969, 51, 635.
4. Hildebrand, J. H., J. Am. Chem. Soc. 1915, 37, 970.
5. Kistiakowsky, W., Z. Phys. Chem. 1923, 107, 65.
6. Nicolas, J. J. Gubbins, K. E., Street, W. B. and Tildesley, D. J., Molecular Physics 1979, 37, 1429.
7. Reid, R. C., Praunitz, J. M., and Sherwood, T. K.,The Properties of Gases and Liquids McGraw-Hill Book Company, Third Edition, (1977).

RECEIVED November 8, 1985

26

Selection and Design of Cubic Equations of State

J.-M. Yu, Y. Adachi, and B.C.-Y. Lu

Department of Chemical Engineering, University of Ottawa, Ottawa, Ontario K1N 9B4, Canada

Fourteen known cubic equations of the van der Waals type, $P = RT/(V-b) - a(T)/(V^2 + ubV + wb^2)$, were evaluated through the calculations of eight properties of normal alkanes. The roles of u and w in these calculations were demonstrated. A new design procedure was developed and a new relationship between u and w was suggested.

Reliable methods for predicting physical properties of pure components and their mixtures (such as single and two-phase properties, and vapor-liquid equilibria) are frequently required in process design and material handling. In view of the wide range of state conditions found in practical applications, and the frequent lack of experimental data, considerable attention has been paid to the development of these methods. In particular, a number of equations of state have been proposed in the literature to meet this demand.

The most popular equations of state are cubic in volume (or density). A generic expression for the currently popular cubic equations of state may be represented in the form of an extended van der Waals (VDW) equation,

$$P = \frac{RT}{V-b} - \frac{a}{V^2+ubV+wb^2} \tag{1}$$

The quadratic expression in volume $(V^2+ubV+wb^2)$ replaces the V^2 term in the denominator of the attractive term of the original VDW equation. When the parameters u and w are assigned certain particular values, Equation 1 can be reduced to the original VDW equation (u=w=0)(1), the Redlich-Kwong (RK) equation (u=1, w=0)(2) and its various modified forms, the Peng-Robinson (PR) equation (u=2, w=-1)(3), the Heyen (H) equation (u+w=1)(4), the Schmidt-Wenzel (SW) equation (u+w=1)(5), and a number of other equations. Some general features of Equation 1 or its equivalent expression have been discussed by Abbott (6).

There is evidence in the literature (7-9) that practically identical vapor-liquid equilibrium (VLE) values (T-P-composition)

0097-6156/86/0300-0537$06.75/0

can be obtained from cubic equations of state containing two to four parameters, and these results are frequently comparable to those obtained from more complex equations of state. It has been further demonstrated (10) that by treating the parameter "a" of Equation 1 temperature dependent, the VDW equation is as capable as other cubic equations for calculating VLE values.

On the other hand, the capabilities of the available cubic equations for representing volumetric properties vary from equation to equation, especially in the calculation of liquid volumes. Kumar and Starling (11) suggested that "at a particular temperature, a higher density dependence leads to a more accurate equation of state". They increased the complexity of Equation 1, and proposed a five-parameter cubic equation of state for predicting liquid densities and low temperature vapor pressures.

One approach is therefore to separate the issues of VLE calculations and volumetric predictions in the application of equations of state. In other words, different equations are used for different purposes. Another approach is to further improve the performance of the available equations or develop new equations to satisfy the requirements of both VLE and volumetric calculations. The task is actually reduced to developing a suitable equation, with a compromise between simplicity and accuracy for representation of volumetric data. In both approaches, guidelines for selecting the appropriate equations are required.

There is still a need to further evaluate the available cubic equations in a systematic manner, so that the utilization of a cubic equation can reach its full potential. The purpose of this study is therefore to evaluate available cubic equations of state of the VDW type as represented by Equation 1 to identify the merits and limitations of these equations for the purpose of selection, and to suggest a suitable procedure for designing new cubic equations of the same type but tailored to specific purposes.

Evaluation of Cubic Equations of State

The techniques used in the improvement of the original VDW equation for physical property predictions, without changing the expression of Equation 1, may be grouped into the following three categories:

1. Modification of the expression used in the denominator of the attractive term.
2. Application of a volume-translation technique to the original VDW equation and the equations of the first category.
3. Introduction of temperature dependence to one or more parameters.

The equations of RK, PR and SW are examples of the first category. The Clausius (C) equation (12) is an example of the second category. The Soave form of the RK equation (SRK)(13) and the modification of the RK equation by Hamam et al. (HCL)(14) are examples of the third category.

In addition to the above mentioned SRK, PR, SW, HCL and H equations, we have included in our consideration the Harmens-Knapp (HK) equation (15), the Patel-Teja (PT) equation (16), the resulting equation from volume translation of the SRK equation (TSRK) proposed by Peneloux et al. (17), the four-parameter equation of Adachi et al. (ALS)(9), and the C1 equation proposed by Peneloux et al. (18).

The original VDW, RK and C equations are not evaluated because their poor performance on volumetric predictions is known. On the other hand, the three-parameter RK equations suggested by Adachi et al. (3RK)(19) and by Fuller (F) (20), and the Martin Equation (21) (a version of the Clausius equation) in two modified versions MMC (Adachi et al. (19)) and KMC (Kubic (22)) are considered. Thus, a total of fourteen cubic equations are evaluated in this work.

Among these equations, the RK equation is one of the most successful two-parameter cubic equations. In the SRK equation, Soave made an effort in 1972 to closely reproduce vapor pressures of pure compounds by assuming the parameter "a" of the original equation to be temperature dependent. This modification enhanced the applicability of the RK equation for calculating VLE values. Many of the equations which have subsequently appeared in the literature have adopted the same or similar modifications. As far as the representation of volumetric data for pure fluids is concerned, Martin (21) concludes that the Clausius-type equation is the best of the simpler cubic equations. However, the calculation of volumetric properties at saturation conditions without considering the equality of fugacity at the same time as applied by Martin could introduce internal inconsistency in the calculated values. As mentioned above, the Clausius equation can be obtained from a volume translation of the VDW equation. It is difficult to envisage that the volume translated VDW equation is superior to the volume-translated RK equation (TSRK). A comparison of the calculated results is definitely of interest. It should be mentioned that in the two modified versions of the Martin equation, MMC (19) and KMC (22), the parameter "a" of Equation 1 was treated as temperature dependent.

Parameters other than "a" of Equation 1 have been assumed to be temperature dependent in some cubic equations. For example, the H equation contains three parameters, two of which are assumed to be temperature dependent. It is designated in this work as a 3P2T equation. The other equations are similarly designated. The investigated equations of state are thus grouped into six types as shown in Table I. The relationships between u and w for these equations, the parameters which are treated temperature dependent as well as the properties (other than T_c and P_c) used to determine the parameters of these equations are also presented in Table I. In the table, the two-parameter equation proposed by Harmens (HA) (23) was also included, because its u and w relationship formed part of the basis in the development of the SW equation.

Representation of Physical Properties

The properties of ten normal alkanes from methane to n-decane, obtained from generalized correlations and tabulations available in the literature were used as the basis for comparing the performance of the fourteen equations. A total of eight properties were considered. The vapor pressures, p^v, were obtained from the generalized correlation proposed by Gomez-Nieto and Thodos (24) for nonpolar substances. The saturated liquid volumes, V^ℓ, were obtained from a modified Rackett equation using the input parameters suggested by Spencer and Alder (25). The saturated vapor volumes, V^v, were obtained from the correlation of Barile and Thodos (26). The second

Table I Features of Some Cubic Equations of State

$$P = RT/(V-b) - a(T)/(V^2 + ubV + wb^2)$$

TYPE	EOS	u	w	a	b	Fitted Properties
	VDW	0	0	a_c	b_c	none
2P1T	RK	1	0	$a_c/T^{\frac{1}{2}}$	b_c	none
	SRK	1	0	a(T)	b_c	P^{ν}
	PR	2	-1	a(T)	b_c	P^{ν}
	HA	3	-2	a(T)	b_c	P^{ν}
3P1T	HK	1-w	$f(\omega)$	a(T)	b_c	P^{ν}, Critical Isotherm
	SW	1-w	$f(\omega)$	a(T)	b_c	$P^{\nu}, V^{\ell}(T_r=0.7)$
	PT	1-w	$f(\omega)$	a(T)	b_c	$P^{\nu}, V^{\ell}(T_r \simeq 0.6$ to $1.0)$
	3RK	$f(\omega)$	0	a(T)	b_c	P^{ν}, V^{ℓ}
	MMC	$f(\omega)$	$u^2/4$	a(T)	b_c	P^{ν}
	TSRK	$f(\omega)$	$(2u^2-u-1)/9$	a(T)	b_c	P^{ν}, V^{ℓ}
	C1	$f(\omega)$	$\simeq(u^2-4u-4)/8$	a(T)	b_c	P^{ν}, Z^{ℓ}
4P1T	ALS	$f(\omega)$	$f(\omega)$	a(T)	b_c	P^{ν}, Critical Isotherm
2P2T	HCL	1	0	a(T)	b(T)	P^{ν}, V^{ℓ}
3P2T	H	1-w	$f(\omega, b)$	a(T)	b(T)	P^{ν}, V^{ℓ}
	KMC	$f(\omega)$	$u^2/4$	a(T)	b(T)	P^{ν}, B
3P3T	F	f(T)	0	a(T)	b(T)	$p^{\nu}, V^{\ell}, V^{\nu}$

virial coefficients, B, were obtained from the correlation of Tsonopoulos (27). For these four properties, points were taken at 0.02 intervals in the T_r range of 0.5 to 0.80 and at T_r equals 0.85, 0.90, 0.95 and 0.98 for a total of 20 T_r values. The correlation of Lee and Kesler (28) was used to obtain the values for liquid compressibility factor (Z^{ℓ}, $0.30 \leq T_r \leq 0.99$, $0.01 \leq P_r \leq 10.0$, 315 points), compressibility factor of vapor (Z^{ν}, $0.55 \leq T_r \leq 0.99$, $0.01 \leq P_r \leq 0.8$, 56 points), compressibility factor of gas above the critical temperature (Z^{sup}, $1.01 \leq T_r \leq 4.00$, $0.01 \leq P_r \leq 10.0$, 240 points) and compressibility factor along the critical isotherm (Z^c, T_r=1.0, $0.01 \leq P_r \leq 10.0$, 15 points).

A summary of the calculation results, in terms of overall average absolute percent deviations, is presented in Table II. The reported values may be slightly different from similar calculations available in the literature due to the difference in the covered ranges of T_r and P_r and in the number of data points selected for the calculation.

It is well known that accurate representation of vapor pressures is essential for vapor-liquid equilibrium calculations. For this reason, vapor pressure values have been used to determine the values of parameter "a", or $\Omega_a (= a\ Pc/R^2\ T_c^2)$. The generalized expressions of the original authors for "a" were used to calculate p^{ν} and practically all equations tested yielded acceptable results.

As far as V^{ℓ} is concerned, the modified Martin (MMC) and the SRK equations yield the largest deviations.

The deviations in the calculated V^{ν} values follow closely to those for p^{ν}, and the deviations in Z^{ℓ} values follow closely to those for V^{ℓ}.

Table II Summary of Overall Average Absolute Percent Deviations in the Calculated Physical Properties for Ten Normal

Property	NT	SRK	PR	HK	SW	PT	3RK	MMC
P^v	200	1.51	2.55	1.02	1.44	1.95	1.29	2.33
V^ℓ	200	13.4	5.47	4.37	2.78	2.56	4.00	12.4
V^v	200	1.23	2.54	1.45	1.22	2.01	1.08	2.51
Z^ℓ	3150	11.2	4.96	4.14	2.96	3.03	3.37	12.6
Z^v	560	0.95	0.58	0.55	0.38	0.40	0.56	0.44
Z^{sup}	2400	2.37	1.54	1.47	1.41	1.28	1.36	1.89
Z^c	150	8.50	4.68	4.05	4.08	4.13	4.92	5.24
B	200	17.2	15.5	15.1	15.6	15.4	15.9	15.4

Property	TSRK	C1	ALS	HCL	H	KMC	F
P^v	1.51	0.96	1.42	1.88	10.5	3.56	1.83
V^ℓ	3.77	3.65	2.61	0.65	2.26	5.41	2.01
V^v	1.20	1.26	1.74	4.10	10.3	4.01	1.97
Z^ℓ	3.67	3.93	3.13	2.61	5.53	13.5	5.28
Z^v	0.64	0.49	0.45	2.24	2.24	0.19	0.99
Z^{sup}	1.35	1.30	1.30	2.98	3.72	2.09	6.16
Z^c	5.42	6.24	4.00	8.37	8.90	5.32	12.6
B	16.1	15.6	15.4	23.2	16.9	1.71	17.4

The differences in the calculated Z^v, Z^{sup} and Z^c values among the 14 equations are not too significant. The KMC equation gives the lowest deviations in the calculated B values, which were used in the determination of the parameters of the equation.

Although some of the findings mentioned above could be envisaged from the features listed in Table I, there are several interesting points revealed by the results of Table II.

1. The currently popular cubic equations of state (SRK and PR) do not yield the best results.
2. The performance of the 3P2T H equation is inferior to that of the 3P1T PT equation. Indeed, the overall performance of equations containing more than one temperature-dependent parameter is generally inferior, indicating that these equations are mainly suitable for representing the physical properties used in the forced fitting procedure.
3. The Martin equation is not suitable for VLE and volumetric calculations simultaneously.
4. A three-parameter cubic equation with only the parameter "a" treated temperature dependent is adequate for the purpose of this study. The performance of the six equations of this type (HK, SW, PT, 3RK, TSRK, C1) seems to be adequate.

Further examination of the deviations obtained for individual alkanes (as shown in Table III) from the 2P1T and 3P1T equations reveals that:

1. The PR equation yields larger deviations in P^v values for larger molecules, and the deviations in V^ℓ increase with increase in molecular weight.

2. For the SRK equation, the deviations in V^{ℓ} also increase with increase in molecular weight.
3. All 2P1T and 3P1T equations yield large deviations in B, and the deviations increase with increase in molecular weight.
4. The TSRK equation yields deviations in p^{v} identical to those of the SRK equation, confirming the fact that volume translation does not affect the p^{v} calculations.
5. As far as the deviations in V^{ℓ} and Z^{ℓ} are concerned, the PT equation appears to be slightly superior in the family of equations (HK SW and PT) which can be represented by the u + w = 1 relationship. Among the equations resulting from volume translation (MMC, TSRK, and C1), the performance of the TSRK equation is better because the individual deviations are about the same for the ten alkanes. The deviations obtained from the C1 equation tend to increase with increase in molecular weight. A comparison of the SW, PT, TSRK and C1 equations reveals that equations with substance-dependent Ω_{ac} values yield better representation of these two properties. All equations appear to represent V^{ℓ} and Z^{ℓ} to the same degree of accuracy.
6. Although nearly all the equations yield low deviations in the calculated Z^{v} and Z^{sup} values, there is a tendency toward larger deviations at larger molecular weights.
7. The family of equations represented by the u + w = 1 relationship yields lower deviations in the calculation of Z^{c} values than the equations obtained from the volume-translation method, and the individual deviations are about the same.

Thus, the results reported in Tables II and III provide some guidance in the selection of equations among the 14 cubic equations for representing physical properties of pure normal fluids. The choice depends on the properties to be emphasized and the molecular weight of the substances to be considered.

Design of Cubic Equations of State

In Equation 1, if u and w are considered as constants for all substances (such as the VDW, RK, SRK, PR and HA equations), the resultant equation would be a two-parameter equation. If u and w are related by an exact mathematic relationship (such as the HK, SW, PT, 3RK, MMC and TSRK equations), Equation 1 would become a three-parameter cubic equation. If u and w are not related to each other through an exact mathematic relationship (such as the ALS equation), Equation 1 would yield a four-parameter equation. Although the C1 equation in its original form (18) appears to be a four-parameter equation, the u and w relationship of this equation can be approximated by the expression shown in Table I. In other words, it may be approximates as a volume-translated PR equation.

If we adopt the classical conditions at the critical point as two of our constraints,

$$(\partial P/\partial V)_{Tc} = 0 \quad (2)$$

$$(\partial^2 P/\partial V^2)_{Tc} = 0 \quad (3)$$

a two-parameter equation yields a constant critical compressibility factor, ζ_c; and a three-, or four-parameter equation may yield substance-dependent ζ_c values.

Table III Comparison of Average Absolute Percentage Deviations Obtained from 14 Cubic Equations of State for Ten Normal Alkanes

Property	SRK	PR	HK	SW	PT	3RK	MMC	TSRK	C1	ALS	HCL	H	KMC	F
Vapor Pressure, p^v $(0.50 < T_r < 0.98)$														
C_1	1.64	1.15	1.00	0.37	1.32	1.60	2.06	1.64	0.97	0.56	0.89	3.44	1.85	1.55
C_2	1.12	2.51	0.67	1.73	2.73	1.01	1.60	1.12	0.66	2.20	0.64	2.32	3.61	1.04
C_3	1.42	2.61	0.85	1.88	2.72	1.21	1.55	1.42	0.78	1.85	0.70	4.02	4.30	1.42
C_4	1.31	2.42	0.75	1.69	2.35	1.04	1.55	1.31	0.79	1.27	0.76	5.93	4.55	1.44
C_5	1.49	0.94	1.00	0.39	0.64	1.54	2.80	1.49	0.69	0.90	1.78	6.70	5.28	1.10
C_6	1.08	2.35	0.68	1.57	1.64	0.88	1.90	1.08	0.96	1.12	0.76	10.3	4.04	1.60
C_7	1.63	1.28	1.41	0.54	1.08	1.73	2.86	1.63	1.13	1.26	0.99	11.5	3.46	0.90
C_8	0.85	2.19	1.37	1.48	1.58	0.95	2.38	0.85	1.18	1.61	1.08	15.0	1.50	1.38
C_9	1.71	4.16	1.10	2.19	2.10	1.11	2.89	1.71	1.02	1.62	3.82	20.0	1.85	3.19
C_{10}	2.81	5.94	1.42	2.59	3.36	1.80	3.66	2.81	1.41	1.89	7.42	25.3	5.15	4.72
Saturated Liquid Volume, V^ℓ $(0.50 < T_r < 0.98)$														
C_1	3.99	9.25	5.11	3.56	3.50	4.13	5.39	3.69	2.33	2.82	0.97	1.42	3.92	0.58
C_2	7.08	7.04	4.97	3.28	3.10	4.60	5.22	3.69	2.14	3.09	0.32	1.15	3.12	0.76
C_3	8.92	5.83	4.96	2.97	2.88	4.49	6.24	3.68	2.47	2.92	0.16	1.49	2.70	1.20
C_4	10.2	4.97	5.13	2.50	2.83	3.93	9.49	3.67	2.55	2.41	0.71	0.84	3.14	1.27
C_5	12.4	3.55	4.89	2.37	2.56	3.62	16.0	3.69	3.45	2.32	0.75	2.18	3.92	1.76
C_6	15.1	2.28	4.00	3.21	2.36	4.22	13.2	3.74	4.66	3.17	0.31	1.53	4.29	3.02
C_7	16.2	2.95	4.26	2.34	2.26	3.32	13.4	3.75	4.46	2.26	0.43	2.71	5.98	2.49
C_8	18.5	4.99	3.57	2.83	2.32	3.33	15.1	3.74	5.12	2.48	0.60	3.74	6.94	3.23
C_9	19.5	5.85	3.74	2.25	2.30	4.00	17.1	4.04	4.34	2.30	0.85	4.17	9.26	2.63
C_{10}	21.8	7.96	3.05	2.46	2.42	4.35	23.0	4.05	4.98	2.36	1.36	3.38	10.8	3.18
Saturated Vapor Volume, V^v $(0.50 < T_r < 0.98)$														
C_1	1.86	2.11	1.36	0.49	1.51	1.80	2.42	1.91	1.47	1.02	2.41	4.14	2.17	2.46
C_2	0.87	3.09	1.10	1.60	2.85	0.86	1.86	1.04	1.07	2.42	2.01	3.95	4.46	1.52
C_3	1.03	2.97	1.19	1.73	2.77	0.98	1.76	1.24	1.18	1.98	2.32	5.83	5.36	1.87
C_4	0.90	2.53	1.01	1.52	2.26	0.67	1.74	0.96	1.13	1.44	2.83	8.16	5.57	1.71
C_5	1.43	1.08	1.40	0.39	0.77	1.54	3.11	1.33	0.89	1.23	4.93	10.8	6.31	1.30
C_6	0.56	2.05	1.11	1.22	1.53	0.52	1.92	0.50	1.19	1.44	3.37	11.5	4.56	1.60
C_7	1.57	1.42	1.96	0.38	1.47	1.72	3.09	1.42	1.30	1.69	4.22	11.0	3.72	1.11
C_8	0.97	2.11	1.93	1.25	2.03	1.15	2.58	0.68	1.63	2.12	4.47	12.9	1.62	1.16
C_9	1.09	3.35	1.60	1.78	2.12	0.57	2.88	1.00	1.22	1.91	5.89	15.6	1.80	2.92
C_{10}	2.04	4.70	1.87	1.87	2.80	0.95	3.77	1.93	1.52	2.19	8.58	19.4	4.55	4.02

Continued on next page

Table III Continued

Property	SRK	PR	HK	SW	PT	3RK	MMC	TSRK	C1	ALS	HCL	H	KMC	F
Liquid Compressibility Factor, Z^{ℓ} ($0.30 < T_r < 0.99$, $0.01 < P_r < 10.0$)														
C_1	3.67	8.35	4.21	3.44	3.16	3.78	3.72	3.58	3.25	3.64	3.60	5.09	5.85	5.12
C_2	5.30	6.64	4.53	3.02	3.11	3.34	5.82	3.43	3.17	3.29	2.26	5.01	7.57	5.19
C_3	6.83	5.46	4.65	2.93	3.09	3.14	7.69	3.40	3.20	3.17	2.23	5.00	9.07	5.13
C_4	8.27	4.52	4.66	2.90	3.06	3.01	9.40	3.38	3.34	3.11	2.76	5.02	10.5	5.09
C_5	10.2	3.52	4.55	2.86	3.02	2.91	11.6	3.43	3.75	3.06	3.09	5.08	12.4	5.12
C_6	11.7	3.03	4.37	2.86	2.99	2.88	13.3	3.46	3.96	3.03	3.09	5.21	13.9	5.17
C_7	13.7	2.96	4.12	2.85	2.92	2.95	15.6	3.81	4.47	3.00	2.77	5.44	16.0	5.36
C_8	15.5	3.52	3.74	2.87	2.94	3.25	17.5	3.85	4.59	2.99	2.26	5.78	17.8	5.40
C_9	17.6	4.97	3.41	2.91	2.98	3.88	19.7	4.08	4.69	2.99	1.83	6.42	20.1	5.57
C_{10}	19.6	6.65	3.18	3.00	3.09	4.54	21.9	4.28	4.90	2.99	2.26	7.25	22.3	5.74
Compressibility Factor of Vapour, Z^{v} ($0.55 < T_r < 0.99$, $0.01 < P_r < 0.8$)														
C_1	0.19	0.88	0.44	0.17	0.20	0.20	0.18	0.18	0.58	0.39	0.76	1.94	0.10	0.95
C_2	0.39	0.71	0.48	0.23	0.27	0.31	0.25	0.29	0.49	0.38	1.20	2.00	0.13	0.95
C_3	0.56	0.60	0.51	0.28	0.31	0.39	0.31	0.38	0.43	0.39	1.50	2.05	0.16	0.95
C_4	0.71	0.52	0.53	0.32	0.35	0.46	0.36	0.47	0.39	0.40	1.78	2.11	0.18	0.95
C_5	0.89	0.45	0.55	0.37	0.39	0.55	0.42	0.60	0.40	0.43	2.12	2.18	0.20	0.96
C_6	1.04	0.43	0.56	0.40	0.42	0.61	0.47	0.68	0.42	0.45	2.39	2.24	0.21	0.98
C_7	1.21	0.45	0.58	0.45	0.46	0.68	0.53	0.80	0.47	0.48	2.72	2.33	0.22	1.01
C_8	1.35	0.49	0.59	0.49	0.49	0.74	0.58	0.89	0.51	0.50	3.00	2.41	0.23	1.03
C_9	1.52	0.57	0.61	0.52	0.52	0.80	0.62	1.00	0.56	0.53	3.32	2.50	0.23	1.06
C_{10}	1.67	0.67	0.62	0.56	0.55	0.84	0.66	1.10	0.63	0.56	3.63	2.60	0.24	1.08
Compressibility Factor of Gas above the Critical Temperature, Z^{sup} ($1.01 < T_r < 4.00$, $0.01 < P_r < 10.0$)														
C_1	1.14	2.13	1.15	1.71	0.83	1.16	0.98	0.98	0.88	0.84	1.05	3.17	1.22	4.80
C_2	1.55	1.66	1.26	1.36	0.87	1.20	1.08	1.06	0.89	0.86	1.52	3.21	1.23	5.15
C_3	1.83	1.48	1.33	1.18	0.92	1.20	1.18	1.11	0.97	0.93	1.92	3.29	1.29	5.41
C_4	2.05	1.37	1.39	1.09	1.01	1.20	1.32	1.17	1.07	1.01	2.30	3.37	1.49	5.67
C_5	2.32	1.31	1.46	1.08	1.14	1.23	1.57	1.24	1.22	1.15	2.79	3.53	1.79	6.00
C_6	2.51	1.31	1.51	1.16	1.27	1.27	1.84	1.32	1.33	1.27	3.17	3.68	2.08	6.27
C_7	2.75	1.36	1.57	1.35	1.44	1.36	2.21	1.44	1.48	1.45	3.65	3.89	2.45	6.62
C_8	2.94	1.44	1.62	1.52	1.59	1.48	2.52	1.55	1.60	1.62	4.03	4.09	2.76	6.91
C_9	3.17	1.57	1.68	1.73	1.78	1.67	2.89	1.71	1.73	1.82	4.45	4.35	3.15	7.24
C_{10}	3.39	1.73	1.73	1.92	1.96	1.89	3.25	1.89	1.88	2.03	4.88	4.62	3.54	7.57

Property	SRK	PR	HK	SW	PT	3RK	MMC	TSRK	C1	ALS	HCL	H	KMC	F
Compressibility Factor along the Critical Isotherm, Z^c (T_r = 1.0, 0.01 < P_r < 10.0)														
C_1	4.33	4.44	3.31	3.59	4.12	3.99	3.84	4.12	3.15	3.30	2.92	7.27	3.81	9.80
C_2	5.30	4.21	3.50	3.72	3.70	4.35	4.11	4.48	3.92	3.46	5.48	7.77	4.10	10.7
C_3	6.19	4.18	3.64	4.12	3.79	4.46	4.34	4.74	5.06	3.74	6.45	8.19	4.33	11.3
C_4	7.02	4.21	3.77	4.23	4.04	4.68	4.65	4.98	5.77	3.71	7.19	8.49	4.61	11.8
C_5	8.07	4.26	3.94	4.32	3.93	4.81	5.01	5.28	6.50	4.00	6.19	8.85	4.99	12.4
C_6	8.89	4.46	4.12	4.21	4.03	4.94	5.29	5.53	6.96	3.98	8.96	9.12	5.30	12.9
C_7	9.93	4.72	4.36	4.09	4.34	5.34	5.66	5.84	7.40	4.31	9.91	9.45	5.70	13.5
C_8	10.8	4.94	4.47	4.12	4.24	5.49	6.06	6.10	7.67	4.45	10.7	9.68	6.20	14.0
C_9	11.8	5.43	4.59	4.19	4.61	5.47	6.50	6.40	7.89	4.37	11.5	9.98	6.78	14.5
C_{10}	12.7	5.99	4.70	4.26	4.49	5.68	6.98	6.69	8.05	4.67	12.4	10.2	7.34	15.0
Second Virial Coefficient, B (0.50 < T_r < 0.98)														
C_1	7.73	7.83	7.16	7.81	7.63	7.77	7.49	7.59	7.59	7.26	8.54	13.4	1.09	8.81
C_2	11.8	10.9	10.6	11.2	11.1	11.4	11.1	11.4	11.1	10.8	14.7	13.7	0.77	12.6
C_3	14.0	12.6	12.4	13.0	12.9	13.3	12.9	13.3	12.9	12.7	18.1	14.6	0.87	14.6
C_4	15.7	14.0	13.8	14.3	14.2	14.7	14.2	14.8	14.3	14.2	20.7	15.5	1.00	16.0
C_5	17.5	15.6	15.5	15.8	15.7	16.2	15.5	16.4	15.7	15.7	23.5	16.5	1.20	17.7
C_6	18.8	16.7	16.4	16.8	16.7	17.3	16.7	17.5	16.9	16.7	25.5	17.4	1.43	18.8
C_7	20.2	18.0	17.5	18.0	17.8	18.4	18.0	18.6	18.0	17.9	27.8	18.3	1.88	20.0
C_8	21.2	18.9	18.4	18.8	18.6	19.1	18.7	19.5	19.0	18.8	29.4	19.0	2.34	20.8
C_9	22.2	19.9	19.2	19.6	19.5	20.0	19.5	20.4	20.0	19.7	31.1	19.8	2.93	21.7
C_{10}	23.1	20.7	19.9	20.3	20.2	20.6	19.9	21.2	20.7	20.4	32.7	20.5	3.56	22.5

Should the parameters of a two-parameter equation be treated temperature dependent up to the critical point by a forced fitting procedure, the selection of p^v and V^ℓ would be desirable for this purpose. However, such an equation can no longer satisfy both Equations 2 and 3. In other words, the representation of the critical isotherm in the critical region cannot be satisfactory. Although a 3P2T equation (having the flexibility of obtaining substance-dependent ζ_c values) might overcome this difficulty, two such equations were evaluated in this work, H and KMC, and both equations yielded poorer overall performance than the evaluated 3P1T equations. Hence, a further examination of the limitations and behavior of 3P1T cubic equations of state would be useful in the design of new cubic equations.

All cubic equations of state suffer certain shortcomings. As discussed above, no equation evaluated could satisfactorily represent all the eight physical properties simultaneously. If one forces an equation to provide acceptable B values, such as the KMC equation, larger deviations would occur in other predictions, such as the V^ℓ and Z^ℓ values. On the other hand, if one ignores the representation of B, there is a chance of obtaining an equation which can simultaneously provide predictions of the other seven physical properties with reasonable results. The parameter values of "a" capable of predicting vapor pressures of pure substances would in general be suitable for vapor-liquid equilibrium calculations. Because some approximation is involved in correlating Ω_a by means of a temperature function, some deviations occur in the calculated p^v values. This shortcoming is not due to the form of the cubic equation of state, but due to the temperature function itself. The deviations in p^v due to this inadequacy are shown in Table III. This inadequacy is also reflected in the calculated V^v values. A plot of p^v and V^v deviations as a function of T_r would reveal that one deviation curve is mirror image of the other along the zero-deviation line for the 2P1T and 3P1T equations. Since the deviations in the calculated Z^c and Z^v values are practically the same among the 2P1T and 3P1T equations and these deviations are of the least concern of this study, the following discussion is mainly centered on the deviations of V^ℓ, Z^ℓ and Z^{sup}.

To illustrate the drawbacks of equations containing two temperature-dependent parameters, the deviation curves obtained for the HCL and H equations are plotted as a function of T_r in Figure 1. The correlation equations originally proposed (4, 14) were used to generate the curves for n-hexane. As both equations used p^v and V^ℓ for determining their parameters, they yield low deviations in these two properties, even though the accuracy of the representation was lost somewhat in the correlations. However, the deviations obtained in the V^v values are too large, due to the shift of errors through the limitations of a cubic equation. The results obtained from a modified PR equation with both parameters a and b simultaneously fitted to p^v and V^ℓ of n-hexane are also included in Figure 1 for comparison. The shift of errors to V^v is also very evident. The difficulty encountered by equations containing two temperature-dependent parameters for representing the critical region is depicted

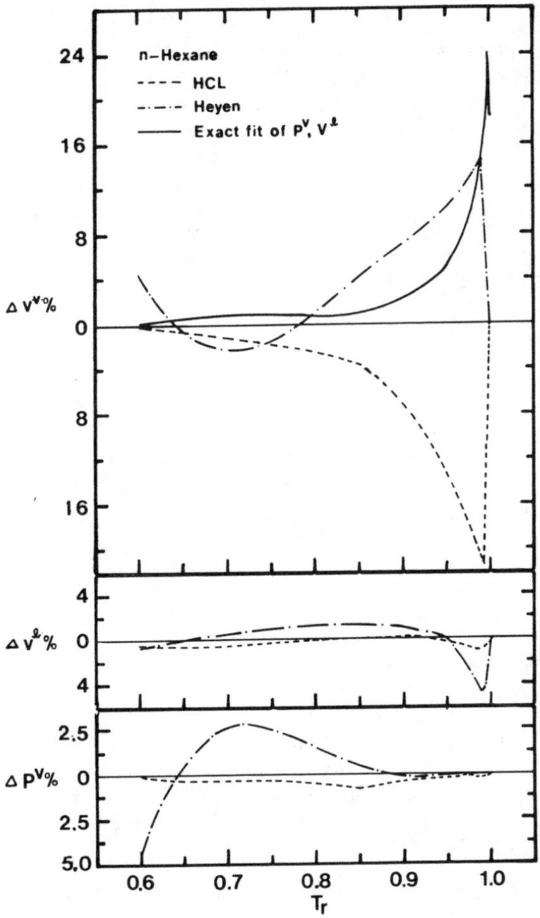

Figure 1 Deviation Curves obtained from equations containing two temperature-dependent parameters.

in Figure 2, using the expression in terms of $\alpha(=a/a_c)$ and $\beta(=b/b_c)$, and plotted in terms of T_r. The shape of the two curves, particularly near the critical point, prevents simple correlation of the parameters in terms of T_r.

The influence of values of u and w on the representation of V^ℓ values by means of the 2P1T equations (SRK, PR and HA) is illustrated in Figure 3. The three equations belong to the u + w = 1 family and their "a" parameters were determined by fitting vapor pressures. In Figure 3, the calculated deviations in V^ℓ and V^ν for methane, n-hexane and n-decane are plotted. None of the three equations yielded really satisfactory results. It is of interest to note that the general patterns of error distribution are about the same. It is impossible to obtain low V^ℓ deviations for substances with different molecular weights from two-parameter equations by simply adjusting the u and w values. This confirms the conclusion reached above that three-parameter equations are desirable.

The deviation contours of V^ℓ obtained from Equation 1 are plotted on u-w diagrams in Figures 4 and 5 for methane, n-hexane and n-decane. The numbers indicated on the diagrams refer to the number of carbons of the alkanes, and the dots represent the minimum deviations. The temperature range covered in Figure 4 ($0.5 \leqslant Tr \leqslant 0.98$) is larger than that in Figure 5 ($0.5 \leqslant Tr \leqslant 0.85$). On the other hand, the contours in Figure 4 represent average absolute deviations of 2.5% in V^ℓ, while in Figure 5, 1%. As pointed out by Schmidt and Wenzel (5), the following constraints must be satisfied to avoid the condition that $V^2 + ubV + wb^2 = 0$ for $V \geqslant b$,

$$w > -1-u \quad \text{for } u \geqslant -2 \tag{4a}$$

$$w > u^2/4 \quad \text{for } u \leqslant -2 \tag{4b}$$

The nonexisting area indicated in Figures 4 and 5 represents the area excluded by Equation 4.

A comparison of the two diagrams indicates that by narrowing the temperature range, the relative positions of the two groups of deviation contours shift towards larger values of w and smaller value of u. In phase equilibrium calculations, it would be desirable to have more volatile components well represented at higher T_r values, and less volatile components well represented at lower T_r values.

The deviation contours for Z^ℓ in the temperature and pressure ranges of $0.30 \leqslant T_r \leqslant 0.99$ and $0.01 \leqslant P_r \leqslant 10.0$ are plotted on the u-w diagram in Figure 6. The contours represent 4% average absolute deviations. Similarly, the deviation contours for Z^{sup} in the temperature and pressure ranges of $1.01 \leqslant T_r \leqslant 4.00$ and $0.01 \leqslant P_r \leqslant 10.0$ are plotted in Figure 7. The contours represent C_1, C_6 and C_{10}, and the average absolute deviations vary from 1.5 to 2.5%.

When the deviation contours of Figures 4-7 are superimposed, the overlapping areas for the alkanes are plotted on the u - w diagram in Figure 8. The deviation contour of V^ℓ for C_{20} is also included in the figure for reference. Any line or curve passing through these contours will yield the following average absolute deviations for the alkanes in the three physical properties:

V^ℓ: 1% ($0.5 \leqslant T_r \leqslant 0.85$); 2.5% ($0.5 \leqslant T_r \leqslant 0.98$)

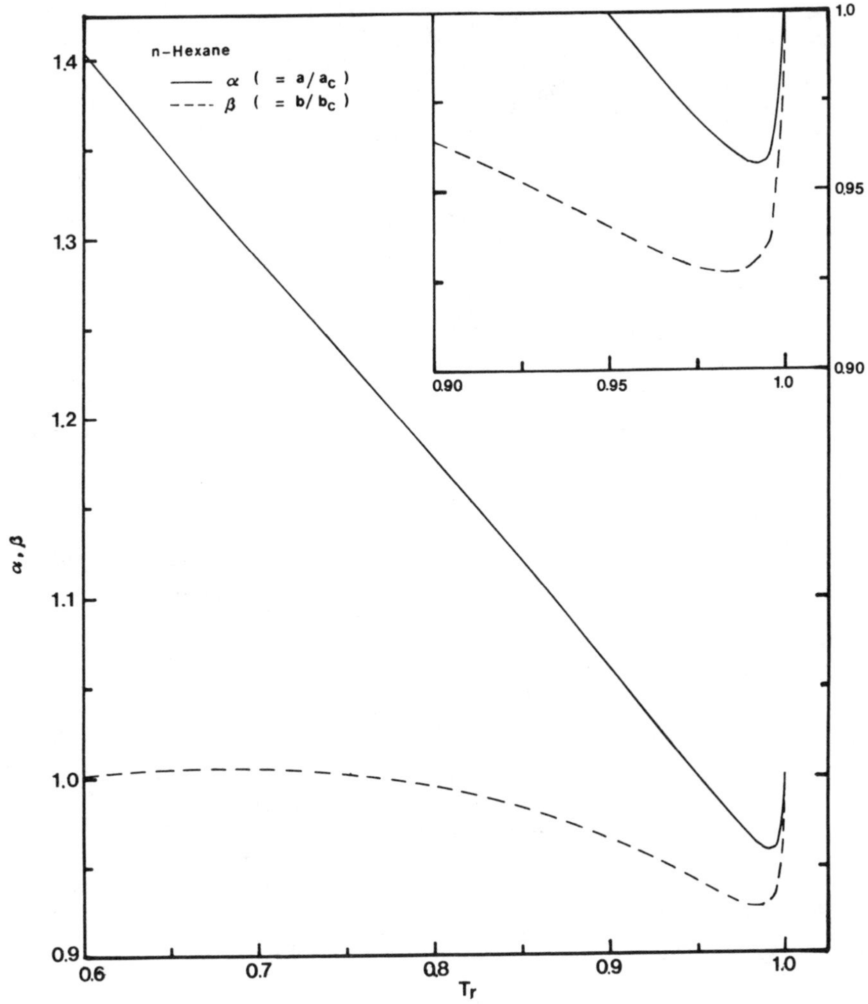

Figure 2 Variation of α and β with T_r for n-hexane from the PR equation

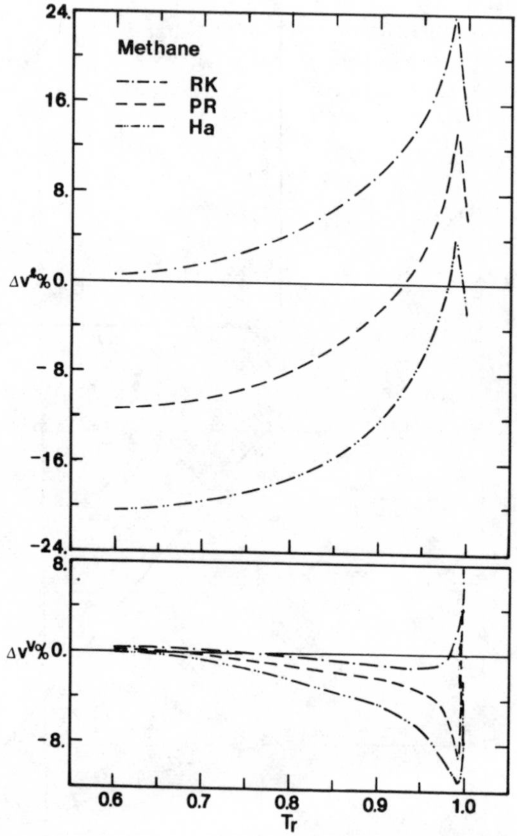

Figure 3a Illustrations of the influence of u and w values on the representation of V^ℓ and V^ν by means of 2P1T equations. -·-·- SRK (u=1, w=0), --- PR (u=2, w=-1), -··-··- HA (u=3, w=-2).

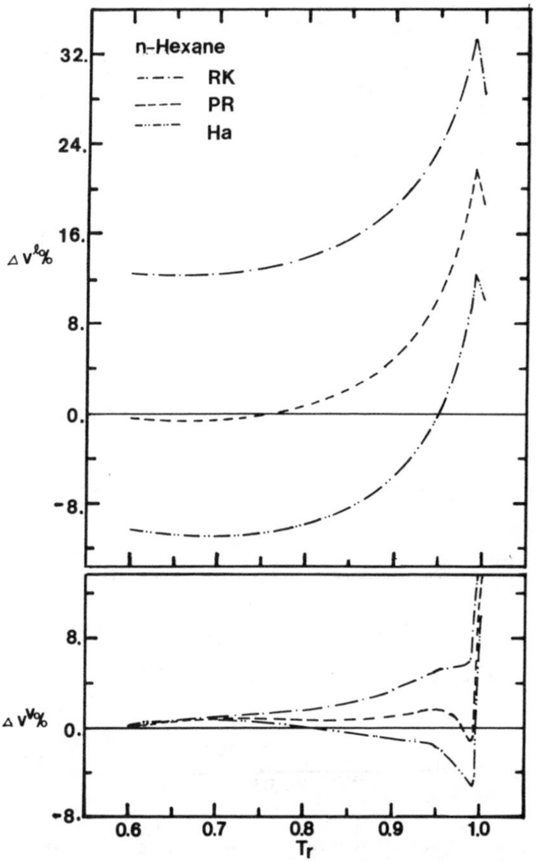

Figure 3b Illustrations of the influence of u and w values on the representation of V by means of 2PIT equations. - - - SRK (u=1, w=0), --- PR (u=2, w=-1), - - - HA (u=3, w=-2).

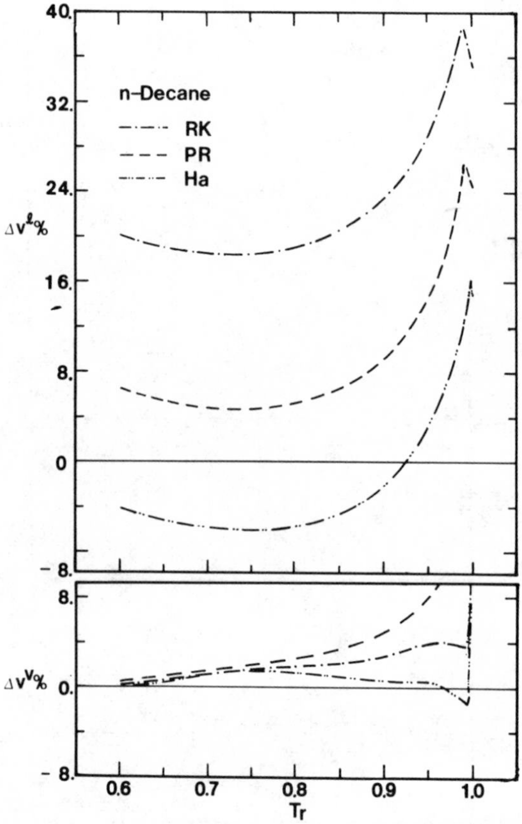

Figure 3c Illustrations of the influence of u and w values on the representation of V by means of 2PIT equations. - - - SRK (u=1, w=0), --- PR (u=2, w=-1), - - - HA (u=3, w=-2).

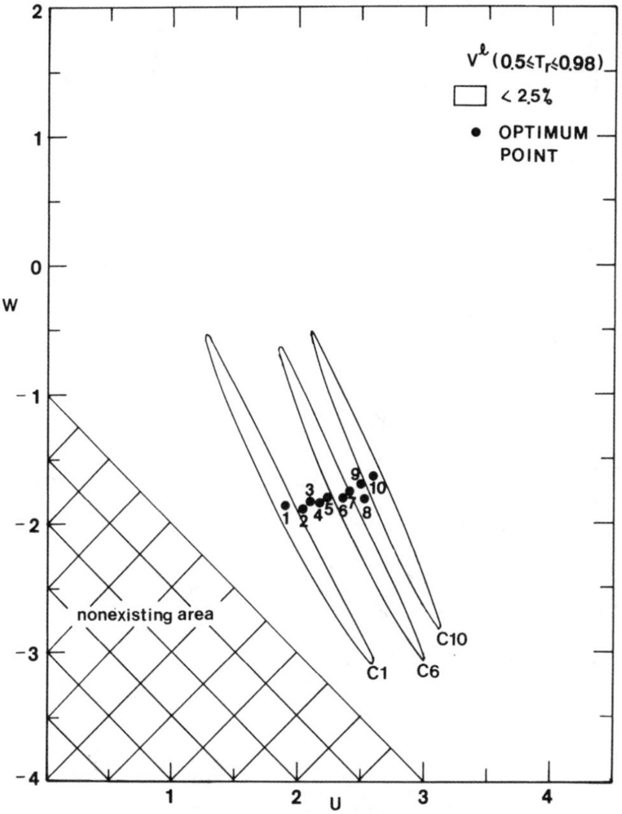

Figure 4 Deviation contours of V^{ℓ} in the temperature range of $0.5 \leq T_r \leq 0.98$. Curves represent deviations of 2.5% in V^{ℓ}.

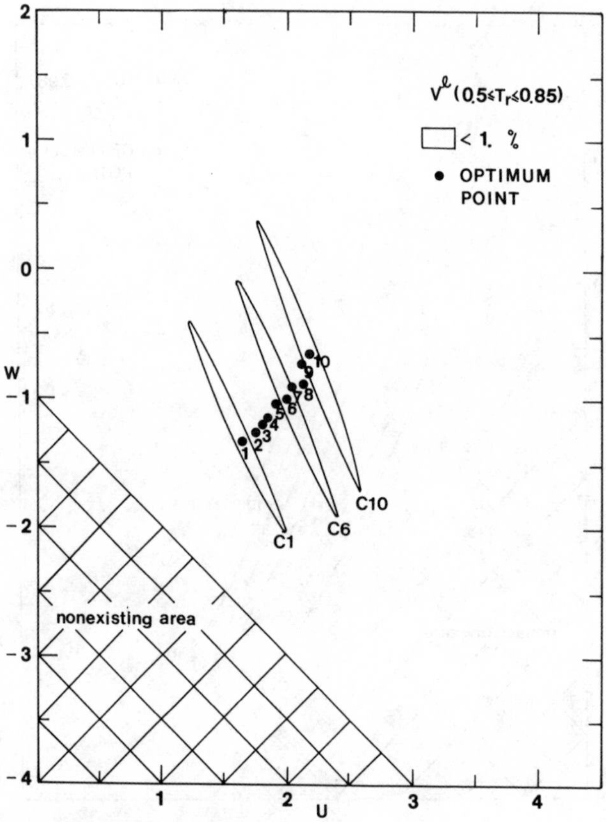

Figure 5 Deviation contours of V^{ℓ} in the temperature range of $0.5 \leq T_r \leq 0.85$. Curves represent deviations of 1% in V^{ℓ}.

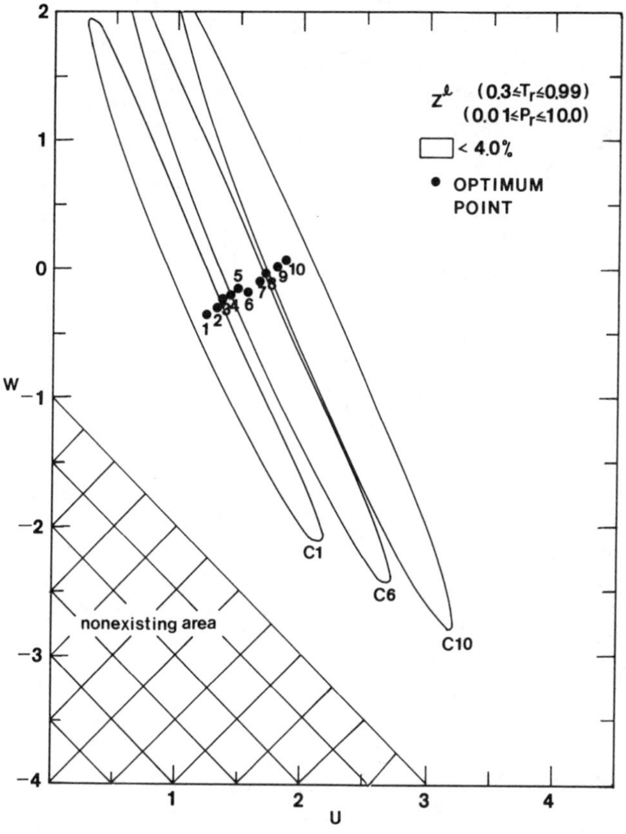

Figure 6 Deviation contours of Z^ℓ. Curves represent deviations of 4% in Z^ℓ.

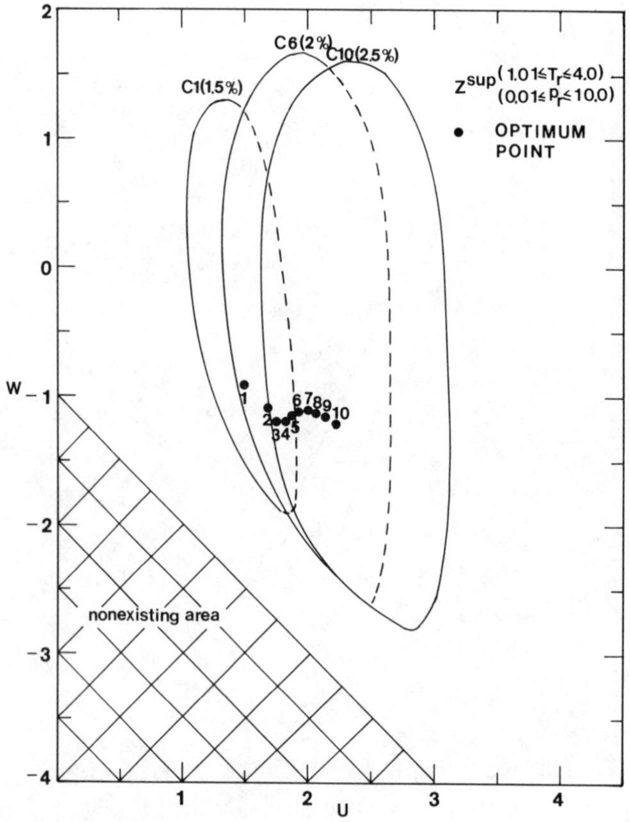

Figure 7 Deviation contours of Z^{sup}. Deviation levels are indicated on each curve.

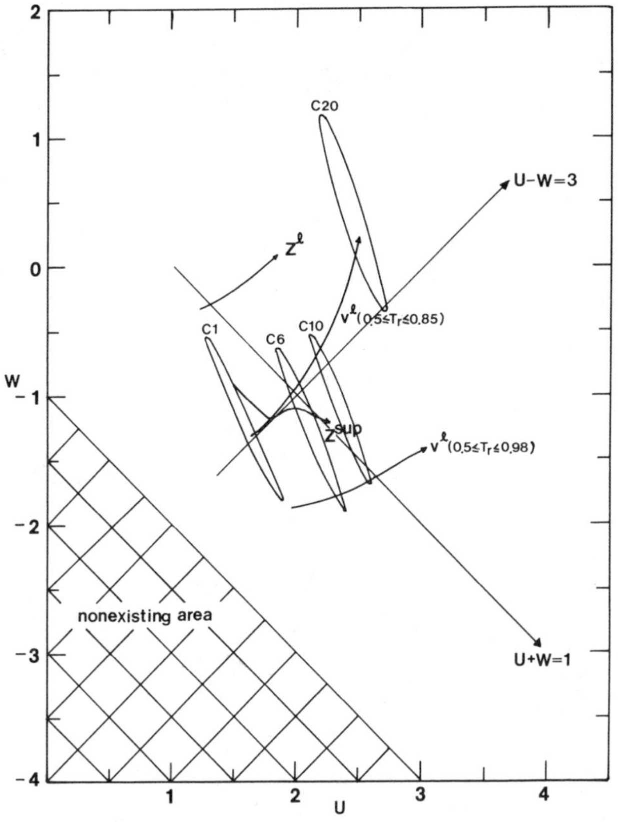

Figure 8 Overlapping areas resulting from superimposing deviation contours of Figures 4-7.

Z^{ℓ}: 4% ($0.3 \leqslant Tr \leqslant 0.99$)
$Z^{sup} \leqslant 2.5\%$ ($1.01 \leqslant T_r \leqslant 4.00$, $0.01 \leqslant P_r \leqslant 10.0$)

The solid curves in Figure 8 represent the loci of the minimum deviation points for the ten alkanes, with the exception of the V^{ℓ} curve, which extends to C_{20}. The family of equations represented by the u + w = 1 relationship on the u-w diagram cannot yield good prediction of V^{ℓ} values, especially for alkanes of larger molecular weights. The desired relationship should have a positive slope on the u-w diagram.

Concluding Remarks

For the purpose of obtaining satisfactory results simultaneously for vapor-liquid equilibrium values and volumetric properties, a three-parameter van der Waals type equation with the parameter "a" fitted to vapor pressure values is desirable.

Using deviation contours on u-w diagrams provides a means for designing a desirable equation for representing specific physical properties of the substances concerned.

For the representation of V^{ℓ}, Z^{ℓ} and Z^{sup} for normal alkanes, the relationship between u and w should yield a positive slope on the u-w diagram, contrary to the suggestion of Schmidt and Wenzel. A possible equation is that represented by the u - w = 3 relationship as shown in Figure 8.

Acknowledgments

The authors are indebted to the Natural Science and Engineering Research Council of Canada for financial support.

Literature Cited

1. van der Waals, J.D. doctoral dissertation, Leiden, Holland, 1873.
2. Redlich, O., Kwong, J.N.S. Chem. Rev. 1949, 44, 239.
3. Peng, D.Y., Robinson, D.B. Ind. Eng. Chem. Fundam. 1976, 15, 59.
4. Heyen, G. Proc. 2nd Int. Conf. on Phase Equilibria and Fluid Properties in the Chemical Industry, 1980, p. 9; In "Chemical Engineering Thermodynamics"; Newman, S.A., Ed.; Ann Arbor Science Publishers, Ann Arbor, 1983; Chap. 15.
5. Schmidt, G., Wenzel, H. Chem. Eng. Sci. 1980, 35, 1503.
6. Abbott, M.M. AIChEJ 1973, 19, 596; in "Equations of State in Engineering and Research", Chao, K.C., Robinson, R.L. Jr., Eds.; Advances in Chemistry Series No. 182, American Chemical Society: Washington, D.C. 1979; Chap. 3.
7. Döring, R., Knapp, H. Proc. 2nd Int. Conf. on Phase Equilibria and Fluid Properties in the Chemical Industry, 1980, p. 34.
8. Lira, R., Malo, J.M., Leiva, M.A., Chapela, G. In "Chemical Engineering Thermodynamics"; Newman, S.A., Ed.; Ann Arbor Science Publishers, Ann Arbor, 1983; Chap. 14.
9. Adachi, Y., Lu, B.C.-Y., Sugie, H. Fluid Phase Equilibria 1983, 11, 29.
10. Adachi, Y., Lu, B.C.-Y. AIChEJ 1984, 30, 991.

11. Kumar, K.H., Starling, K.E. Ind. Eng. Chem. Fundam. 1982, 21, 255.
12. Clausius, R. Ann Phys. Chem. 1880, IX, 337.
13. Soave, G. Chem. Eng. Sci. 1972, 27, 1197.
14. Hamam, S.E.M., Chung, W.K., Elshayal, I.M., Lu, B.C.-Y. Ind. Eng. Chem. Process Des. Dev. 1977, 16, 51.
15. Harmens, A., Knapp, H. Ind. Eng. Chem. Fundam. 1980, 19, 291.
16. Patel, N.C., Teja, A.S. Chem. Eng. Sci. 1982, 37, 463.
17. Peneloux, A., Rauzy, E., Freze, R. Fluid Phase Equilibria 1982, 8, 7.
18. Freze, R., Chevalier, J.-L., Peneloux, A., Rouzy, E. Fluid Phase Equilibria, 1983, 15, 33.
19. Adachi, Y., Lu, B.C.-Y., Sugie, H. Fluid Phase Equilibria 1983, 13, 133.
20. Fuller, G.G. Ind. Eng. Chem. Fundam. 1976, 15, 254.
21. Martin, J.J. Ind. Eng. Chem. Fundam. 1979, 18, 81.
22. Kubic, W.L. Fluid Phase Equilibria 1982, 9, 79.
23. Harmens, A. Cryogenics 1977, 17, 519.
24. Gomez-Nieto, M., Thodos, G. Ind. Eng. Chem. Fundam. 1978, 17, 45.
25. Spencer, C.F., Alder, S.B. J. Chem. Eng. Data 1978, 43, 137.
26. Barile, R.G., Thodos, G. Can. J. Chem. Eng. 1965, 43, 137.
27. Tsonopoulos, C. AIChEJ 1974, 20, 263.
28. Lee, B.I., Kesler, M.G. AIChEJ 1975, 21, 510.

RECEIVED November 8, 1985

27

An Improved Cubic Equation of State

R. Stryjek[1] and J. H. Vera

Department of Chemical Engineering, McGill University, Montreal, Quebec H3A 2A7, Canada

The Peng-Robinson equation of state has been modified to extend its use to low reduced temperatures. The modified form, called the PRSV equation, is able to reproduce pure compound vapor pressures down to 1.5 kPa with accuracy comparable to Antoine's equation for hydrocarbons, polar and associated compounds. Hydrocarbon/hydrocarbon and aromatic hydrocarbon/polar compound vapor-liquid equilibria have been correlated using a single binary parameter with an accuracy similar to the γ-ψ approach. The binary parameter is slightly temperature dependent. Two parameters are required to correlate vapor-liquid equilibria of saturated hydrocarbon/polar compound systems. Water/methanol system is well represented with one binary parameter model. Representation of the systems water/ethanol and water/propanols requires the use of two binary parameters.

The use of cubic equations of state to correlate vapor-liquid equilibrium data has received increased attention in the last few years. For the purposes of this study, important contributions have been presented by Mathias (1), Mathias and Copeman (2), Soave (3) and Gibbons and Laughton (4).

Mathias (1) proposed an improved temperature dependence for the attractive term of the Redlich-Kwong equation of state. Mathias and Copeman (2) suggested a form for the temperature dependence of the attractive term of the Peng-Robinson equation of state. Soave (3) showed that even the simple van der Waals' equation of state may be used to correlate vapor-liquid equilibria with an appropriate temperature dependence of the attractive term for pure compounds and the use of the mixing rules proposed by Huron and Vidal (5). Gibbons and Laughton (4) presented a modified form of the Redlich-Kwong equation of state including two adjustable parameters per pure compound.

---

[1]Current address: Institute of Physical Chemistry, PAN, Warsaw, Poland.

0097-6156/86/0300-0560$06.00/0

Pure Compounds

In this work we have selected the cubic equation of state proposed by Peng and Robinson (6), namely,

$$P = \frac{RT}{v-b} - \frac{a}{v^2 + 2bv - b^2} \tag{1}$$

with

$$a = (0.45724\ R^2T_c^2/P_c)\alpha \tag{2}$$

and

$$b = 0.0778\ R\ T_c/P_c \tag{3}$$

where α is given by

$$\alpha = [1 + \kappa\ (1 - T_R^{0.5})]^2 \tag{4}$$

Peng and Robinson (6) considered κ to be a function of the acentric factor ω only. This equation performs well for hydrocarbons and slightly polar compounds at reduced temperatures of the order of 0.7 and above. At low reduced temperatures, however, it fails to reproduce the vapor pressures of pure compounds even for nonpolar substances.

Preliminary studies at reduced temperatures below 0.7 showed that different values of κ could be obtained for the same compound depending on the reduced temperature range of the vapor pressure data used in the fitting procedure. In addition, for different compounds with a large variation in critical temperature, values of κ had a smooth variation with the acentric factor only when all compounds were considered in the same reduced temperature range. These results suggested that it was necessary to consider a fixed reduced temperature for all compounds in order to generate a self consistent set of κ values. In this work we selected $T_R = 0.7$. Since this temperature is close to the normal boiling point for most substances, reliable vapor pressure data are available in its vicinity for most compounds. Critical pressures and critical temperatures reported in the literature were carefully examined and the most precise values were selected for the evaluation of both ω and κ. The acentric factor ω was obtained from its definition. The value of κ, here called κ_o, was obtained so as to reproduce exactly the experimental vapor pressure when the fugacity of the vapor phase is equal to the fugacity of the liquid phase at $T_R = 0.7$. Values of κ_o so obtained fall over a single curve when plotted against ω, irrespective of the polarity, degree of association or geometrical complexity of the molecules. The following correlation reproduces well the variation of κ_o with ω for over eighty compounds.

$$\kappa_o = 0.378893 + 1.4897153\omega - 0.17131848\omega^2 + 0.0196554\omega^3 \tag{5}$$

The use of equation (5) instead of the form for the dependence of κ on ω proposed by Peng and Robinson (6) produced a first improvement in the results. Not only the reproduction of vapor pressures was improved in the reduced temperature range from 0.7 to 1.0, but also deviations between calculated and experimental vapor pressures for different compounds presented a more regular variation at reduced temperatures below 0.7. However, for compounds with $\omega \approx 0.1$ and $\omega \approx 0.4$ both forms give the same value of κ and thus, the same results. Maximum differences between both correlations are produced for low and high acentric factors.

In order to reproduce saturation pressures at reduced temperatures below 0.7 it is necessary to introduce a temperature dependence on κ with at least one adjustable parameter characteristic of each compound. After some trials, the following form was found to give a good fit of vapor pressures down to low reduced temperatures

$$\kappa = \kappa_o + \kappa_1 \, (1 + T_R^{0.5})(0.7 - T_R) \qquad (6)$$

Although values of κ_1 may be obtained from a single vapor pressure datum at reduced temperature below 0.7, in this work values of κ_1 were obtained by the fitting of low reduced temperature vapor pressure data. Some representative values of κ_1 are given in Table I together with critical properties and acentric factors used in this work. For water and lower alcohols there is a small advantage in using equation (6) in all the range from low reduced temperature up to the critical temperature. For all other compounds we recommend to use $\kappa = \kappa_o$ for $T_\kappa > 0.7$.

Table I. Pure Compound Parameters

Compound	T_c,K	P_c, kPa	ω	κ_1
Ethane	302.43	4879.76	0.0978	0.02814
Cyclohexane	553.64	4074.96	0.2088	0.07023
Benzene	562.16	4897.95	0.2093	0.07019
Biphenyl	769.15	3120.78	0.3810	0.11487
Acetone	508.10	4695.95	0.3067	-0.00888
Ammonia	405.55	11289.52	0.2517	0.00100
Water	647.29	22089.75	0.3438	-0.06635
Methanol	515.58	8095.79	0.5653	-0.16816
Ethanol	513.92	6148.33	0.6444	-0.03374
1-Propanol	536.71	5169.55	0.6201	0.21419
2-Propanol	508.40	4764.25	0.6637	0.23264
Pyridine	620.00	5595.26*	0.2372	0.06946
Thianaphthene	752.00	3880.71	0.2936	0.06043

* estimated by group contribution method.

Figure 1 compares the performance of the modified Peng-Robinson equation, from here on called the PRSV equation, with the modified cubic equations of state of Soave (3) and of Gibbons and Laughton (4). It should be observed the forms of Soave (3) and of

Gibbons and Laughton (4) contain two adjustable parameters per compound while the PRSV equation contains only one. Table II presents a comparison of results obtained with the PRSV equation and those reported by Mathias (1) using a modified Redlich-Kwong equation of state. Clearly, the PRSV equation gives a better representation of pure compound vapor pressures than any other modified cubic equation of state. For water, for example, vapor pressures are correlated with a percent deviation of less than 0.03 from its triple point to a reduced temperature of 0.75. Slightly higher deviations are obtained in the high reduced temperature range but, in any case, deviations are smaller than those obtained with other cubic equations of state. Similar results are obtained for ammonia. Heavy organic compounds, are equally well represented by the PRSV equation down to very low vapor pressures. Figure 2 presents the performance of the PRSV equation for some compounds. Observe that for biphenyl vapor-pressures are well represented by the PRSV equation down to a pressure of 0.004 kPa. Results for pyridine present an average pressure deviation of about 0.1%.

Table II. Comparison of Results Obtained with the PRSV Equation and Those Reported by Mathias (1)

		$\lvert\overline{\Delta P}\rvert$%	
Compound	T,K	Mathias	This work
Water	273-647	0.3	0.1
Acetone	259-508	0.4	0.2
Methanol	288-513	0.4	0.6
Ethanol	293-514	0.7	0.5
1-Pentanol	348-512	0.7	0.7
1-Octanol	386-554	2.2	1.6

Binary Mixtures

As for the Peng-Robinson equation of state, the expression for the fugacity coefficient of a component i in a mixture obtained from the PRSV equation has the general form

$$\ell n\psi_i = \frac{\bar{b}_i}{b}(z-1) - \ell n(z-B) - \frac{A}{2\sqrt{2}\,B}\left(\frac{\bar{a}_i}{a} + 1 - \frac{\bar{b}_i}{b}\right)\ell n\,\frac{z+(1+\sqrt{2})B}{z+(1-\sqrt{2})B} \quad (7)$$

where $z = Pv/RT$, $A = Pa/(RT)^2$, $B = Pb/RT$ and $\bar{a}_i$ and $\bar{b}_i$ are obtained as

$$\bar{a}_i = \left(\frac{\partial na}{\partial n_i}\right)_{n_{j\neq i}} \quad (8) \qquad \bar{b}_i = \left(\frac{\partial nb}{\partial n_i}\right)_{n_{j\neq i}} \quad (9)$$

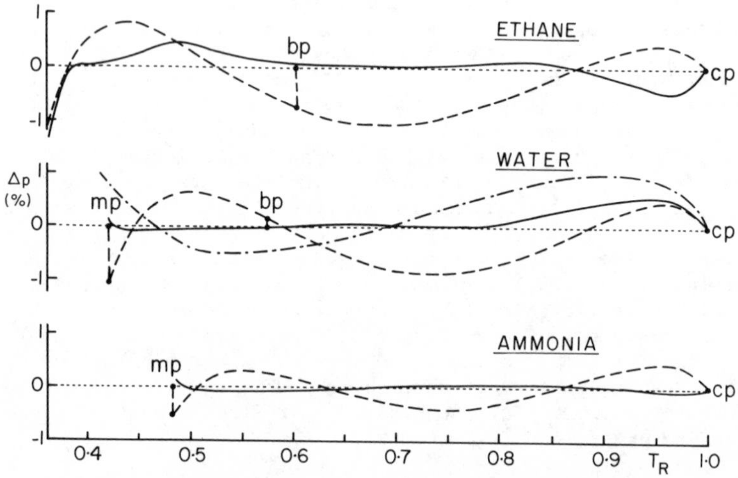

Figure 1. Per Cent Deviations in Calculated Pure Compound Vapor Pressures for Ethane, Water and Ammonia. —— PRSV; ----- Soave (1984); -•-•- Gibbons and Laughton (1984).

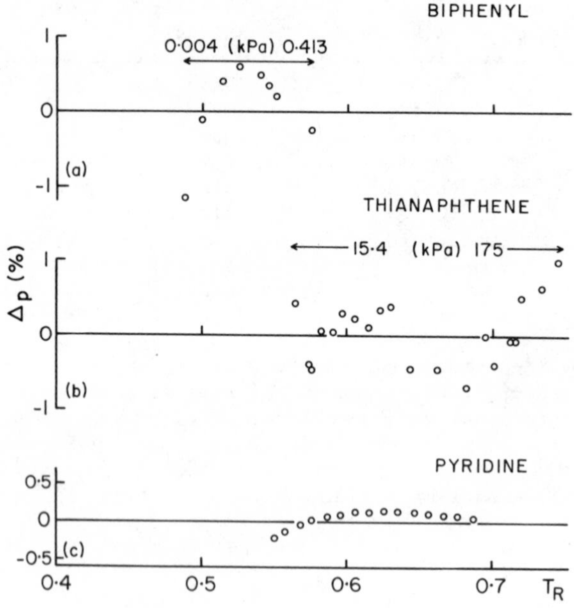

Figure 2. Per Cent Deviations in Calculated Pure Compound Vapor Pressures for Biphenyl, Thianaphtene and Pyridine using the PRSV Equation of State.

The particular expressions of $\bar{a}_i$ and $\bar{b}_i$ depend on the mixing rules chosen for a and b, respectively. In this work we have used the conventional mixing rule for b in all cases

$$b = \sum_j x_j b_j \tag{10}$$

which gives

$$\bar{b}_i = b_i(\text{pure compound}) \tag{11}$$

For a, we tested first the conventional mixing rule

$$a = \sum_{i,j} x_i x_j a_{ij} \tag{12}$$

with

$$a_{ij} = (a_i a_j)^{0.5} (1-k_{ij}) \tag{13}$$

which gives

$$\bar{a}_i = 2 \sum_j x_j a_{ij} - a \tag{14}$$

These mixing rules proved to be satisfactory for mixtures of the type hydrocarbon/hydrocarbon and aromatic hydrocarbon/polar compound. However they gave only fair results for systems of the type saturated hydrocarbon/polar compound. For binary systems the binary parameters were fitted minimizing the objective function: $Q = \sum(\Delta P/P)^2$, where P is the bubble pressure. Results presented in the tables are those with the optimum value of the parameter at each temperature. Results for the binary system benzene/biphenyl are presented in Table III. For comparison we have also included in Table III results obtained by Gmehling et al. (7) using standard γ-ψ approach. It is interesting to observe that at the lowest isotherm biphenyl is below its normal melting point. Table IV presents a similar comparison for different isotherms of the system benzene/acetone. Again here the one parameter mixing rule produces surprisingly good results. Notably, values of the k_{12} parameter are almost independent of temperature for the system benzene/acetone and present a regular temperature dependence for the system benzene/biphenyl. No attempt was done to smooth the temperature dependence of k_{12} since γ-ψ approaches use also particular values of two adjustable parameters at each temperature. For systems of the type saturated hydrocarbon/polar compound such as cyclohexane/acetone the use of the one-parameter mixing rule given by equation (13) produced larger deviations in the calculated pressures. These deviations were not random, they presented a systematic trend with respect to composition for different systems studied. The use of two binary parameters was clearly required.

In this work we have used the two-parameter mixing rule

Table III. Vapor-Liquid Equilibria for the System Benzene/Biphenyl

$\overline{\Delta P}$, mm Hg

	PRSV-1			Wilson*		NRTL*		UNIQUAC*	
T,K	$k_{12}\cdot10^3$	$\overline{\Delta P}$	$\overline{\Delta y}\cdot10^3$	$\overline{\Delta P}$	$\overline{\Delta y}\cdot10^3$	$\overline{\Delta P}$	$\overline{\Delta y}\cdot10^3$	$\overline{\Delta P}$	$\overline{\Delta y}\cdot10^3$
318.15	4.420	0.60	0.2	7.73	0.1	3.16	0.0	12.57	0.1
328.15	5.785	0.50	0.3	15.74	0.3	4.50	0.1	18.78	0.4
338.15	7.100	0.73	0.4	18.68	0.4	5.84	0.1	24.75	0.5

*As reported by Gmehling et al. (1980)

Table IV. Vapor-Liquid Equilibria for the System Benzene/Acetone

$\overline{\Delta P}$, mm Hg

	PRSV-1			Wilson*		NRTL*		UNIQUAC*	
T,K	$k_{12}\cdot10^2$	$\overline{\Delta P}$	$\overline{\Delta y}\cdot10^3$	$\overline{\Delta P}$	$\overline{\Delta y}\cdot10^3$	$\overline{\Delta P}$	$\overline{\Delta y}\cdot10^3$	$\overline{\Delta P}$	$\overline{\Delta y}\cdot10^3$
298.15	2.716	0.75	5.0	0.50	2.8	0.51	2.8	0.48	2.7
303.15	2.836	1.23	9.9	1.78	6.4	1.78	6.4	1.78	6.4
313.15	2.877	2.46	6.8	2.14	4.2	2.13	4.2	2.08	4.1
318.15	2.512	0.68	4.2	0.71	3.4	0.76	3.5	0.60	3.4
323.15	2.985	2.28	7.2	3.19	7.6	3.43	7.8	3.06	7.5

*As reported by Gmehling et al. (1980)

proposed by Huron and Vidal (5) for the a term, which for the PRSV equation takes the form,

$$a = b\ (\sum x_i\ a_i/b_i - c\ g^E_\infty) \quad (15)$$

with $c = 2\sqrt{2}/\ln\left[(2 + \sqrt{2})/(2 - \sqrt{2})\right]$. With this mixing rule for a, the term of equation (7) containing a_i takes the form

$$\frac{A}{2\sqrt{2}B}\left(\frac{\bar{a}_i}{a} + 1 - \frac{\bar{b}_i}{b}\right) = \frac{1}{2\sqrt{2}}\left[\frac{a_i}{b_i RT} - c\ \ln \gamma_{i,\infty}\right] \quad (16)$$

For the "infinite pressure" excess function we have used in equation (15) the expression for the excess Gibbs energy proposed by Renon and Prausnitz (8) as suggested by Huron and Vidal (5). Thus,

$$\ln \gamma_{i,\infty} = \frac{1}{RT}\frac{\sum_j \Delta g_{ji} G_{ji} x_j}{\sum_k G_{ki}\ x_k} + \sum_j \frac{x_j G_{ij}}{\sum_k G_{kj}\ x_k}\left(\Delta g_{ij} - \frac{\sum_k x_k \Delta g_{kj} G_{kj}}{\sum_k G_{kj}\ x_k}\right) \quad (17)$$

with

$$G_{ij} = \exp\left[-\frac{0.3\ \Delta g_{ij}}{RT}\right] \quad (18)$$

Water/alcohol systems present complex molecular interactions and thus they are a severe test for correlating methods. Table V shows results obtained with the one-binary parameter PRSV-1 form of equation (13) for the system water/methanol. These results are perfectly comparable with those obtained by Gmehling et al. (7) using the γ-ψ approach. For water/ethanol, water/1-propanol and water/2-propanol systems, the PRSV-1 equation gave poor results.

Table VI shows the results obtained with the PRSV-HV model of equations (15) to (18), for the systems water/ethanol and water/propanols. A dramatic improvement in the results is obtained with the two-binary-parameter PRSV-HV approach in comparison with the previous treatment. For these highly non ideal systems the PRSV equation of state is very sensitive to the mixing rule applied. Values of the binary parameters Δg_{12} and Δg_{21} are also reported in Table VI.

Conclusions

The PRSV equation is a valuable tool for computation of vapor-liquid equilibria for systems that previously could only be treated by the γ-ψ approach. In this preliminary presentation of the PRSV equation we have attempted to give a general overview of its potentialities. More detailed studies will be presented in forthcoming publications.

Table V. Vapor-Liquid Equilibria for the System Water/Methanol

$\overline{\Delta P}$, mm Hg

T	N	PRSV-1			
		$k_{12}\cdot10^2$	$\overline{\Delta P}$	$\overline{\Delta P\%}$	$\overline{\Delta y}\cdot10^3$
298.15	10	8.866	0.86	1.39	7.6
298.15	10	9.450	1.56	3.56	13.9
308.15	14	8.555	1.62	1.95	14.4
312.91	21	9.172	3.49	2.85	42.5
313.05	10	8.240	1.27	0.75	7.0
322.91	12	9.033	6.12	2.44	18.8
323.15	13	8.500	3.12	1.59	11.4
333.15	7	7.680	2.16	0.49	-
333.15	12	8.385	6.45	2.37	18.2
338.15	12	8.225	5.19	1.29	9.5
373.15	16	8.360	15.59	0.91	10.4

VLE data were taken from Gmehling et al. (1980)

Table VI. Vapor-Liquid Equilibria for the Systems Water/Ethanol, Water/1-Propanol, and Water/2-Propanol with the PRSV-HV Model+

T,K	Δg_{12}	Δg_{21}	$\overline{\Delta P\%}$	$\overline{\Delta y}\cdot10^3$	T,K	Δg_{12}	Δg_{21}	$\overline{\Delta P\%}$	$\overline{\Delta y}\cdot10^3$
		Water/Ethanol					Water/1-Propanol		
283.15	15.389	37.330	1.30	-	333.15	30.025	113.080	0.63	12.19
288.15	17.069	38.054	0.78	-	333.15	26.341	111.025	1.42	12.02
293.15	12.873	45.070	0.96	-	333.15	30.479	111.773	0.89	10.97
298.15	12.242	49.783	1.02	6.78	363.15	27.979	122.111	0.58	8.38
303.15	13.956	49.483	1.07	-	Σ			0.88	10.89
303.15	11.757	52.320	0.72	-					
313.15	12.463	54.900	2.43	8.18			Water/2-Propanol		
323.15	7.649	64.892	1.33	12.39					
323.15	9.465	65.629	0.50	-	308.15	22.282	85.950	0.43	4.18
323.15	9.339	64.331	0.64	-	318.15	25.538	93.873	0.87	22.23
323.15	10.002	64.692	0.68	-	318.20	23.154	91.787	0.46	4.45
323.15	9.890	64.534	0.49	-	328.18	23.502	97.766	0.44	5.19
328.15	9.876	67.724	0.57	8.11	333.15	24.862	101.854	0.87	27.06
333.15	9.242	66.508	0.60	14.37	338.15	24.109	103.587	0.50	4.64
343.15	10.290	73.020	0.33	8.14	348.14	24.138	109.184	0.52	4.09
343.15	9.394	72.342	0.40	-	423.15	22.444	144.766	0.94	15.50
363.15	10.818	76.827	0.15	-	473.15	24.068	152.649	0.64	6.10
423.15	6.330	94.911	1.31	5.88	523.15	21.999	159.728	0.25	7.92
523.15	7.367	98.509	0.81	-	548.15	22.386	165.561	0.41	16.96
548.15	26.043	73.547	1.99	9.13	Σ			0.58	10.76
573.15	44.183	68.324	1.51	6.22					
598.15	54.582	68.014	0.59	6.28					
Σ			0.92	8.55					

+For Water/Ethanol, sets 1-17 from Gmehling et al. (1980) and 18-22 from Barr-David and Dodge (1959), for Water/1-Propanol from Gmehling et al. (1980), and for Water/2-Propanol, sets 1-7 from Gmehling et al. (1980) and 8-11 from Barr-David and Dodge (1959).

In general, the dramatic improvement in the reproduction of the pure compound vapor pressures allows to obtain less temperature dependent binary parameters. A one-parameter mixing rule suffices to represent systems with symmetric excess Gibbs energy curves and a two-parameter mixing rule is required when a system is highly asymmetric.

Acknowledgements

The authors are grateful to NSERC and McGill University for financial support, and one of us (R.S.) to the Polish Academy of Sciences, for a leave of absence.

Legend of Symbols

a,b	equation of state parameters
$\bar{a}_i, \bar{b}_i$	for definition see equations (8) and (9), respectively
A,B	dimensionless terms: $A = Pa/(RT)^2$; $B = Pb/RT$
c	numerical constant, $2\sqrt{2}/\ln\left[(2+\sqrt{2})/(2-\sqrt{2})\right]$
g^E_∞	excess Gibbs energy at infinite pressure
Δg_{ij}	binary parameter
G_{ij}	temperature dependent binary parameter
k	binary parameter
n	number of moles
N	number of data points
P	pressure
R	gas constant
T	absolute temperature
T_R	reduced temperature
v	molar volume
x	mole fraction
y	vapor, phase mole fraction
z	compressibility factor
α	function of reduced temperature and acentric factor value
$\overline{\Delta P}$	average deviation in pressure
$\overline{\Delta y}$	average deviation in vapor phase composition
κ	see equation (6)
κ_o	see equation (5)
κ_1	pure compound parameter. See Table I.
ψ	fugacity coefficient
ω	acentric factor

Subscripts

c	critical property
i,j	compounds
∞	value at infinite pressure

Literature Cited

1. Mathias, P.M., Ind. Eng. Chem., Proc. Des. Develop., 22, 385-391 (1983).
2. Mathias, P.M. and Copeman, T.W., Fluid Phase Equil., 13, 91-108 (1983).
3. Soave, G., Chem. Eng. Sci., 39, 357-369 (1984).
4. Gibbons, R.M. and Laughton, A.P., J. Chem. Soc., Faraday Trans. 2, 80, 1019-1038 (1984).
5. Huron, M.J. and Vidal, J., Fluid Phase Equil., 3, 255-271 (1979).
6. Peng, D.Y. and Robinson, D.B., Ind. Eng. Chem. Fundam., 15, 59-64 (1976).
7. Gmehling, J., Onken, V. and Arlt, W., "Vapor Liquid Equilibrium Data Collection", Chemistry Data Series, Dechema, Frankfurt/Main 1980.
8. Renon, H. and Prausnitz, J.M., AIChE J., 14, 135-144 (1968).
9. Barr-David, F. and Dodge, B.F., J. Chem. Eng. Data, 4, 107 (1959).

RECEIVED November 5, 1985

New Mixing Rule for Cubic Equations of State for Highly Polar, Asymmetric Systems

A. Z. Panagiotopoulos and R. C. Reid

Department of Chemical Engineering, Massachusetts Institute of Technology, Cambridge, MA 02139

A new two-parameter mixing rule for van der Waals-type cubic equations of state was developed by making the normally used single binary interaction parameter k_{ij} a linear function of composition. A significant improvement was observed in the representation of binary and ternary phase equilibrium data for highly polar and asymmetric systems. Results are presented for systems with water and supercritical fluids at high pressures, as well as for low-pressure non-ideal systems. Ternary phase equilibrium data at high pressures, including LLG three-phase equilibria, were successfully correlated using parameters regressed from binary data only.

Cubic equations of state (EOS) have become important tools in the area of phase equilibrium modelling, especially for systems at pressures close to or above the critical pressure of one or more of the system components. Among the more common of the currently used cubic EOS, are the Soave modification of the Redlich-Kwong (1) and the Peng-Robinson EOS (2). The functional form of both equations, as well as several other proposed cubics, can be represented in a general manner as shown in Equation 1 (3) :

$$P = \frac{RT}{V - b_m} - \frac{a_m}{V^2 + uVb_m + wb_m^2} \tag{1}$$

where u and w are numerical constants. Table I lists the values of u and w for some common EOS.

For a mixture, parameters a_m and b_m are related to the pure component parameters and the mixture composition through a mixing rule. Equations 2 and 3 show one common choice for the mixing rule, the van der Waals 1-fluid mixing rule :

$$a_m = \sum_i \sum_j x_i x_j a_{ij} \tag{2}$$

$$b_m = \sum_i x_i b_i \tag{3}$$

0097-6156/86/0300-0571$06.00/0

Table I. Parameters for a few cubic Equations of State

Equation of State	u	w
van der Walls (1873)	0	0
Redlich-Kwong (1949)	1	0
Soave (1972)	1	0
Peng-Robinson (1976)	2	-1

The cross-parameters a_{ij} are related in turn to the pure-component parameters by a "combining rule". Equation 4 shows a common form of the combining rule for a_{ij} :

$$a_{ij} = \sqrt{a_i \, a_j} \; (1-k_{ij}) \tag{4}$$

In Equation 4, k_{ij} is called a binary interaction parameter, and was originally introduced so that the equation can better reproduce experimental composition data in systems that contain components other than the light hydrocarbons.

There has been criticism directed toward the oversimplicity of the cubic equation form, and rightly so. Nevertheless this representation does describe, at least qualitatively, all the important characteristics of vapor-liquid equilibrium behavior. Alternative equations of state have (and are) being suggested, but, to date, none have been widely used and tested. Also, too often, alternate EOS are significantly more complex and bring with them additional pure-component and mixture parameters which must be evaluated by regressing experimental data.

We feel that the key to success in employing the cubic equation form to model phase equilibria is in the choice of the mixing and combining rules. Our goal, then, was to select the simplest forms, with the fewest interaction parameters that could be used to correlate complex phase behavior in both non-polar and highly polar systems.

Proposed Method

We propose an empirical modification of the combining rule shown in Equation 4. In our approach, we relax the assumption $k_{ij} = k_{ji}$ thus introducing a second interaction parameter per binary :

$$a_{ij} = \sqrt{a_i \, a_j} \; [1-k_{ij}+(k_{ij}-k_{ji})x_i] \tag{5}$$

Equation 5 has the following characteristics :

If $k_{ij} = k_{ji}$, the original mixing rule given by Equation 4 is recovered.

- The "effective" interaction parameter between components i and j approaches k_{ij} as x_i, the mole fraction of component i, approaches zero. It also approaches k_{ji} if x_i approaches unity. The apparent asymmetry under an interchange of i and j is corrected by the fact that both a_{ij} and a_{ij} enter in the calculation of the mixture parameter a_m symmetrically.

- Application of the rule given by Equation 2 for the calculation of the mixture parameter a_m results in a cubic expression for the mole fraction dependence of the mixture parameter a_m. This is different from the case of the conventional mixing rule (Equation 4), which results in a quadratic expression for a_m.

Using this mixing rule with Equation 1, we can obtain the fugacity coefficient of a component in a mixture as:

$$\ell n \phi_k = \ell n \frac{\hat{f}_k}{x_k P} = \frac{b_k}{b_m}\left(\frac{PV}{RT} - 1\right) - \ell n \frac{P(V-b_m)}{RT} +$$

$$+ \left[\frac{\sum_i x_i (a_{ik} + a_{ki}) - \sum_{ij}\sum x_i^2 x_j (k_{ij} - k_{ji})\sqrt{a_i a_j} + x_k \sum_i x_i (k_{ki} - k_{ik})\sqrt{a_k a_i}}{a_m} - \frac{b_k}{b_m} \right] \times$$

$$\times \frac{a_m}{\sqrt{u^2 - 4w}\, b_m RT} \ell n \frac{2V + b_m (u - \sqrt{u^2 - 4w})}{2V + b_m (u + \sqrt{u^2 - 4w})} \qquad (6)$$

An algorithm for the determination of the phase equilibrium properties using the equations described above was developed. The main characteristics of the method used are :

- Pure component parameter evaluation : A recently proposed technique (7) was used for the subcritical components of the mixtures under consideration. This technique, similar to the Joffe and Zudkevitch method (8), provides pure component parameters that exactly reproduce vapor pressure and liquid volume of a compound at a temperature of interest. The reason this approach was preferred over the conventional acentric factor correlation is that the latter does not work well for highly polar or associated components. For supercritical components though, the normal acentric factor correlation was used.

- Multicomponent equilibria were calculated as follows: We start by assuming that the component and interaction parameters are known at a given temperature, and postulate the existence of a given number of phases (2 or 3 for the examples given below). Then, Newton's method is applied for the solution of the system of non-linear equations given by Equation 7.

$$\hat{f}_i^{\alpha} = \hat{f}_i^{\beta} \;\; (= \hat{f}_i^{\gamma} \;) \;, \quad i = 1,2 \ldots N \qquad (7)$$

where the superscript specifies the phase and i refers to the component number.

For the calculations presented in this paper, we elected to use the Peng - Robinson EOS. The values of the two binary interaction parameters were regressed from experimental binary phase equilibrium data on a given isotherm. It was found that although the temperature variation of the parameters is weak, it must be taken into account if a quantitative agreement with experimental data over a wide temperature range is desired.

Results

Polar - Supercritical Fluid Systems. As suggested earlier, the principal motivation behind the development of the new mixing rule, has been the representation of phase equilibrium in systems that contain a supercritical component and one or more highly polar compounds, such as water.

The inadequacy of the conventional method for such systems is demonstrated in Figure 1, that shows the predicted composition of the two phases for the system carbon dioxide - water at 323 K, for both the one-parameter and the two-parameter mixing rules. Parameter k_{12} was fitted to the supercritical fluid phase concentration of water (Y2) for both models. Parameter k_{21} for the two parameter model was fitted to the liquid phase composition of CO_2 (X1). It is clear that when the adjustable parameter in the single-parameter correlation is fitted to the composition of one phase, the results for the other phase are very poor. In contrast, the two-parameter correlation predicts the composition of both phases essentially within experimental accuracy throughout a pressure range from atmospheric to 1000 bar.

An important feature of the new method is that the two parameters are essentially uncorrelated in many cases, as shown in the previous example, in which the parameters were determined from data in different phases. This is generally true only for systems in which the compositions of the coexisting phases are very different. In this respect, the proposed method is similar to a previously suggested technique for highly asymmetric systems (9) in which a different interaction parameter is used for each phase. The advantage of the proposed method is best seen in systems that approach a critical point (for example a liquid-liquid phase split), and therefore cannot be adequately modelled if a different equation is used for different phases.

In Figure 2, a comparison is made between model predictions and the experimentally observed behavior for the CO_2 - water system over a temperature range from 298 K to 348 K (pure CO_2 is subcritical below 304.2 K). Very good agreement is obtained for all pressures for the composition of both phases.

An interesting question arises if we consider the lower pressure range examined. At those conditions, the density of the gas phase is low and deviations from ideality are small. One of the reasons for the complexity of previously proposed modifications of the mixing rules for cubic EOS (4,5,6) is the requirement that the mixing rule should reduce to a quadratic form for a mixture second virial coefficient B_m at low densities, whereas the observed behavior at the high density limit can only be modelled with a higher order mixing rule. For a cubic EOS of the type shown in Equation 1, the expression

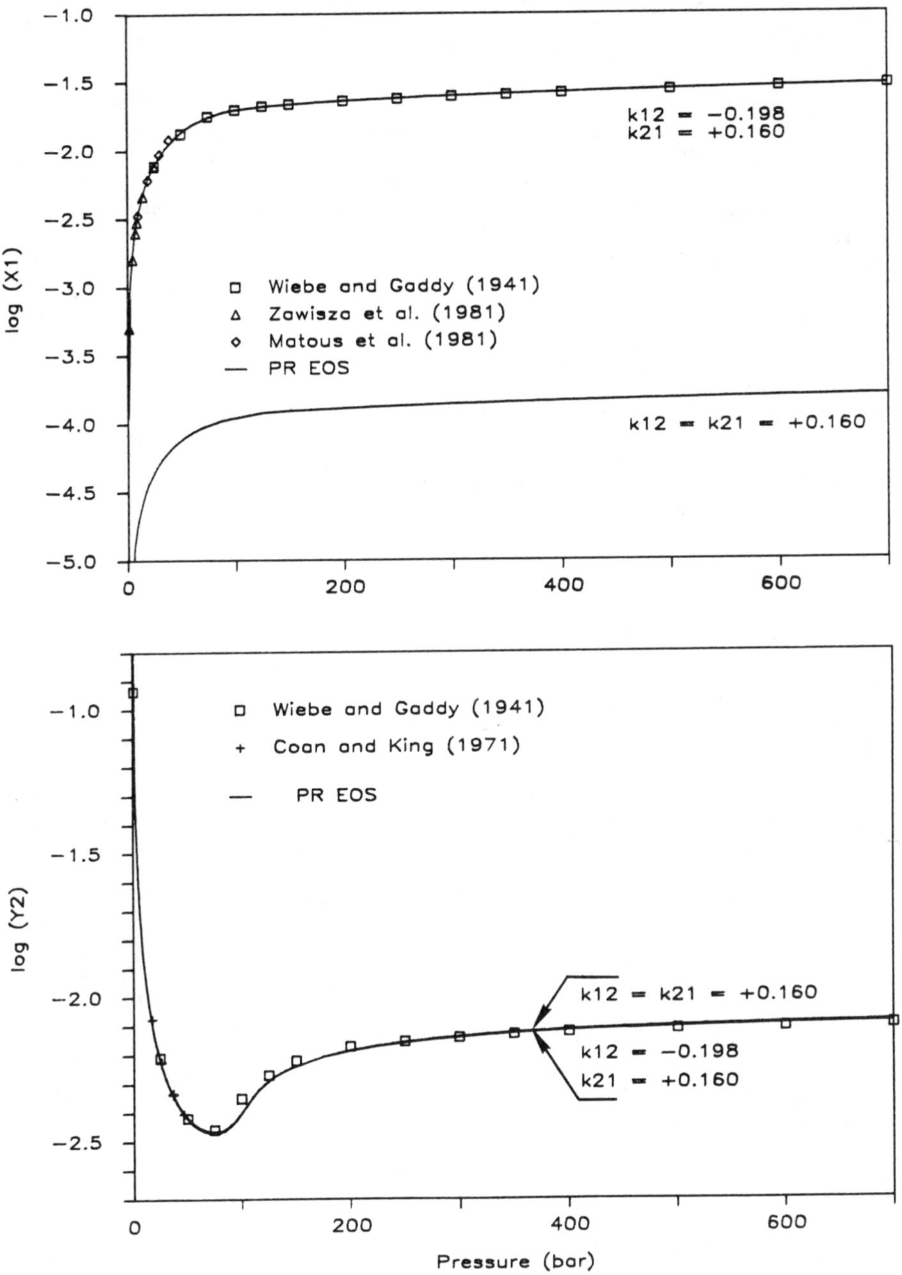

Figure 1. Experimental and predicted phase equilibrium behavior for the system carbon dioxide - water at 323 K. X1 is the mole fraction of CO_2 in the liquid phase and Y2 the mole fraction of water in the fluid phase.

for the mixture second virial coefficient is

$$B_m = b_m - \frac{a_m}{RT} \tag{9}$$

The proposed mixing rule does not obey the requirement for a quadratic dependence of B_m, since the dependence of a_m on the mole fractions of the components of a mixture is cubic. Because of this, it was surprising at first that such a good agreement was found at the low pressure range between model predictions and experimental results. To compare the performance of the proposed mixing rule for the mixture volumetric properties, we calculated the mixture second virial coefficient at 323 K using Equation 7 with the values of k_{12} and k_{21} calculated earlier, as well as using a common interaction parameter equal to their arithmetic average. In the latter case, the mixing rule results in a quadratic functionality for B_m. As can be seen from Figure 3, the difference in the calculated dependence of B_m on the mixture composition between the two cases is indeed small.

Additional examples of the model performance for polar - supercritical fluid systems are shown in Figures 4 and 5 for the carbon dioxide - ethanol and the carbon dioxide - acetone systems. Again, agreement between model predictions and experiment is excellent, even quite close to mixture critical points. It is interesting to note, that for the system carbon dioxide - acetone, the optimal values of the two parameters are quite close to each other. Using the conventional mixing rule for this particular system would have resulted in almost as good agreement between experiment and prediction as for the two - parameter correlation. It is encouraging to note that the proposed correlation naturally leads itself to the elimination of the second parameter if the systems being modelled are relatively simple.

Low Pressure VLE. Up to this point, we have focused our discussion on systems at relatively high pressures. For such systems, the equation of state approach has already been tested extensively. If we are interested in modelling ternary phase behavior , we also need interaction parameters between the less volatile components of a mixture that do not form two-phase systems at high pressures. The easiest way to obtain such parameters is to utilize VLE data at low pressures that are widely available.

As a test of the applicability of the method for low pressure systems, we correlated isothermal VLE data for the binary system ethanol - water at a range of temperatures (18). The results are plotted in Figure 6. Agreement is again very good, even close to the azeotropic region, and the accuracy of the predictions is clearly comparable to the accuracy of excess Gibbs Free Energy models with the same number of adjustable parameters.

Ternary Systems One of the most stringent tests for a proposed correlation is its ability to predict ternary behavior when only the binary behavior is known. In Figure 7 the model predictions for the ternary system carbon dioxide - ethanol - water are presented. The model predictions are based solely on values of the interaction parameters regressed from binary data at the same temperature (the values of the pure component and mixture parameters used are also shown on the same figure).

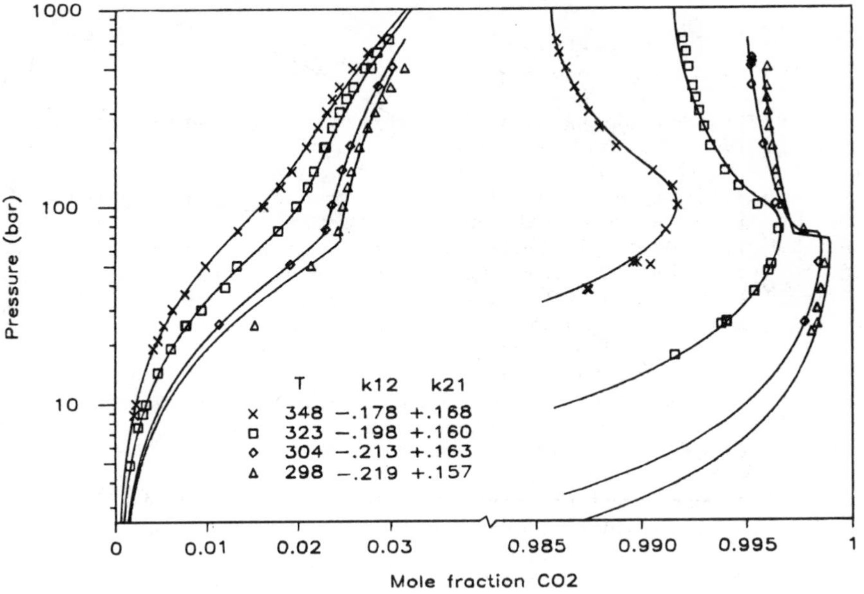

Figure 2. Experimental and predicted phase equilibrium behavior for the system carbon dioxide - water at several temperatures. Experimental data are from references 12-15.

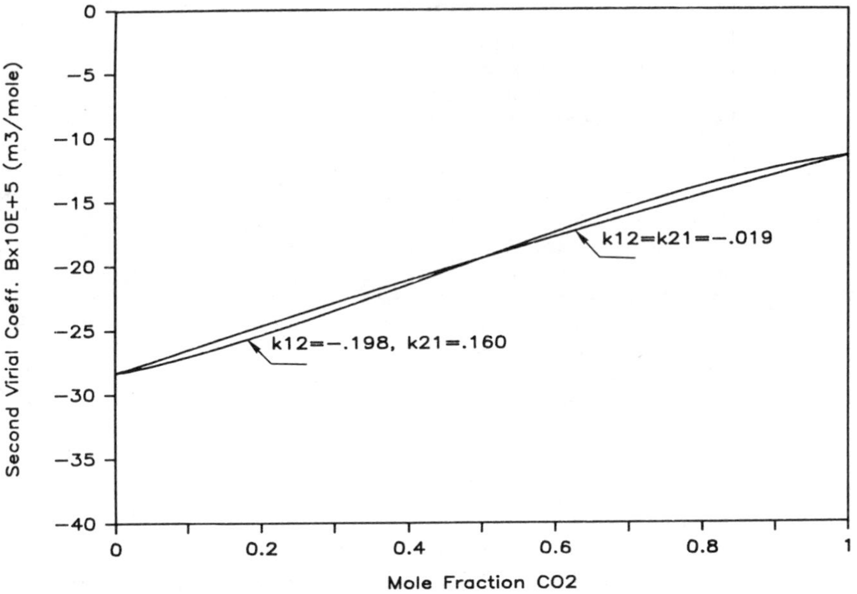

Figure 3. Predicted second virial coefficient versus composition for the carbon dioxide - water system at 323 K.

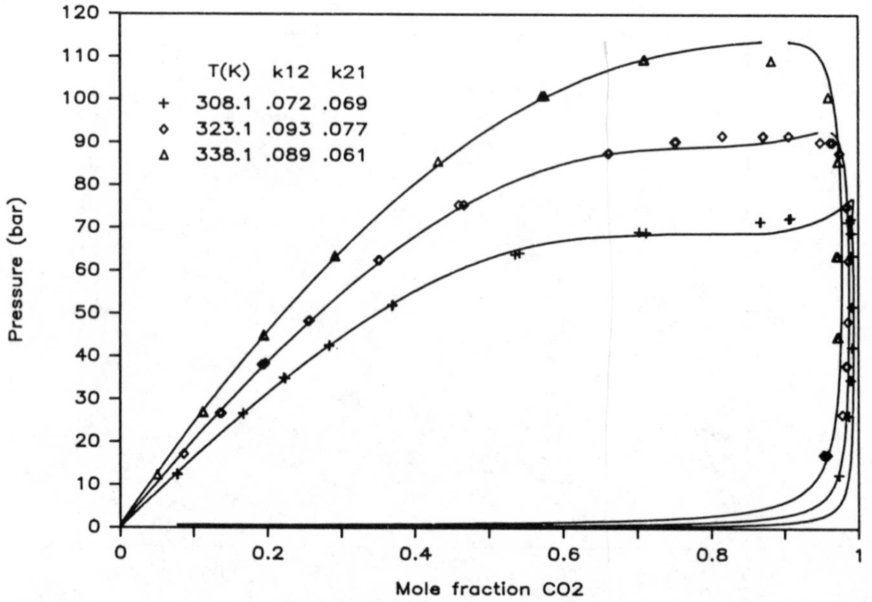

Figure 4. Experimental (10) and predicted phase equilibrium behavior for the system carbon dioxide - ethanol.

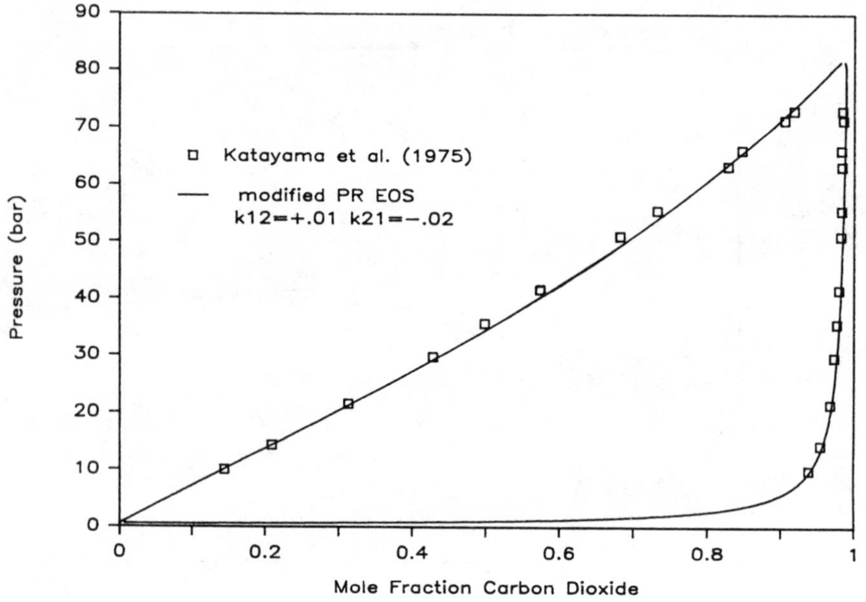

Figure 5. Experimental (16) and predicted phase equilibrium behavior for the system carbon dioxide - acetone at 313 K.

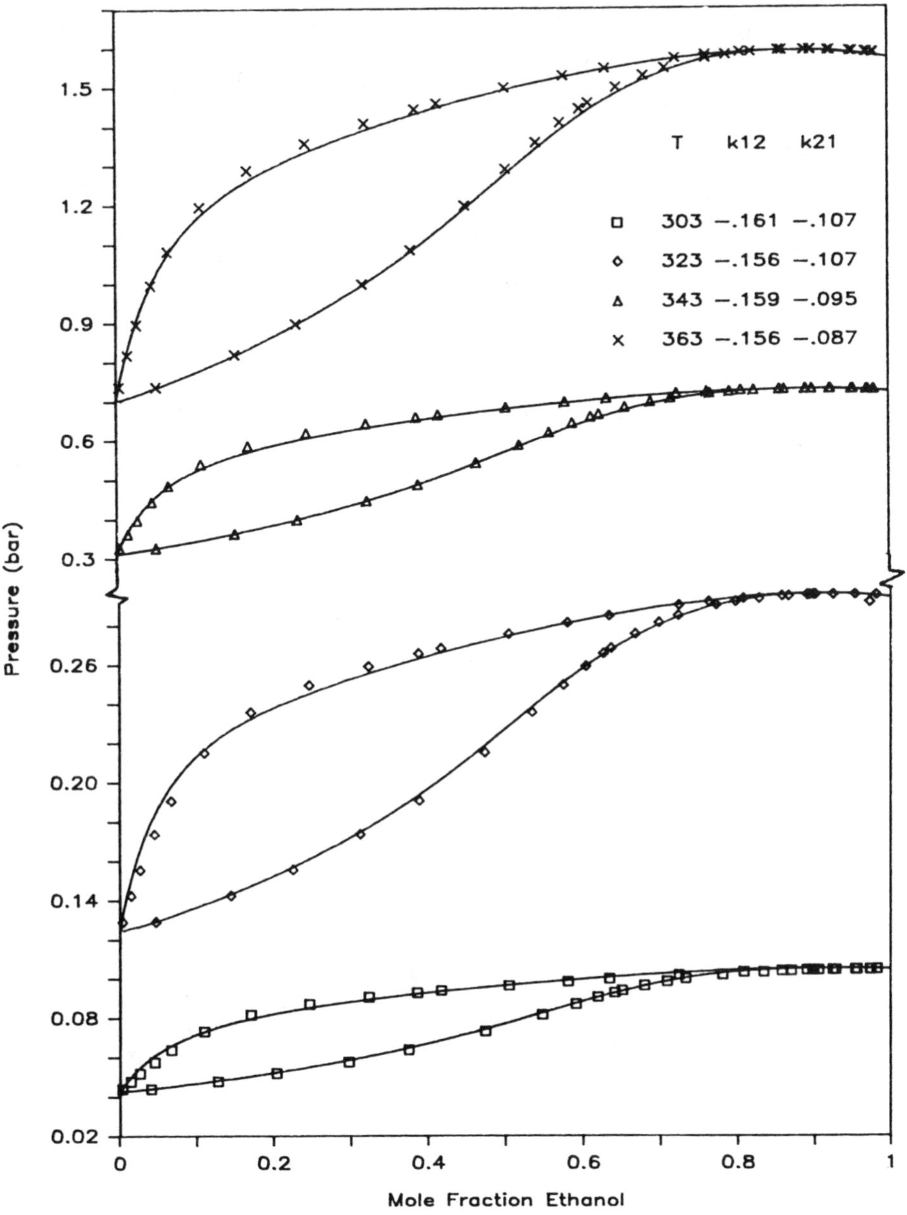

Figure 6. Experimental (18) and predicted phase equilibrium behavior for the system ethanol - water.

As can be seen from Figure 7, the accuracy of the model predictions at the low ethanol concentration range is very good. The calculated tie-lines deviate from the experimental data as the plait point is approached, but still the correct qualitative behavior is predicted. One more illustrative example of the capabilities of the model is given in Figure 8 for the ternary system carbon dioxide - acetone - water at 313 K. The most prominent characteristic of the model predictions for the system is an extensive three-phase region. At this temperature, the solubility of carbon dioxide in the acetone phase is very high even at moderate pressures. This can be seen already in Figure 8a that corresponds to a pressure of 20 bar. As pressure is increased to approx. 22 bar (Figure 8b), the liquid phase undergoes a phase split, into a lower liquid phase rich in water and a middle liquid phase rich in acetone, which coexist with a fluid phase rich in CO_2. The three phase region increases in size as pressure is increased (Figures 8c,8d) and then gradually shrinks again (Figure 8e). At approx. 82 bar (not shown on Figure 8), the system passes through a critical state again, and the middle liquid phase becomes identical to the supercritical phase. At even higher pressures, only two phases coexist, and the effect of pressure on the composition of the phases is much smaller (Figure 8f).

This complex behavior was observed earlier for the system ethylene - acetone - water by Elgin and Weinstock (11). Their " Type 2 " qualitative phase diagrams bear a striking resemblance to our model predictions. The model predictions are fully supported by experimental evidence from our laboratory as indicated on Figures 8c,8d and 8e, in which the measured compositions of the three phases are shown in addition to the model results. The calculated phase compositions are not in exact agreement with the experimental data, but the correct prediction of the appearance and disappearance of the third phase strongly implies that the model captures the substantial features of the physical reality. A more complete presentation of the pertinent experimental results and comparison with model predictions is given elsewhere (10).

Conclusions

A new two-parameter mixing rule is proposed for use in cubic Equations of State and is shown to be especially useful in correlating the phase equilibrium behavior in highly polar systems that cannot be correctly represented by a conventional one-parameter mixing rule. The introduction of Equation 5, is at this point a purely empirical correction to the original form of the mixing rule. The modification is related, but is not directly derived from, the idea of local compositions, that has been shown in the past to result in improved representation of the phase equilibrium behavior in highly polar and asymmetric mixtures.

Among the advantages of the proposed mixing rule, are the relative simplicity of the resulting expressions for the derived thermodynamic properties, together with the fact that the model can be easily reduced to the conventional one-parameter mixing rule for which a substantial amount of regressed interaction parameters exists. The model is shown to accurately reproduce data for both high

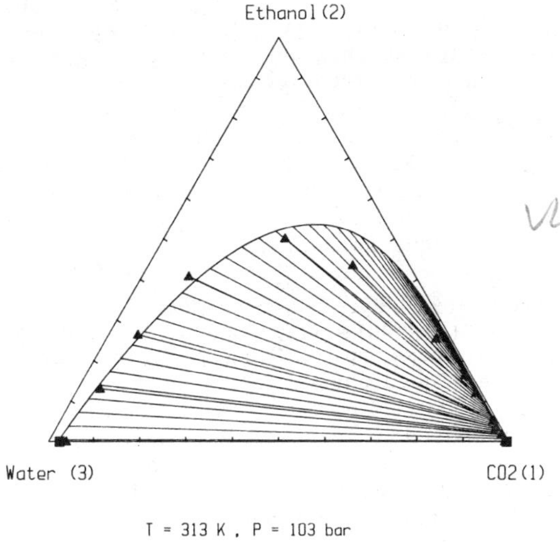

Figure 7. Ternary phase equilibrium behavior for the system carbon dioxide - ethanol - water (experimental data from <u>17</u>,<u>12</u>).

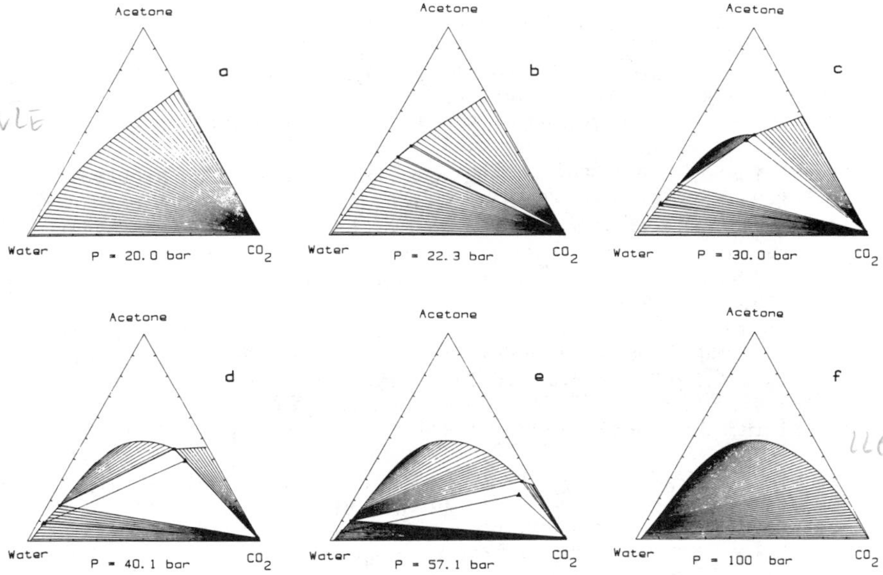

Figure 8. Ternary phase equilibrium behavior for the system carbon dioxide - acetone - water at 313 K (experimental data (<u>10</u>) for the three-phase equilibrium compositions in c,d and e are shown as filled triangles).

pressure polar - supercritical fluid, and low pressure polar - polar binary phase equilibrium. In addition, predictions for ternary systems based on coefficients regressed from binary data only, are qualitatively correct for the systems studied.

One possible weakness of the proposed model, is that it does not reproduce the correct functional dependence of a mixture second virial coefficient on the composition.

Acknowledgments

Financial support for this research was provided by the National Science Foundation (CPE-8318494). One of the authors (AZP) was also supported by a fellowship from Halcon Inc. We gratefully acknowledge both sponsors.

Literature Cited

1. Soave, G., Chem. Eng. Sci. 1972, 27, 1197-1203.
2. Peng, D.-Y. and Robinson, D.B., Ind. Eng. Chem. Fundam. 1976, 15(1), 59-64.
3. Schmidt G. and Wenzel H., Chem. Eng. Sci., 1980, 35, 1503-1512.
4. Whiting, W.B. and Prauznitz, J.M., Fluid Phase Equil. 1982, 9, 119-147.
5. Mollerup, J., Fluid Phase Equil. 1981, 7, 121-138.
6. Mathias, P.M. and Copeman, T.W., Fluid Phase Equil. 1983, 13, 91-108.
7. Panagiotopoulos, A.Z. and Kumar, S.K., Fluid Phase Equil. 1985, 22, 77-88.
8. Joffe, J. and Zudkevitch, D., Ind. Eng. Chem. Fundam. 1970, 9(4), 545-548.
9. Robinson, D.B., Peng, D.-Y. and Chung C.Y.-K., " The Development of the Peng - Robinson Equation and its Application to Phase Equilibrium in a System containing Methanol", paper presented at the Annual AIChE meeting in San Fransisco, CA, November 1984. See also Peng, D.-Y. and Robinson, D. B., ACS Symposium Series, 1980, 133, 393-414.
10. Panagiotopoulos, A.Z. and Reid, R.C., 1985, " High Pressure Phase Equilibria in Ternary Fluid Mixtures with a Supercritical Fluid ", ACS Division of Fuel Chemistry Preprints, vol.30 No 3, 46-56.
11. Elgin, J.C. and Weinstock, J.J., J. Chem. Eng. Data 1959, 4(1), 3-12.
12. Wiebe, R. and Gaddy, V.L., J. Am. Chem. Soc. 1941, 63,475-77 and Wiebe, R., Chem. Rev. 1941, 29, 475-81.
13. Coan, C.R. and King, A.P., J. Am. Chem. Soc. 1971, 93, 1857-62.
14. Matous, J. et al., Coll. Czech. Chem. Commun. 1969, 34, 3982-85.
15. Zawisza, A. et al., J. Chem. Eng. Data 1981, 26, 388-391.
16. Katayama, T., et al., J. Chem. Eng. Jpn. 1975, 8(2) , 89-92.
17. Kuk, M.S. and Montagna, J.C., Chapter 4 in Paulaitis et al. (ed.), " Chemical Engineering at Supercritical Fluid Conditions ", Ann Arbor Science Publishers, 1983.
18. Pemberton, R.C. and Mash, C.J., J. Chem. Thermodynamics 1978, 10, 867-888.

RECEIVED November 5, 1985

Author Index

Subject Index

N

Production and indexing by Keith B. Belton
Jacket design by Pamela Lewis

Elements typeset by Hot Type Ltd., Washington, D.C.
Printed and bound by Maple Press Co., York, Pa.